Nutritional Biochemistry of the Vitamins

SECOND EDITION

The vitamins are a chemically disparate group of compounds whose only common feature is that they are dietary essentials that are required in small amounts for the normal functioning of the body and maintenance of metabolic integrity. Metabolically, they have diverse functions, such as coenzymes, hormones, antioxidants, mediators of cell signaling, and regulators of cell and tissue growth and differentiation. This book explores the known biochemical functions of the vitamins, the extent to which we can explain the effects of deficiency or excess, and the scientific basis for reference intakes for the prevention of deficiency and promotion of optimum health and well-being. It also highlights areas in which our knowledge is lacking and further research is required. This book provides a compact and authoritative reference volume of value to students and specialists alike in the field of nutritional biochemistry, and indeed all who are concerned with vitamin nutrition, deficiency, and metabolism.

David Bender is a Senior Lecturer in Biochemistry at University College London. He has written seventeen books, as well as numerous chapters and reviews, on various aspects of nutrition and nutritional biochemistry. His research has focused on the interactions between vitamin B_6 and estrogens, which has led to the elucidation of the role of vitamin B_6 in terminating the actions of steroid hormones. He is currently the Editor-in-Chief of *Nutrition Research Reviews*.

Nutritional Biochemistry of the Vitamins

SECOND EDITION

DAVID A. BENDER
University College London

CAMBRIDGE
UNIVERSITY PRESS

CAMBRIDGE UNIVERSITY PRESS
Cambridge, New York, Melbourne, Madrid, Cape Town, Singapore,
São Paulo, Delhi, Dubai, Tokyo

Cambridge University Press
The Edinburgh Building, Cambridge CB2 8RU, UK

Published in the United States of America by Cambridge University Press, New York

www.cambridge.org
Information on this title: www.cambridge.org/9780521122214

First published 2003
This digitally printed version 2009

A catalogue record for this publication is available from the British Library

ISBN 978-0-521-80388-5 Hardback
ISBN 978-0-521-12221-4 Paperback

Contents

List of Figures

List of Tables

Preface

In the preface to the first edition of this book, I wrote that one stimulus to write it had been teaching a course on nutritional biochemistry, in which my students had raised questions for which I had to search for answers. In the intervening decade, they have continued to stimulate me to try to answer what are often extremely searching questions. I hope that the extent to which helping them through the often conflicting literature has clarified my thoughts is apparent to future students who will use this book and that they will continue to raise questions for which we all have to search for answers.

The other stimulus to write the first edition of this book was my membership of United Kingdom and European Union expert committees on reference intakes of nutrients, which reported in 1991 and 1993, respectively. Since these two committees completed their work, new reference intakes have been published for use in the United States and Canada (from 1997 to 2001) and by the United Nations Food and Agriculture Organization/World Health Organization (in 2001). A decade ago, the concern of those compiling tables of reference intakes was on determining intakes to prevent deficiency. Since then, the emphasis has changed from prevention of deficiency to the promotion of optimum health, and there has been a considerable amount of research to identify biomarkers of optimum, rather than minimally adequate, vitamin status. Epidemiological studies have identified a number of nutrients that appear to provide protection against cancer, cardiovascular, and other degenerative diseases. Large-scale intervention trials with supplements of individual nutrients have, in general, yielded disappointing results, but these have typically been relatively short-term (typically 5–10 years); the obvious experiments would require lifetime studies, which are not technically feasible.

The purpose of this book is to review what we know of the biochemistry of the vitamins, and to explain the extent to which this knowledge explains the clinical signs of deficiency, the possible benefits of higher intakes than are obtained from average diets, and the adverse effects of excessive intakes.

In the decade since the first edition was published, there have been considerable advances in our knowledge: novel functions of several of the vitamins have been elucidated; and the nutritional biochemist today has to interact with structural biochemists, molecular, cell, and developmental biologists and geneticists, as well as the traditional metabolic biochemist. Despite the advances, there are still major unanswered questions. We still cannot explain why deficiency of three vitamins required as coenzymes in energy-yielding metabolism results in diseases as diverse as fatal neuritis and heart disease of thiamin deficiency, painful cracking of the tongue and lips of riboflavin deficiency, or photosensitive dermatitis, depressive psychosis, and death associated with niacin deficiency.

This book is dedicated in gratitude to those whose painstaking work over almost 100 years since the discovery of the first accessory food factor in 1906 has established the basis of our knowledge, and in hope to those who will attempt to answer the many outstanding questions in the years to come.

David A. Bender

August 2002 London

ONE

The Vitamins

The vitamins are a disparate group of compounds; they have little in common either chemically or in their metabolic functions. Nutritionally, they form a cohesive group of organic compounds that are required in the diet in small amounts (micrograms or milligrams per day) for the maintenance of normal health and metabolic integrity. They are thus differentiated from the essential minerals and trace elements (which are inorganic) and from essential amino and fatty acids, which are required in larger amounts.

The discovery of the vitamins began with experiments performed by Hopkins at the beginning of the twentieth century; he fed rats on a defined diet providing the then known nutrients: fats, proteins, carbohydrates, and mineral salts. The animals failed to grow, but the addition of a small amount of milk to the diet both permitted the animals to maintain normal growth and restored growth to the animals that had previously been fed the defined diet. He suggested that milk contained one or more "accessory growth factors" – essential nutrients present in small amounts, because the addition of only a small amount of milk to the diet was sufficient to maintain normal growth and development.

The first of the accessory food factors to be isolated and identified was found to be chemically an amine; therefore, in 1912, Funk coined the term *vitamine*, from the Latin vita for "life" and amine, for the prominent chemical reactive group. Although subsequent accessory growth factors were not found to be amines, the name has been retained – with the loss of the final "-e" to avoid chemical confusion. The decision as to whether the word should correctly be pronounced "vitamin" or "veitamin" depends in large part on which system of Latin pronunciation one learned – the *Oxford English Dictionary* permits both.

During the first half of the twentieth century, vitamin deficiency diseases were common in developed and developing countries. At the beginning of the twenty-first century, they are generally rare, although vitamin A deficiency (Section 2.4) is a major public health problem throughout the developing world, and there is evidence of widespread subclinical deficiencies of vitamins B_2 (Section 7.4) and B_6 (Section 9.4). In addition, refugee and displaced populations (some 20 million people according to United Nations estimates in 2001) are at risk of multiple B vitamin deficiencies, because the cereal foods used in emergency rations are not usually fortified with micronutrients [Food and Agriculture Organization/World Health Organization (FAO/WHO, 2001)].

1.1 DEFINITION AND NOMENCLATURE OF THE VITAMINS
In addition to systematic chemical nomenclature, the vitamins have an apparently illogical system of accepted trivial names arising from the history of their discovery (Table 1.1). For several vitamins, a number of chemically related compounds show the same biological activity, because they are either converted to the same final active metabolite or have sufficient structural similarity to have the same activity.

Different chemical compounds that show the same biological activity are collectively known as *vitamers*. Where one or more compounds have biological activity, in addition to individual names there is also an approved generic descriptor to be used for all related compounds that show the same biological activity.

When it was realized that milk contained more than one accessory food factor, they were named A (which was lipid-soluble and found in the cream) and B (which was water-soluble and found in the whey). This division into fat- and water-soluble vitamins is still used, although there is little chemical or nutritional reason for this, apart from some similarities in dietary sources of fat-soluble or water-soluble vitamins. Water-soluble derivatives of vitamins A and K and fat-soluble derivatives of several of the B vitamins and vitamin C have been developed for therapeutic use and as food additives.

As the discovery of the vitamins progressed, it was realized that "Factor B" consisted of a number of chemically and physiologically distinct compounds. Before they were identified chemically, they were given a logical series of alphanumeric names: B_1, B_2, and so forth. As can be seen from Table 1.2, a number of compounds were assigned vitamin status, and were later shown either not to be vitamins, or to be compounds that had already been identified and given other names.

Table 1.1 The Vitamins

Vitamin		Functions	Deficiency Disease
A	Retinol β-Carotene	Visual pigments in the retina; regulation of gene expression and cell differentiation; (β-carotene is an antioxidant)	Night blindness, xerophthalmia; keratinization of skin
D	Calciferol	Maintenance of calcium balance; enhances intestinal absorption of Ca^{2+} and mobilizes bone mineral; regulation of gene expression and cell differentiation	Rickets = poor mineralization of bone; osteomalacia = bone demineralization
E	Tocopherols Tocotrienols	Antioxidant, especially in cell membranes; roles in cell signaling	Extremely rare – serious neurological dysfunction
K	Phylloquinone Menaquinones	Coenzyme in formation of γ-carboxyglutamate in enzymes of blood clotting and bone matrix	Impaired blood clotting, hemorrhagic disease
B_1	Thiamin	Coenzyme in pyruvate and 2-oxo-glutarate dehydrogenases, and transketolase; regulates Cl^- channel in nerve conduction	Peripheral nerve damage (beriberi) or central nervous system lesions (Wernicke–Korsakoff syndrome)
B_2	Riboflavin	Coenzyme in oxidation and reduction reactions; prosthetic group of flavoproteins	Lesions of the corner of the mouth, lips, and tongue; sebhorreic dermatitis
Niacin	Nicotinic acid Nicotinamide	Coenzyme in oxidation and reduction reactions, functional part of NAD and NADP; role in intracellular calcium regulation and cell signaling	Pellagra-photosensitive dermatitis; depressive psychosis
B_6	Pyridoxine Pyridoxal Pyridoxamine	Coenzyme in transamination and decarboxylation of amino acids and glycogen phosphorylase; modulation of steroid hormone action	Disorders of amino acid metabolism, convulsions
	Folic acid	Coenzyme in transfer of one-carbon fragments	Megaloblastic anemia

(continued)

Table 1.1 (*continued*)

Vitamin		Functions	Deficiency Disease
B$_{12}$	Cobalamin	Coenzyme in transfer of one-carbon fragments and metabolism of folic acid	Pernicious anemia = megaloblastic anemia with degeneration of the spinal cord
	Pantothenic acid	Functional part of coenzyme A and acyl carrier protein: fatty acid synthesis and metabolism	Peripheral nerve damage (nutritional melalgia or "burning foot syndrome")
H	Biotin	Coenzyme in carboxylation reactions in gluconeogenesis and fatty acid synthesis; role in regulation of cell cycle	Impaired fat and carbohydrate metabolism; dermatitis
C	Ascorbic acid	Coenzyme in hydroxylation of proline and lysine in collagen synthesis; antioxidant; enhances absorption of iron	Scurvy – impaired wound healing, loss of dental cement, subcutaneous hemorrhage

NAD, nicotinamide adenine dinucleotide; NADP, nicotinamide adenine dinucleotide phosphate.

For a compound to be considered a vitamin, it must be shown to be a dietary essential. Its elimination from the diet must result in a more-or-less clearly defined deficiency disease, and restoration must cure or prevent that deficiency disease.

Demonstrating that a compound has pharmacological actions, and possibly cures a disease, does not classify that compound as a vitamin, even if it is a naturally occurring compound that is found in foods.

Equally, demonstrating that a compound has a physiological function as a coenzyme or hormone does not classify that compound as a vitamin. It is necessary to demonstrate that endogenous synthesis of the compound is inadequate to meet physiological requirements in the absence of a dietary source of the compound. Table 1.3 lists compounds that have clearly defined functions, but are not considered vitamins because they are not dietary essentials; endogenous synthesis normally meets requirements. However, there is some evidence that premature infants and patients maintained on long-term total parenteral nutrition may be unable to meet their requirements for carnitine (Section 14.1.2), choline (Section 14.2.2), and taurine (Section 14.5.3) unless they are provided in the diet, and these are sometimes regarded as

Table 1.2 Compounds that Were at One Time Assigned Vitamin Nomenclature, But Are Not Considered to Be Vitamins

B_3	Assigned to a compound that was probably pantothenic acid, also sometimes used (incorrectly) for niacin
B_4	Later identified as a mixture of arginine, glycine, and cysteine, possibly also riboflavin and vitamin B_6
B_5	Assigned to what was later assumed to be either vitamin B_6 or nicotinic acid; also sometimes used for pantothenic acid
B_7	A factor that prevented digestive disturbance in pigeons (also called vitamin I)
B_8	Later identified as adenylic acid
B_9	Never assigned
B_{10}	A factor for feather growth in chickens, probably folic acid and thiamin
B_{11}	Later identified as a mixture of folic acid and thiamin
B_{13}	A growth factor in rats; orotic acid, intermediate in pyrimidine synthesis
B_{14}	An unidentified compound isolated from urine that increases bone marrow proliferation in culture
B_{15}	Pangamic acid, reported to enhance oxygen uptake
B_{16}	Never assigned
B_{17}	Amygdalin (laetrile), a cyanogenic glycoside with no physiological function
B_c	Obsolete name for folic acid
B_p	Chicken antiperosis factor; can be replaced by choline and manganese salts
B_T	Carnitine, a growth factor for insects
B_W	A growth factor, probably biotin
B_x	Obsolete name for p-aminobenzoic acid (intermediate in folate synthesis); also used at one time for pantothenic acid
C_2	A postulated antipneumonia factor (also called vitamin J)
F	Essential fatty acids (linoleic, linolenic, and arachidonic acids)
G	Obsolete name for riboflavin
H_3	"Gerovital," novocaine (procaine hydrochloride) promoted without evidence as alleviating aging, not a vitamin
I	A factor that prevented digestive disturbance in pigeons (also called vitamin B_7)
J	A postulated antipneumonia factor (also called vitamin C_2)
L	Factor isolated from yeast that was claimed to promote lactation
M	Obsolete name for folic acid
N	Extracts from the brain and stomach, purported to have anticancer activity
P	Bioflavonoids
PP	Pellagra-preventing factor, obsolete name for niacin
Q	Ubiquinone (also called Q_{10})
R	Bacterial growth factor, probably folic acid
S	Bacterial growth factor, probably biotin
T	Growth factor in insects, and reported to increase protein uptake in rats, later identified as a mixture of folic acid, vitamin B_{12}, and nucleotides
U	Methylsulfonium salts of methionine
V	Bacterial growth factor, probably NAD
W	Bacterial growth factor, probably biotin
X	Bacterial growth factor, probably biotin
Y	Probably vitamin B_6

NAD, nicotinamide adenine dinucleotide.

Table 1.3 Marginal Compounds that Are Probably Not Dietary Essentials

Carnitine	Required for transport of fatty acids into mitochondria
Choline	Constituent of phospholipids; acetylcholine is a neurotransmitter
Inositol	Constituent of phospholipids; inositol trisphosphate acts as second messenger in transmembrane signaling
Pyrroloquinoline quinone	Coenzyme in redox reactions
Taurine	Osmotic agent in retina and used for conjugation of bile acids; dietary essential for cats
Ubiquinone (coenzyme Q)	Redox coenzyme in mitochondrial electron transport chain

"marginal compounds," for which there is no evidence to estimate requirements.

The rigorous criteria outlined here would exclude niacin (Chapter 8) and vitamin D (Chapter 3) from the list of vitamins, because under normal conditions endogenous synthesis does indeed meet requirements. Nevertheless, they are considered to be vitamins, even if only on the grounds that each was discovered as the result of investigations into once common deficiency diseases, pellagra and rickets.

In addition to the marginal compounds listed in Table 1.3, there are a number of compounds present in foods of plant origin that are considered to be beneficial, in that they have actions that may prevent the development of atherosclerosis and some cancers, although there is no evidence that they are dietary essentials, and they are not generally considered as nutrients.

These compounds are listed in Table 1.4 and discussed in Section 14.7.

1.1.1 Methods of Analysis and Units of Activity

Historically, the vitamins, like hormones, presented chemists with a considerable challenge. They are present in foods, tissues, and body fluids in very small amounts, of the order of μmoles, nmoles, or even pmoles per kilogram, and cannot readily be extracted from the multiplicity of other compounds that might interfere in chemical analyses. Being organic, they are not susceptible to determination by elemental analysis as are the minerals. In addition, for several vitamins, there are multiple vitamers that may have the same biological activity on a molar basis (e.g., the vitamin B_6 vitamers, Section 9.1), or may have very different biological activity (e.g., the vitamin E vitamers, Section 4.1).

The original methods of determining vitamins were biological assays, initially requiring long-term depletion experiments in animals, and later using a

Table 1.4 Compounds that Are Not Dietary Essentials, But May Have Useful Protective Actions

Anthocyanins	Plant (flower) pigments, antioxidants
Bioflavonoids	Polyphenolic compounds with antioxidant action, at one time known as vitamin P
Glucosinolinates	Modify metabolism of foreign compounds and reduce yield of active carcinogens from procarcinogens
Glycosides	Modify metabolism of foreign compounds and reduce yield of active carcinogens from procarcinogens
Polyterpenes	Inhibit cholesterol synthesis
Squalene	Final acyclic intermediate in cholesterol synthesis, acts as feedback inhibitor of cholesterol synthesis
Phytoestrogens	Weak estrogenic and antiestrogenic actions, potentially protective against estrogen- and androgen-dependent tumors and osteoporosis
Polyphenols	Antioxidants
Ubiquinone (coenzyme Q)	Redox coenzyme in mitochondrial electron transport chain, coantioxidant with vitamin E
Vitamin A inactive carotenoids	Antioxidants

variety of microorganisms with more or less defined requirements. Microbiological assays are still commonly used for many of the vitamins; problems of both overestimation and underestimation may occur:

1. Overestimation of the vitamin content of foods will occur if the test organism can use chemical forms and derivatives of the vitamin that are not biologically active in, or available to, human beings.
2. Underestimation will occur if the test organism is unable to use some vitamers, although human beings have appropriate enzymes for interconversion.

Before some of the vitamins had been purified, they were determined in terms of units of biological activity. All should now be expressed in mass or, preferably, molar terms, although occasionally the (now obsolete) international units (iu) are still used for vitamins A (Section 2.1.3), D (Section 3.1), and E (Section 4.1). Where different vitamers differ greatly in biological activity (e.g., the eight tocopherol and tocotrienol vitamers of vitamin E, Section 4.1), it is usual to express total vitamin activity in terms of milligram equivalents of the major vitamer or that with the highest biological activity.

Many of the methods that have been devised for vitamin analysis are now of little more than historical interest, and, in general, unless there is some reason, no analytical methods are listed in this book. A number of recommended methods for vitamin analysis in foods were published as the outcome of a European Union (EU) COST-91 project (Brubacher et al., 1985); since then, the development of ligand binding assays (radioimmunoassays) and high-performance liquid chromatography techniques has meant that individual chemical forms of most of the vitamins can now be determined with great precision and specificity, often with only a minimal requirement for extraction from complex biological materials. Nevertheless, microbiological assays are still sometimes the method of choice, and biological assay is still essential to determine the relative biological activity of different vitamers.

Although modern analytical techniques have considerable precision and sensitivity, food composition tables cannot be considered to give more than an approximation to vitamin intake. Apart from the problems of biological availability (Section 1.1.2), there is considerable variation in the vitamin content of different samples of the same food, depending on differences between varieties, differences in growing conditions (even of the same variety), losses in storage, and losses in food preparation.

When foods have been enriched with vitamins, because of the requirement for the food to contain the stated amount of vitamin after normal storage, manufacturers commonly add more than the stated amount – so-called overage. One of the problems in the debate concerning folate enrichment of flour (Section 10.12) is the relatively small difference between the amount that is considered desirable and the amount that may pose a hazard to vulnerable population groups, and the precision to which manufacturers can control the amount in the final products. In pharmaceutical preparations, considerable latitude is allowed; the U.S. Pharmacopeia permits preparations to contain from 90% to 150% of the declared amount of water-soluble vitamins and from 90% to 165% of the fat-soluble vitamins.

1.1.2 Biological Availability

The biological availability of a nutrient is the proportion of the nutrient present in a food that can be used by the body. It is determined by the extent to which the nutrient is digested, the extent to which the products of digestion are absorbed, and the metabolism of the products of digestion. A number of factors affect digestion, absorption, and metabolism, and hence biological availability. These factors include the physical properties of the food matrix (e.g., nutrients may be inside intact cells of plant foods, and the plant cell wall is not digested); the chemical nature of the vitamin in the food; and the presence of inhibitors that may be present in the food, taken with food, or taken as drugs or medications (Bates and Heseker, 1994; Ball, 1998).

1. Many vitamins are absorbed by active transport; this is a saturable process, and, therefore, the percentage that is absorbed will decrease as the intake increases.

2. The fat-soluble vitamins (A, D, E, and K) are absorbed dissolved in lipid micelles, and, therefore, absorption will be impaired when the meal is low in fat. Gastrointestinal pathology that results in impaired fat absorption and steattorhea (e.g., untreated celiac disease) will also impair the absorption of fat-soluble vitamins, because they remain dissolved in the unabsorbed lipid in the intestinal lumen. Lipase inhibitors used for the treatment of obesity and fat replacers (e.g., sucrose polyesters such as Olestra^{TM}) will similarly impair the absorption of fat-soluble vitamins.

3. Many of the water-soluble vitamins are present in foods bound to proteins, and their release may require either the action of gastric acid (as for vitamin B_{12}, Section 10.7.1) or specific enzymic hydrolysis [e.g., the action of conjugase to hydrolyze folate conjugates (Section 10.2.1) and the hydrolysis of biocytin to release biotin (Section 11.2.3)].

4. The state of body reserves of the vitamin may affect the extent to which it is absorbed (by affecting the synthesis of binding and transport proteins) or the extent to which it is metabolized after uptake into the intestinal mucosa [e.g., the oxidative cleavage of carotene to retinaldehyde is regulated by vitamin A status (Section 2.2.1)].

5. Compounds naturally present in foods may have antivitamin activity. Many foods contain thiaminases and compounds that catalyze nonenzymic cleavage of thiamin to biologically inactive products (Section 6.4.7).

6. Both drugs and compounds naturally present in foods may compete with vitamins for absorption. Chlorpromazine, tricyclic antidepressants, and some antimalarial drugs inhibit the intestinal transport and metabolism of riboflavin (Section 7.4.4); carotenoids lacking vitamin A activity compete with β-carotene for intestinal absorption and metabolism (Section 2.2.2.2); and alcohol inhibits the active transport of thiamin across the intestinal mucosa (Section 6.2).

7. Some vitamins are present in foods in chemical forms that are not susceptible to enzymic hydrolysis during digestion, although they are released during the preparation of foods for analysis. Much of the vitamin B_6 in plant foods is present as pyridoxine glycosides (Section 9.1), which are only partially available, and may also antagonize the metabolism of free pyridoxine (Gregory, 1998); excessive heating can lead to nonenzymic formation of pyridoxyllysine in foods, rendering both the vitamin and the lysine unavailable (Section 9.1); and most of the niacin in cereals

is present as niacytin (nicotinoyl-glucose esters in oligosaccharides and nonstarch polysaccharides), which is only hydrolyzed to a limited extent by gastric acid (Section 8.2.1.1).

Occasionally, protein binding of a vitamin on foods increases its absorption and hence its biological availability. For example, folate from milk is considerably better absorbed than that from either mixed food folates or free folic acid (Section 10.2.1). Folate bound to a specific binding protein in milk is absorbed in the ileum, whereas free folate monoglutamate is absorbed in the (smaller) jejunum.

1.2 VITAMIN REQUIREMENTS AND REFERENCE INTAKES

A priori, it would appear to be a simple matter to determine requirements for vitamins. In practice, a number of problems arise. The first of these is the definition of the word *requirement*. The U.S. usage (Institute of Medicine, 1997) is that the requirement is the lowest intake that will "maintain a defined level of nutriture in an individual" – i.e., the lowest amount that will meet a specified criterion of adequacy. The WHO (1996) defines both a *basal requirement* (the level of intake required to prevent pathologically relevant and clinically detectable signs of deficiency) and a *normative requirement* (the level of intake to maintain a desirable body reserve of the nutrient).

We have to define the purpose for which we are determining the requirement (the criteria of adequacy), then determine the intake required to meet these criteria.

1.2.1 Criteria of Vitamin Adequacy and the Stages of Development of Deficiency

For any nutrient, there is a range of intakes between that which is clearly inadequate, leading to clinical deficiency disease, and that which is so much in excess of the body's metabolic capacity that there may be signs of toxicity. Between these two extremes is a level of intake that is adequate for normal health and the maintenance of metabolic integrity, and a series of more precisely definable levels of intake that are adequate to meet specific criteria and may be used to determine requirements and appropriate levels of intake. These follow.

1. Clinical deficiency disease, with clear anatomical and functional lesions, and severe metabolic disturbances, possibly proving fatal. Prevention of deficiency disease is a minimal goal in determining requirements and is the criterion of the WHO basal requirement (WHO, 1996).

2. Covert deficiency, where there are no signs of deficiency under normal conditions, but any trauma or stress reveals the precarious state of the body reserves and may precipitate clinical signs. For example, as discussed in Section 13.7.1, an intake of 10 mg of vitamin C per day is adequate to prevent clinical deficiency, but at least 20 mg per day is required for healing of wounds.

3. Metabolic abnormalities under normal conditions, such as impaired carbohydrate metabolism in thiamin deficiency (Section 6.5) or excretion of methylmalonic acid in vitamin B_{12} deficiency (Section 10.10.3).

4. Abnormal response to a metabolic load, such as the inability to metabolize a test dose of histidine in folate deficiency (Section 10.10.4), or tryptophan in vitamin B_6 deficiency (Section 9.5.4), although at normal levels of intake there may be no metabolic impairment.

5. Inadequate saturation of enzymes with (vitamin-derived) coenzymes. This can be tested for three vitamins, using red blood cell enzymes: thiamin (Section 6.5.3), riboflavin (Section 7.5.2), and vitamin B_6 (Section 9.5.3).

6. Low plasma concentration of the nutrient, indicating that there is an inadequate amount in tissue reserves to permit normal transport between tissues. For some nutrients, this may reflect failure to synthesize a transport protein rather than primary deficiency of the nutrient itself.

7. Low urinary excretion of the nutrient, reflecting low intake and changes in metabolic turnover.

8. Incomplete saturation of body reserves.

9. Adequate body reserves and normal metabolic integrity. This is the (possibly untestable) goal. Both immune function and minimization of DNA damage offer potential methods of assessing optimum micronutrient status, but both are affected by a variety of different nutrients and other factors (Fenech, 2001).

10. Possibly beneficial effects of intakes more than adequate to meet requirements: the promotion of optimum health and life expectancy. There is evidence that relatively high intakes of vitamin E and possibly other antioxidant nutrients (Section 4.6.2) may reduce the risk of developing cardiovascular disease and some forms of cancer. High intake of folate during early pregnancy reduces the risk of neural tube defects in the fetus (Section 10.9.4).

11. Pharmacological (druglike) actions at very high levels of intake. This is beyond the scope of nutrition, and involves using compounds that happen to be vitamins for the treatment of diseases other than deficiency disease.

12. Abnormal accumulation in tissues and overloading of normal metabolic pathways, leading to signs of toxicity and possibly irreversible lesions. Niacin (Section 8.7.1), and vitamins A (Section 2.5.1), D (Section 3.6.1), and B_6 (Section 9.6.4) are all known to be toxic in excess (see Section 1.2.4.3 for a discussion of tolerable upper levels of intake).

Problems arise in interpreting the results, and therefore defining requirements, when different markers of adequacy respond to different levels of intake. This explains the difference in the tables of reference intakes published by different national and international authorities (see Tables 1.5–1.8).

1.2.2 Assessment of Vitamin Nutritional Status
The same criteria used to define requirements can also be used to assess vitamin nutritional status.

Although vitamin deficiencies give rise to more-or-less clearly defined signs and symptoms, diagnosis is not always easy, so biochemical assessment is frequently needed to confirm a presumptive diagnosis. Furthermore, whereas experimental studies may involve feeding diets deficient in one nutrient, but otherwise complete, it is unlikely that under normal conditions an individual would have such a diet. Undernutrition is likely to lead to deficiency or depletion of several vitamins, with the signs of one deficiency predominating. Biochemical assessment will permit more specific diagnosis. There is an obvious advantage in being able to detect biochemical signs of early or marginal deficiency.

An individual who shows biochemical evidence of deficiency or inadequacy may be metabolically stable, and adequately adapted to his or her current intake, or may be in the early stages of developing clinically significant deficiency disease. In population studies, whereas the number of people with clear clinical deficiency signs gives some indication of the scale of the problem, detection of the larger number who show biochemical signs of deficiency gives a better indication of the number of people at risk of developing deficiency, and hence a more realistic estimate of the true scale of the problem.

Biochemical criteria of vitamin adequacy and methods for biochemical assessment of nutritional status can be divided into the following two distinct groups:

1. Determination of plasma, urine, or tissue concentrations of vitamins and their metabolites. These methods depend on comparison of an individual or group with the population reference range, which is normally taken as the 95% confidence interval: ± twice the standard deviation about the mean value. By definition, 5% of the normal healthy population will lie outside the 95% reference range.

Table 1.5 Reference Nutrient Intakes of Vitamins, U.K., 1991

Age	Vitamin A (µg)	Vitamin D (µg)	Vitamin B$_1$ (mg)	Vitamin B$_2$ (mg)	Niacin (mg)	Vitamin B$_6$ (mg)	Vitamin B$_{12}$ (µg)	Folate (µg)	Vitamin C (mg)
0–3 m	350	8.5	0.2	0.4	3	0.2	0.3	50	25
4–6 m	350	8.5	0.2	0.4	3	0.2	0.3	50	25
7–9 m	350	7	0.2	0.4	4	0.3	0.4	50	25
10–12 m	350	7	0.3	0.4	5	0.4	0.4	50	25
1–3 y	400	7	0.5	0.6	8	0.7	0.5	70	30
4–6 y	500	—	0.7	0.8	11	0.9	0.8	100	30
7–10 y	500	—	0.7	1.0	12	1.0	1.0	150	30
Males									
11–14 y	600	—	0.9	1.2	15	1.2	1.2	200	35
15–18 y	700	—	1.1	1.3	18	1.5	1.5	200	40
19–50 y	700	—	1.0	1.3	17	1.4	1.5	200	40
50+ y	700	10	0.9	1.3	16	1.4	1.5	200	40
Females									
11–14 y	600	—	0.7	1.1	12	1.0	1.2	200	35
15–18 y	600	—	0.8	1.1	14	1.2	1.5	200	40
19–50 y	600	—	0.8	1.1	13	1.2	1.5	200	40
50+ y	600	10	0.8	1.1	12	1.2	1.5	200	40
Pregnant	+100	10	+0.1	+0.3	—	—	—	+100	+10
Lactating	+350	10	+0.1	+0.5	+2	—	+0.5	+60	+30

Source: Department of Health, 1991.

Table 1.6 Population Reference Intakes of Vitamins, European Union, 1993

Age	Vitamin A (µg)	Vitamin B_1 (mg)	Vitamin B_2 (mg)	Niacin (mg)	Vitamin B_6 (mg)	Folate (µg)	Vitamin B_{12} (µg)	Vitamin C (mg)
6–12 m	350	0.3	0.4	5	0.4	50	0.5	20
1–3 y	400	0.5	0.8	9	0.7	100	0.7	25
4–6 y	400	0.7	1.0	11	0.9	130	0.9	25
7–10 y	500	0.8	1.2	13	1.1	150	1.0	30
Males								
11–14 y	600	1.0	1.4	15	1.3	180	1.3	35
15–17 y	700	1.2	1.6	18	1.5	200	1.4	40
18+ y	700	1.1	1.6	18	1.5	200	1.4	45
Females								
11–14 y	600	0.9	1.2	14	1.1	180	1.3	35
15–17 y	600	0.9	1.3	14	1.1	200	1.4	40
18+ y	600	0.9	1.3	14	1.1	200	1.4	45
Pregnant	700	1.0	1.6	14	1.3	400	1.6	55
Lactating	950	1.1	1.7	16	1.4	350	1.9	70

Source: Scientific Committee for Food, 1993.

Table 1.7 Recommended Dietary Allowances (RDAs) and Acceptable Intakes for Vitamins, U.S./Canada, 1997–2001

Age	Vitamin A (µg)	Vitamin D (µg)	Vitamin E (mg)	Vitamin K (µg)	Vitamin B$_1$ (mg)	Vitamin B$_2$ (mg)	Niacin (mg)	Vitamin B$_6$ (mg)	Folate (µg)	Vitamin B$_{12}$ (µg)	Vitamin C (mg)
0–6 m	400	5	4	2.0	0.2	0.3	2	0.1	65	0.4	40
7–12 m	500	5	5	2.5	0.3	0.4	4	0.3	80	0.5	50
1–3 y	300	5	6	30	0.5	0.5	6	0.5	150	0.9	15
4–8 y	400	5	7	55	0.5	0.6	8	0.6	200	1.2	25
Males											
9–13 y	600	5	11	60	0.9	0.9	12	1.0	300	1.8	45
14–18 y	900	5	15	75	1.2	1.3	16	1.3	400	2.4	75
19–30 y	900	5	15	120	1.2	1.3	16	1.3	400	2.4	90
31–50 y	900	5	15	120	1.2	1.3	16	1.3	400	2.4	90
51–70 y	900	10	15	120	1.2	1.3	16	1.7	400	2.4	90
>70 y	900	15	15	120	1.2	1.3	16	1.7	400	2.4	90
Females											
9–13 y	600	5	11	60	0.9	0.9	12	1.0	300	1.8	45
14–18 y	700	5	15	75	1.0	1.0	14	1.2	400	2.4	65
19–30 y	700	5	15	90	1.1	1.1	14	1.3	400	2.4	75
31–50 y	700	5	15	90	1.1	1.1	14	1.3	400	2.4	75
51–70 y	700	10	15	90	1.1	1.1	14	1.5	400	2.4	75
>70 y	700	15	15	90	1.1	1.1	14	1.5	400	2.4	75
Pregnant	770	5	15	90	1.4	1.4	18	1.9	600	2.6	85
Lactating	900	5	16	90	1.4	1.6	17	2.0	500	2.8	120

Figures for infants under 12 months are Adequate Intakes, based on the observed mean intake of infants fed principally on breast milk; for nutrients other than vitamin K, figures are RDA, based on estimated average requirement +2 SD; figures for vitamin K are Adequate Intakes, based on observed average intakes.

Source: Institute of Medicine, 1997, 1998, 2000, 2001.

Table 1.8 Recommended Nutrient Intakes for Vitamins, FAO/WHO, 2001

Age	Vitamin A (µg)	Vitamin D (µg)	Vitamin K (µg)	Vitamin B$_1$ (mg)	Vitamin B$_2$ (mg)	Niacin (mg)	Vitamin B$_6$ (mg)	Folate (µg)	Vitamin B$_{12}$ (µg)	Vitamin C (mg)	Panto (mg)	Biotin (µg)
0–6 m	375	5	5	0.2	0.3	2	0.1	80	0.4	25	1.7	5
7–12 m	400	5	10	0.3	0.4	4	0.3	80	0.5	30	1.8	6
1–3 y	400	5	15	0.5	0.5	6	0.5	160	0.9	30	2.0	8
4–6 y	450	5	20	0.6	0.6	8	0.6	200	1.2	30	3.0	12
7–9 y	500	5	25	0.9	0.9	12	1.0	300	1.8	35	4.0	20
Males												
10–18 y	600	5	35–55	1.2	1.3	16	1.3	400	2.4	40	5.0	30
19–50 y	600	5	65	1.2	1.3	16	1.3	400	2.4	45	5.0	30
50–65 y	600	10	65	1.2	1.3	16	1.7	400	2.4	45	5.0	30
>65 y	600	15	65	1.2	1.3	16	1.7	400	2.4	45	5.0	30
Females												
10–18 y	600	5	35–55	1.1	1.0	16	1.2	400	2.4	40	5.0	25
19–50 y	600	5	55	1.1	1.1	14	1.3	400	2.4	45	5.0	30
50–65 y	600	10	55	1.1	1.1	14	1.5	400	2.4	45	5.0	30
>65 y	600	15	55	1.1	1.1	14	1.5	400	2.4	45	5.0	30
Pregnant	800	5	55	1.4	1.4	18	1.9	600	2.6	55	6.0	30
Lactating	850	5	55	1.5	1.6	17	2.0	500	2.8	70	7.0	35

Source: FAO/WHO, 2001.

2. Metabolic loading tests and the determination of enzyme saturation with cofactor measure the ability of an individual to meet his or her idiosyncratic requirements from a given intake, and, therefore, give a nearly absolute indication of nutritional status, without the need to refer to population reference ranges. A number of factors other than vitamin intake or adequacy can affect responses to metabolic loading tests. This is a particular problem with the tryptophan load test for vitamin B_6 nutritional status (Section 9.5.4); a number of drugs can have metabolic effects that resemble those seen in vitamin deficiency or depletion, whether or not they cause functional deficiency.

1.2.3 Determination of Requirements

Having decided an appropriate criterion of adequacy, which will differ from one vitamin to another, the problem is to determine what are adequate intakes to meet those criteria. Studies of vitamin requirements can be divided into the following four groups.

1.2.3.1 Population Studies of Intake
In areas in which deficiency diseases are common, it is possible to estimate requirements to prevent the development of deficiency by comparing the intakes of people with and without specific signs. This permits determination of minimally adequate intakes (basal requirements), subject to the considerable problems of determining nutrient intake with adequate accuracy.

There have been a small number of large-scale studies of healthy populations, comparing nutrient intake with specific biochemical indices of nutritional adequacy. Such studies generally rely on seven-day weighed food intakes (and make the assumption that recording of intake will not alter habitual diets significantly), with estimation of nutrient intakes from tables of food composition; only very rarely are duplicate portions of foods analyzed for their vitamin content in such studies. The study of almost 2,000 adults in Great Britain (Gregory et al., 1990) was of this type and permitted revision of estimates of requirements for some of the vitamins.

Such population studies also permit the definition of a range of acceptable intakes. Quite apart from determining average requirements, and then reference intakes (Section 1.2.4), it is useful to know the range of intakes that is compatible with normal health.

The lower acceptable intake will often be the intake of the 5th percentile of the healthy population, although a lower intake may be classified as acceptable on the basis of other information. Thus, for example, average requirements for vitamin B_{12} are of the order of 0.1 to 1 μg per day, but the intake of the 2.5th percentile in Great Britain is 2.4 μg per day (Gregory et al., 1990). In this case, average intakes are well above the known lower acceptable level. Conversely, if the intake of the 5th percentile is lower than that at which signs of deficiency are known to occur, then this would not be classified as an acceptable intake.

Similarly, upper acceptable intakes are defined on the basis of the intake of the 95th percentile of the healthy population, unless this is known to be above the toxic threshold; higher intakes may still be within the acceptable range.

1.2.3.2 **Depletion/Repletion Studies**   There have been a number of studies to determine vitamin requirements by deliberate depletion of initially healthy subjects, following the development of biochemical and clinical signs of deficiency, then determining the intake required to reverse those signs. Such studies permit reasonably precise estimation of requirements to meet different criteria of adequacy and give some indication of the extent of individual variation. The number of subjects studied in this way has been relatively small, and estimation of requirements has generally been by interpolation into the results of experiments using a relatively wide range of intakes and is thus subject to considerable possible error. Such studies, however, provide most of the experimental evidence on which current estimates of vitamin requirements are based.

Depletion/repletion studies may give a false indication of requirements, because they are based on a nonphysiological experiment – more-or-less complete deprivation of the vitamin under test, but the provision of an otherwise completely adequate diet. More important, they also measure requirements in a state of changing nutritional status, during the acute development and cure of deficiency, rather than measuring the requirements for maintenance of normal body reserves and metabolic integrity against the background of a relatively constant habitual intake.

1.2.3.3 **Replacement of Metabolic Losses**   An alternative approach to determining vitamin requirements is to measure the loss from the body pool in a steady state. This requires estimation of the total body pool, and measurement of the fractional rate of loss from that pool, generally using radioactive or stable isotope tracers. Three problems can arise in such studies.

1. The measured total body pool may not be appropriate or desirable, because it will reflect the state of the subjects' nutrition on a self-selected diet (see Section 13.7.3 for a discussion of the problem of the desirable body pool of vitamin C).

2. There may be multiple metabolic pools of the vitamin, with very different rates of turnover. In this case, short-term and long-term studies will give very different estimates of the fractional rate of turnover of the total body pool. As discussed in Section 9.6.1, this is known to be a problem with vitamin B_6, because some 80% of the total body pool is associated with muscle glycogen phosphorylase and has a much lower fractional turnover rate than the remaining 20%.

3. The fractional rate of turnover of the body pool may well change with changes in intake; as discussed in Section 13.7.3, this is known to be the case with vitamin C.

1.2.3.4 Studies in Patients Maintained on Total Parenteral Nutrition

Subjects who are maintained for prolonged periods by total parenteral nutrition are obviously wholly dependent on what is provided in the nutrient mixture, normally with no contribution from intestinal bacteria. A great deal has been learned from such patients, including the essentiality of the amino acid histidine, and evidence that endogenous synthesis of taurine (Section 14.5.3) and carnitine (Section 14.1.2) may not be adequate to meet requirements without some dietary provision. However, for obvious ethical reasons, such patients have not been subjected to trials of graded intakes of vitamins, but are generally provided with amounts calculated to be adequate and in excess of minimum requirements.

A further problem with studies in patients maintained on long-term total parenteral nutrition is that they are not normal healthy subjects – there is some good medical reason for their treatment. Furthermore, they will have little or no enterohepatic recirculation of vitamins, and hence may have considerably higher requirements than normal; there is considerable enterohepatic circulation of folate (Section 10.2.1) and vitamin B_{12} (Section 10.7.1).

1.2.4 Reference Intakes of Vitamins

Notwithstanding the problems involved in determining requirements for vitamins, most national authorities (as well as the United Nations FAO/WHO and the European Commission) publish, and periodically revise, tables of recommended intakes of nutrients or dietary reference values.

As shown in Tables 1.5–1.8, reference intakes published by different authorities show considerable differences. Some of the reasons for this are apparent from the discussion above; different criteria of adequacy may be applied by the members of different expert committees, and the estimation of average requirements and, hence, reference intakes requires a considerable exercise of judgment to interpret the relatively small body of scientific literature. Historically, tables of reference intakes were based on requirements to prevent deficiency disease, or subclinical signs of inadequacy; increasingly, as evidence accumulates, the emphasis is on requirements to promote optimum health rather than to prevent deficiency.

As shown in Table 1.9, a number of terms are used: Recommended Daily (or Dietary) Intake (RDI), Recommended Dietary (or Daily) Amount (RDA), Reference or Recommended Nutrient Intake (RNI), and Population Reference Intake (PRI). All have the same statistical basis, and all are defined as an intake of the nutrient that is adequate to ensure that the requirements of essentially all healthy people in the specified population group are met. The 2001 FAO/WHO report introduced the term *protective nutrient intake* – an amount greater than the reference intake that may be protective against specified health risks of public health importance.

There is considerable individual variation in nutrient requirements. It is generally assumed that requirements follow a more or less statistically normal (Gaussian) distribution, as shown in the upper curve in Figure 1.1. This means that 95% of the population has a requirement for a given nutrient within the range of ±2 SD about the observed mean requirement. Therefore, an intake at the level of the observed (or estimated) mean requirement plus $2 \times$ SD will be more than enough to meet the requirements of 97.5% of the population. This is the level that is generally called the RDI, RDA, RNI, or PRI.

There is, in fact, little evidence that requirements do follow a Gaussian distribution; the U.S./Canadian tables (Institute of Medicine, 1997, 1998, 2000, 2001) note this, and state that when the distribution is skewed the 97.5th percentile can be estimated by transforming the data to a normal distribution. Where the standard deviations from different studies are inconsistent, the U.S./Canadian tables determine the RDA on the basis of $1.2 \times$ average requirement. This assumes a coefficient of variation of 10%, which is based on the known variance in basal metabolic rate.

It is apparent from this discussion that reference intake figures are intended for use in populations and communities, and do not apply to individuals. An individual might have a requirement anywhere within the range, and therefore

Table 1.9 Terms that Have Been Used to Describe Reference Intakes of Nutrients

RDA	Recommended Dietary Allowances	U.S., 1941	The name was deliberately chosen to allow the possibility of future modification of the values and was not intended to carry any connotation of minimum or optimal requirements.
RDI	Recommended Dietary Intakes	U.K., 1969	... to emphasize that the recommendations related to foodstuffs as actually eaten
RDA	Recommended Daily Amounts	U.K., 1979	... to make it clear that the amounts referred to averages for a group of people and not to amounts that individuals must meet, as implied by the term "allowances"
	Safe levels of intake	UN agencies	Means safe and adequate, but does not imply that higher intakes are unsafe
RNI	Reference Nutrient Intakes	U.K., 1991	By parallel with clinical chemistry reference ranges, which encompass 95% of *normal* values; to emphasize that they are not recommendations for individuals, nor are they amounts to be consumed daily; see Table 1.5
RNI	Recommended Nutrient Intake	FAO, 2001	
PRI	Population Reference Intakes	EU, 1993	By parallel with RNI, but emphasizing that these are population ranges, and not applicable to individuals; see Table 1.6
U.S.-RDA		U.S., 1973	Reference intakes for labeling purposes, the highest RDA value for any population group; see Table 1.11
RDI	Reference Daily Intakes	U.S., 1990	Reference intakes for labeling purposes, numerically equal to U.S.-RDA; see Table 1.11
DRV	Daily Reference Values	U.S., 1990	Reference values for fat, carbohydrate, sodium, potassium, and protein; for labeling purposes
AI	Adequate Intake	U.S./Canada, 1997	
UL	Tolerable Upper Intake Level	U.S./Canada, 1997	

EU, European Union; FAO, Food and Agriculture Organization; UN, United Nations.

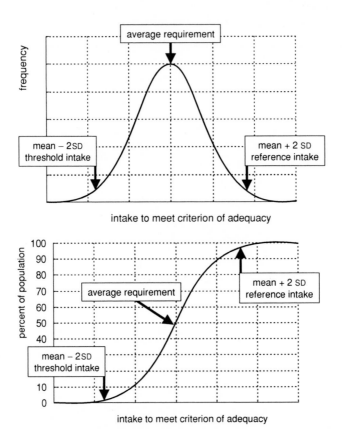

Figure 1.1. Derivation of reference intakes of nutrients from the distribution around the observed mean requirement; plotted below as a cumulative distribution curve, permitting estimation of the probability that a given level of intake is adequate to meet an individual's requirement.

might satisfy requirements with an intake considerably below the reference intake. In the United Kingdom, the RNI is regarded as a goal for planning and evaluating the intake of population groups and communities, rather than applying to an individual (Department of Health, 1991); in contrast, in the United States and Canada, the RDA is regarded as a goal for individuals to achieve (Institute of Medicine, 1997, 1998, 2000, 2001).

The lower curve in Figure 1.1 shows the population distribution of nutrient requirements plotted as a cumulative percentage. This can then be reinterpreted as indicating the statistical probability that a given level of intake will be adequate for an individual.

1. At an intake equal to the mean requirement minus twice the standard deviation, only 2.5% of the population have been included. Therefore, there is only a 2.5% probability that this intake is adequate for an individual.
2. At an intake equal to the mean observed requirement, 50% of the population have been included, and there is thus a 50% probability that this level of intake will be adequate for an individual.
3. At an intake equal to the observed mean requirement plus twice the standard deviation (RDA or RNI), 97.5% of the population have been included, and there is thus a 97.5% probability that this intake will be adequate for an individual.

1.2.4.1 **Adequate Intake** For some vitamins, notably biotin (Section 11.5) and pantothenic acid (Section 12.6), dietary deficiency is more-or-less unknown, and there are no data from which to estimate average requirements or derive reference intakes. In such cases, the observed range of intakes is obviously more than adequate to meet requirements, and the average intake is used to calculate an adequate intake figure.

1.2.4.2 **Reference Intakes for Infants and Children** For obvious ethical reasons, there have been almost no experimental studies of the vitamin requirements of infants and children. For infants, it is conventional to use the nutrient yield of breast milk and assume that this is equal to or greater than requirements. Although this is termed an RNI in U.K. tables (Table 1.5), in the U.S./Canadian tables (Table 1.7), it is more correctly referred to as an acceptable intake.

Most authorities have estimated reference intakes for children by linear interpolation between the experimental data for young adults and the nutrient yield of breast milk (Figure 1.2). The EU expert group (Table 1.6; Scientific Committee for Food, 1993) took a different approach and extrapolated backward from the experimentally determined reference intakes for young adults on the basis of energy requirement (for which there are good experimental data), with the possibly unjustified assumption that the nutrient density of adequate diets should be essentially constant through childhood. The advantage of this approach was that it takes into account the higher nutrient requirements at times of rapid growth (because energy requirement increases in growth). This backward extrapolation gave figures for vitamin requirements in infancy that were the same as those based on the composition of breast milk.

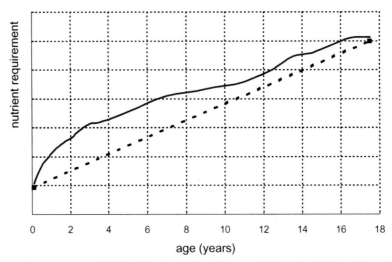

Figure 1.2. Derivation of requirements or reference intakes for children. Dotted line shows linear interpolation between the assumed acceptable intake at age 3 months (based on breast milk composition) and the experimentally derived reference intake at age 17 to 18 years. Solid line shows extrapolation backward from the experimentally derived reference intake at age 17 to 18 years on the basis of energy requirements.

1.2.4.3 Tolerable Upper Levels of Intake A number of the vitamins are known to be toxic in excess. For most, there is a considerable difference between reference intakes that are more than adequate to meet requirements and the intake at which there may be adverse effects, although for vitamins A (Section 2.5.1) and D (Section 3.6.1) there is only a relatively small margin of safety.

For food additives and contaminants, an acceptable level of intake is calculated from the highest intake at which there is no detectable adverse effect – the *no adverse effect level* (NOAEL) – by dividing by a factor of 100, thus ensuring a very wide margin of safety. This approach is not appropriate for compounds that are dietary essentials, and indeed in many cases would result in a (toxicologically calculated) acceptable intake below the reference intake or even below the requirement for metabolic integrity. The U.K. (Department of Health, 1991) and EU (Scientific Committee for Food, 1993) tables give "guidance on higher intakes," suggesting upper safe levels of habitual intake from supplements. The U.S./Canadian tables (Institute of Medicine, 1997, 1998, 2000, 2001) give tolerable upper levels of intake derived from the NOAEL divided by appropriate safety factors. The upper level of intake is defined as the maximum level of habitual intake that is unlikely to pose any risk of adverse health effects

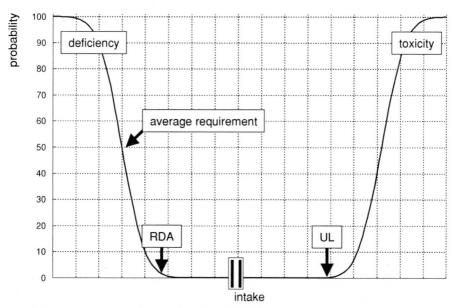

Figure 1.3. Derivation of reference intake [Recommended Dietary (or Daily) Amount (RDA)], and tolerable upper level (UL) for a nutrient. Curve shows the probability that a subject will show signs of deficiency (left) or toxicity (right) at any given level of intake.

to almost all individuals in the (stated) population group. It is a level of intake that can (with a high degree of probability) be tolerated biologically, but is not a recommended level. "There is no established benefit for healthy individuals consuming more than the RDA." As shown in Figure 1.3, the RDA is set at the 97.5th percentile of the distribution of requirements, and is thus adequate to meet the requirements of "essentially all" of the population group, whereas the upper tolerable intake is set below the level at which *any* of the population might be expected to show adverse effects.

Table 1.10 shows the NOAEL for the vitamins, the upper limits for supplements available over the counter proposed by the European Federation of Health Product Manufacturers Associations (Shrimpton, 1997), the U.S./Canadian tolerable upper levels, and the prudent upper levels of consumption from the EU tables.

The upper levels for over-the-counter supplements shown Table 1.10 are voluntary, but because the report (Shrimpton, 1997) was commissioned by the European Federation of Health Product Manufacturers, it is likely that most manufacturers of nutritional supplements will abide by them. The problem is that in most countries nutritional supplements are covered by food

Table 1.10 Toxicity of Vitamins: Upper Limits of Habitual Consumption and Tolerable Upper Limits of Intake

Vitamin	NOAEL		Upper Limit for Supplements[a]	Tolerable Upper Levels U.S.[b]	Tolerable Upper Levels EU[c]
A	μg	3,000	3,000	2,800, 3,000[d]	7,500, 9,000
Carotene	μg	—	25	—	—
D	μg	20	20	50	50
E	mg	800	800	1,000	>2,000
K	mg	30	—	—	—
B$_1$	mg	50	50	—	>500
B$_2$	mg	200	200	—	—
Niacin	mg	500, 250[e]	—	—	—
Nicotinic acid	mg	—	500, 250[e]	35	—
Nicotinamide	mg	—	1,500	—	—
B$_6$	mg	200	200	100	25[f]
Folate	μg	1,000	1,000	1,000	1,000
^{3}B$_{12}$	μg	3,000	3,000	—	200[g]
Biotin	mg	2,500	2,500	—	—
Pantothenate	mg	1,000	1,000	—	—
C	mg	>1,000	1,000	2,000	10,000

EU, European Union; NOAEL, no adverse effect level, the highest level of intake at which no adverse effects are observed.
Sources: [a]Shrimpton, 1997; [b]Institute of Medicine, 1997, 1998, 2000, 2001; [c]Scientific Committee for Food, 1993; [d]where two figures are shown for vitamin A, the lower is for women and the higher is for men (Table 2.5). [e]for niacin and nicotinic acid, the lower values are for sustained release preparations; [f]the EU upper level of 25 mg of vitamin B$_6$ was proposed by the Scientific Committee for Food Opinion, 2000; and [g]the EU upper level of 200 μg of vitamin B$_{12}$ was set because of the possible presence of inactive corrinoids in pharmaceutical preparations, not because of toxicity of the vitamin itself.

legislation rather than regulations covering medicines. A report to the U.S. Food and Drug Administration (FDA; Department of Health and Human Services, 2001) noted the lack of surveillance, and the lack of an adequate system for reporting or investigating adverse effects of nutritional supplements. Indeed, in one-third of cases in which adverse effects were reported, the FDA was unable to discover from the manufacturers precisely what was included in the supplements. In some cases, they could not find the manufacturer of products that were implicated. The report recommended the following:

- a requirement for manufacturers to report serious adverse effects;
- provision of information to health care professionals and consumers about the procedure for reporting adverse effects;

Table 1.11 Labeling Reference Values for Vitamins

	U.S. Reference Daily Intake[a]	European Union Proposed by Scientific Committee for Food[b]	European Union Required by Directive[c]
Vitamin A, μg	1,500	500	800
Vitamin D, μg	10	5	5
Vitamin E, mg	30	—	10
Vitamin C, mg	60	30	60
Thiamin, mg	1.5	0.8	1.4
Riboflavin, mg	1.7	1.3	1.6
Niacin, mg	20	15	18
Vitamin B$_6$, mg	2.0	1.3	2.0
Folate, μg	400	140	200
Vitamin B$_{12}$, μg	6.0	1.0	1.0
Biotin, μg	300	—	150
Pantothenic acid, mg	10	—	6

Sources: [a]National Research Council, 1989; [b]Scientific Committee for Food, 1993; [c]European Commission, 1990.

- a requirement for manufacturers to register themselves and their products with the FDA;
- a requirement for guidance to manufacturers on safety information to be provided.

1.2.4.4 Reference Intake Figures for Food Labeling Tables of reference intakes provide figures for different age groups, separate figures for men and women, and additional figures for pregnancy and lactation. For nutritional labeling of foods, it is obviously desirable to have a single figure that will permit comparison of different foods and pharmaceutical preparations. In the United States, the figure that is used for labeling (the U.S. RDA or RDI) is the highest RDA for any population group for that nutrient (National Research Council, 1989). The EU Scientific Committee for Food (1993) noted that to use the highest reference intake might lead to excess nutrient intake by a substantial proportion of the population and that those unable to achieve this high level might be tempted to take (unnecessary) supplements. It might also lead to a loss of confidence in traditional foods that would appear to be of low nutritional value. They proposed that the labeling reference value should be the average requirement for men (which in many cases equals the reference intake for women). However, at present, the law in the EU requires that the figures shown in column 3 of Table 1.11 be used for labeling.

FURTHER READING

Ames BN (2001) DNA damage from micronutrient deficiencies is likely to be a major cause of cancer. *Mutation Research* **475,** 7–20.

Ames BN, Elson-Schwab I, and Silver EA (2002) High dose vitamin therapy stimulates variant enzymes with decreased coenzyme binding affinity (increased K_m): relevance to genetic disease and polymorphisms. *American Journal of Clinical Nutrition* **75,** 616–58.

Anonymous (1997) The development and use of dietary reference intakes. *Nutrition Reviews* **55,** 319–51.

Ball GFM (1998) *Bioavailability and Analysis of Vitamins in Foods.* London: Chapman and Hall.

Bates C and Heseker H (1994) Human bioavailability of vitamins. *Nutrition Research Reviews* **7,** 93–128.

Beaton G (1988) Nutrient requirements and population data. *Proceedings of the Nutrition Society* **47,** 63–78.

Bhaskaram P (2001) Immunobiology of mild micronutrient deficiencies. *British Journal of Nutrition* **85**(Suppl 2), S75–80.

Brubacher G, Muller-Mulot W, and Southgate DAT (1985) *Methods for the Determination of Vitamins in Food: Recommended by COST 91.* London: Elsevier Applied Science Publishers.

Crews H, Alink G, Andersen R, Braesco V, Holst B, Maiani G, Ovesen L, Scotter M, Solfrizzo M, van den Berg R, Verhagen H, and Williamson G (2001) A critical assessment of some biomarker approaches linked with dietary intake. *British Journal of Nutrition* **86**(Suppl 1), S5–35.

Donnelly JG (2001) Nutrient requirements in health and disease. In *Food and Nutritional Supplements: Their Role in Health and Disease,* JK Ransley, JK Donnelly, and NW Read (eds.), pp. 29–44. Berlin: Springer.

Fairfield KM and Fletcher RH (2002) Vitamins for chronic disease prevention in adults: scientific review. *JAMA* **287,** 3116–26.

Fenech M (2001) Recommended dietary allowances (RDAs) for genomic stability. *Mutation Research* **480,** 51–4.

Gibson R (1990) *Principles of Nutritional Assessment.* Oxford: Oxford University Press.

Hathcock JN (1997) Vitamins and minerals: efficacy and safety. *American Journal of Clinical Nutrition* **66,** 427–37.

Powers H (1997) Vitamin requirements for term infants: considerations for infant formula. *Nutrition Research Reviews* **10,** 1–33.

Ransley JK (2001) The rise and rise of food and nutritional supplements: an overview of the market. In *Food and Nutritional Supplements: Their Role in Health and Disease,* JK Ransley, JK Donnelly, and NW Read (eds.), pp. 1–16. Berlin: Springer.

van den Berg H,van der Gaag M, and Hendriks H (2002) Influence of lifestyle on vitamin bioavailability. *International Journal of Vitamin and Nutrition Research* **72,** 53–9.

Various authors (1996) New approaches to define nutrient requirements. *American Journal of Clinical Nutrition* **63,** 983s–1001s.

Walter P, Stähelin H, and Brubacher G (1989) *Elevated Dosages of Vitamins: Benefits and Hazards.* Toronto: Hans Huber Publishers.

References cited in the text are listed in the Bibliography.

TWO

Vitamin A: Retinoids and Carotenoids

Vitamin A deficiency is a serious problem of public health nutrition, second only to protein-energy malnutrition worldwide, and is probably the most important cause of preventable blindness among children in developing countries. Marginal deficiency is a significant factor in childhood susceptibility to infection, and hence morbidity and mortality, in developing countries; even in developed countries, vitamin A (along with iron) is the nutrient most likely to be supplied in marginal amounts. In addition to primary deficiency of the vitamin, secondary (functional) vitamin A deficiency results from protein-energy malnutrition, because of impaired synthesis of the plasma retinol binding protein (RBP) that is required for transport of the vitamin from liver reserves to its sites of action.

The main physiologically active forms of vitamin A are retinaldehyde and retinoic acid, both of which are derived from retinol. Retinaldehyde functions in the visual system as the prosthetic group of the opsins, which act as the signal transducers between reception of light in the retina and initiation of the nervous impulse.

Retinoic acid modulates gene expression and tissue differentiation, acting by way of nuclear receptors. Historically, there was confusion between the effects of deficiency of vitamins A and D; by the 1950s, it was believed that the confusion had been resolved. Elucidation of the nuclear actions of the two vitamins has shown that, in many systems, the two act in concert, forming retinoid–vitamin D heterodimeric receptors; hypervitaminosis A can antagonize the actions of vitamin D.

In vitro, and in experimental animals, vitamin A has anticancer action related to its role in modulating gene expression and tissue differentiation. It retards the initiation and growth of some experimental tumors. However, it only shows these effects at toxic levels, and a number of synthetic analogs, collectively known as retinoids, have been developed for use as anticancer drugs and in dermatology.

Preformed vitamin A is found only in animals and a small number of bacteria. A number of the carotenoid pigments in plants can be cleaved oxidatively to yield retinol; β-carotene is quantitatively the most important of these provitamin A carotenoids. Although preformed retinol is both acutely and chronically toxic in excess, carotene is not, because there is only a limited capacity to cleave it to retinol.

In addition to its provitamin A role, β-carotene is a radical trapping antioxidant and may be nutritionally important in its own right both as an antioxidant and possibly also through direct actions that are independent of retinoids. Other carotenoids that occur in foods, and circulate in the bloodstream, also have free radical trapping activity, and, hence, potential metabolic significance, whether or not they are metabolic precursors of vitamin A.

2.1 VITAMIN A VITAMERS AND UNITS OF ACTIVITY

The term vitamin A can include any compound with the biological activity of the vitamin: provitamin A carotenoids, retinol, and its active metabolites.

2.1.1 Retinoids

The term retinoid is used to include retinol and its derivatives and analogs, either naturally occurring or synthetic, with or without the biological activity of the vitamin. The main biologically active retinoids are shown in Figure 2.1; until the late 1990s, only retinol, retinaldehyde, all-*trans*-retinoic acid, and 9-*cis*-retinoic acid were known to be biologically active. However, a number of other retinoids are now also known or believed to have important functions, including 4-oxo-retinol, 4-oxo-retinoic acid, and a variety of retroretinoids. The term *rexinoid* has been introduced to include only those retinoids that bind to the retinoid X receptor (RXR) and not the retinoic acid receptor (RAR) (Section 2.3.2.1).

Free retinol is chemically unstable and does not occur to any significant extent in foods or tissues. Rather, it is present as a variety of esters, mainly retinyl palmitate. Retinyl acetate is generally used as an analytical standard and in pharmaceutical preparations. Dehydroretinol (vitamin A_2) is found in freshwater fishes and amphibians; it has about half the biological activity of retinol.

Retinoic acid occurs in foods in only small amounts. In conventional biological assays, it has lower potency than retinol or retinyl esters because it is not stored, but is metabolized rapidly. Furthermore, because retinoic acid

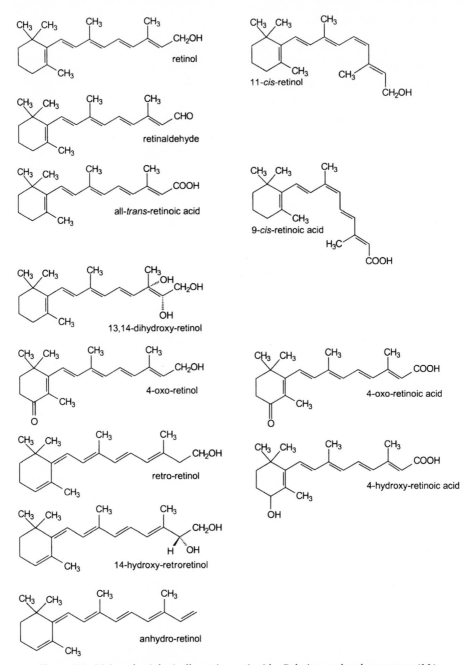

Figure 2.1. Major physiologically active retinoids. Relative molecular masses (M_r): retinol, 286.5; retinaldehyde, 284.4; retinoic acid, 300.4; 13,14-dihydroxy-retinol, 318.5; 4-oxoretinol, 301.5; retroretinol, 286.5; 14-hydroxy-retroretinol, 302.5; anhydroretinol, 269.6; 4-oxo-retinoic acid, 315.4; 4-hydroxyretinoic acid, 316.4; dehydroretinol, 284.4; retinyl acetate, 328.5; and retinyl palmitate, 524.9.

cannot be reduced to retinaldehyde or retinol, it cannot support vision (or fertility) in deficient animals.

Anhydroretinol binds to plasma and intracellular RBPs, but not to the cellular retinoic acid binding proteins (CRABPs) or retinoid receptors. In experimental animals, it protects against the development of chemically induced tumors while showing none of the toxic effects of other retinoids.

4-Oxoretinol is synthesized from canthaxin (see Figure 2.2) and occurs in plasma. It binds to the retinoic acid nuclear receptor (but not the RXR) and is active in tissue differentiation. In the early embryo, the main biologically active retinoid is 4-oxoretinaldehyde, which both activates RARs and also acts as a precursor of oxo-retinoic acid and oxoretinol.

2.1.2 Carotenoids

A number of carotenoids that have an unsubstituted β-ionone ring can be cleaved oxidatively to yield retinaldehyde, which is then reduced to retinol or oxidized to retinoic acid. Such compounds are known collectively as provitamin A carotenoids. Because relatively few foods are especially rich sources of retinol – and these are all animal foods – they are nutritionally important, accounting for a significant proportion of total vitamin A intake in most countries. Even in developed countries, with a relatively high intake of animal foods and fortified products, 25% to 30% of dietary vitamin A is derived from carotenoids; in developing countries, 80% or more of the potentially available vitamin A is from carotenoids. Only a small number of the several hundred carotenoid pigments found in plants have an unsubstituted β-ionone ring and are therefore capable of cleavage to yield retinaldehyde. The major dietary carotenoids are shown in Figure 2.2.

Carotenoids are classified in two ways:

1. Nutritionally, on the basis of whether or not they have an unsubstituted β-ionone ring and can therefore act as precursors of retinol (the provitamin A carotenoids)
2. On the basis of their chemistry, as:
    (a) hydrophobic hydrocarbon carotenoids, of which α- and β-carotene and lycopene are the most important in human tissues; lycopene is not a precursor of vitamin A in mammals, although it is the acyclic precursor of β-carotene in plants
    (b) monohydroxycarotenoids, of which β-cryptoxanthin is the most important in human tissues
    (c) dihydroxycarotenoids, of which lutein (dihydroxy α-carotene) and zeaxanthin (dihydroxy β-carotene) are the most important in human tissues.

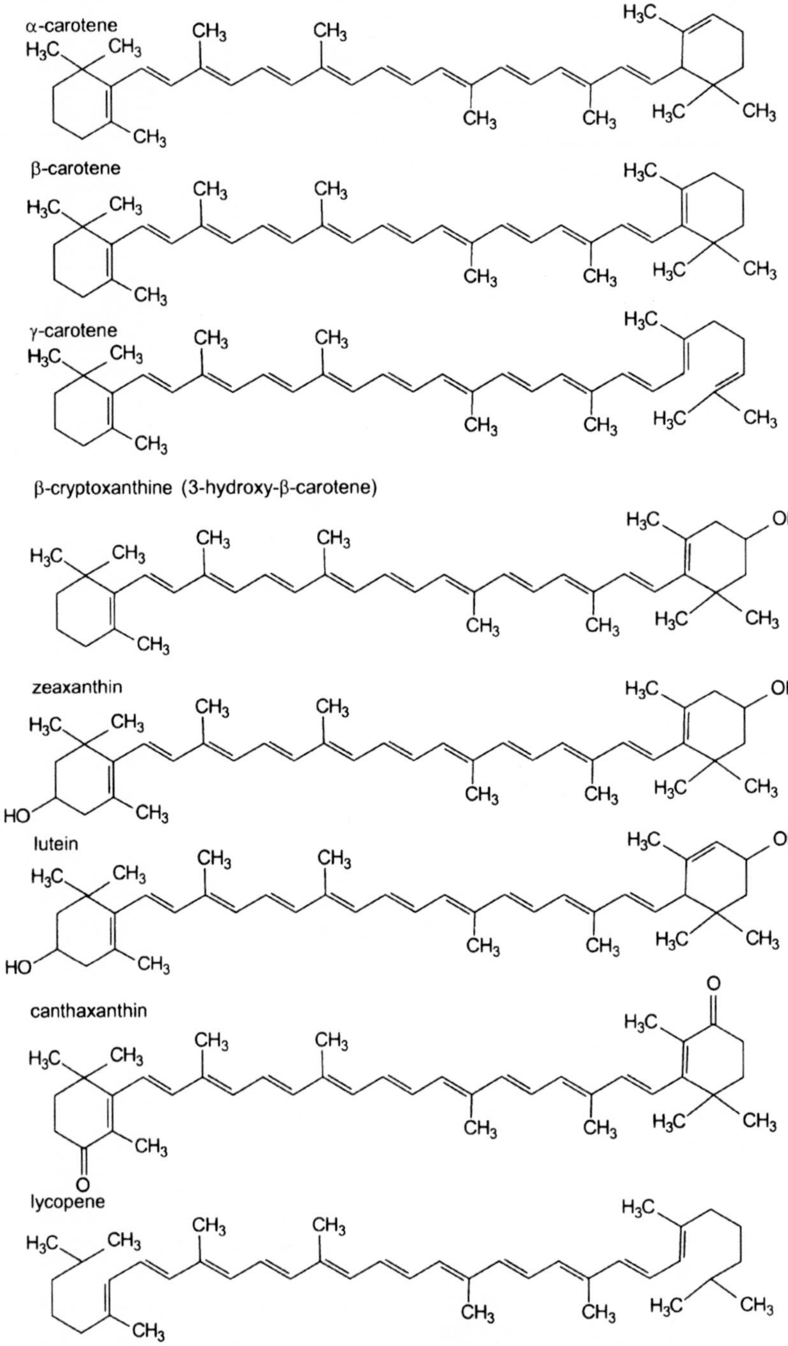

Figure 2.2. Major dietary carotenoids. α-, β-, and γ-carotenes and β-cryptoxanthine have vitamin A activity; zeaxanthin, lutein, canthaxanthin, and lycopene do not. Relative molecular masses (M_r): α-, β-, and γ-carotenes, 536.9; β-cryptoxanthine, 552.9; zeaxanthin, 568.9; lutein (xanthophyll), 568.85; canthaxanthin, 564.8; and lycopene, 536.9.

In plasma, the more hydrophobic carotenoids are deep within chylomicrons or very low-density lipoproteins, whereas the more polar hydroxycarotenoids are at the surface, and therefore potentially available for transfer between plasma lipoproteins and uptake into tissues.

2.1.3 International Units and Retinol Equivalents

The obsolete International Unit (iu) of vitamin A activity was based on biological assay of the ability of the test compound to support growth in deficient animals (1 iu = 10.47 nmol of retinol = 0.3 μg of free retinol or 0.344 μg of retinyl acetate).

13-*Cis*-retinol has 75% of the biological activity of all-*trans*-retinol, and retinaldehyde has 90%. Food composition tables give total preformed vitamin A as the sum of all-*trans*-retinol + 0.75 × 13-*cis*-retinol + 0.9 × retinaldehyde (Holland et al., 1991).

To take account of the contribution from carotenoids, the total vitamin A content of foods is expressed as micrograms of retinol-equivalents – the sum of that provided by retinoids and from carotenoids. Because of the relatively low absorption of carotenes and incomplete metabolism to yield retinol (Section 2.2.2), 6 μg of β-carotene is 1 μg of retinol-equivalent – a molar ratio of 3.2 mol of β-carotene equivalent to 1 mol of retinol. β-Carotene is absorbed much better from milk than from other foods; in milk, 2 μg of β-carotene is 1 μg of retinol-equivalent (1.07 mol equivalent to 1 mol of retinol). This is still far from the theoretical yield of 2 mol of retinol per mol of β-carotene. Other provitamin A carotenoids yield at most half the retinol of β-carotene, and 12 μg of these compounds = 1 μg of retinol-equivalent. On this basis, 1 iu of vitamin A activity = 1.8 μg of β-carotene or 3.6 μg of other provitamin A carotenoids.

The U.S./Canadian Dietary Reference Values report (Institute of Medicine, 2001) introduced the term retinol activity equivalent to take account of the incomplete absorption and metabolism of carotenoids; 1 RAE = 1 μg of all-*trans*-retinol, 12 μg of β-carotene, and 24 μg of α-carotene or β-cryptoxanthin. On this basis, 1 iu of vitamin A activity = 3.6 μg of β-carotene or 7.2 μg of other provitamin A carotenoids.

2.2 ABSORPTION AND METABOLISM OF VITAMIN A AND CAROTENOIDS

2.2.1 Absorption and Metabolism of Retinol and Retinoic Acid

About 70% to 90% of dietary retinol is absorbed, and, even at high intakes, this falls only slightly. Retinyl esters are hydrolyzed by pancreatic lipase and carboxyl ester lipase in lipid micelles in the intestinal lumen, and also by one or more retinyl ester hydrolases in the intestinal mucosal brush border membrane. At physiological levels of intake, retinol uptake into enterocytes is by facilitated diffusion from the lipid micelles. When the transport protein in the intestinal mucosal brush border cells is saturated, there is also passive uptake of retinol.

Within the enterocyte, retinol is bound to cellular retinol binding protein (CRBP II) and is esterified by lecithin:retinol acyltransferase (LRAT), which uses phosphatidylcholine as the fatty acid donor, mainly yielding retinyl palmitate, although small amounts of stearate and oleate are also formed. At unphysiologically high levels of retinol, when CRBP II is saturated, acyl coenzyme A (CoA):retinol acyltransferase (ARAT) esterifies the free retinol that accumulates in intracellular membranes. Then the retinyl esters enter the lymphatic circulation and then the bloodstream (in chylomicrons), together with dietary lipid and carotenoids (Norum et al., 1986; Olson, 1986; Blomhoff et al., 1991; Green et al., 1993; Harrison and Hussain, 2001).

A small proportion of dietary retinol is oxidized to retinoic acid, which is absorbed into the portal circulation and bound to serum albumin. Some retinyl esters are also transferred into the portal circulation. Patients with abetalipoproteinemia, who are unable to synthesize chylomicrons, can nevertheless maintain adequate vitamin A status if they are provided with relatively high intakes of retinol.

2.2.1.1 **Liver Storage and Release of Retinol**   Tissues can take up retinyl esters from chylomicrons, but most is left in the chylomicron remnants that are taken up into the liver by endocytosis. The retinyl esters are hydrolyzed at the hepatocyte cell membrane, and free retinol is transferred to the rough endoplasmic reticulum, where it binds to apo-RBP. Holo-RBP then migrates through the smooth endoplasmic reticulum to the Golgi and is secreted as a 1:1 complex with the thyroid hormone binding protein, transthyretin (Section 2.2.3).

Studies in vitamin A replete animals suggest that most of the retinol is transferred from hepatocytes to the perisinusoidal stellate cells of the liver. Here, it is again esterified by LRAT to form mainly retinyl palmitate (76% to 80%), with smaller amounts of stearate (9% to 12%), oleate (5% to 7%), and linoleate (3% to 4%). The stellate cells contain 90% to 95% of hepatic vitamin A, as cytoplasmic lipid droplets that consist of between 12% to 65% retinyl esters (Batres and Olson, 1987). Studies with [^{13}C]retinyl palmitate show that much of the recently ingested retinol appears more or less immediately in the circulation, bound to RBP, and is only sequestered in the liver reserves sub-sequently. This suggests that there may be little or no direct transfer of retinol from hepatocytes to stellate cells, but rather retinol is cleared from the circu-lation into stellate cells for storage. LRAT is induced by retinoic acid (and by dietary vitamin A) and is down-regulated in vitamin A depletion, when the need is to transfer vitamin A to tissues rather than to store it in the liver (Zolfaghari and Ross, 2000; Wolf, 2001).

Release of retinol from stellate cells into the circulation may occur either directly, as free retinol bound to RBP, or indirectly as a result of the transfer of retinol from stellate cells to hepatocytes. The release of retinol from stores is impaired in iron deficiency, as is the absorption of dietary vitamin A (Jang et al., 2000).

The concentration of retinol in most tissues is between 1 to 5 μmol per kg; in the liver, the mean concentration is 500 μmol per kg, with a very wide range of individual variation. In a number of studies of *postmortem* tissue, between 10% to 30% of the population of the United States had liver retinol below 140 μmol per kg, and about 5% had reserves in excess of 1,700 μmol per kg. Five percent to 10% of samples analyzed in Canada showed undetectably low liver reserves of retinol, although similar studies in Britain did not show any significant proportion of the population with extremely low liver reserves of vitamin A (Sauberlich et al., 1974; Huque, 1982). Abnormally low liver reserves of retinol may result not only from prolonged low intake, but also from the in-duction by barbiturates of cytochrome P_{450}, which catalyzes the catabolism of retinol (Section 2.2.1.2). Chlorinated hydrocarbons, as in many agricultural pesticides, also deplete liver retinol by effects on the metabolism of RBP (Section 2.2.3).

Opinions differ as to what constitutes an adequate concentration of retinol in the liver. When the concentration rises above 70 μmol per kg, there is increased catabolism of retinol (Section 2.2.1.2). Estimates of requirements based on the fractional catabolic rate of whole body retinol and using liver reserves of 70 μmol per kg as a basis for calculation are generally in agree-ment with estimates based on the very few depletion/repletion studies that have been performed (Section 2.4). However, from the observed range of liver reserves in healthy subjects, it can be argued that a more appropriate level is 140 μmol per kg, which gives a higher estimate of requirements (Sauberlich et al., 1974; Hodges et al., 1978; Olson, 1987a).

Although the major storage of vitamin A is in the liver (50% to 80% of the total body content), adipose tissue may contain 15% to 20% of total body vita-min A. Much of this is taken up from chylomicrons; retinyl esters are hydrolyzed by lipoprotein lipase (Blaner et al., 1994), but some vitamin A is also taken up from circulating vitamin bound to RBP. Release of retinol from adipose tissue is by hydrolysis of stored retinyl esters, catalyzed by (cAMP-stimulated) hormone-sensitive lipase, bound to RBP, which is synthesized by both white and brown adipose tissue (Wei et al., 1997).

A variety of other tissues synthesize RBP; this provides a mechanism for return to the liver of retinol in excess of requirements that has been taken up from chylomicrons by the action of lipoprotein lipase. Because these tissues do not synthesize transthyretin, the binding of holo-RBP to transthyretin must occur in the circulation after release.

2.2.1.2 Metabolism of Retinoic Acid   Retinoic acid is the normal major metabolite of physiological amounts of retinol. However, it is not a catabolic product of retinol, but the ligand for nuclear retinoid receptors involved in modulation of gene expression (Section 2.3.2). It may be formed in the liver, although there is no hepatic storage, and is then transported bound to serum albumin rather than RBP. Other tissues are also able to form retinoic acid from retinol. The rate-limiting step is the dehydrogenation of retinol to retinaldehyde; the K_m of the dehydrogenase is high, so that a major determinant of the rate of formation of retinoic acid will be the concentration of retinol in the cell (Napoli, 1996).

Cytosolic alcohol dehydrogenases only act on free retinol, not retinol bound to CRBP, so they are unlikely to be involved in formation of retinaldehyde and retinoic acid. Furthermore, inhibition of cytosolic alcohol dehydrogenases does not inhibit the oxidation of retinol to retinoic acid (Boerman and Napoli, 1996). CRBP-bound retinol is a substrate for at least three microsomal $NADP^+$-dependent dehydrogenases; but, given the intracellular $NADP^+$:NADPH ratio (0.01, compared with an NAD^+:NADH ratio of the order of 10^3), it is likely that these microsomal enzymes will act mainly to reduce retinaldehyde to retinol and not to oxidize retinol.

A microsomal retinol dehydrogenase catalyzes the oxidation of CRBP-bound all-*trans*-retinol to retinaldehyde; it also acts as a 3α-hydroxysteroid dehydrogenase. A similar enzyme catalyzes the oxidation of 9-*cis*- and 11-*cis*-retinol, but not all-*trans*-retinol; again, it has 3α-hydroxysteroid dehydrogenase activity. In the eye, the major product of this enzyme is 11-*cis*-retinaldehyde, whereas in other tissues it is 9-*cis*-retinaldehyde, which is then oxidized to 9-*cis*-retinoic acid (Section 2.3.2.1; Chen et al., 2000; Duester, 2000, 2001; Gamble et al., 2000; Napoli, 2001). Although there is known to be an isomerase in the eye for the formation of 11-*cis*-retinaldehyde as a substrate for the dehydrogenase, there is no information concerning an iso-merase in other tissues to produce 9-*cis*-retinol (Wang et al., 1999; Gamble et al., 2000; McBee et al., 2000).

Intracellular concentrations of retinoic acid are controlled not only by the rate of synthesis, but also by catabolism. At least in culture, prior exposure of cells to retinoic acid induces the enzymes of retinoic acid catabolism (Chytil, 1984; Napoli and Race, 1987). The major metabolite of retinoic acid is the glucuronide (Section 2.2.1.3).

All-*trans*-retinoic acid (but apparently not 9-*cis*-retinoic acid) undergoes microsomal oxidation to yield a variety of polar metabolites. Retinoic acid hy-droxylase is a retinoic acid-induced cytochrome P_{450} (CYP26) – from its amino acid sequence, it appears to represent a novel family of cytochrome P_{450}. 4-Hydroxyretinoic acid then undergoes further oxidation to yield 4-oxo-retinoic acid. The same enzyme also catalyzes 18-hydroxylation and 5,6-epoxidation of retinoic acid. 4-Hydroxy- and 4-oxo-retinoic acids were originally consid-ered to be inactivation products of all-*trans*-retinoic acid; however, they bind to, and activate, the RAR and show high activity as modulators of positional specificity in Xenopus embryogenesis. Furthermore, CYP26 is expressed dif-ferently through the process of embryogenesis (Sonneveld et al., 1998).

In addition to oxidation of retinol, retinoic acid may be formed by the β-oxidation of apo-carotenals arising from the asymmetric cleavage of β-carotene (Section 2.2.2.1).

Retinoic acid regulates its own synthesis from retinol in a variety of ti-ssues by induction of LRAT; this increases the rate of esterification of re-tinol, thereby decreasing the amount available for oxidation to retinoic acid (Kurlandsky et al., 1996). Retinoic acid also induces the cytochrome P_{450} that catalyzes oxidation to 4-oxo-retinoic acid, and regulates both its own syn-thesis and catabolism.

2.2.1.3 Retinoyl Glucuronide and Other Metabolites The main excre-tory product of both retinol and retinoic acid is retinoyl glucuronide, which is secreted in bile. Some retinoyl taurine is also secreted in bile. This suggests that, in addition to regulated synthesis of retinoic acid for its biological activity, oxidation to the acid is also a significant pathway for catabolism of retinol. Un-like retinoyl glucuronide, which has biological activity, retinoyl taurine seems to be solely an excretory product.

The plasma concentration of retinoyl glucuronide is between 5 and 14 nmol per L, and the activity of retinoic acid UDP-glucuronyltransferase increases in vitamin A deficiency, suggesting that glucuronidation may be important other than as a pathway for inactivation and excretion of retinoic acid (Miller and DeLuca, 1986). Retinoyl glucuronide has biological activity; it is not clear whether or not this is as a result of hydrolysis to retinoic acid. In some experimental systems, the glucuronide appears to act without undergoing hydrolysis, although it binds to neither cellular retinoic acid binding protein nor nuclear retinoid receptors. However, glucuronidases in the liver and kidney hydrolyze retinoyl glucuronide, and activity of the glucuronidase, like that of the UDP-glucuronyltransferase, increases in vitamin A deficiency (Barua, 1997; Sidell et al., 2000). It has also been suggested that retinoyl glucuronide, rather than retinoic acid, may be the precursor of retinoyl CoA for retinoylation of proteins (Section 2.3.3.1).

Small amounts of a number of other metabolites, including epoxy-retinoic acid glucuronide and a number of products of side-chain oxidation of retinol and retinoic acid, are also formed, some of which are excreted in the urine as well as bile. As the intake of retinol increases, and the liver concentration rises above 70 μmol per kg, a different catabolic pathway becomes increasingly important for the catabolism of retinol in liver parenchymal cells. This is a microsomal cytochrome P_{450}-dependent oxidation, leading to a number of polar metabolites, including 4-hydroxyretinol, which are excreted in the urine and bile. Thus, there is a catabolic mechanism that allows excretion of excess retinol. However, at high intakes, the microsomal pathway is saturated, and this may be one of the factors in the toxicity of excess retinol, because there is no further capacity for its catabolism and excretion. Stored retinyl esters in the stellate cells of the liver are only slowly released to the parenchymal cells for catabolism, and retinol is chronically toxic (Section 2.5.1). Induction of cytochrome P_{450} enzymes by chronic administration of barbiturates can result in depletion of liver reserves of retinol and may be a factor in drug-induced vitamin A deficiency (Leo and Lieber, 1985; Olson, 1986; Leo et al., 1989).

Anhydroretinol may arise by nonenzymic isomerization of all-*trans*-retinol under acidic conditions and can act as a precursor for the synthesis of other biologically active retroretinoids (McBee et al., 2000).

2.2.2 Absorption and Metabolism of Carotenoids
Carotenoids are absorbed passively, dissolved in lipid micelles; various studies have estimated the biological availability and absorption of dietary carotene as between 5% to 60%, depending on the nature of the food, whether it is cooked or raw, and the amount of fat in the meal. In addition, much of the carotene in foods is present as crystals that may not dissolve to any significant extent in intestinal contents (Parker, 1989; 1996; Parker et al., 1999; Hickenbottom et al., 2002; Ribaya-Mercado, 2002; Tanumihardjo, 2002; Yeum and Russell, 2002).

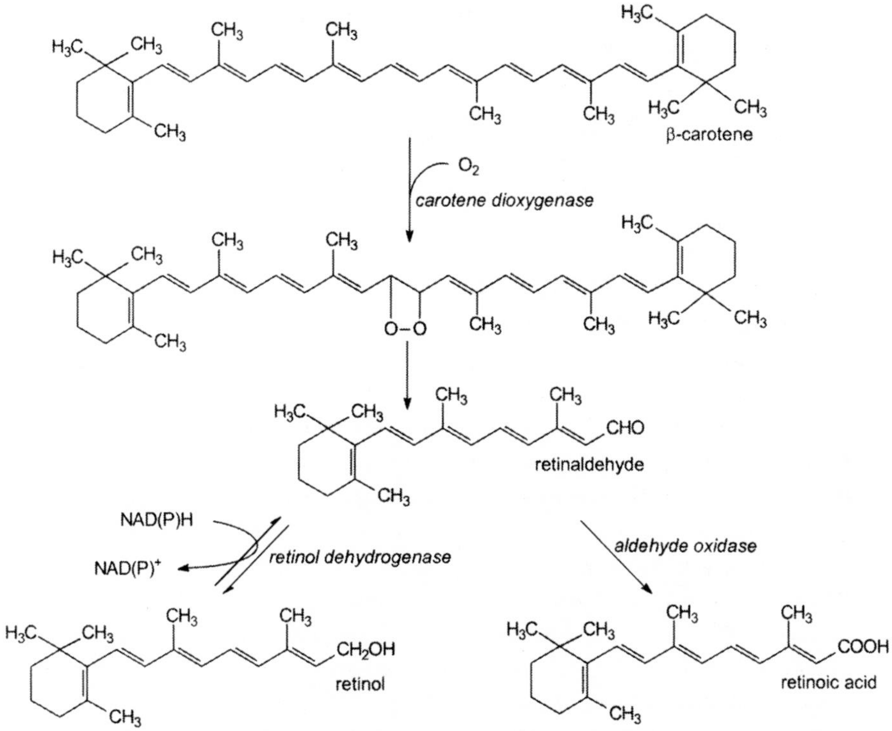

Figure 2.3. Oxidative cleavage of β-carotene by carotene dioxygenase (EC 1.14.99.36), and onward metabolism of retinaldehyde catalyzed by retinol dehydrogenase (EC 1.1.1.105) and retinaldehyde oxidase (EC 1.2.3.11).

Although the generally accepted factors for calculating retinol equivalence of dietary carotenoids are that 1 μg of retinol is provided by 6 μg of β-carotene or 12 μg of other provitamin A carotenoids, feeding studies suggest that 1 μg of retinol is provided by 26 μg of β-carotene from dark green leafy vegetables (with a range of 13 to 76 μg), or 12 μg from yellow and orange fruits (with a range of 6 to 29 μg) (Castenmiller and West, 1998; West and Castenmiller, 1998). A number of studies have shown that increasing the consumption of dark green leafy vegetables as a means of increasing vitamin A intake in children in developing countries is unlikely to provide a significant improvement in nutritional status. These studies provide the rationale for the retinol activity equivalents that are half the traditional retinol equivalents (Section 2.1.3).

2.2.2.1 **Carotene Dioxygenase**   As shown in Figure 2.3, β-carotene and other provitamin A carotenoids undergo oxidative cleavage to retinaldehyde in the intestinal mucosa, catalyzed by carotene dioxygenase. Retinaldehyde binds to the intracellular retinoid binding protein (CRBP II), and is reduced to retinol by a microsomal dehydrogenase, then esterified and secreted in chylomicrons together with retinyl esters formed from dietary retinol.

As discussed in Section 2.2.2.2, only a relatively small proportion of carotene undergoes oxidation in the intestinal mucosa, and a significant amount of carotene enters the circulation in chylomicrons. Novotny and coworkers (1995) reported a study in one subject given an oral dose of $[^2H]\beta$-carotene dissolved in oil; 22% was absorbed – 17.8% as carotene and 4.2% as retinoids. Their results suggest that nonintestinal carotene dioxygenase is important in retinoid formation, because there was a late disappearance of labeled carotene from the circulation and the appearance of labeled retinoids.

There is some hepatic cleavage of carotene taken up from chylomicron remnants, again giving rise to retinaldehyde and retinyl esters; the remainder is secreted in very low-density lipoproteins and may be taken up and cleaved by carotene dioxygenase in extrahepatic tissues. During and coworkers (1996) reported the activity of carotene dioxygenase in various tissues: liver (25% of the specific activity in intestinal mucosa), brain (2.5% of intestinal activity), and lungs (1% of intestinal activity), with some activity also in the kidneys. The carotene 15,15′-dioxygenase gene is expressed in duodenal villi, the liver, lungs, and kidney tubules (Wyss et al., 2001).

Central oxidative cleavage of β-carotene, as shown in Figure 2.3, gives rise to two molecules of retinaldehyde, which can be reduced to retinol. However, as noted previously, the biological activity of β-carotene, on a molar basis, is considerably lower than that of retinol, not two-fold higher as might be expected. Three factors may account for this: limited absorption of carotenoids from the intestinal lumen, limited activity of carotene dioxygenase, and excentric (asymmetric) cleavage.

2.2.2.2 Limited Activity of Carotene Dioxygenase   The intestinal activity of carotene dioxygenase is relatively low, so that in many species (including human beings) a relatively large proportion of ingested β-carotene may appear in the circulation unchanged. In general, herbivores have higher activity of carotene dioxygenase than omnivores. In some carnivores, such as the cat, there is virtually no carotene dioxygenase activity, and cats are unable to meet their vitamin A requirements from carotene (Lakshmanan et al., 1972). Species with high intestinal activity of carotene dioxygenase have white body fat, whereas in species with lower activity, body fat has a yellow tinge. Although the activity of carotene dioxygenase in most species is probably adequate to meet vitamin A requirements solely from dietary carotene, it is low enough to ensure that even very high intakes of carotene will not result in the formation of potentially toxic amounts of retinol (Section 2.5.1).

In animals fed a vitamin A-deficient diet, the activity of intestinal carotene dioxygenase is significantly higher than in animals fed a high intake of carotene or preformed retinol. Dietary protein also affects the intestinal conversion of carotene to retinol, resulting in increased liver retinol stores in animals fed a high-protein diet. In both human beings and experimental animals, feeding a high-protein diet results in increased activity of intestinal mucosal carotene dioxygenase. By contrast, protein deficiency in experimental animals and protein-energy malnutrition in human beings lead to reduced cleavage of carotene to vitamin A (van Vliet et al., 1996; Parvin and Sivakumar, 2000).

Other carotenoids in the diet, which are not substrates, such as canthaxanthin and zeaxanthin, may inhibit carotene dioxygenase and reduce the proportion that is converted to retinol (Grolier et al., 1997). Similarly, a variety of antioxidants that occur in foods together with carotenoids, including flavonoids (Section 14.7.2), also inhibit carotene dioxygenase.

A number of studies have reported low serum concentrations of retinol and high concentrations of β-carotene in patients with insulin-dependent diabetes mellitus. Krill and coworkers (1997) showed that up to one-third of nondiabetic first-degree relatives of patients with diabetes also showed a low serum retinol:carotene ratio, implying a genetic predisposition to low activity of carotene dioxygenase, possibly associated with insulin-dependent diabetes.

2.2.2.3 The Reaction Specificity of Carotene Dioxygenase

Whereas the principal site of carotene dioxygenase attack is the 15,15′-central bond of β-carotene, there is evidence that asymmetric cleavage also occurs, leading to formation of 8′-, 10′-, and 12′-apo-carotenals, as shown in Figure 2.4. These apo-carotenals are metabolized by oxidation to apo-carotenoic acids, which are substrates for β-oxidation to retinoic acid and a number of other metabolites.

Early studies of the reaction specificity of carotene dioxygenase in intestinal mucosal homogenates suggested that it catalyzed both central and asymmetric cleavage (Wang et al., 1991, 1992), although there was evidence that excentric cleavage was nonenzymic. Devery and Milborrow (1994) suggested that there are two enzymes: a cytosolic dioxygenase that acts centrally and a membrane-associated enzyme that catalyzes asymmetric cleavage. Using intestinal homogenates and under conditions to minimize nonenzymic action, there was a near stoichiometric yield of retinaldehyde from β-carotene

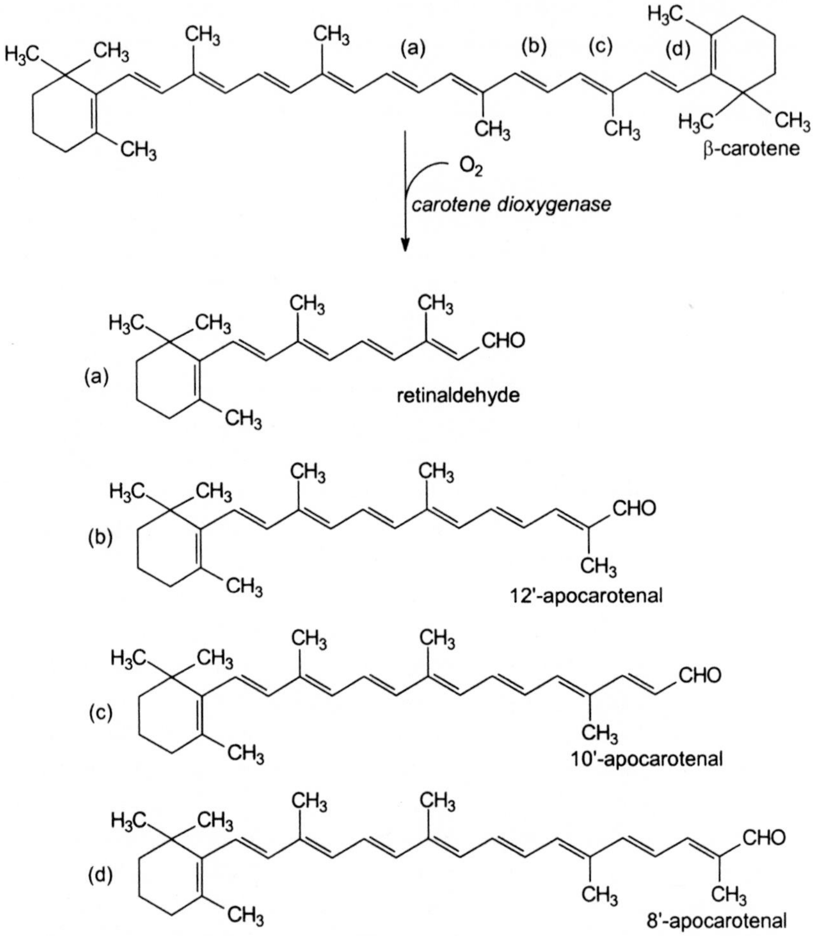

Figure 2.4. Potential products arising from enzymic or nonenzymic symmetrical (a) or asymmetric (b to d) oxidative cleavage of β-carotene. Apocarotenals can undergo side chain oxidation to yield retinoic acid, but cannot form retinaldehyde or retinol.

(Nagao et al., 1996); later studies suggested that excentric cleavage did not occur in the presence of antioxidants, such as α-tocopherol (Yeum et al., 2000). Genetic cloning of an enzyme that catalyzed specifically central cleavage supported the view that excentric cleavage was an artifact (Barua and Olson, 2000; Redmond et al., 2001); however, there is also a mammalian enzyme that catalyzes C9′ to 10′ cleavage of β-carotene, yielding apo-10-carotenal and β-ionone, an enzyme that also catalyzes cleavage of lycopene (Kiefer et al., 2001). There is some evidence that apocarotenoic acids arising from asymmetric cleavage have effects on cell proliferation independently of actions mediated by RARs (Tibaduiza et al., 2002).

2.2.3 Plasma Retinol Binding Protein (RBP)

Retinol is released from the liver bound to an α-globulin, retinol binding protein (RBP); this serves to maintain the vitamin in aqueous solution, protects it against oxidation, and also delivers the vitamin to target tissues. RBP binds 1 mol of retinol per mol of protein.

RBP forms a 1:1 complex with the tetrameric thyroxine-binding prealbumin, transthyretin. This is important to prevent urinary loss of retinol bound to the relatively small RBP (M_r 21,000), which would be filtered by the glomerulus; transthyretin has an M_r of 54,000; hence, the complex will not normally be filtered. However, moderate renal damage, or the increased permeability of the glomerulus in infection, may result in considerable loss of vitamin A bound to RBP-transthyretin.

The transthyretin tetramer could theoretically bind 2 mol of holo-RBP, but does not, because holo-RBP is limiting. In vitamin A deficiency, the ratio of RBP:transthyretin falls, indicating that although binding to transthyretin is essential for secretion of holo-RBP from the liver, vitamin A is not essential for secretion of transthyretin. Other tissues secrete holo-RBP, but not transthyretin; it is assumed that this binds to transthyretin in the circulation.

In the liver, the apo-RBP-transthyretin complex is formed in the rough endoplasmic reticulum and then migrates through the smooth endoplasmic reticulum to the Golgi on binding retinol. Calnexin, an integral membrane protein, coprecipitates with the apo-RBP-transthyretin complex, suggesting that migration of the apo-protein into the Golgi is prevented by membrane binding. Binding of retinol to the complex displaces calnexin, so that the holo-RBP-transthyretin complex is now free to migrate to the Golgi for secretion (Bellovino et al., 1996).

Metabolites of polychlorinated biphenyls bind to the thyroxine binding site of transthyretin and, in doing so, impair the binding of RBP. As a result of this, there is free RBP-bound retinol in plasma, which is filtered at the glomerulus and hence lost in the urine. This may account for the vitamin A depleting action of polychlorinated biphenyls (Brouwer and van den Berg, 1986).

RBP is relatively rich in aromatic amino acids, which create a deep hydrophobic pocket that is specific for the β-ionone ring, polyene side chain, and polar end group. In addition to all-*trans*-retinol, RBP binds retinaldehyde, retinoic acid, and 13-*cis*-retinol, but not retinyl esters or β-carotene. RBP shows considerable structural homology with β-lactoglobulin from milk and other binding proteins for lipophilic compounds. β-Lactoglobulin also binds retinol and may be important in the absorption of the vitamin in young animals.

Cell surface receptors in target tissues take up retinol from the RBP-transthyretin complex, esterifying it externally, then transferring free retinol by esterase activity onto an intracellular RBP. Part of the function of the receptor is to catalyze a conformational change in RBP so that the retinol held in the hydrophobic pocket can be released. There is no endocytosis of the RBP-transthyretin complex.

The cell surface receptors also remove the carboxy terminal arginine residue from RBP, thus inactivating it by reducing its affinity for both transthyretin and retinol. As a result, apo-RBP is filtered at the glomerulus. Some may be lost in the urine, but most is resorbed in the proximal renal tubules and is then catabolized by lysosomal hydrolases. This seems to be the main route for catabolism of RBP; the apo-protein is not recycled (Peterson et al., 1974).

During the development of vitamin A deficiency in experimental animals, the plasma concentration of RBP falls, while the liver content rises. The administration of retinol to deficient animals results in a considerable release of holo-RBP from the liver. This is a rapid effect on the release of preformed apo-RBP in response to the availability of retinol, rather than an increase in the synthesis of the protein. There is no evidence that retinol controls the synthesis of RBP (Soprano et al., 1982). This provides the basis of the relative dose response (RDR) test for liver stores of vitamin A (Section 2.4.1.3); administration of a test dose of retinol gives a considerably greater increase in plasma retinol, bound to RBP, in deficient subjects than in those with adequate liver reserves, because of the accumulation of apo-RBP in the liver.

As well as protecting retinol against oxidation, the binding of retinol to RBP may also serve to protect the body against the general membrane-seeking and potentially membrane lytic effects of retinol. In tissue culture, the addition of retinol nonspecifically bound to albumin results in lysosomal membrane damage and the release of lysosomal hydrolases. Retinol bound to RBP does not have this effect, suggesting that the high affinity of RBP for retinol protects tissues against nonspecific uptake of the vitamin. Vitamin A toxicity occurs when there is such an excess that RBP is saturated and retinol circulates bound to other proteins and as esters in plasma lipoproteins (Meeks et al., 1981).

Protein-energy malnutrition results in functional vitamin A deficiency, with very low circulating levels of the vitamin and development of clinical signs of xerophthalmia (Section 2.4). The condition is unresponsive to the administration of vitamin A and often occurs despite adequate liver reserves of retinol. The problem is one of impaired synthesis of RBP in the liver and hence a seriously impaired ability to release retinol from liver stores. During rehabilitation of protein-energy malnourished children, there is a rapid increase in plasma retinol as a result of increased synthesis of RBP. Deficiency of zinc also impairs synthesis of RBP.

2.2.4 Cellular Retinoid Binding Proteins: CRBPs and CRABPs

There are five intracellular retinoid binding proteins that show considerable sequence homology not only with each other, but also with a variety of other intracellular binding proteins for hydrophobic metabolites, including the intracellular fatty acid binding protein (Li and Norris, 1996; Noy, 2000; Vogel et al., 2001).

The two cellular retinol binding proteins bind all-*trans*- and 13-*cis*-retinol, but not 9-*cis*- or 11-*cis*-retinol, or retinoic acid. They also bind retinaldehyde, although there is a distinct retinaldehyde binding protein in the eye:

1. CRBP(I) occurs in almost all tissues, apart from skeletal and cardiac muscle; it is especially abundant in tissues that contain large amounts of retinol.
2. CRBP(II) occurs only in the small intestinal mucosal cells.
3. CRBP(III) occurs in skeletal and cardiac muscle.

There are two cellular retinoic acid binding proteins:

1. CRABP(I) is widely distributed.
2. CRABP(II) occurs primarily in the skin, uterus, ovary, choroid plexus, and in fetal cells.

All five proteins have a high affinity and are present in excess of their ligands, with CRBP 1.4-fold higher than intracellular retinol, and CRABP 20-fold higher than intracellular retinoic acid. This means that, under normal conditions, essentially all of the intracellular retinoids will be protein-bound.

The intracellular retinoid binding proteins function as a passive reservoir of retinoids, permitting accumulation within the cell while both protecting the ligands against oxidative damage and also protecting cells against the membrane lytic effects of retinoids.

They are also important in intracellular trafficking and transport of retinoids. CRBP(II) interacts directly with the enterocyte membrane retinol transporter, and CRBP(I) with the cell surface RBP receptor, thus permitting direct uptake and accumulation of retinol from the intestinal lumen and circulation respectively. CRBP(I) is present in large amounts in cells that synthesize and secrete RBP, suggesting that it also functions to transport retinol into the lumen of the endoplasmic reticulum and present it to apo-RBP.

CRABP(I) and CRABP(II) function to transport retinoic acid into the nucleus for binding to retinoid receptors. CRABP(II), with retinoic acid bound, also interacts directly with the liganded RAR-RXR heterodimer bound to hormone response elements on DNA and enhances the activity of the nuclear receptor (Section 2.3.2.1; Delva et al., 1999).

Both CRBP and CRABP are also important in regulating the metabolism of retinoids:

1. In the enterocyte, CRBP(II) regulates
   (a) reduction to retinol of the retinaldehyde formed by carotene dioxygenase;
   (b) esterification of retinol by LRAT.
2. In the liver and other tissues, CRBP(I) regulates
   (a) esterification of retinol and hydrolysis of retinyl esters;
   (b) oxidation of retinol to retinaldehyde; and
   (c) oxidation of retinaldehyde to retinoic acid.
3. CRABP is required for the microsomal oxidation of retinoic acid to more polar compounds.

In each case, the binding protein is required for binding of the substrate to the enzyme, and protein binding protects retinol from other enzymes that can act on free, but not protein-bound, substrate. It is likely that this requirement to interact with not only the ligand, but also the catalytic sites of enzymes, explains the very high degree of conservation of the cellular retinoid binding proteins across species (Napoli et al., 1995; Boerman and Napoli, 1996). After relatively large amounts of retinol, there is significant formation of (potentially teratogenic) all-*trans*-retinoic acid as a result of nonspecific (and hence unregulated) oxidation in enterocytes of excess retinol that is not bound to CRBP(II) (Arnhold et al., 2002).

In most tissues, apo-CRBP does not bind to enzymes; only the holo-CRBP binds. However, in the liver, apo-CRBP(I) does bind to LRAT and acts to reduce the rate of esterification of retinol when there is a significant excess of apo-CRBP. This serves to reduce the esterification of retinol for storage at times of low retinol availability and will presumable direct retinol into the endoplasmic reticulum for binding to apo-RBP for export. Apo-CRBP(II) in the intestinal mucosa does not bind to LRAT.

Apo-CRBP(I) stimulates the hydrolysis of retinyl esters, thus releasing retinol from stores for transfer to RBP. This means that the esterification and hydrolysis of retinyl esters is regulated to a considerable extent by the ratio of apo-CRBP:holo-CRBP.

2.3 METABOLIC FUNCTIONS OF VITAMIN A

Vitamin A has four metabolic roles:

1. as the prosthetic group of the visual pigments;
2. as a nuclear modulator of gene expression;
3. as a carrier of mannosyl units in the synthesis of hydrophobic glyco-proteins; and
4. in the retinoylation of proteins.

Ahmed and coworkers (1990) suggested that the development of clinical signs of vitamin A deficiency may require the additional stress of infection. They showed that vitamin A-depleted mice were more sensitive to rotavirus infection, and the infected animals showed clinical signs of vitamin A deficiency, whereas uninfected animals that had stopped growing as a result of vitamin A depletion did not. The increased susceptibility to infection was associated with reduced differentiation and formation of a specific subpopulation of intestinal, mucus-secreting goblet cells in the crypts of Leberkühn, suggesting the importance of retinol in mannosyltransfer reactions in mucopolysaccharide synthesis (Section 2.3.3).

2.3.1 Retinol and Retinaldehyde in the Visual Cycle

Binding of retinaldehyde to the protein opsin in the rods, and related proteins in the cones, of the retina gives a highly sensitive signal transduction and amplification system, such that a single photon results in a measurable change in the current across the outer section membrane, and hence the propagation of a nerve impulse. In the outer segment of rod cells, opsin may constitute more than 90% of the total protein; it is present at a concentration of approximately 3 mmol per L.

Photoexcited rhodopsin activates transducin, a G-protein, which in turn stimulates cyclic GMP phosphodiesterase; this leads to closing of an ion channel, hyperpolarization of the membrane, and a decreased rate of neurotransmitter release (Wald, 1968; Stryer, 1986; Chabré and Deterre, 1989).

The pigment epithelium of the retina receives all-*trans*-retinol from plasma RBP. It is then isomerized to 11-*cis*-retinol, which may either be stored as 11-*cis*-retinyl esters or oxidized to 11-*cis*-retinaldehyde, which is transported to the photoreceptor cells bound to an interphotoreceptor retinoid binding protein.

As shown in Figure 2.5, within the rods and cones of the retina, 11-*cis*-retinaldehyde forms a protonated Schiff base to the ε-amino group of lysine[296] in opsin, forming the holo-protein rhodopsin. Lysine[296] is within the membrane, in one of the transmembrane helical regions of the protein. Opsins are cell type-specific. They serve to shift the absorption of 11-*cis*-retinaldehyde from ultraviolet (UV) light into what we call, in consequence, the visible range – either a relatively broad spectrum of sensitivity for vision in dim light (in the rods, with an absorbance peak at 500 nm) or more defined spectral peaks for differentiation of colors in stronger light (in the cones). The absorption maxima are at 425 (blue), 530 (green), or 560 nm (red), depending on the cell type.

Any one cone cell contains only one type of opsin and is sensitive to only one color of light. Color blindness results from loss or mutation of one or the other of the cone opsins. The combination of 11-*cis*-retinaldehyde with cone opsin is sometimes called iodopsin, with rhodopsin meaning more specifically the holo-protein of rod opsin. Most studies of the mechanisms of vision shown in Figure 2.5 have been performed using rods; by extrapolation, it is assumed that the same mechanisms are involved in cone vision.

Opsin can be considered to be a retinaldehyde receptor protein, functioning in the same way as cell surface receptor G-proteins (Sakmar, 1998). Like receptor proteins, opsin is a transmembrane protein with seven α-helical regions in the transmembrane domain; the difference is that opsin spans the intracellular disk membrane of the rod or cone cell, whereas hormone and neurotransmitter receptors span the plasma membrane of the cell. The response time of rhodopsin is considerably faster than that of cell surface receptor proteins.

The absorption of light by rhodopsin results in a change in the configuration of the retinaldehyde from the 11-*cis* to the all-*trans* isomer, together with a conformational change in opsin. This results in both the release of retinaldehyde from the Schiff base and the initiation of a nerve impulse. The overall process is known as bleaching, because it results in the loss of the color of rhodopsin.

The formation of the initial excited form of rhodopsin – bathorhodopsin – depends on the isomerization of 11-*cis*-retinaldehyde to a strained form of all-*trans*-retinaldehyde. This occurs within picoseconds of illumination and is the only light-dependent step in the visual cycle. Thereafter, there is a series of conformational changes leading to the formation of metarhodopsin II. In metarhodopsin II, the Schiff base is unprotonated, and the retinaldehyde is in the unstrained all-*trans* configuration.

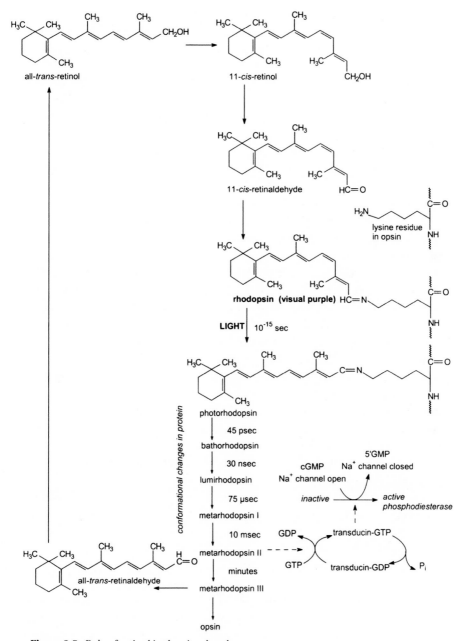

Figure 2.5. Role of retinol in the visual cycle.

The conversion of metarhodopsin II to metarhodopsin III is relatively slow, with a time course of minutes. It is the result of phosphorylation of serine residues in the protein catalyzed by rhodopsin kinase. The final step is hydrolysis to release all-*trans*-retinaldehyde and opsin.

Under conditions of low light intensity, the all-*trans*-retinaldehyde released from rhodopsin is reduced to all-*trans*-retinol, which is then transported to the retinal pigment epithelium bound to the interphotoreceptor RBP. This protein also binds fatty acids, including palmitate and docosahexaenoic acid (C22:6 ω3), which is known to be essential for vision and which comprises some 50% of the phospholipid of photoreceptor cells.

In the retinal pigment epithelium, palmitate is bound to the fatty acid binding site of the interphotoreceptor RBP, and the retinoid binding site has a high affinity for 11-*cis*-retinaldehyde, which is to be transported to the photoreceptor cells. In the photoreceptor cells, the palmitate is displaced by docosahexaenoic acid, which causes a conformational change in the protein, so that it no longer binds 11-*cis*-retinaldehyde, which is delivered to the photoreceptor cells and binds all-*trans*-retinol for transport back to the pigment epithelium. Here, the docosahexaenoic acid is displaced by palmitate, and the affinity of the protein for 11-*cis*-retinaldehyde is restored (Palczewski and Saari, 1997; Tschanz and Noy, 1997).

Under conditions of high light intensity, all-*trans*-retinaldehyde binds to the retinal G-protein–coupled receptor (RGR), which catalyzes photoisomerization to 11-*cis*-retinaldehyde – the reverse of the photoisomerization catalyzed by rhodopsin. Retinaldehyde dehydrogenase binds to the RGR and reduces the 11-*cis*-retinaldehyde to 11-*cis*-retinol, which then enters the pool available to undergo oxidation to the aldehyde and reform rhodopsin. Knockout mice lacking RGR have impaired responses to light under conditions of continuous intense illumination, but normal responses under conditions of low-light intensity when all-*trans*-retinaldehyde is reduced to all-*trans*-retinol and transported by the interphotoreceptor retinoid binding protein (Hao and Fong, 1999; Chen et al., 2001).

Metarhodopsin II is the excited form of rhodopsin that initiates the guanine nucleotide amplification cascade that causes nerve stimulation. The final event is a hyperpolarization of the outer section membrane of the rod or cone caused by the closure of sodium and calcium channels through the membrane – excitation of a single molecule of rhodopsin, the action of a single photon, causes a drop of 1 pA in the normal dark current across this membrane in rods. In cones, the response to a single photon is only 1% to 10% of that in the rods.

In the dark, the sodium channels are kept open, and there is a dark current because they bind GMP. They are closed by the loss of this bound GMP as it is hydrolyzed to 5'-GMP by phosphodiesterase. The phosphodiesterase is activated by the guanine nucleotide binding protein (G-protein) transducin.

Transducin, in its inactive form in the dark, has bound GDP; interaction with metarhodopsin II causes it to release this GDP and bind GTP. Transducin-GTP has a low affinity for metarhodopsin II, which is therefore free to interact with another molecule of transducin-GDP. Thus, for as long as metarhodopsin II remains in its active state (i.e., until it has been fully phosphorylated and converted to metarhodopsin III, which does not interact with transducin-GDP), it will continue to activate transducin molecules.

Metarhodopsin II activates transducin, leading to an exchange of bound GDP for GTP; several hundred molecules of transducin are activated by a single molecule of metarhodopsin II within a fraction of a second. Transducin-GTP binds to, and activates, GMP phosphodiesterase, lowering the intracellular concentration of cGMP. As cGMP falls, a cation channel in the membrane closes, thus interrupting the steady inward current of sodium and calcium ions. This leads to hyperpolarization of the membrane and reduced secretion of neurotransmitter (Baylor, 1996).

Like other G-proteins, transducin has innate GTPase activity, and over a time course of seconds or less is autocatalytically converted to transducin-GDP, which does not interact with phosphodiesterase. This restores the normal inhibition of phosphodiesterase, permitting cGMP concentrations to rise again, reopening the sodium channels and restoring the dark current.

Metarhodopsin II is inactivated by phosphorylation of three serine residues at the carboxyl terminal of the protein, catalyzed by rhodopsin kinase. In transgenic mice with carboxyl terminal-truncated rhodopsin, lacking the phosphorylation sites, there is a prolonged response to a single photon. Rhodopsin kinase is activated by its substrate, metarhodopsin II, and is inhibited by calcium bound to the protein recoverin, which thus prolongs the photoresponse.

Phosphorylation is a necessary, but not sufficient, condition for quenching metarhodopsin II; it also has to bind the protein arrestin before it loses the bound retinaldehyde and is converted to metarhodopsin III. Then, it is dephosphorylated by protein phosphatase 2A and a calcium-activated protein phosphatase. It is this dephosphorylation of metarhodopsin III that is correlated with dark adaptation and regeneration of active rhodopsin by binding to

11-*cis*-retinaldehyde (Palczewski and Saari, 1997; Hurley et al., 1998; Kennedy et al., 2001).

The rate-limiting step in initiation of the visual cycle is the regeneration of 11-*cis*-retinaldehyde. In vitamin A deficiency, when there is little 11-*cis*-retinyl ester in the pigment epithelium, both the time taken to adapt to darkness and the ability to see in poor light will be impaired.

2.3.2 Genomic Actions of Retinoic Acid

Apart from the effects on vision, most of the effects of vitamin A deficiency (Section 2.4) involve derangements of cell proliferation and differentiation (squamous metaplasia and keratinization of epithelia), dedifferentiation, and loss of ciliated epithelia. Retinoic acid has both a general role in growth and a specific morphogenic role in development and tissue differentiation. These functions are the result of genomic actions, modulating gene expression by activation of nuclear receptors. Both deficiency and excess of retinoic acid cause severe developmental abnormalities.

Retinoic acid has a specific morphogenic role in limb development. There is a small concentration gradient of retinoic acid across the developing limb bud and a gradient of retinoic acid binding protein in the opposite direction, suggesting that the resultant relatively steep gradient of free retinoic acid may be the important factor determining the pattern of development (Thaller and Eichele, 1987, 1990). It may also be important in the development of the central nervous system. CRABP has a strictly delimited anatomical localization in the developing mouse brain and is only expressed transiently, between days 11 to 14 of gestation (Momoi et al., 1990).

At pharmacological levels, retinoic acid enhances the expression of uncoupling protein 1 (thermogenin) in brown adipose tissue and decreases the expression of leptin in white adipose tissue, suggesting that it may have an effect on energy homeostasis, but it is not known whether or not the effects are relevant at physiological levels (Kumar et al., 1999; Villarroya et al., 1999). Retinoic acid also induces synthesis of glucokinase in pancreatic β-islet cells. Increased metabolism of glucose as a result of glucokinase activity is responsible for initiating insulin secretion in response to a rise in blood glucose concentration, and retinoic acid increases the secretion of insulin by pancreatic islets in culture (Cabrera-Valladares et al., 1999).

Retinoic acid may either enter the target cell from the circulation or may be formed intracellularly by oxidation of retinol. A number of tissues – but not muscle, kidneys, small intestines, liver, lungs, or spleen – have a cellular retinoic acid binding protein (CRABP) that is distinct from CRBP. Testis and uterus contain CRBP and CRABP; both retinol and retinoic acid are essential in the functions of these organs. Although retinoic acid will support testosterone synthesis, it will not support spermatogenesis, nor will it support placental development in vitamin A-deficient animals (Appling and Chytil, 1981). Similarly, retinol and retinoic acid have different actions on bone cells in culture, suggesting that both have functions in normal bone development, in addition to antagonizing the actions of vitamin D when retinoic acid is present in excess. Retinol inhibits collagen synthesis, whereas retinoic acid stimulates the synthesis of noncollagen bone proteins (Dickson et al., 1989).

The role of these intracellular binding proteins seems to be to transport retinoic acid into the nucleus. Unlike receptor proteins, which bind their ligand in the nucleus and then interact with regulatory elements on DNA, the CRBPs do not enter the nucleus or interact with DNA and nucleoproteins.

2.3.2.1 **Retinoid Receptors and Response Elements**   There are two families of nuclear retinoid receptors: the retinoic acid receptors (RARs), which bind all-*trans*-retinoic acid and the retinoid X receptors (RXRs), so-called because their physiological ligand was unknown when they were first discovered. It is now known to be 9-*cis*-retinoic acid, which also binds to and activates RARs. As well as being the activating ligand for RXR, 9-*cis*-retinoic acid also increases the rate of catabolism of its receptor, which may be important in the regulation of the various hormonal responses that require formation of RXR heterodimers (Figure 2.6).

14-Hydroxyretroretinol, 4-oxoretinol, and anhydroretinol, as well as possibly other retinoids, also bind to and activate the RAR family of receptors at physiological concentrations, but do not bind to the RXR family. 4-Oxoretinol is formed from all-*trans*-retinol in differentiating cells; 10% to 15% of intracellular retinol may be oxidized to 4-oxoretinol during an 18-hour period, whereas there is no formation of all-*trans*-retinoic acid or 9-*cis*-retinoic acid. 4-Oxoretinol induces differentiation of cells in culture (Achkar et al., 1996). In the developing Xenopus embryo, 4-oxoretinaldehyde is the major retinoid, acting as a precursor of both 4-oxoretinol and 4-oxoretinoic acid, both of which activate the RAR (Blumberg et al., 1996). This developmental role of 4-oxoretinoids may explain the observation that pregnant vitamin A-deficient rats resorb the fetuses around day 15 of gestation if they are provided with retinoic acid, but not if they are provided with retinol (Wellik and DeLuca, 1995). 14-Hydroxyretroretinol and 13,14-dihydroxyretroretinol act as growth promoters for retinol-deficient cells in culture, but do not induce differentiation; by contrast, anhydroretinol has growth-inhibiting activity.

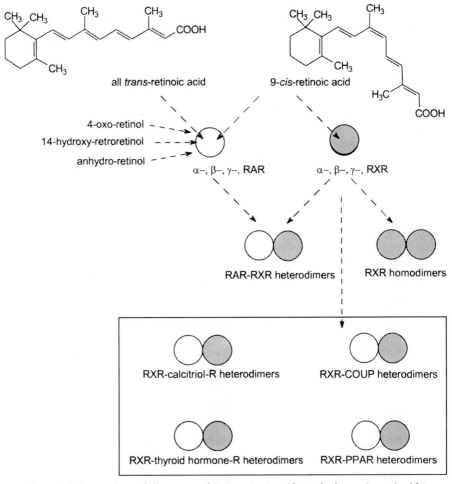

Figure 2.6. Interactions of all-*trans*- and 9-*cis*-retinoic acids (and other active retinoids) with retinoid receptors. COUP, chicken ovalbumin upstream promoter.

The two families of receptors differ from each other considerably, and RARs show greater sequence homology with thyroid hormone receptors than with RXRs. RARs act only as heterodimers with RXRs; homodimers of RARs have poor affinity for retinoid response elements on DNA. The liganded CRABP(II) enhances the binding of the RAR-RXR heterodimer to response elements on DNA and amplifies the effect of the receptor dimer (Delva et al., 1999).

RXR forms active homodimers and also form heterodimers with the calcitriol (vitamin D) receptor (Section 3.3.1), the thyroid hormone receptor, the peroxisome proliferation-activated receptor (PPAR, whose physiological ligand is a long-chain polyunsaturated fatty acid or an eicosanoid derivative), and the chicken ovalbumin upstream promoter (COUP) receptor – an orphan receptor whose physiological ligand is unknown (Mangelsdorf and Evans, 1995; Glass, 1996; Glass et al., 1997). Formation of RXR heterodimers seems to be required for DNA binding of calcitriol, thyroid hormone, and PPAR receptors. Interestingly, there is no requirement for a ligand bound to the RXR for this to occur. An unliganded RXR can form active heterodimers (Wendling et al., 1999).

In the presence of all-*trans*- or 9-*cis*-retinoic acid, the receptor heterodimers are transcriptional activators. However, the heterodimers will also bind to DNA in the absence of retinoic acid, in which case they act as repressors of gene expression (Fujita and Mitsuhashi, 1999).

There are three isoforms of each receptor type: RARα, RARβ, RARγ, RXRα, RXRβ, and RXRγ. They are encoded by different genes, with different tissue-specific expression, and different expression at different times during development. There is a greater conservation of amino acid sequence between species for any one type of retinoid receptor than between the different receptor types in the same species, suggesting that the receptor types evolved separately a considerable time ago. In addition, there are two different subforms of each (RARα, RARγ, RXAα, RXRβ, and RXRγ) and four subforms of RARβ. These arise by differential splicing of the RNA transcript (Rowe and Brickell, 1993).

RARα and RXRβ have widespread distribution in tissues; RARβ and RARγ are expressed to different extents in different tissues during development. RARα and RXRγ have tissue-specific distribution.

Studies with knockout mutant mice lacking one or another of the retinoid receptors suggest that there is some degree of redundancy or overlap between the receptor subtypes, and that RARγ is especially important in the teratogenic actions of retinoids (Section 2.5.1.1; Mark et al., 1999; Maden, 2000):

- RARα^0 mice show no congenital abnormalities, but have a high postnatal mortality.
- RARβ^0 mice show no detectable effects.
- RARγ^0 mice have widespread congenital abnormalities, as seen in severe vitamin A deficiency.
- RARα^0 mice show the same teratogenic effects of retinoic acid excess as wild-type mice.
- RARβ^0 mice show the same teratogenic effects of retinoic acid excess as wild-type mice.
- RARγ^0 mice show some, but not all, of the teratogenic effects of retinoic acid excess.

- $RXR\alpha^0$ mice show abnormalities of the eye and heart.
- $RXR\beta^0$ mice are morphologically normal, but the males are sterile.

The multiplicity of possible combinations of homodimers and heterodimers of RAR and RXR subtypes, and the various possible RXR heterodimers with other receptors, permits a wide variety of active retinoid receptor complexes that bind to different response elements on DNA. Unlike most hormone response elements on DNA, which are palindromic and bind a symmetrical receptor homodimer, the most common type of retinoid response element is a direct repeat: purine-G-(G or T)-T-C-A-(X_n)-purine-G-(G or T)-T-C-A, in which the spacer (X_n) is commonly 5 base pairs, but may be 1 or 2. There are also more complex retinoid response elements, including palindromic and inverted palindromic repeats, as well as hexameric motifs with variable spacing. This means that a wide variety of different genes may be regulated differently in response to retinoids.

2.3.3 Nongenomic Actions of Retinoids

In addition to its genomic functions, retinoic acid also has a number of nongenomic functions. It enhances the stability of keratin mRNA, leading to increased synthesis of this protein (Crowe, 1993). Retinoic acid acts also as an effector in response to transmembrane signaling by retinoylation of target proteins.

In the synthesis of most glycoproteins containing mannose, the intermediate carrier of the mannosyl moiety is the polyene dolichol phosphate. However, in some systems, retinyl phosphate can act as the intermediate carrier between UDP-mannose and the acceptor glycoprotein. Retinyl phosphate mannose seems to be involved especially in the synthesis of hydrophobic regions of glycoproteins (DeLuca, 1977; Frot-Coutaz et al., 1985).

2.3.3.1 Retinoylation of Proteins
Studies with knockout mice, lacking nuclear retinoid receptors, suggest that retinoic acid has physiological functions unrelated to its genomic actions. Label from [^{3}H]retinoic acid is incorporated into cells in culture in a form that is not extractable by organic solvents, suggesting that there is covalent binding of retinoic acid to proteins – retinoylation.

There are two main routes involved in the retinoylation of proteins:

1. Formation of retinoyl CoA, followed by formation of an ester with the hydroxyl group of tyrosine, threonine or serine, or a thio-ester with the sulfhydryl group of cysteine (Figure 2.7). The source of retinoic acid for

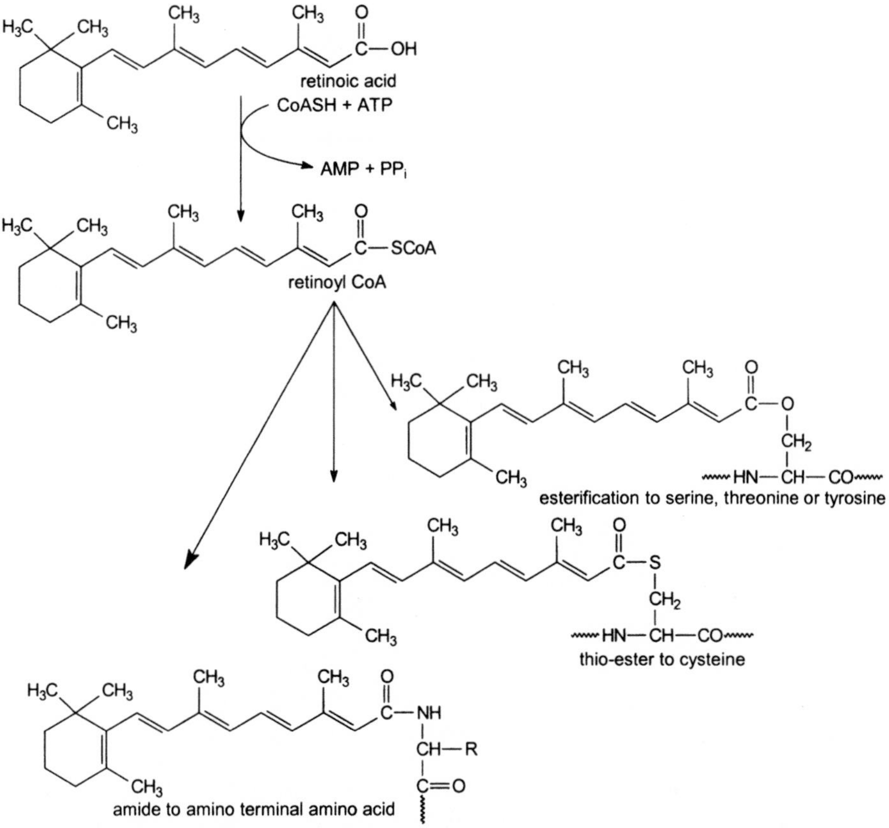

Figure 2.7. Retinoylation of proteins by retinoyl CoA.

this reaction may be retinoyl glucuronide (Section 2.2.1.3) rather than free retinoic acid.

2. Cytochrome P_{450}-catalyzed 4-hydroxylation, followed by formation of an ether bond to the hydroxyl group of tyrosine, threonine or serine, or a thio-ether bond to the sulfhydryl group of cysteine (Figure 2.8).

Both in cells in culture and in vivo, the major targets for retinoylation are the regulatory subunits of cAMP-dependent protein kinases, suggesting a role for retinoic acid in modulation of the actions of cell surface acting hormones and neurotransmitters (Myhre et al., 1996). In a variety of cell types, cAMP-dependent protein kinase activity increases after exposure to retinoic acid. In a number of experimental situations, retinoic acid and cAMP act synergistically in cell differentiation. Takahashi et al. (1997) reported that 40 different proteins

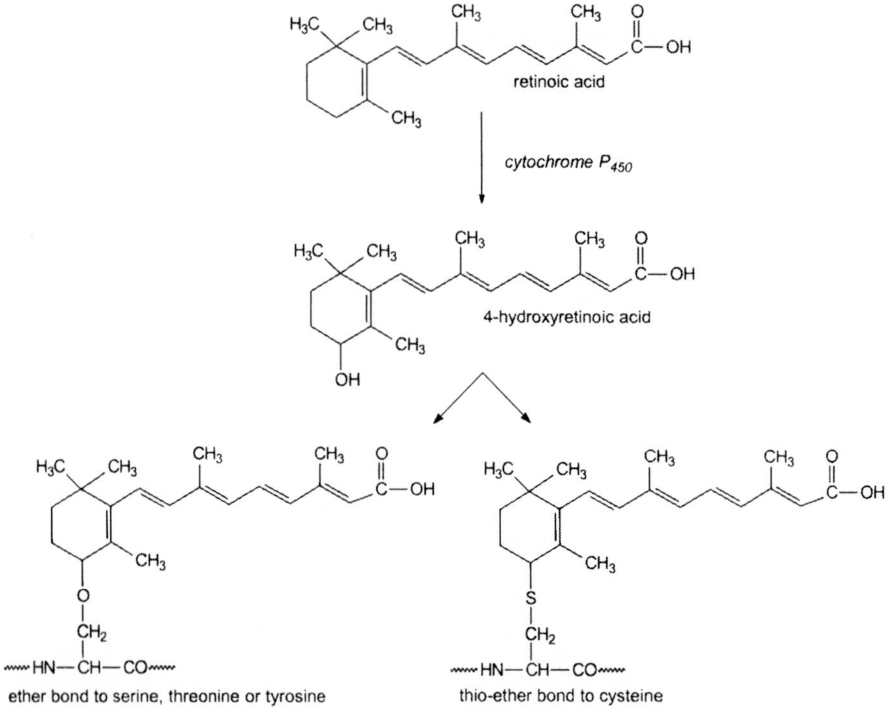

Figure 2.8. Retinoylation of proteins by 4-hydroxyretinoic acid.

are retinoylated in cells in culture, whereas cells in which the *ras* oncogene has been activated and that are insensitive to growth inhibition by retinoic acid, only 15 proteins are retinoylated (Takahashi et al., 1997).

2.3.3.2 Retinoids in Transmembrane Signaling Neutrophils treated with physiological concentrations of all-*trans*-retinoic acid show a dose-dependent increase in synthesis of superoxide. Inhibitor studies suggest that retinoic acid acts via an inositol trisphosphate cascade rather than calcium and protein kinase C (Koga et al., 1997). There is also evidence that all-*trans*-retinoic acid leads to increased formation of cADP-ribose and nicotinic acid adenine dinucleotide phosphate as second messengers (Section 8.4.4; Dousa et al., 1996; Mehta and Cheema, 1999).

Some of the retroretinoids also have cell signaling functions at a cell surface or a cytoplasmic receptor. 14-Hydroxyretroretinol is required for lymphocyte proliferation, whereas anhydroretinol is a growth inhibitor; the two compounds act antagonistically. Treatment of T lymphocytes with anhydroretinol in the absence of 14-hydroxyretroretinol leads to rapid cell death, with widespread morphological changes but little or no nuclear abnormality. This suggests a cytoplasmic mechanism of apoptosis (O'Connell et al., 1996).

There is also evidence that retinoic acid directly modulates transmission at electrical synapses of retinal cells. This is independent of G-proteins and second messengers, and involves a nonnuclear RAR-like binding site associated with ion channels (Zhang and McMahon, 2000).

2.4 VITAMIN A DEFICIENCY (XEROPHTHALMIA)

Vitamin A-deficient experimental animals fail to grow; adults are blind and sterile, with testicular degeneration in males and keratinization of the uterine epithelium in females. Although deficient female animals will conceive, and the fetuses will implant, formation of the placenta is impaired and the fetuses are resorbed. Epithelia in general are hyperplastic and keratinized, and there is impaired cellular immunity with increased susceptibility to infection. Both retinol and retinoic acid are required for gestation in the rat; in deficient animals, retinoic acid alone will not prevent fetal resorption after about day 10 of gestation (Wellik and DeLuca, 1995; Wellik et al., 1997).

Vitamin A deficiency is a major problem of children under five in developing countries, being the single most common preventable cause of blindness. Table 2.1 shows the prevalence of vitamin A deficiency in different regions of the world. The increased susceptibility to infection and impairment of immune responses in vitamin A deficiency causes significant childhood mortality. A number of trials of vitamin A supplementation in areas of endemic deficiency show a 20% to 35% reduction in child mortality.

Table 2.1 Prevalence of Vitamin A Deficiency among Children under Five

WHO Region	Subclinical Deficiency		Clinical Deficiency	
	Millions	% Prevalance	Millions	% Prevalance
Africa	49	45.8	1.08	1.0
Americas	17	21.5	0.06	0.1
Southeast Asia	125	70.2	1.3	0.7
Europe	—	—	—	—
Eastern Mediterranean	23	31.5	0.16	0.3
Western Pacific	42	30.0	0.1	0.1
Total	256	40.3	2.7	0.1

WHO, World Health Organization.

Functional vitamin A deficiency may occur despite adequate liver reserves of retinol, as a result of impaired synthesis of RBP in protein-energy malnutrition, and possibly also in zinc deficiency (Smith et al., 1973; Solomons and Russell, 1980; Rahman et al., 2002).

A mild infection, such as measles, commonly triggers the development of xerophthalmia in children whose vitamin A status is marginal. In addition to functional deficiency as a result of impaired synthesis of RBP (Section 2.2.3) and transthyretin in response to infection, there may be a considerable urinary loss of vitamin A because of increased renal epithelial permeability and proteinuria, permitting loss of retinol bound to RBP-transthyretin. The American Academy of Pediatrics Committee on Infectious Diseases (1993) recommended vitamin A supplements for all children who have been hospitalized with measles.

In adults, excessive alcohol consumption reduces liver reserves of vitamin A, both as a result of alcoholic liver damage and also by induction of cytochrome P_{450} enzymes that catalyze the oxidation of retinol to retinoic acid (as also occurs with chronic use of barbiturates). However, chronic consumption of alcohol can also potentiate the toxicity of retinol (Section 2.5.1).

Even marginal vitamin A status leads to significantly impaired resistance to infection, and deficient children are significantly more prone to infection. A number of studies show beneficial effects of vitamin A supplementation, and adverse effects of marginal status in measles, diarrheal and respiratory diseases, malaria, human immunodeficiency virus (HIV) infection, and tuberculosis (Semba, 1999; Semba and Tang, 1999). The vitamin has multiple effects on the immune system, including modulating the expression of cytokines, and the differentiation, function, and apoptosis of macrophages, T and B lymphocytes, neutrophils, and other cells.

Mild deficiency results in impaired dark adaptation; as the deficiency progresses, there is inability to see in the dark (night blindness). As discussed in Section 2.3.1, the recycling of retinaldehyde to reform rhodopsin is the rate-limiting step in the visual cycle. If reserves of retinaldehyde in the pigment epithelium are inadequate, then there will be slow dark adaptation and impaired vision in poor light. More prolonged and severe deficiency leads to conjunctival xerosis – squamous metaplasia and keratinization of the epithelial cells of the conjunctiva with loss of goblet cells in the conjunctival mucosa, leading to dryness, wrinkling, and thickening of the cornea (xerophthalmia). As the deficiency progresses, there is keratinization of the cornea. At this stage, the condition is still reversible, although there may be residual scarring of the

Table 2.2 WHO Classification of Xerophthalmia

Classification Code	Clinical Description	Prevalence among Preschool Children to Indicate Significant Public Health Problems
XN	Night blindness	>1%
X1A	Conjunctival xerosis	—
X1B	Bitot's spots	>0.5%
X2	Corneal xerosis	
X3A	Corneal ulceration/keratomalacia involving less than one-third of the corneal surface	>0.01%
X3B	Corneal ulceration/keratomalacia involving more than one-third of the corneal surface	>0.01%
XS	Corneal scar	>0.05%
XF	Xerophthalmic fundus	
Biochemical	Plasma retinol <0.35 μmol/L	>5%

cornea. In advanced xerosis, yellow-gray foamy patches of keratinized cells and bacteria (Bitot's spots) may accumulate on the surface of the conjunctiva. The next stage is ulceration of the cornea from increased proteolytic action, thus causing irreversible blindness (Pirie et al., 1975). Table 2.2 shows the World Health Organization classification of xerophthalmia.

As well as the conjunctiva, other epithelia are affected by moderate or mild vitamin A deficiency, with increased intestinal permeability to disaccharides, later a reduction in the number of goblet cells per villus, and then a reduction in mucus secretion (McCullough et al., 1999). There is also atrophy of the respiratory epithelium, again with loss of goblet cells, followed by keratinization, resulting in increased susceptibility to infection. These changes in intestinal and respiratory epithelium develop earlier than the more readily observed diagnostic changes in the eye.

In adults maintained on vitamin A-deficient diets for a period of months, there are a number of early signs, apparent before the impairment of dark adaptation: impairment of the senses of taste, smell, and balance and distortion of color vision, with impaired sensitivity to green light. With the exception of the effects on color vision, these can all be attributed to dedifferentiation of ciliated epithelia (Sauberlich et al., 1974; Hodges et al., 1978).

Early studies showed impaired gluconeogenesis and low hepatic stores of glycogen in vitamin A-deficient animals. Synthesis of one of the key regulatory enzymes of glycolysis, the GTP-dependent isoenzyme of phosphoenolpyruvate carboxykinase, is regulated by all-*trans*-retinoic acid. Both gene expression and gluconeogenesis fall in vitamin A deficiency (Shin and McGrane, 1997).

2.4.1 Assessment of Vitamin A Nutritional Status

An early sign of vitamin A deficiency is impaired dark adaptation – an increase in the time taken to adapt to seeing in dim light. The apparatus required is not suitable for use in field studies, or for use with children (the group most at risk from deficiency), and the dark adaptation test is largely of historical interest. Balance, color vision, and the senses of taste and smell are also affected in early deficiency, but none of these provides a sensitive or specific test of status.

Liver reserves of vitamin A can be estimated by isotope dilution after a test dose of isotopically labeled retinol, but this is not a suitable technique for assessment of status in population studies.

Assessment of vitamin A nutritional status depends on the biochemical criteria shown in Table 2.3.

2.4.1.1 Plasma Concentrations of Retinol and β-Carotene Because RBP is released from the liver only as the holo-protein and apo-RBP is cleared from the circulation rapidly after tissue uptake of retinol (Section 2.2.3), the fasting plasma concentration of retinol remains constant over a wide range of intakes. It is only when liver reserves are nearly depleted that it falls significantly, and it only rises significantly at the onset of toxic signs. Therefore, although insensitive to changes within the normal range, measurement of plasma retinol provides a convenient means of detecting people whose intake of vitamin A is inadequate to permit normal liver reserves to be maintained.

Interpretation of plasma concentrations of retinol is confounded by the fact that both RBP and transthyretin are negative acute phase proteins, and their synthesis falls, and hence the plasma concentration of retinol fall, in response to infection. Similarly, both protein-energy malnutrition and zinc deficiency result in a low plasma concentration, despite possibly adequate liver reserves as a result of impaired synthesis of RBP.

Carotene in plasma is mainly in lipoproteins; thus, as with vitamin E (Section 4.5), measurements of plasma concentrations of carotene should be related to either cholesterol or total plasma lipids. Only 10% to 20% of total plasma carotenoids is β-carotene, with a very wide range of individual variation. There are no reliable determinations of β-carotene or total provitamin A carotenoids

Table 2.3 Biochemical Indices of Vitamin A Status

Liver Retinyl Esters (as Retinol)	μmol/kg	mg/kg
Adequate	>70	>20
Marginal	35–70	10–20
Poor	17.5–35	5–10
Deficient	<17.5	<5
Plasma Retinol	**μmol/L**	**μg/L**
Elevated	>1.75	>500
Normal	0.7–1.75	200–500
Unsatisfactory	0.35–0.7	100–200
Liver stores depleted/deficient	<0.35	<100
Plasma Total Carotenoids[a]	**μmol/L**	**μg/L**
Adult reference range	0.4–4.0	240–2,200
Acceptable	>0.75	>400
Hypercarotinemia	>5.6	>3,000
Plasma Retinoic Acid	**nmol/L**	**μg/L**
Adults	10–13	3–4
Plasma Retinol Binding Protein	**μmol/L**	**μg/L**
Adults	1.9–4.28	40–90
Preschool children	1.19–1.67	25–35
Relative Dose Response		
Normal	<20%	
Marginal deficiency	>20%	
Modified Dose Response	Dehydroretinol/retinol	
Normal	<0.03	
Marginal deficiency	>0.03	

[a] β-Carotene is 10% to 20% of total plasma carotenoids.
Sources: International Vitamin A Consultative Group, 1983; Underwood, 1990.

in appropriate populations to permit plasma concentrations of carotene to be related to vitamin A nutritional status.

2.4.1.2 **Plasma Retinol Binding Protein** Measurement of plasma concentrations of RBP may give some additional information. Indeed, it has been suggested that because retinol is susceptible to oxidation on storage of blood samples, measurement of RBP may be a better indication of the state of vitamin A status. In adequately nourished subjects, about 13% of immunologically reactive RBP in plasma is present as the apo-protein, whereas in vitamin A-deficient children, the proportion of apo-protein may rise to 50% to 90% of circulating RBP. Measurement of the ratio of plasma retinol:RBP may provide a sensitive index of status (Thurnham and Northrop-Clewes, 1999).

2.4.1.3 The Relative Dose Response (RDR) Test The RDR test is a test of the ability of a dose of vitamin A to raise the plasma concentration of retinol several hours later, after chylomicrons have been cleared from the circulation. What is being tested is the ability of the liver to release retinol into the circulation. In subjects who are retinol deficient, a test dose will produce a large increase in plasma retinol, because of the accumulation of apo-RBP in the liver in deficiency (Section 2.2.3). In those whose problem is due to lack of RBP, then little of the dose will be released into the circulation. An RDR greater than 20% indicates depletion of liver reserves of retinol to less than 70 μmol per kg (Underwood, 1990).

The test requires two samples of blood, taken before and 5 hours after the test dose of retinol. A modified RDR test involves giving a test dose of dehydroretinol, then determining the ratio of dehydroretinol:retinol in a single plasma sample taken 30 hours later. Again, because of the accumulation of RBP in the liver in deficiency and because in deficiency there is less dilution of dehydroretinol with liver pools of retinyl esters, the ratio is inversely proportional to the liver stores of retinol (Tanumihardjo et al., 1987).

2.4.1.4 Conjunctival Impression Cytology Early changes in vitamin A deficiency include loss of the mucus-secreting goblet cells from the conjunctival epithelium, and the appearance of enlarged, flattened, and partially keratinized epithelial cells. An impression of the conjunctiva is taken by blotting onto cellulose acetate, then fixing and staining prior to histological examination. The technique detects children who do not yet show any clinical signs and whose serum retinol is within the normal range (Wittpenn et al., 1986; Natadisastra et al., 1987).

2.5 VITAMIN A REQUIREMENTS AND REFERENCE INTAKES
Very few direct studies have been performed to determine human vitamin A requirements. In the Sheffield study (Hume and Krebs, 1949), 16 subjects were depleted of vitamin A for 2 years; only three subjects showed clear signs of impaired dark adaptation. One of these subjects was repleted with 390 μg of retinol per day, which resulted in a gradual restoration of dark adaptation; the other two subjects received β-carotene. On this basis, the minimum requirement was presumed to be 390 μg, and the reference intake was set at 750 μg.

Table 2.4 Reference Intakes of Vitamin A (μg/day)

Age	U.K. 1991	EU 1993	U.S./Canada 2001	FAO 2001
0–6 m	350	—	400	375
7–12 m	350	350	500	400
1–3 y	400	400	300	400
4–6 y	500	400	400	450
7–8 y	500	500	400	500
Males				
9–10 y	500	500	600	600
11–13 y	600	600	600	600
>14 y	700	700	900	600
Females				
9–10 y	500	500	600	600
11–13 y	600	600	600	600
>14 y	600	600	700	600
Pregnant	700	700	770	800
Lactating	950	950	900	850

EU, European Union; FAO, Food and Agriculture Organization; WHO, World Health Organization.
Sources: Department of Health, 1991; Scientific Committee for Food, 1993; Institute of Medicine, 2001; FAO/WHO, 2001.

Since then, eight more subjects have been studied (Sauberlich et al., 1974; Hodges et al., 1978). From these studies, the reference intake for adult men was set at 1,000 μg of retinol equivalent, with a minimum physiological requirement of 600 μg per day. Because the signs of deficiency only resolve slowly, it is possible that depletion/repletion studies overestimate requirements.

An alternative approach to determining requirements is to measure the fractional rate of catabolism of the vitamin by use of a radioactive tracer, then determine the intake that would be required to maintain an appropriate level of liver reserves. As discussed in Section 2.2.1.1, when the liver concentration rises above 70 μmol per kg, there is increased activity of the microsomal oxidation of vitamin A and biliary excretion of retinol metabolites. The fractional catabolic rate is 0.5% per day; assuming 50% efficiency of storage of dietary retinol, this gives a mean requirement of 6.7 μg per kg of body weight and a reference intake of 650 to 700 μg for adult men (Olson, 1987a). Reference intakes for vitamin A are shown in Table 2.4.

Although there is some evidence that β-carotene and other carotenoids may have actions in their own right, apart from their provitamin A activity (Section 2.6.3), there is no evidence on which to base any recommendations or suggestions of requirements for carotene other than as a precursor of retinol. There is no evidence of any carotene deficiency disease in depletion studies of people provided with an adequate intake of retinol (Institute of Medicine, 2001). The epidemiological evidence that shows a high intake of carotenoids to be associated with a lower incidence of cancer (Section 2.6.3) may reflect intake of (carotene-rich) fruits and vegetables, which are sources of other potentially protective compounds (Section 14.7), rather than carotene intake per se.

2.5.1 Toxicity of Vitamin A

Vitamin A is both acutely and chronically toxic. Acutely, large doses of vitamin A (in excess of 300 mg in a single dose to adults) cause nausea, vomiting, and headache, with increased pressure in the cerebrospinal fluid – signs that disappear within a few days. After a very large dose, there may also be drowsiness and malaise, with itching and exfoliation of the skin; extremely high doses can prove fatal. Single doses of 60 mg of retinol are given to children in developing countries as a prophylactic against vitamin A deficiency – an amount adequate to meet the child's needs for 4 to 6 months. About 1% of children so treated show transient signs of toxicity, but this is considered to be an acceptable adverse effect in view of the considerable benefit of preventing xerophthalmia.

The chronic toxicity of vitamin A is a more general cause for concern; prolonged and regular intake of more than about 7,500 to 9,000 μg per day by adults (and significantly less for children) causes signs and symptoms of toxicity affecting:

1. The skin: excessive dryness, scaling and chapping of the skin, desquamation, and alopecia.
2. The central nervous system: headache, nausea, ataxia, and anorexia, all associated with increased cerebrospinal fluid pressure.
3. The liver: hepatomegaly, hyperlipidemia, and histological changes in the liver, including increased collagen formation. Alcohol potentiates the hepatotoxicity of vitamin A.
4. Bones: joint pains, thickening of the long bones, hypercalcemia, and calcification of soft tissues, but with reduced bone mineral density. High intakes of vitamin A are associated with an increased rate of loss of bone mineral density with age, and some studies have shown that intakes above 1,500 μg per day are associated with increased incidence of osteoporosis and hip fracture, although other studies have not shown any relationship between vitamin A intake and osteoporosis (Institute of Medicine, 2001). At high levels of intake, vitamin A both stimulates bone

Table 2.5 Prudent Upper Levels of Habitual Intake (μg of preformed vitamin A/day)

Age (y)	Upper Limit of Safe Habitual Intake (U.K., 1991)	Tolerable Upper Intake Level (U.S./Canada, 2001)
Birth–1	900	600
1–3	1,800	600
4–6	3,000	
4–8		900
7–10	4,500	
Males		
9–13		1,700
11–18	6,000	
14–18		2,800
Adult females	9,000	3,000
9–13		600
11–18	6,000	
14–18		2,800
Adult	7,500	3,000
Pregnant	3,300	2,800–3,000
Lactating	—	2,800–3,000

Sources: Department of Health, 1991; Institute of Medicine, 2001.

resorption and inhibits bone formation (Binkley and Krueger, 2000), largely as a result of antagonism of the actions of vitamin D (Section 3.3.5). As the tissue concentration of 9-*cis*-retinoic acid increases, so there is increased formation of RXR homodimers, leaving an inadequate amount of RXR to dimerize with the vitamin D receptor.

Prudent upper levels of habitual intake of retinol are shown in Table 2.5.

As the intake of vitamin A increases, there is an increase in the excretion of metabolites in bile, once adequate liver reserves have been established. However, the biliary excretion of retinol metabolites reaches a plateau at relatively low levels, and it seems likely that this explains the relatively low toxic threshold (Olson, 1986). Vitamin A intoxication is associated with the appearance of both retinol and retinyl esters bound to albumin and in plasma lipoproteins, which can be taken up by tissues in an uncontrolled manner; the amount of circulating retinol bound to RBP does not increase. Retinol has a membrane lytic action; it was noted in Section 2.2.2.3 that one of the functions of RBP binding seems to be to protect tissues against retinol, as well as to protect retinol against oxidation (Meeks et al., 1981).

At relatively high levels of intake, vitamin A induces synthesis of glycine *N*-methyltransferase. This can lead to depletion of methyl groups and undermethylation of DNA, a potential factor in carcinogenesis (Section 10.9.5; Rowling et al., 2002).

Carotenoids do not cause hypervitaminosis A. As discussed in Section 2.2.2.1, the conversion of provitamin A carotenoids to retinol is limited; therefore, vitamin A intoxication is unlikely to occur even at high intakes of carotene. Accumulation of even abnormally large amounts of carotene seems to have no short-term adverse effects, although plasma, body fat, and skin can have a strong orange-yellow color (hypercarotinemia) after prolonged high intakes of carotenoids. A small number of people lack carotene dioxygenase and suffer from (asymptomatic) carotinemia with normal modest intakes. However, as discussed in Section 2.6.3, two large-scale intervention trials have shown an increased incidence of cancers in subjects receiving supplements of β-carotene.

2.5.1.1 **Teratogenicity of Retinoids** 13-*Cis*-retinoic acid and etretinate (Section 2.6.2) are highly teratogenic. After women have been treated with retinoids for dermatological problems, it is generally recommended that contraceptive precautions be continued for 12 months, because of the retention of retinoids in the body. With embryonic limb bud cells in culture, it is those retinoids that bind to RAR that are teratogenic; those that bind to RXR are not. However, retinoids that bind to RXR potentiate the teratogenicity of those that bind to RAR (Soprano and Soprano, 1995; Collins and Mao, 1999). There is considerable species variation in sensitivity to the teratogenic effects of 13-*cis*-retinol; in species such as rats and mice, it has low teratogenicity because it is metabolized rapidly to the glucuronide. In primates, it is mainly oxidized to 13-*cis*-4-oxoretinoic acid, which is transported across the placenta (Nau, 2001).

By extrapolation, it has been assumed that retinol is also teratogenic, although there is little evidence; in case control studies, intakes between 2,400 to 3,300 μg per day during pregnancy have been associated with birth defects. Other studies have not demonstrated any teratogenic effect at this level of intake, and it has been suggested that the plasma concentration associated with teratogenic effects is unlikely to be reached with intakes below 7,500 μg per day (Miller et al., 1998; Ritchie et al., 1998; Wiegand et al., 1998). Arnhold and coworkers (2002) demonstrated formation of all-*trans*-retinoic acid in enterocytes after feeding relatively large amounts of retinol that saturated the intestinal CRBP (Section 2.2.4) and suggested that there is indeed a threshold intake of retinol above which teratogenic metabolites are formed. Pregnant women are variously advised not to consume more than 2,800 to 3,000 μg per day (Institute of Medicine, 2001) or 3,300 μg per day (Department of Health, 1991).

2.5.2 Pharmacological Uses of Vitamin A, Retinoids, and Carotenoids

2.5.2.1 Retinoids in Cancer Prevention and Treatment Since the discovery of vitamin A, the observation that the main effects of deficiency are hyperplasia and loss of differentiation of squamous epithelium has raised speculation that the vitamin may be associated with carcinogenesis. Either deficiency may be a risk factor for cancer or increased intake may be protective. Deficient animals develop more spontaneous tumors and are more sensitive to chemical carcinogens, whereas liver reserves of vitamin A are lower in patients with cancer than in controls. One of the genes repressed by retinoic acid is the *myc*-oncogene.

The addition of relatively high concentrations of retinol to organ culture media produces changes that are apparently the opposite of those seen in deficiency; chick epidermis, which is normally keratinized, becomes mucus producing and in some cases ciliated. Studies of experimentally induced and transplanted tumors in experimental animals given very high intakes of retinol or retinoic acid, and of tumors in tissue culture with very high concentrations of retinol and retinoic acid, suggest that there is a potentially beneficial effect of very high intakes in inhibiting the initiation and growth of epithelial tumors.

The doses of retinol that are protective in animals are in the toxic range (Section 2.5.1) and are unlikely to be useful in cancer therapy or prevention. A number of synthetic retinoids have been developed, in a search for compounds that show anticancer activity, but are metabolized, stored, and transported differently, or bind to different subtypes of retinoid receptor and are less toxic. RXR-selective ligands are less toxic and more active in animal cancer models than RAR ligands (Lippman and Lotan, 2000). Fenretinamide, and possibly other retinoids that have antitumor activity, exerts at least part of its action by induction of apoptosis by a receptor-independent mechanism (Wu et al., 2001).

In addition to the regression of established tumors, a number of retinoids show apparent inhibition of the chemical induction of the bladder and other epithelial tumors in experimental animals. The effect is not in fact inhibition of carcinogenesis, but rather a lengthening of the latent period between the initiation step of carcinogenesis and the development of tumors. Although perhaps not as exciting as compounds that prevent the development of cancers, such a delaying action may be useful. If the results can be scaled from experimental animals to man, then it seems that the recurrence of bladder tumors after initial surgery or chemotherapy might be delayed by 5 to 10 years, a clinically useful effect (Hicks and Turton, 1986).

2.5.2.2 Retinoids in Dermatology   13-*Cis*-retinoic acid (isotretinoin, Accutane®) is used orally, and all-*trans*-retinoic acid (Tretinoin®) topically, for treatment of severely disfiguring cystic acne. Etretinate (the trimethoxyphenyl analog of retinoic acid) and tazarotene (a receptor-specific retinoid) are used topically for the treatment of psoriasis. They are effective in cases in which other therapy has failed, and at lower levels than are required for the control of tumor development in experimental animals, although they have been associated with birth defects (Section 2.5.1.1; Johnson and Chandraratna, 1999).

2.5.2.3 Carotene   Average daily intakes of carotenoids in western countries are of the order of 7 to 8 mg per day: α-carotene, 0.7 mg; β-carotene, 3 mg; lutein and zeaxanthin, 2.5 mg each; and lycopene, 1 mg. The major importance of dietary carotenoids is as precursors of vitamin A; even among omnivores in western countries, some 25% to 30% of vitamin A is provided by carotenes rather than preformed retinol. In plants and microorganisms, carotenoids function not only as pigments (e.g., in flowers), but also as energy-transferring molecules in photosynthesis, broadening the spectrum of light that can activate chlorophyll. A number of carotenoids of natural origin are used as food colors – those that have an unsubstituted β-ionone ring will also have provitamin A activity.

Most carotenoids are stored in adipose tissue. Lutein and zeaxanthin (but not other carotenoids) are specifically accumulated in the pigment layer of the retina, and there is epidemiological evidence that they are protective against the development of age-related macular degeneration. There is also epidemiological evidence that lutein and zeaxanthin may provide protection against the development of cataracts. Lycopene is accumulated in the adrenal glands and testes at concentrations 20-fold higher than occur in adipose tissue, suggesting that there is active accumulation in these two tissues. Epidemiological evidence suggests that it may be protective against prostate cancer (Stahl and Sies, 1996; Gann et al., 1999; Handelman, 2001).

In addition to their importance as precursors of vitamin A, carotenes can also act, at least in vitro (and under conditions of low oxygen tension), as antioxidants, trapping singlet oxygen generated by photochemical reactions or lipid peroxidation of membranes (Burton and Ingold, 1984). Studies with

β-carotene and other carotenoids have not shown any consistent effect on in vivo markers of oxidative damage (Institute of Medicine, 2000).

Epidemiological and case-control studies show a negative association between β-carotene intake and a number of cancers (Peto et al., 1981), suggesting that β-carotene may have a protective effect against some forms of cancer, and hence a function in its own right, not simply as a precursor of retinol. This has generally been assumed to be due to the antioxidant activity of β-carotene, although it is noteworthy that different dietary carotenoids induce different isoenzymes of cytochrome P_{450} and might be predicted to have positive or negative effects on (chemical) carcinogenesis (Jewell and O'Brien, 1999). In addition, there is evidence that some carotenoids may have a genomic action in their own right, inducing synthesis of connexin 43, one of the proteins involved in maintaining tissue integrity and cell-to-cell communication. Teicher and coworkers (1999) reported that the action of apo-carotenoic acids is to increase the stability of connexin mRNA (by binding to the 3′-untranslated region), rather than enhancement of gene expression. Increased synthesis of connexin might retard the growth of a tumor by maintaining an outer layer of tightly connected normal cells or stimulating intercell gap-junction communication (Stahl and Sies, 1996; Bertram, 1999; Stahl et al., 2000).

On the basis of the epidemiological evidence, there have been a number of intervention studies using supplements of β-carotene. They have typically used supplements of 20 to 30 mg per day β-carotene in a highly available form, compared with average intakes from foods of 7 to 8 mg of mixed carotenoids of generally low biological availability. In the Linxian study in China (Blot et al., 1993), supplements of β-carotene, vitamin E, and selenium to a marginally malnourished population led to a reduction in mortality from a variety of cancers, especially gastric cancer. The Physicians' Health Study (Hennekens et al., 1996) was a 12-year trial in the United States in which β-carotene supplements showed no effect on the incidence of cardiovascular disease or cancer.

Two major intervention trials among people at risk of lung cancer were the Alpha-Tocopherol Beta-Carotene Study in Finland, in which heavy smokers were given supplements of 20 mg per day of β-carotene and/or 50 mg of vitamin E (Alpha-Tocopherol Beta-Carotene Cancer Prevention Study Group, 1994), and the CARET Study (Carotene and Retinol Efficacy Lung Cancer Chemoprevention Trial; Omenn et al., 1996a, 1996b) in the United States involving people who had been exposed to asbestos dust, who received 30 mg of β-carotene and 7,500 μg of retinyl palmitate per day. Both studies showed a significant increase in death from lung cancer among people taking the supposedly protective carotene supplements. A number of hypotheses have been advanced to explain this unexpected finding:

1. The studies were performed on established heavy smokers and people who had been exposed to asbestos dust some years previously. It has been suggested that whereas β-carotene may inhibit induction of cancers by reactive oxygen species, it may also enhance later stages in tumor development.

2. The plasma concentrations of β-carotene in the intervention trials were considerably higher than those observed to be protective in epidemiological studies; as with any antioxidant, β-carotene is also potentially a prooxidant. This is especially likely in the lung, where the partial pressure of oxygen is high; the radical-trapping antioxidant action is observed at low partial pressures of oxygen (Burton and Ingold, 1984). Whereas β-carotene may be an antioxidant at low levels of intake, higher intakes may lead to the formation of oxidized metabolites that are prooxidants (Wang and Russell, 1999; Young and Lowe, 2001).

3. It is possible that whereas asymmetric cleavage of β-carotene at low concentrations leads to the formation of apocarotenals that can be oxidized to retinoic acid (Section 2.2.2.1), at higher levels apocarotenals and apocarotenoic acids may antagonize retinoic acid. Ferrets (which metabolize carotene similarly to human beings) show lower concentrations of retinoic acid in the lung when they are given high intakes of β-carotene, as well as lower expression of RARβ, although RARα and RARγ are unaffected. In lung cancer cells in vitro, RARβ has tumor-suppressing activity (Houle et al., 1993).

4. What the epidemiological studies have actually shown is a negative association between various types of cancer and the consumption of fruits and vegetables that are rich in carotenoids and a great many other potentially protective compounds (Section 14.7). It may be that β-carotene is simply a marker for some other protective factor.

FURTHER READING

Albert AD and Yeagle PL (2000) Structural aspects of the G-protein receptor, rhodopsin. *Vitamins and Hormones* **58,** 27–51.

Bachmair F, Hoffmann R, Daxenbichler G, and Langer T (2000) Studies on structure-activity relationships of retinoic acid receptor ligands by means of molecular modeling. *Vitamins and Hormones* **59,** 159–215.

Bauernfeind J (1986) *Vitamin A Deficiency and Its Control.* New York: Academic Press.

Chambon P (1996) A decade of molecular biology of retinoic acid receptors. *FASEB Journal* **10,** 940–54.

Clagett-Dame M and DeLuca HF (2002) The role of vitamin A in mammalian reproduction and embryonic development. *Annual Reviews of Nutrition* **22,** 347–81.

Evans T and Kaye S (1999) Retinoids, present role and future potential. *British Journal of Cancer* **80,** 1–8.

Gerster H (1997) Vitamin A: functions, dietary requirements and safety in humans. *International Journal of Vitamin and Nutrition Research* **67,** 71–90.

Giguère V (1994) Retinoic acid receptors. In *Vitamin Receptors: Vitamins and Ligands in Cell Communication*, K Dakshinamurti (ed.), Cambridge, UK: Cambridge University Press.

Goodman D (1984) Vitamin A and retinoids in health and disease. *New England Journal of Medicine* **310,** 1023–31.

Gottesman ME, Quadro L, and Blaner WS (2001) Studies of vitamin A metabolism in mouse model systems. *Bioessays* **23,** 409–19.

Green M and Green J (1996) Quantitative and conceptual contributions of mathematical modelling to current views on vitamin A metabolism, biochemistry and nutrition. *Advances in Food and Nutrition Research* **40,** 3–24.

Hathcock J, Hattan D, Jenkins M, McDonald J, Sundaresan P, and Wilkening V (1990) Evaluation of vitamin A toxicity. *American Journal of Clinical Nutrition* **52,** 183–202.

Hill D and Grubbs C (1992) Retinoids and cancer prevention. *Annual Reviews of Nutrition* **12,** 61–181.

Means AL and Gudas L (1995) The roles of retinoids in vertebrate development. *Annual Reviews of Biochemistry* **64,** 201–33.

Morriss-Kay GM and Ward SJ (1999) Retinoids and mammalian development. *International Review of Cytology* **188,** 73–131.

Pfahl M and Chytil F (1996) Regulation of metabolism by retinoic acid and its nuclear receptors. *Annual Reviews of Nutrition* **16,** 257–83.

Russell RM (2000) The vitamin A spectrum: from deficiency to toxicity. *American Journal of Clinical Nutrition* **71,** 878–84.

Smith S, Dickman E, Power S, and Lancman J (1998) Retinoids and their receptors in vertebrate embryogenesis. *Journal of Nutrition* **128,** 467s–70s.

Soprano DR (1994) Serum and cellular retinoid binding proteins. In *Vitamin Receptors: Vitamins and Ligands in Cell Communication*, K Dakshinamurti (ed.), pp. 1–27. Cambridge, UK: Cambridge University Press.

Stephensen CB (2001) Vitamin A, infection, and immune function. *Annual Reviews of Nutrition* **21,** 167–92.

Sun SY and Lotan R (2002) Retinoids and their receptors in cancer development and chemoprevention. *Critical Reviews in Oncology and Haematology* **41,** 41–55.

Thurnham D and Northrop-Clewes C (1999) Optimum nutrition: vitamin A and the carotenoids. *Proceedings of the Nutrition Society* **58,** 449–57.

Tomkins A and Hussey G (1989) Vitamin A, immunity and infection. *Nutrition Research Reviews* **2,** 17–28.

Underhill T, Kotch L, and Linney E (1995) Retinoids and mouse embryonic development. *Vitamins and Hormones* **51**.

Various authors (1994). Vitamin A, from molecular biology to public health. *Nutrition Reviews* **52,** 1s–90s (14th Maribou Symposium).

von Reinersdorff D, Green MH, and Green JB (1998) Development of a compartmental model describing the dynamics of vitamin A metabolism in men. *Advances in Experimental Medicine and Biology* **445,** 207–23.

West K, Howard G, and Sommer A (1989) Vitamin A and infection: public health implications. *Annual Reviews of Nutrition* **9,** 63–86.

Zile MH (2001) Function of vitamin A in vertebrate embryonic development. *Journal of Nutrition* **131,** 705–8.

References cited in the text are listed in the Bibliography.

THREE

Vitamin D

Vitamin D is not strictly a vitamin, rather it is the precursor of one of the hormones involved in the maintenance of calcium homeostasis and the regulation of cell proliferation and differentiation, where it has both endocrine and paracrine actions. Dietary sources are relatively unimportant compared with endogenous synthesis in the skin by photolysis of 7-dehydrocholesterol; problems of deficiency arise when there is inadequate exposure to sunlight. The deficiency diseases (rickets in children and osteomalacia in adults) are therefore largely problems of temperate and subarctic regions, although cultural factors that result in little exposure to sunlight may also cause problems in subtropical and tropical areas. There are few foods that are rich sources of vitamin D. It is generally accepted that, for people with inadequate exposure to sunlight (young children and the house-bound elderly), supplements are necessary to maintain adequate status. Excessively high intakes of vitamin D are associated with hypercalcemia and calcinosis.

Although the pioneering studies of Chick and others during the 1920s clarified the dual roles of sunlight exposure to promote endogenous synthesis and dietary sources of the vitamin, it was not until high specific activity [^{3}H]vitamin D became available in the 1960s that the onward metabolism of vitamin D to the active metabolite, calcitriol, was discovered, and its mechanism of action elucidated, largely by Kodicek and coworkers in Cambridge and DeLuca and coworkers in Wisconsin. Calcitriol acts as a steroid hormone, binding to a nuclear receptor protein in target tissues and regulating gene expression. As a result of studies of the distribution of calcitriol receptors and the induced proteins, a number of functions have been discovered for the vitamin other than in the maintenance of calcium balance, including roles in cell proliferation and differentiation, in the modulation of immune system responses, and in the secretion of insulin and thyroid and parathyroid hormones.

More recent studies, during the 1990s, have shown that calcitriol also has rapid actions, acting via cell surface G-protein receptors linked to both adenylate cyclase and phospholipase cascade systems.

3.1 VITAMIN D VITAMERS, NOMENCLATURE, AND UNITS OF ACTIVITY

Two compounds have the biological activity of vitamin D: cholecalciferol, which is the compound formed in the skin, and ergocalciferol, which is synthesized by ultraviolet (UV) irradiation of ergosterol (see Figure 3.1). The name vitamin D_1 was originally given to the crude product of irradiation of ergosterol, which contained a mixture of ergocalciferol with inactive lumisterol (an isomer of ergosterol) and suprasterols. When ergocalciferol was identified as the active compound, it was called vitamin D_2. Later, when cholecalciferol was identified as the compound formed in the skin and found in foods, it was called vitamin D_3. Vitamin D is a secosteroid – i.e., a steroid in which the B-ring has undergone cleavage, followed by rotation of the A-ring (see Figures 3.1 and 3.2). The numbering of carbon atoms in the vitamin follows that of the parent steroid nucleus, and, more confusingly, the assignation of positions of substituents above or below the plane of the ring also follows that of the parent steroid. This means that the 1-hydroxylated derivative, which actually has the β-configuration, is correctly referred to as 1α-hydroxy. As discussed in Section 3.2, vitamin D undergoes hydroxylation to the metabolically active 1,25-dihydroxy derivative, and a number of abbreviations for the various hydroxylated derivatives are used in the literature. The recommended nomenclature for the metabolites is shown in Table 3.1.

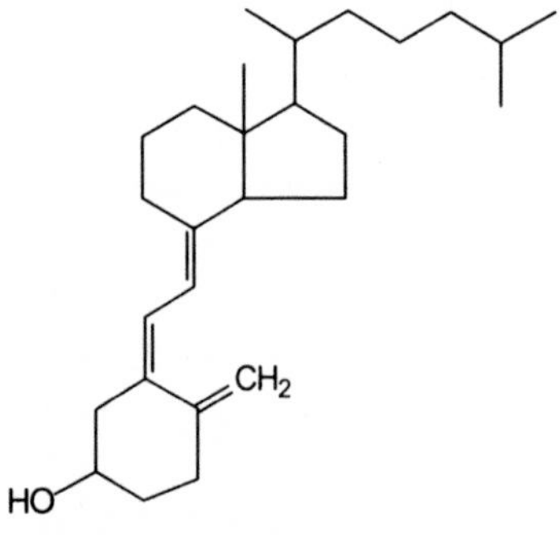

vitamin D_3
calciol (cholecalciferol)

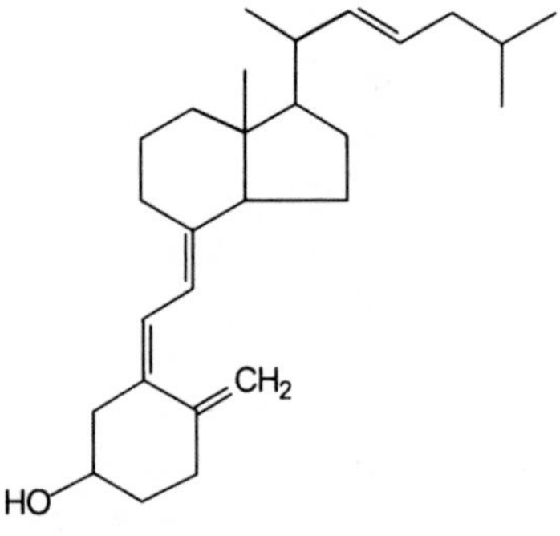

vitamin D_2
ercalciol (ergocalciferol)

Figure 3.1. Vitamin D vitamers. Relative molecular masses (M_r): calciol, 384.6; ercalciol, 396.6.

Table 3.1 Nomenclature of Vitamin D Metabolites

Trivial Name	Recommended Name	Abbreviation	M_r
Vitamin D_3			
Cholecalciferol	Calciol	—	384.6
25-Hydroxycholecalciferol	Calcidiol	$25(OH)D_3$	400.6
1α-Hydroxycholecalciferol	1(S)-Hydroxycalciol	$1α(OH)D_3$	400.6
24,25-Dihydroxycholecalciferol	24(R)-Hydroxycalcidiol	$24,25(OH)_2D_3$	416.6
1,25-Dihydroxycholecalciferol	Calcitriol	$1,25(OH)_2D_3$	416.6
1,24,25-Trihydroxycholecalciferol	Calcitetrol	$1,24,25(OH)_3D_3$	432.6
Vitamin D_2			
Ergocalciferol	Ercalciol	—	396.6
25-Hydroxyergocalciferol	Ercalcidiol	$25(OH)D_2$	412.6
24,25-Dihydroxyergocalciferol	24(R)-Hydroxyercalcidiol	$24,25(OH)_2D_2$	428.6
1,25-Dihydroxyergocalciferol	Ercalcitriol	$1,25(OH)_2 D_2$	428.6
1,24,25-Trihydroxyergocalciferol	Ercalcitetrol	$1,24,25(OH)_3D_2$	444.6

Abbreviations shown in column 3 are not recommended, but are frequently used in the literature.

Before the preparation of crystalline cholecalciferol, the standard for biological activity of vitamin D was a solution of irradiated ergosterol (and hence ergocalciferol). The (obsolete) international unit (iu) of vitamin D activity is equivalent to 25 ng (65 pmol) of cholecalciferol. One microgram of cholecalciferol is equivalent to 40 iu (1 nmol is 104 iu). Cholecalciferol and ergocalciferol are not equipotent, and the relative biological activities of the two vitamers differ in different species. In most species (including human beings), cholecalciferol causes a greater increase in the circulating concentration of the 25-hydroxy-derivative than does ergocalciferol, because of faster metabolic clearance of ergocalciferol than cholecalciferol. In the rat, by contrast, there is metabolic discrimination against cholecalciferol in favor of ergocalciferol. As far as is known, the active metabolites (calcitriol and ercalcitriol) are equipotent and bind to calcitriol receptors in target tissues (Section 3.3.3.1) with equal affinity (Horst et al., 1982; Trang et al., 1998).

3.2 METABOLISM OF VITAMIN D

Synthetic ergocalciferol is used for enrichment and fortification of foods; its metabolic fate is the same as that of dietary cholecalciferol. Except where there are known to be differences between the two vitamers, it is assumed that all of the following discussion applies equally to ergocalciferol and cholecalciferol. There are few rich dietary sources of vitamin D, and the major source is usually photosynthesis in the skin. Dietary vitamin D is absorbed in chylomicrons and taken up rapidly by the liver as chylomicron remnants are cleared from

Table 3.2 Plasma Concentrations of Vitamin D Metabolites

		nmol/L
Cholecalciferol		1.3–156
24-Hydroxycalcidiol		2–20
Calcitriol		0.038–0.144
Calcidiol	Adults, summer	37–87
	Adults, winter	20–45
	Adults with osteomalacia	<10
	Children, summer	50–100
	Children, winter	27–52
	Children with rickets	<20
	Risk of hypercalcemia	>400

the circulation. By contrast, vitamin D synthesized in the skin is bound to plasma vitamin D binding protein (Section 3.3.2.7) and is metabolized more gradually.

Both dietary and endogenously synthesized vitamin D undergo 25-hydroxylation in the liver to yield calcidiol (25-hydroxycholecalciferol), which is the main circulating form of the vitamin. This undergoes 1-hydroxylation in the kidney to produce the active hormone calcitriol (1,25-dihydroxy-cholecalciferol) or 24-hydroxylation in the kidney and other tissues to yield 24-hydroxycalcidiol (24,25-dihydroxycholecalciferol).

Unlike the other fat-soluble vitamins, there is little or no storage of vitamin D in the liver, except in oily fish. In human liver, concentrations of vitamin D do not exceed about 25 nmol per kg. Significant amounts may be present in adipose tissue, but this is not really storage of the vitamin, because it is released into the circulation as adipose tissue is catabolized, rather than in response to demand for the vitamin. The main storage of the vitamin seems to be as plasma calcidiol, which has a half-life of the order of 3 weeks (Holick, 1990). In temperate climates, there is a considerable seasonal variation, with plasma concentrations at the end of winter as low as half those seen at the end of summer (see Table 3.2). The major route of vitamin D excretion is in the bile, with less than 5% as a variety of water-soluble conjugates in urine. Calcitroic acid (see Figure 3.3) is the major product of calcitriol metabolism; but, in addition, there are a number of other hydroxylated and oxidized metabolites.

3.2.1 Photosynthesis of Cholecalciferol in the Skin
Cholecalciferol is formed nonenzymatically in the skin by UV irradiation of 7-dehydrocholesterol, as shown in Figure 3.2. 7-Dehydrocholesterol is an

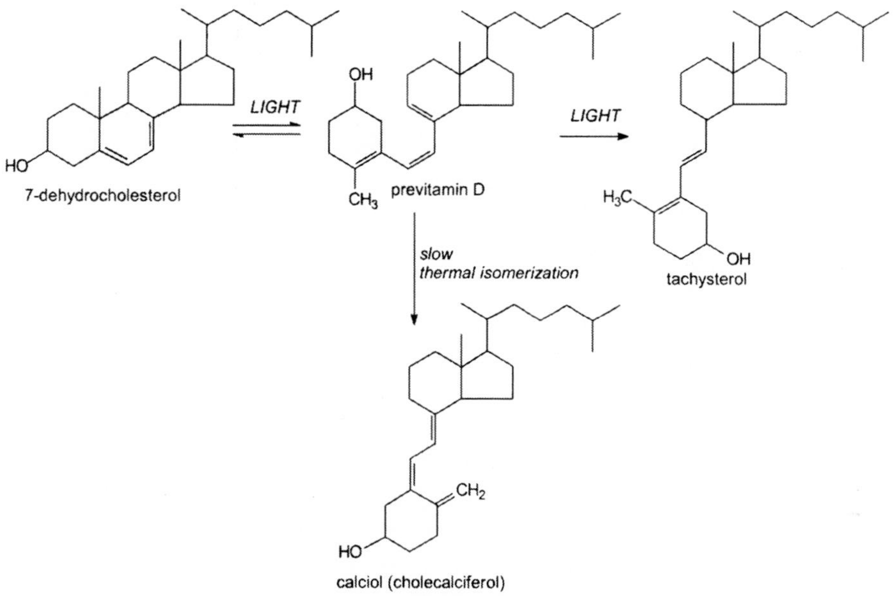

Figure 3.2. Synthesis of calciol from 7-dehydrocholesterol in the skin.

intermediate in the synthesis of cholesterol that accumulates in the skin, but not other tissues. It is synthesized in the sebaceous glands, secreted onto the surface of the skin, and then absorbed into the epidermis. It is found throughout the epidermis and dermis, with the highest concentration per unit surface area in the stratum basale and stratum spinosum, which therefore have the highest capacity for cholecalciferol synthesis. One of the possible causes of vitamin D deficiency in the elderly, in addition to low exposure to sunlight, is an age-dependent decrease in the concentration of 7-dehydrocholesterol in the epidermis – hence a reduction in the capacity for endogenous cholecalciferol synthesis.

On exposure to UV light, 7-dehydrocholesterol undergoes photolysis, with cleavage of the B-ring and inversion of the A-ring, to yield precalciferol (previtamin D or tacalciol). The peak wavelength for this photolysis is 296.5 nm; for practical purposes, the useful range of solar radiation is the UV-B range, between 290 nm (the lowest wavelength transmitted by ozone) and 320 nm. At 310 nm, however, the yield of precalciferol is only 1% of that at 296.5 nm. Precalciferol undergoes thermal isomerization to cholecalciferol. This is a slow process; at 37°C, there is 50% isomerization within 48 hours, and, after 4 days, equilibrium is reached in vitro with about 83% cholecalciferol. In vivo, isomerization is somewhat more rapid because the equilibrium is shifted by the removal of cholecalciferol bound to plasma vitamin D binding protein.

As discussed in Section 3.6.1, excess oral vitamin D results in hypercalcemia; toxicity is associated with plasma concentrations of calcidiol above 400 nmol per L. However, even excessive exposure to sunlight does not result in vitamin D intoxication, and the plasma concentration of calcidiol does not rise above 100 to 200 nmol per L. During prolonged UV irradiation of the skin, the concentration of precalciferol does not rise above 10% to 15% of the initial concentration of 7-dehydrocholesterol. This is because photolysis of 7-dehydrocholesterol is reversible; light-catalyzed closure of the B-ring can result in formation of either 7-dehydrocholesterol or lumisterol, in which the 19-methyl group has the opposite configuration. In addition, precalciferol undergoes photoisomerization to tachysterol, which is biologically inactive; and cholecalciferol is also sensitive to photodegradation, yielding 5,6-*trans*-cholecalciferol and biologically inactive suprasterols. There is no evidence that the cutaneous synthesis of cholecalciferol is regulated by vitamin D status, and the administration of calcitriol has no effect on the increase in serum calcidiol after exposure to UV irradiation. Because of both the slow isomerization of precalciferol to cholecalciferol and photoisomerization to inactive compounds, skin pigmentation seems not to affect the total amount of cholecalciferol formed to any significant extent.

Sunlight is not strictly essential for cutaneous synthesis of cholecalciferol, because UV-B penetrates clouds reasonably well; complete cloud cover reduces the available intensity by about 50%. It also penetrates light clothing. However, low-intensity irradiation (below 20 mJ per cm^2 in vitro) does not result in significant photolysis of 7-dehydrocholesterol to previtamin D. Acute whole-body exposure to UV-B irradiation below 18 mJ per cm^2 does not result in any detectable increase in plasma cholecalciferol or calcidiol. In temperate regions (beyond about 40°N or S), the intensity of UV-B is below this threshold in winter, so there is unlikely to be any significant cutaneous synthesis of the vitamin in winter, and plasma concentrations of calcidiol show a marked seasonal variation in temperate regions (Holick, 1995; see Table 3.2).

3.2.2 Dietary Vitamin D
There are few dietary sources of cholecalciferol. The richest sources are oily fish (especially fish liver oils), although eggs also contain a relatively large amount, and there is a modest amount in milk fat and animal liver. In many countries, margarine is fortified with vitamin D. No common plant foods contain vitamin D, although some tropical plants contain calciferol glucuronides that are hydrolyzed in the intestinal lumen and are a source of the vitamin. Indeed, this can be a cause of hypervitaminosis and calcinosis in grazing animals.

Dietary cholecalciferol and ergocalciferol are absorbed from the small intestine in lipid micelles and are transferred into chylomicrons. Some is then transferred in the circulation onto the plasma vitamin D binding protein (Section 3.2.7), but much enters the liver in chylomicron remnants. There is evidence that cholecalciferol arriving in the liver in chylomicron remnants is more susceptible to catabolism, rather than 25-hydroxylation, than is cholecalciferol bound to the plasma vitamin D binding protein. In addition, intermittent and relatively high dietary intakes of cholecalciferol will lead to high concentrations in the liver and hence increased catabolism, compared with the more gradual release from the skin. Even in subjects consuming foods fortified with ergocalciferol, and in winter when the intensity of UV-B irradiation is below the threshold for cutaneous synthesis of calciferol, 80% to 90% of circulating 25-hydroxyvitamin D is calcidiol rather than ercalcidiol. In elderly subjects, the plasma concentration of calcidiol may fall to below 15 to 20 nmol per L in winter, suggesting that normal dietary intake does not make a significant contribution (Lawson et al., 1979). An intake of at least 5 μg per day is required to avoid this seasonal variation in plasma calcidiol and the associated rise in parathyroid hormone secretion (Section 3.2.8.2), and maintenance of plasma calcidiol above 20 nmol per L requires intakes of the order of 10 μg per day (Krall et al., 1989). This is considerably above what can be achieved from a normal diet and suggests that supplements are necessary when sunlight exposure and endogenous synthesis are inadequate.

3.2.3 25-Hydroxylation of Cholecalciferol

There are two separate cytochrome P_{450}-dependent mixed-function oxidases in the liver that catalyze the 25-hydroxylation of cholecalciferol (see Figure 3.3). The activity of both enzymes is higher in tissue from vitamin D-deficient animals, and there is some evidence that calcitriol either inhibits or represses them. The mitochondrial enzyme (CYP27A), which has a K_m of 10^{-5} M, catalyzes the hydroxylation of cholecalciferol twice as fast ergocalciferol. It also acts on a number of C-27 steroids and is involved in bile acid synthesis. This enzyme requires ferredoxin and ferredoxin reductase for activity. The microsomal enzyme (CYP2D25), with a K_m of 10^{-7} M, only acts on cholecalciferol, and not ergocalciferol, although it also acts on a number of C-27 steroids and will catalyze the 25-hydroxylation of 1α-hydroxycholecalciferol (hydroxycalciol) to calcitriol. As discussed in Section 3.4.2, this has been exploited to treat and prevent problems of vitamin D-resistant rickets and osteomalacia in cases of renal failure (renal osteodystrophy) using 1α-hydroxycholecalciferol.

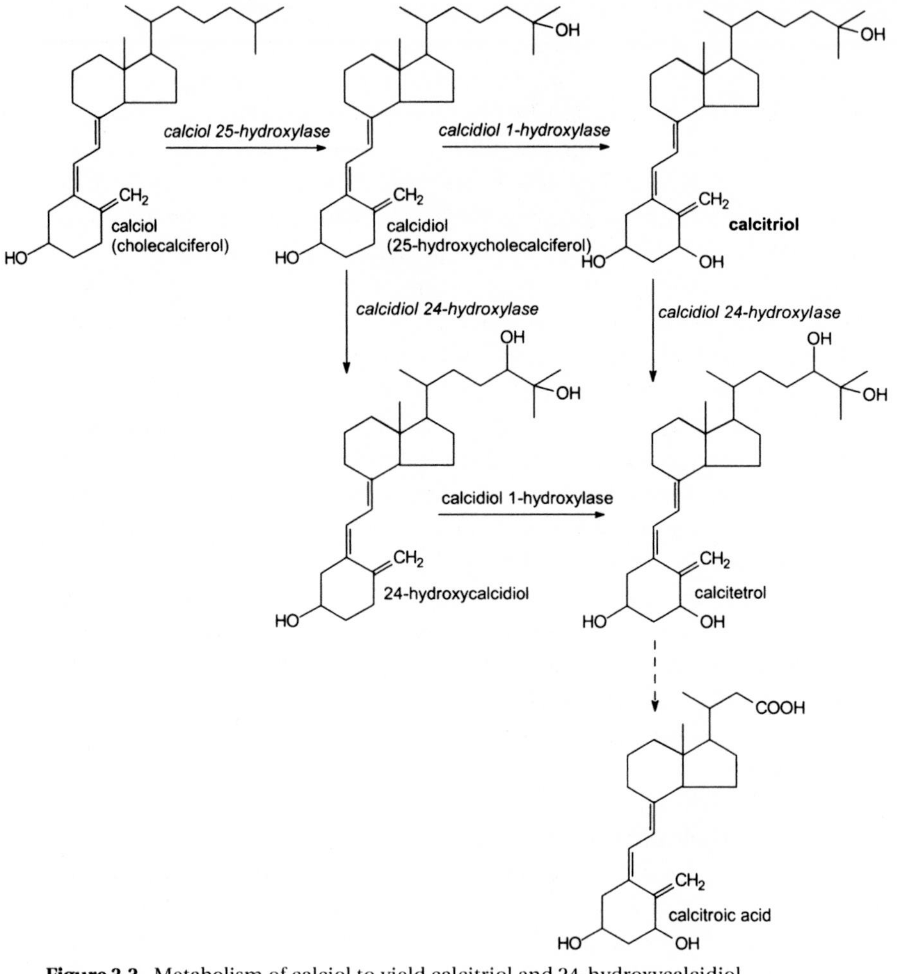

Figure 3.3. Metabolism of calciol to yield calcitriol and 24-hydroxycalcidiol.

Cholecalciferol 25-hydroxylase is not restricted to the liver; kidneys, skin, and gut microsomes also have a cytochrome P_{450}-dependent enzyme that catalyzes the 25-hydroxylation of cholecalciferol and 1α-hydroxycholecalciferol, but not ergocalciferol. Although there is some evidence that calcitriol can reduce the activity of calciferol 25-hydroxylase, it is not known whether this is physiologically important; the major factor controlling 25-hydroxylation is the rate of uptake of cholecalciferol into the liver. It is the fate of calcidiol in the kidneys that provides the most important regulation of vitamin D metabolism (Wikvall, 2001).

3.2.4 Calcidiol 1α-Hydroxylase

The active metabolite of vitamin D, calcitriol, is formed in the proximal tubules of the kidneys from calcidiol. There are three cytochrome P_{450}-dependent enzymes in kidneys that catalyze 1-hydroxylation of calcidiol: CYP27A and CYP27 in mitochondria and a microsomal 1α-hydroxylase, which is ferredoxin-dependent. It is likely that the microsomal enzyme is the most important; its synthesis is induced by cAMP in response to parathyroid hormone (Section 3.2.8.2) and repressed by calcitriol (Omdahl et al., 2001; Wikvall, 2001).

Calcidiol 1α-hydroxylase is not restricted to the kidney, but is also found in placenta, bone cells (in culture), mammary glands, and keratinocytes. The placental enzyme makes a significant contribution to fetal calcitriol, but it is not clear whether the calcidiol 1-hydroxylase activity of other tissues is physiologically significant or not. Acutely nephrectomized animals given a single dose of calcidiol do not form any detectable calcitriol, but there is some formation of calcitriol in anephric patients, which increases on the administration of cholecalciferol or calcidiol. However, this extrarenal synthesis is not adequate to meet requirements, so that osteomalacia develops in renal failure (Section 3.4.1). The enzyme is inhibited, or possibly repressed, by strontium ions; this is the basis of strontium-induced vitamin D-resistant rickets, which responds to the administration of calcitriol or 1α-hydroxycalciol, but not calciferol or calcidiol (Omdahl and DeLuca, 1971).

Calcidiol 1α-hydroxylase also acts on 24-hydroxycalcidiol, yielding calcitetrol; indeed, it has a relatively low specificity and will act on any secosterol with hydroxyl groups at C-3 and C-25. Calcitriol has a short metabolic half-life after injection of the order of 4 to 6 hours (Holick, 1990). But, under normal conditions, the regulation of its synthesis means that the plasma concentration remains fairly constant, depending on the state of calcium balance (Hewison et al., 2000).

3.2.5 Calcidiol 24-Hydroxylase

Both calcidiol and calcitriol are substrates for 24-hydroxylation, catalyzed by a cytochrome P_{450}-dependent enzyme in kidneys, intestinal mucosa, cartilage, and other tissues that contain calcitriol receptors. This enzyme is induced by calcitriol; the activities of calcidiol 1-hydroxylase and 24-hydroxylase in the kidney are subject to regulation in opposite directions, so that decreased requirement for, and synthesis of, calcitriol results in increased formation of 24-hydroxycalcidiol. Kidney epithelial cells in culture show increased formation of 24-hydroxycalcidiol, and decreased formation of calcitriol, after the addition of calcitriol or high concentrations of calcium to the culture medium.

Conversely, the addition of parathyroid hormone results in decreased 24-hydroxylation and increased 1-hydroxylation (Juan and DeLuca, 1977; Omdahl et al., 2001; Wikvall, 2001). There is evidence that the high prevalence of vitamin D deficiency among people from the Indian subcontinent may be because of genetically determined high activity of calcidiol 24-hydroxylase, rather than cultural and dietary factors (Awumey et al., 1998).

Early studies suggested that 24-hydroxylation of calcidiol was a pathway for inactivation of the vitamin, a conclusion that is supported by the observation that calcidiol 24-hydroxylase is activated and induced by calcitriol. Fish-eating mammals, such as seals, that have a very high intake of cholecalciferol, do not show vitamin D intoxication. Although they have plasma concentrations of calcitriol similar to those seen in other mammals, after the administration of [^{3}H]cholecalciferol, label is found in calcidiol and 24-hydroxycalcidiol, not calcitriol, suggesting that 24-hydroxylation provides a means of inactivating excess calciferol and hence avoiding vitamin D intoxication (Keiver et al., 1988).

There is evidence that 24-hydroxycalcidiol has physiological functions distinct from those of calcitriol, and the regulation of the 24-hydroxylase suggests that it functions to provide a metabolically active product, as well as diverting calcidiol away from calcitriol synthesis (Henry, 2001). Studies of knockout mice lacking the 24-hydroxylase show that 24-hydroxycalcidiol has a role in both intramembranous bone formation during development and the suppression of parathyroid hormone secretion (St-Arnaud, 1999; van Leeuwen et al., 2001).

3.2.6 Inactivation and Excretion of Calcitriol

Most vitamin D is excreted in the bile; less than 5% is excreted as water-soluble metabolites in urine. Some 2% to 3% of the vitamin D in bile is cholecalciferol, calcidiol, and calcitriol, but most is a variety of polar metabolites and their glucuronide conjugates. In most tissues, the major pathway for inactivation of calcitriol is by way of 24-hydroxylation to calcitetrol, then onward oxidation by way of the 24-oxo-derivative, 23-hydroxylation, and oxidation to calcitroic acid (see Figure 3.3). In addition, a variety of hydroxylated and other polar metabolites have been identified in bile, and many of these onward oxidation products also undergo glucuronide conjugation in the liver (Reddy and Tserng, 1989).

Compounds that induce cytochrome P_{450}-dependent hydroxylases, such as barbiturates and the anticonvulsants primidone and diphenylhydantoin, cause increased output of vitamin D metabolites in the bile, and increase the rate of inactivation of calcidiol by liver microsomes. As a result of this, long-term use of these anticonvulsants may be associated with the development of osteomalacia, although barbiturates also cause some induction of calciferol 25-hydroxylase, thus increasing the hydroxylation of calciferol to calcidiol.

3.2.7 Plasma Vitamin D Binding Protein (Gc-Globulin)

Cholecalciferol, calcidiol, calcitriol, and 24-hydroxycalcidiol are all transported bound to the same plasma binding protein – Gc-globulin, also known as the group-specific component or transcalciferin. There are three major forms of Gc-globulin, with differing primary structures, and a number of minor variants of each because of differences in postsynthetic glycosylation. There is considerable polymorphism among human populations; because of this, Gc-globulin has been investigated both for its interest in population genetics and also its potential value in forensic medicine. All the variants bind vitamin D and its metabolites with similar affinity. It is noteworthy that the absence of Gc-protein has never been detected, suggesting that a deletion of this protein may be fatal. Cholecalciferol is also transported in plasma lipoproteins, so that about 60% is normally bound to Gc-globulin and 40% to lipoproteins. It is only that fraction bound to lipoproteins that is taken up by the liver for 25-hydroxylation (Haddad et al., 1988). In addition to its role in the plasma transport of vitamin D, and control over tissue uptake, Gc-globulin represents the major storage site for the vitamin, mainly as calcidiol.

The plasma binding protein has a higher affinity for calcidiol and 24-hydroxycalcidiol than for calcitriol or cholecalciferol. The plasma concentration of Gc-globulin is about 6 mmol per L – considerably higher than the concentrations of other hormone binding proteins, such as thyroxine binding globulin (300 μmol per L), cortisol binding globulin (800 μmol per L), or sex hormone binding globulin (40 μmol per L in males and 80 μmol per L in females) and far in excess of circulating vitamin D. As a result of this, whereas the other hormone binding globulins are about 50% saturated under normal conditions, the vitamin D binding protein is only about 2% saturated. This means that changes in the circulating concentration of the protein are unlikely to have any significant effect on the small proportion of vitamin D metabolites that is free, rather than protein-bound. Again, unlike other hormone binding globulins, the plasma concentration of Gc-globulin is not affected by vitamin D status or other factors that affect calcium homeostasis and vitamin D metabolism (Cooke and Haddad, 1989; Haddad, 1995).

3.2.8 Regulation of Vitamin D Metabolism

The main physiological function of vitamin D is in the control of calcium homeostasis, and vitamin D metabolism is regulated largely by the state of calcium balance. The main regulation of vitamin D metabolism is by control of the activities of calcidiol 1-hydroxylase and 24-hydroxylase, and hence the fate of calcidiol. In general, factors that increase the activity of one of the hydroxylases simultaneously reduce the activity of the other. Although plasma calcitriol is relatively constant throughout the year, 24-hydroxycalcidiol shows a seasonal fluctuation that reflects that of calcidiol.

3.2.8.1 **Calcitriol**   The major determinant of the relative activities of calcidiol 1-hydroxylase and 24-hydroxylase is the availability of calcitriol. In vitamin D-deficient animals, with low circulating concentrations of calcitriol, the activity of 1-hydroxylase in the kidneys is maximal. There is little or no detectable 24-hydroxylase activity. Both in vivo and in isolated kidney cells in culture, the addition of calcitriol results in induction of the 24-hydroxylase and repression of 1-hydroxylase; removal of calcitriol from the culture medium results in induction of 1-hydroxylase and repression of 24-hydroxylase.

3.2.8.2 **Parathyroid Hormone**   Parathyroid hormone raises plasma calcium by direct effects on bone resorption and renal reabsorption of calcium, and indirectly by regulating the metabolism of vitamin D. It is a peptide and acts via cell surface G-protein receptors linked to adenylate cyclase. The parathyroid glands have G-protein cell surface calcium receptors linked to phospholipase C, and parathyroid hormone is secreted in response to hypocalcemia. Magnesium is required for secretion of the hormone, which may explain the development of hypocalcemia in premature infants who are magnesium deficient.

In the kidneys, parathyroid hormone increases 1-hydroxylation of calcidiol and reduces 24-hydroxylation. This is not the result of de novo enzyme synthesis, but an effect on the activity of the preformed enzymes, mediated by cAMP-dependent protein kinases. In turn, calcitriol has a direct role in the control of parathyroid hormone, acting to repress expression of the gene. In chronic renal failure, there is reduced synthesis of calcitriol, leading to the development of secondary hyperparathyroidism that results in excess mobilization of bone mineral, hypercalcemia, hypercalciuria, hyperphosphaturia, and the development of calcium phosphate renal stones.

3.2.8.3 **Calcitonin**   Calcitonin is secreted by the C cells of the thyroid gland in response to hypercalcemia. Its primary action is to oppose the actions of parathyroid hormone by suppressing osteoclast actions. It also stimulates calcidiol 1-hydroxylation in the kidney. Two separate mechanisms seem to be involved: (1) a rapid increase in the activity of 1-hydroxylase, mediated by a cAMP-dependent protein kinase, and (2) a slower response that involves de novo enzyme synthesis. Isolated kidney cells in culture do not respond to calcitonin. The effect is not seen when calcitonin is given to thyro-para-thyroidectomized animals. This suggests that calcitonin may act indirectly, via actions on the parathyroid gland and parathyroid hormone secretion, rather than directly on calcidiol hydroxylases.

3.2.8.4 **Plasma Concentrations of Calcium and Phosphate**   Although the main response to changes in plasma calcium is a change in the secretion of parathyroid hormone, the activity of calcidiol 1-hydroxylase in kidney slices is decreased directly by high concentrations of calcium in the incubation medium. Calcium has no direct effect on the activity of calcidiol 24-hydroxylase under these conditions. Strontium and cadmium also inhibit calcidiol 1-hydroxylase.

The serum concentration of calcitriol varies inversely with phosphate throughout the day. Feeding subjects on low phosphate diets leads to a fall in serum phosphate and an increase in circulating calcitriol. It is not clear whether or not this is a direct effect of phosphate on the kidney hydroxylases.

3.3 METABOLIC FUNCTIONS OF VITAMIN D

The principal physiological role of vitamin D is in the maintenance of the plasma concentration of calcium. Calcitriol acts to increase intestinal absorption of calcium, to reduce its excretion by increasing reabsorption in the distal renal tubule, and to mobilize the mineral from bone – of the 25 mol of calcium in the adult body, 99% is in bone. The daily intake of calcium is around 25 mmol, and intestinal secretions add an additional 7 mmol to the intestinal contents; 10 to 14 mmol of this is normally absorbed, with 18 to 22 mmol excreted in feces. Bone turnover accounts for exchange of 10 mmol of calcium between bone and plasma daily. The kidneys filter some 240 mmol of calcium daily, almost all of which is reabsorbed; urinary excretion of calcium is about 3 to 7 mmol per day.

Calcitriol acts like a steroid hormone, binding to, and activating, nuclear receptors that modulate gene expression. More than 50 genes are known to be regulated by calcitriol (see Table 3.3), but vitamin D response elements have only been identified in a relatively small number, including: calcidiol 1-hydroxylase and 24-hydroxylase; calbindin, a calcium binding protein in the

Table 3.3 Genes Regulated by Calcitriol

	Increased Expression	Decreased Expression
Vitamin D metabolism	Calcitriol receptor Calcidiol 1-hydroxylase Calcidiol 24-hydroxylase	
Mineral metabolism	Calbindin D Osteocalcin Osteopontin Plasma membrane calcium pump Metallothionein	Preproparathyroid hormone Transferrin receptor
Energy metabolism	Glyceraldehyde 3-phosphate dehydrogenase ATP synthase NADH dehydrogenase subunit I NADH dehydrogenase subunit IV Cytochrome oxidase Protein kinase C	Fatty acid binding protein NADH dehydrogenase subunit II Cytochrome b Protein kinase inhibitor Ferredoxin
Regulatory peptides	Nerve growth factor Interleukin I Interleukin 6 Interleukin III receptor Cachexin (tumor necrosis factor-α) Monocyte-derived neutrophil-activating peptide	Histone H_4 Interleukin II γ-Interferon Granulocyte-macrophage colony stimulating factor
Cytoskeleton	Fibronectin Osteoclast integrin	α-tubulin
Oncogenes	c-fms, c-fos, c-ki-ras, c-myc	c-myc Type I collagen

Source: From data reported by Hannah and Norman, 1994.

intestinal mucosa and other tissues; the vitamin K-dependent protein osteo-calcin in bone (Section 5.3.3); and osteopontin, which permits the attachment of osteoclasts to bone surfaces and the osteoclast cell membrane isoform of integrin. In addition, calcitriol affects the secretion of insulin and the synthesis and secretion of parathyroid and thyroid hormones – these actions may be secondary to changes in intracellular calcium concentrations resulting from induction of calbindin.

Calcitriol also has a role in the regulation of cell proliferation and differentiation. In addition to genomic actions, it has a variety of actions that are because of interaction with cell surface G-protein receptors.

24-Hydroxycalcidiol is also biologically active. In hypocalcemic vitamin D-deficient chicks, calcitriol alone does not reverse the hypertrophy of the parathyroid gland; 24-hydroxycalcidiol, together with calcitriol, does – although alone it has no effect (Henry et al., 1977). In hens raised to maturity with calcitriol as the sole source of vitamin D, although the fertility of the eggs is unimpaired, hatchability is greatly reduced and can be restored by feeding 24-hydroxycalcidiol, together with calcitriol (Henry and Norman, 1978). 24-Hydroxycalcidiol also has biological activity in cartilage. Isolated chondrocytes show increased formation of proteoglycans in response to both calcitriol and 24-hydroxycalcidiol. Studies of knockout mice lacking the 24-hydroxylase show that 24-hydroxycalcidiol has a role in intramembranous bone formation during development (St-Arnaud, 1999; van Leeuwen et al., 2001).

3.3.1 Nuclear Vitamin D Receptors

The nuclear vitamin D receptor was originally studied in intestinal mucosa, but has subsequently been found in a variety of other tissues that have therefore been shown to be vitamin D-responsive, including kidneys, bone, parathyroid gland, β-islet cells of the pancreas, pituitary, placenta, uterus, mammary glands, skin, thymus, monocytes, macrophages, and activated T lymphocytes. Like other steroid hormone receptors, it is a zinc finger protein; it has the same high affinity (of the order of 10^{-11} M) for both calcitriol and ercalcitriol.

The vitamin D receptor acts mainly as a heterodimer with the retinoid X receptor (RXR; Section 2.3.2.1). Binding of calcitriol induces a conformational change in the receptor protein, permitting dimerization with occupied or unoccupied RXR, followed by phosphorylation to activate binding to the vitamin D response element on DNA (DeLuca and Zierold, 1998). Abnormally high concentrations of 9-*cis*-retinoic acid result in sequestration of RXR as the homodimers, meaning that it is unavailable to form heterodimers with the vitamin D receptor (or other receptors); excessive vitamin A can therefore antagonize the nuclear actions of vitamin D (Haussler et al., 1995; Rohde et al., 1999).

The vitamin D receptor-RXR heterodimer binds in 5′RXR-VDR3′ polarity to a direct repeat hormone response element. However, the vitamin D receptor also forms heterodimers with the retinoic acid receptor and the thyroid hormone receptor. All three vitamin D receptor dimers can interact with either direct repeat or inverted palindromic hormone response elements. In heterodimers, the vitamin D receptor may be at the 5′-position or 3′-position, resulting in six types of activated vitamin D receptor dimers that can bind to two types of response elements, raising the possibility of multiple signaling pathways (Carlberg, 1996; Carlberg et al., 2001; Yamada et al., 2001b).

Synthesis of the vitamin D receptor is increased in response to both parathyroid hormone (Section 3.2.8.2) and calcitriol. It is not clear whether or not the response to calcitriol is from increased transcription. There is evidence that the ligand-occupied receptor protein is more resistant to degradation than the empty receptor and therefore survives longer (Christakos et al., 1996).

Type II vitamin D-resistant rickets (Section 3.4.2) is associated with target tissue resistance to calcitriol. Most cases are from either a lack of calcitriol receptors or impaired binding of calcitriol to the receptor. Thus, higher than normal concentrations of calcitriol are required to saturate the receptor. Some affected families show normal binding of calcitriol to the receptor, with an apparent defect in the DNA binding domain (Griffin and Zerwekh, 1983).

3.3.2 Nongenomic Responses to Vitamin D

With the isolated perfused duodenum, there is a rapid increase in calcium transport in response to the addition of calcitriol to the perfusion medium. Isolated enterocytes and osteoblasts also show a rapid increase in calcium uptake in response to calcitriol. It is not associated with changes in mRNA or protein synthesis, but seems to be because of recruitment of membrane calcium transport proteins from intracellular vesicles to the cell surface. It is inhibited by the antimicrotubule compound colchicine. It can only be demonstrated in tissues from animals that are adequately supplied with vitamin D; in vitamin D-deficient animals, the increase in intestinal calcium absorption occurs only more slowly, together with the induction of calbindin.

In osteoblasts, keratinocytes, and colonocytes, and possibly other cells, calcitriol acts via cell surface receptors linked to phospholipase C, resulting in release of diacylglycerol and inositol trisphosphate (Section 14.4.1), followed by opening of intracellular calcium channels and activation of protein kinase C and mitogen-activated protein (MAP) kinases. The effect of this is inhibition of cell proliferation and induction of differentiation. A variety of analogs of calcitriol that do not bind to the nuclear receptor do bind to, and activate, the cell surface receptor, including 1,25-dihydroxy-7-dehydrocholesterol and 1,25-dihydroxylumisterol. The rapid nongenomic responses to vitamin D can be demonstrated in knockout mice that lack the vitamin D nuclear receptor (Farach-Carson and Ridall, 1998; Nemere and Farach-Carson, 1998).

Calcitriol modulates the maturation of chondrocytes via a cell surface receptor linked to phospholipase and protein kinase C; in response to calcitriol, there are rapid changes in arachidonic acid release from, and reincorporation into, membrane phospholipids, and increased synthesis of prostaglandins E_1 and E_2 (Boyan et al., 1999). 24-Hydroxycalcidiol also modulates the maturation of chondrocytes, acting via cell surface receptors linked to phospholipase D, causing inactivation of both protein kinase C and MAP kinases, thus resulting in both genomic and nongenomic actions. Chondrocytes form 24-hydroxycalcidiol themselves, and this may represent both endocrine and autocrine actions (Boyan et al., 2001).

In thyroid cells in culture, calcitriol reduces production of cAMP in response to thyroid stimulating hormone by a nuclear action on the synthesis of G-protein subunits. However, it also reduces the responsiveness to cAMP, and attenuates cell growth and iodide uptake in response to thyroid stimulating hormone, with a rapid time course from direct action on protein kinase A (Berg and Haug, 1999).

3.3.3 Stimulation of Intestinal Calcium and Phosphate Absorption

Early studies showed that, after the administration of [³H]cholecalciferol or ergocalciferol to vitamin D-deficient animals, there is marked accumulation of [³H]calcitriol in the nuclei of intestinal mucosal cells. Physiological doses of vitamin D cause an increase in the intestinal absorption of calcium in deficient animals; the response is faster after the administration of calcidiol and faster still after calcitriol. There are two separate responses of intestinal mucosal cells to calcitriol: a rapid increase in calcium uptake that is due to recruitment of calcium transporters to the cell surface (Section 3.3.2) and a later response from the induction of a calcium binding protein, calbindin-D.

3.3.3.1 Induction of Calbindin-D   In response to calcitriol administration, there is an increase in mRNA synthesis and then in the synthesis of calbindin-D in intestinal mucosal cells, which is correlated with the later and more sustained increase in calcium absorption. In vitamin D-deficient animals, there is no detectable calbindin in the intestinal mucosa, whereas in animals adequately provided with vitamin D, it may account for 1% to 3% of soluble protein in the cytosol of the columnar epithelial cells. Although the rapid response to calcitriol is an increase in the permeability of the brush border membrane to calcium, the induction of calbindin permits intracellular accumulation and transport of calcium. The rapid increase in net calcium transport in tissue from vitamin D-replete animals is presumably dependent on the calbindin that is already present; in deficient animals, there can be no increase in calcium transport until sufficient calbindin has accumulated to permit intracellular accumulation, despite the increased permeability of the brush border.

Calbindin is a relatively small protein (the chick protein has an M_r of 28,000, whereas those from mammalian intestinal mucosa have M_r between 8,000 to 11,000) and binds calcium with high affinity (K_{diss} $1 - 10 \times 10^{-7}$ M). The mammalian intestinal protein (calbindin-D9k) has two calcium binding domains, whereas the chick protein (calbindin-D28k) has four. As well as being the intestinal calcium binding protein in birds, calbindin-D28k is also found in mammalian kidneys, liver, and pancreas. There is little sequence homology between calbindin-D9k and calbindin-D28k, despite the fact that there is considerable interspecies homology in calbindin-D28k. The calcium binding domain of calbindin-D is distinct from the lower affinity calcium binding domains of the γ-carboxyglutamate-containing calcium binding proteins (Section 5.3) and is similar to that of other high-affinity calcium binding proteins, such as calmodulin, parvalbumin, troponin C, and brain S-100 protein, none of which is calcitriol dependent. It consists of an octahedral calcium binding region formed from a helix–loop–helix structure; the loop region has side-chain oxygen atoms that chelate calcium. Other metal ions are also bound, with lower affinity: cadmium > strontium > manganese > zinc > barium > cobalt > magnesium.

3.3.4 Stimulation of Renal Calcium Reabsorption

Almost all of the 240 mmol of calcium filtered daily in the kidney is reabsorbed by three mechanisms (Friedman, 2000):

1. In the proximal tubules, calcium absorption is mainly by a paracellular route that is not regulated by hormones.
2. In the thick ascending limbs, there are both paracellular and transcellular routes: the active transcellular route is regulated by parathyroid hormone (increasing reabsorption) and calcitonin (reducing reabsorption), the paracellular route by cotransport with sodium.
3. In the distal tubule, reabsorption is entirely transcellular, and is regulated by parathyroid hormone and calcitriol (increasing reabsorption) and calcitonin (reducing reabsorption).

Although calcitriol is synthesized only in the proximal renal tubule, after the administration of [^{3}H]calcidiol, radioactivity in the kidney accumulates only in the distal and collecting tubules. This is the region in which selective resorption of calcium from the urine occurs and, in response to calcitriol, there is induction of calbindin-D28k. As in the intestinal mucosa, calbindin in the kidney is a cytosolic protein and is presumably involved in the intracellular accumulation and transport of calcium.

3.3.5 The Role of Calcitriol in Bone Metabolism

In addition to the role of bone mineral as a structural component of bone, it can be regarded as a major reserve of calcium for the body. In an adult, the skeleton contains 25 mol of calcium, whereas the total extracellular fluids contain only about 25 mmol. Parathyroid hormone, calcitonin, and calcitriol regulate the intestinal absorption and renal excretion of calcium, and the mobilization and deposition of bone calcium, maintaining the plasma concentration in a narrow range between 2.2 to 2.55 mmol per L. Of this, 0.7 mmol per L is bound to albumin and therefore not readily diffusible; a further 0.25 mmol per L is chelated, for example, by citrate. The metabolically important fraction is the remaining 1.3 mmol per L that is present as free Ca^{2+} ions.

Bone mineral is largely calcium phosphate, in the form of hydroxyapatite $[Ca_{10}(PO_4)_6(OH)_2]$, although it also contains carbonate and citrate, as well as magnesium and traces of fluoride and strontium. The mineral has a very fine crystal structure, and hence a large surface area; as discussed in Section 5.3.3, the function of osteocalcin, the γ-carboxyglutamate-containing calcium binding protein in the bone matrix, is to modify crystallization of bone mineral. The maintenance of bone structure is because of the balanced activity of osteoclasts, which erode existing bone mineral and organic matrix. Osteoblasts synthesize and secrete the proteins of the bone matrix and also have resorptive activity. Mineralization of the organic matrix seems to be largely controlled by the availability of adequate concentrations of calcium and phosphate, modulated by osteocalcin.

Osteoblasts also synthesize and secrete into the bone matrix a variety of compounds that modify the responsiveness of osteoclasts to inhibition by calcitonin and prostaglandins. They have both nuclear and cell surface receptors for calcitriol, as well as receptors for parathyroid hormone, glucocorticoids, epidermal growth factor, prostaglandins, estrogens, and androgens. They are susceptible to multiple hormonal modulation of activity. In response to calcitriol, they show decreased synthesis of collagen and alkaline phosphatase, and increased synthesis of osteocalcin and osteopontin. These are genomic responses associated with nuclear receptors. There are also rapid (nongenomic) responses mediated by cell surface receptors that include activation of voltage-gated calcium channels, induction of phospholipid and sphingolipid turnover, an increase in intracellular calcium, and priming of parathyroid hormone-sensitive ion channels, as well as second messenger cascades. There is also a slower, nongenomic response to calcitriol, with a time course of 1 to 3 hours and phosphorylation of a variety of secreted proteins, including osteopontin (Farach-Carson and Ridall, 1998).

Physiologically, the response of bone to calcitriol is resorption of bone mineral and matrix protein. Both calcitriol and parathyroid hormone increase bone resorption in vivo and in bone organ culture, but have no effect on the activity of isolated osteoblasts. In addition to direct stimulation of the resorptive activity of osteoblasts, calcitriol increases osteoclastic activity. This is not because of a direct effect of calcitriol on osteoclasts, which lack calcitriol receptors, but rather an increase in the differentiation of osteoclast precursor cells into mature osteoclasts. It is not known whether osteoclast precursors respond directly to calcitriol, or whether the effect is indirect, mediated by calcitriol-responsive leukocytes or osteoblasts. Osteoblasts stimulated by calcitriol or parathyroid hormone secrete one or more small proteins that increase the activity of osteoclasts (Bar-Shavit et al., 1983).

Osteoclastic activity is inhibited by calcitonin and prostaglandins I_2, E_1, and E_2, all of which act directly on the osteoclast; there is some evidence that osteoblasts may synthesize and secrete some of the osteoclast inhibitory prostaglandins. Calcitriol and parathyroid hormone stimulation of osteoblast resorptive activity also cause the synthesis and release of a variety of growth factors from the osteoblasts. These accumulate in the bone and act as delayed activators of osteoblast proliferation and activation. Although the immediate response of osteoblasts to calcitriol is repression of the synthesis of collagen and alkaline phosphatase (Rowe and Kream, 1982), between 24 to 48 hours after calcitriol administration there is increased collagen synthesis, with both new mRNA synthesis and an increased rate of translation of the existing mRNA, and induction of alkaline phosphatase and osteocalcin (Franceschi et al., 1988; Boyan et al., 1989). Alkaline phosphatase may have an important role in mineralization by hydrolyzing pyrophosphate and ATP in the bone matrix, both of which are inhibitors of mineralization; inhibition of alkaline phosphatase inhibits calcification of cartilage in culture. The combined effect of the delayed autocrine activators of osteoblast proliferation released by parathyroid hormone or calcitriol-stimulated osteoblasts, and the delayed induction of collagen, osteocalcin, and alkaline phosphatase synthesis is thus to promote the formation and mineralization of new bone matrix to replace that resorbed.

3.3.6 Cell Differentiation, Proliferation, and Apoptosis
Not only is calcitriol an important determinant for the differentiation of osteoclast precursor cells, it also directs the differentiation and maturation of normal and leukemic cells into monocytes, and potentiates apoptosis induced by 9-cis-retinoic acid, although it does not induce apoptosis itself. This suggests the possibility of what has been called maturation therapy for leukemia rather than conventional chemotherapy (James et al., 1999).

Calcitriol induces terminal differentiation of skin keratinocytes in culture, an action that has been exploited in the treatment of psoriasis (Section 3.6.2). In keratinocytes in culture, both calcium and calcitriol are required for differentiation. Cells lacking the calcium-sensing receptor or phospholipase C-γ 1 fail to differentiate in response to calcium or calcitriol, suggesting that in addition to nuclear actions, calcitriol also modulates differentiation via cell surface receptors, phospholipase, and a protein kinase C cascade (Bikle et al., 2001; Bollinger and Bollag, 2001). Keratinocytes not only form cholecalciferol from 7-dehydrocholesterol (Section 3.2.1), but also hydroxylate it to calcitriol. This endogenous formation of calcitriol is regulated, and changes as the cells differentiate, so there is both endocrine and autocrine regulation of differentiation by calcitriol (Bikle, 1995).

Hair follicles also have calcitriol receptors and type II vitamin D-resistant rickets (Section 3.4.2), which is caused by lack of calcitriol receptor function, is associated with total alopecia, suggesting that calcitriol has a role in their development.

Vitamin D receptors have been identified in a variety of tumor cells. At low concentrations, calcitriol is a growth promoter, whereas at higher concentrations it inhibits proliferation of a variety of tumor cells in culture, including breast and prostate tumor cells. There is an epidemiological association between low vitamin D status and prostate cancer. Calcitriol has both antiproliferative and proapoptotic actions in cancer cells in culture. The antiproliferative effect is by suppression of growth stimulatory factors and the potentiation of growth inhibitory signals, and serves to introduce a block in the cell cycle at the transition from the G1 phase to the S phase. Apopotosis is induced by increased translocation of the proapoptotic *Bax* protein into mitochondria, leading to increased formation of cytotoxic reactive oxygen species (Blutt and Weigel, 1999; Narvaez et al., 2001; Ylikomi et al., 2002).

Adipocytes have vitamin D receptors, and there is evidence that vitamin D may act as a suppressor of adipocyte development (Kawada et al., 1996). It has been suggested that vitamin D inadequacy may be a factor in the development of the metabolic syndrome ("syndrome X," the combination of insulin resistance, hyperlipidemia, and atherosclerosis associated with abdominal obesity). Sunlight exposure, and hence vitamin D status, may be a factor in the difference in incidence of atherosclerosis and myocardial infarction between northern and southern European countries; in addition to effects on adipocyte development, calcitriol also enhances insulin secretion through induction of calbindin-D (Section 3.3.7.1), and there is some evidence vitamin D supplements can improve glucose tolerance (Boucher, 1998).

3.3.7 Other Functions of Calcitriol

Calcitriol receptors have been identified in a variety of tissues; in some of these, the effect of calcitriol is to induce the synthesis of calbindin-D; in others, it is regulation of cell proliferation and differentiation.

Calbindin-D9k has been identified in both the placenta and yolk sac of rats and mice, and increases in the later stages of gestation when there is considerable fetal uptake of calcium for mineralization of the skeleton. In both birds and mammals, calcitriol is also required for ovulation (Halloran, 1989).

Calbindin-D28k is found in both the central and peripheral nervous systems. In the central nervous system, it is apparently a constitutive, calcitriol-independent protein, whereas in peripheral nerves, it is induced by calcitriol.

3.3.7.1 **Endocrine Glands**   Calcium is known to be important in the secretion of insulin; in vitamin D deficiency, there is impairment of secretion. Calbindin-D28k in the β-islet cells of the pancreas is believed to be calcitriol dependent, unlike similar calcium binding proteins in other pancreatic cells that are constitutive proteins and independent of calcitriol.

3.3.7.2 **The Immune System**   Calcitriol has effects on the proliferation, differentiation, and immune function of lymphocytes and monocytes. Lymphocytes from vitamin D-deficient mice show impaired inflammatory and phagocytic responses. There is a correlation between plasma concentrations of calcidiol (and hence vitamin D status) and circulating concentrations of immunoglobulins (Sedrani, 1988). Peripheral monocytes and macrophages have a constitutive calcitriol receptor at all stages of development and activation. Calcitriol promotes the differentiation of monocyte precursor cells to form monocytes and macrophages, and enhances monocyte function (Manolagas et al., 1985). Activated macrophages have calcidiol 1-hydroxylase and can synthesize calcitriol from calcidiol, suggesting that, in addition to the endocrine role of calcitriol, it may have a paracrine role in the immune system (Casteels et al., 1995). Resting lymphocytes do not have calcitriol receptors, although the receptor is induced within 24 hours of activation. Calcitriol is a potent inhibitor of interleukin-2, and suppresses the effector functions of T and B lymphocytes; it is a potent inhibitor of immunoglobulin production by peripheral blood monocytes in culture, apparently as a result of its antiproliferative effect on immunoglobulin-producing B cells and/or T-helper cells (Lemire et al., 1984; Manolagas et al., 1985). Thus, calcitriol acts at the site of inflammation both to limit T-lymphocyte action and to enhance or activate macrophage cytotoxicity.

3.4 **VITAMIN D DEFICIENCY – RICKETS AND OSTEOMALACIA**
Rickets is a disease of young children and adolescents, resulting from a failure of the mineralization of newly formed bone. In infants, epiphyseal cartilage continues to grow, but is not replaced by bone matrix and mineral. The earliest

sign of this is craniotabes – the occurrence of unossified areas in the skull, accompanied by late closure of the fontanelles. At a later stage, there is enlargement of the epiphyses, initially at the costachondral junction of the ribs, giving a beading effect – the rachitic rosary. This may lead to deformity of the chest, and in severe cases collapse of the rib cage, with consequent obstruction of respiration. Other epiphyseal junctions also become enlarged. When the child begins to walk, the weight of the body deforms the undermineralized long bones, leading to bow legs or knock knees and deformity of the pelvis. Similar problems may develop during the adolescent growth spurt. In severe deficiency, the plasma concentration of calcium may fall to the level at which intracellular calcium in nerves and muscles cannot be maintained, and tetany occurs.

Rickets was more or less eradicated as a nutritional deficiency disease during the 1950s, as a result of widespread enrichment of infant foods with vitamin D. The level of supplementation was reduced as a result of the development of hypercalcemia caused by vitamin D intoxication (Section 3.6.1) in a small number of especially susceptible infants. As a result, rickets has reemerged, especially in northern cities in temperate countries.

There have been a number of reports of rickets, especially among African-Americans in the southern United States. Rickets and osteomalacia are problems among Indians living in the United Kingdom and elsewhere. Although dietary and cultural factors may be involved, there is evidence of a genetic predisposition from high activity of calcidiol 24-hydroxylase (Section 3.2.5) (Dunnigan and Henderson, 1997; Awumey et al., 1998; Shaw and Pal, 2002).

Osteomalacia is the defective remineralization of bone during normal bone turnover in adults, so that there is a progressive demineralization, but with adequate bone matrix, leading to bone pain and skeletal deformities, with muscle weakness. Women with inadequate vitamin D status are especially at risk of osteomalacia after repeated pregnancies, as a result of the considerable drain on calcium reserves for fetal bone mineralization and lactation.

Elderly people are at risk of osteomalacia, because of both decreased synthesis of 7-dehydrocholesterol in the skin with increasing age and low exposure to sunlight. Plasma concentrations of calcidiol below 10 nmol per L are commonly seen in people over 75 years of age, not rising above 20 nmol per L at any time of the year. Histologically proven osteomalacia is observed in 2% to 5% of elderly people presenting to the hospital in Britain.

3.4.1 Nonnutritional Rickets and Osteomalacia

Induction of cytochrome P_{450} enzymes by barbiturates and other anticonvulsants can result in increased catabolism of calcidiol, and hence secondary, drug-induced osteomalacia. The antituberculosis drug isoniazid inhibits cholecalciferol 25-hydroxylase, and again prolonged administration can cause osteomalacia.

Three conditions associated with defective 1-hydroxylation of calcidiol can all be treated by the administration of either calcitriol itself or 1α-hydroxycholecalciferol, which is a substrate for 25-hydroxylation in the liver forming calcitriol:

1. Strontium intoxication can cause vitamin D-resistant rickets because strontium is a potent inhibitor of calcidiol 1-hydroxylase (Omdahl and DeLuca, 1971).
2. Renal failure is associated with an osteomalacia-like syndrome, renal osteodystrophy, as a result of the loss of calcidiol 1-hydroxylase activity. The condition may be complicated by defective reabsorption of calcium and phosphate from the urine. Furthermore, the half-life of parathyroid hormone is increased, because the principal site of its catabolism is the kidney, so there is increased parathyroid hormone-stimulated osteoclastic action without the compensatory action of calcitriol (Mawer et al., 1973).
3. Hypoparathyroidism is also associated with a failure of calcidiol 1-hydroxylation, in this case because the major stimulus for induction of 1-hydroxylase is parathyroid hormone.

3.4.2 Vitamin D-Resistant Rickets

There are a number of rachitic syndromes that do not respond to normal amounts of vitamin D:

1. X-linked hypophosphatemic rickets is caused by abnormal reabsorption of phosphate in the proximal renal tubule, resulting in excessive excretion of phosphate and hence hypophosphatemia. There may also be blunting of the normal increase in calcidiol 1-hydroxylase activity in response to hypophosphatemia. The gene responsible for the condition has been identified (the PHEX gene); its product is a membrane-bound endopeptidase that normally acts to clear the hormone phosphatonin from the circulation. Phosphatonin acts to decrease the activity of the sodium/phosphate cotransporter in the kidney (Drezner, 2000).
2. Tumor-induced osteomalacia (oncogenic hypophosphatemic osteomalacia) is also characterized by excessive urinary excretion of phosphate, and hence hypophosphatemia and low circulating calcitriol. Removal of the tumor results in normalization of phosphate excretion and

recovery from osteomalacia. The tumors that cause osteomalacia secrete abnormally large amounts of phosphatonin (Kumar, 2000).

3. Type I vitamin D-resistant rickets is due to a genetic defect in calcidiol 1-hydroxylase, so that little or no calcitriol is formed. Patients respond well to the administration of 1α-hydroxycholecalciferol, which is a substrate for 25-hydroxylation in the liver, leading to normal circulating concentrations of calcitriol.

4. Type II vitamin D-resistant rickets is characterized by a lack of responsiveness of target tissues to calcitriol and is caused by a genetic defect in the calcitriol receptor. Affected children develop more or less normally until about 9 months of age, then develop severe rickets with alopecia and a wide variety of disorders, including immune system dysfunction. Three variants are known:

   (a) Complete absence of calcitriol receptor, presumably the result of a deletion or early nonsense mutation in the receptor gene.
   (b) Poor affinity of the receptor for calcitriol, presumably the result of a mutation affecting the calcitriol binding site.
   (c) Normal receptor binding of calcitriol with impaired responsiveness of target tissues, presumably the result of a mutation affecting the DNA binding domain of the receptor.

3.4.3 Osteoporosis

Osteoporosis is a condition involving loss of bone mineral and matrix in elderly people, and may affect 40% of women and 12% of men as they age. The loss of bone mineral is associated with inappropriate calcification of other tissues, especially arteries and the kidneys. This may be more dangerous than the loss of bone, in that the majority of deaths among women suffering from osteoporosis are from cardiovascular disease.

Unlike osteomalacia, there is no defect of bone mineralization in osteoporosis. The lower density of the bone renders it more susceptible to fracture, whereas in osteomalacia the incompletely mineralized bone matrix is liable to deformation rather than fracture.

Type I osteoporosis, also known as postmenopausal osteoporosis, involves loss of trabecular bone in the vertebrae, leading to crush fracture with minimal trauma. It is essentially a condition affecting postmenopausal women, with a female:male ratio of 10:1. Type II osteoporosis (senile osteoporosis) is osteoporotic hip fracture. It shows only a 2:1 excess of females over males and a geometric increase in incidence with increasing age. The two types of osteoporosis are not exclusive, and type I patients are more susceptible to hip fracture, whereas many hip fracture patients have asymptomatic vertebral crush fractures.

The principal cause of osteoporosis seems to be the loss of estrogen and androgen secretion with increasing age. It was noted in Section 3.3.5 that osteoblasts have both estrogen and androgen receptors. Although the mechanism of action of the sex steroids is not clear, it seems likely that they act by reducing the osteoclastogenesis resulting from osteoblast activation. They may also antagonize the release of osteoblast-derived resorption factors following calcitriol and parathyroid hormone action. Thus, loss of estrogen at menopause and loss of testosterone with increasing age in men result in loss of some of the normal modulation of bone resorption stimulated by calcitriol.

Osteoporosis is probably an inevitable consequence of aging. The peak bone mass is achieved between the ages of 20 to 30; thereafter, there is a progressive loss of bone, becoming more marked postmenopausally. The condition is considerably less severe in women who enter menopause with greater bone mass, which is largely genetically determined. Polymorphism of the vitamin D receptor gene is one factor, but variants of the estrogen receptor gene and the gene for type I collagen are also involved (Wood and Fleet, 1998; Audi et al., 1999; Eisman, 1999).

A lifetime low intake of calcium is a risk factor, and there is some evidence that a moderately high intake of calcium during early life, while the skeleton is being formed, is protective. Postmenopausal hormone replacement therapy is beneficial in reducing the rate of bone loss.

Although vitamin D status declines with increasing age and there is reduced activity of calcidiol 1-hydroxylase, osteoporosis is not caused by vitamin D deficiency. Equally, although there is negative calcium balance in osteoporosis, this is a result of bone loss, not a cause. Two consensus statements (National Institutes of Health, 2000; North American Menopause Society, 2001) note that, although supplementary calcium is less effective than hormone replacement therapy or treatment with antiresorptive agents (such as bisphosphonates), it is – together with vitamin D – an essential component of treatment for osteoporosis. An intake of at least 1,200 mg of calcium (but not more than 2,500 mg), together with 10 to 15 μg of vitamin D, is recommended.

3.4.3.1 Glucocorticoid-Induced Osteoporosis
Therapeutic doses of glucocorticoid hormones, and even high physiological levels associated with chronic stress, can cause or exacerbate osteoporosis. The gene for osteocalcin has an inhibitory glucocorticoid response element that overlaps the TATA box and hence impairs the induction of osteocalcin in response to calcitriol (Christakos et al., 1996).

Analysis of a number of controlled trials shows that, whereas bisphosphonates are most effective in preventing or treating glucocorticoid-induced osteoporosis, calcium and vitamin D together are superior to either no treatment or calcium supplements alone (Amin et al., 1999). It is recommended that all patients being treated with glucocorticoids should receive 20 μg of vitamin D, 1 μg of 1α-hydroxycalcidiol, or 0.5 μg of calcitriol per day to normalize calcium balance, in addition to bisphosphonates or estrogen replacement therapy as appropriate (American College of Rheumatology Ad Hoc Committee on Glucocorticoid-Induced Osteoporosis, 2001).

3.5 ASSESSMENT OF VITAMIN D STATUS

Before gross anatomical deformities are apparent in vitamin D-deficient children, bone density is lower than normal, and this can be detected by radiography. This preclinical condition is known as radiological rickets. At an earlier stage of deficiency, there is a marked elevation of plasma alkaline phosphatase released by osteoclastic activity; the reference range for alkaline phosphatase in children is 75 to 250 units per L. For many years, this stage of biochemical rickets was used as a means of detecting children with preclinical rickets.

Osteocalcin is induced in osteoblasts by calcitriol, and circulating osteocalcin can be used as an index of calcitriol action and metabolic bone disease. In rachitic children, the plasma concentration of osteocalcin is lower than in controls, and rises on therapy, remaining high until there is radiological evidence of cure. However, plasma osteocalcin can be undetectably low in normal subjects with adequate vitamin D status, so this does not provide a useful indication of deficiency (Greig et al., 1989).

The plasma concentration of calcidiol is the most sensitive and useful index of vitamin D status, and is correlated with elevated plasma parathyroid hormone and alkaline phosphatase activity (Table 3.4). As shown in Table 3.2, the reference range of plasma calcidiol is between 20 to 150 nmol per L, with a twofold seasonal variation in temperate regions. Concentrations below 20 nmol per L are considered to indicate impending deficiency, and osteomalacia is seen in adults when plasma calcidiol falls below 10 nmol per L. In children, clinical signs of rickets are seen when plasma calcidiol falls below 20 nmol per L. The plasma concentration of calcitriol does not give a useful indication of vitamin D status. The reference range is between 38 to 144 pmol per L and is maintained because of the stimulation of calcidiol 1-hydroxylation by parathyroid hormone secreted in response to falling concentrations of calcium (Holick, 1990).

Table 3.4 Plasma Concentrations of Calcidiol, Alkaline Phosphatase, Calcium, and Phosphate as Indices of Nutritional Status

	Calcidiol (nmol/L)	Alkaline Phosphatase (units/L)	Calcium (mmol/L)	Phosphate (mmol/L)
Infants	27–100	100–300	2.5	1.6–2.6
Children with rickets	<20	>390	2.0–2.25	1.0
Adults	20–87	57–100	2.5	1.0–1.4
Adults with osteomalacia	<10	300	2.25	0.6–1.0
Adults with osteoporosis	20–87	40	2.5–3	1.3–1.6

3.6 REQUIREMENTS AND REFERENCE INTAKES

Dietary vitamin D makes little contribution to status, and the major factor is exposure to sunlight, a conclusion that is supported by the two-fold seasonal variation in plasma calcidiol in temperate regions (see Table 3.2). There are no reference intakes for young adults in the United Kingdom and Europe; for house-bound elderly people, the reference intake is 10 μg per day, based on the intake required to maintain a plasma concentration of calcidiol of 20 nmol per L (see Table 3.5). This will almost certainly require supplements of the vitamin, because average intakes are less than half this amount. The U.S./Canadian adequate intake is 5 μg per day up to age 50, increasing to 10 μg between 51 to 70, and 15 μg over 70 years of age (Institute of Medicine, 1997).

A number of studies have suggested that the cholecalciferol content of human milk is inadequate to meet the requirements of breast-fed infants without exposure to sunlight, especially during the winter, when the mother's reserves of the vitamin are low. Infant formulae normally provide 10 μg of cholecalciferol per day, and a similar amount is recommended for breast-fed infants. Supplements of 10 μg per day are also recommended for children between 3 months and 3 years, because of the relatively high requirement during the phase of maximum bone development and the limited exposure to sunlight in temperate regions. Such supplements maintain the plasma concentration of calcidiol above 20 nmol per L.

The U.S./Canadian report (Institute of Medicine, 1997) discussed requirements only in terms of bone density and maintenance of a plasma concentration of calcitriol above that associated with elevated parathyroid hormone and alkaline phosphatase. Vieth (1999) noted that intakes above 5 μg per day are required to prevent osteoporosis (Section 3.4.3) and secondary hyperparathyroidism, and suggested that normal sunlight exposure may provide the

Table 3.5 Reference Intakes of Vitamin D (μg/day)

Age	U.K. 1991	U.S./Canada 1997	FAO 2001
0–6 m	8.5	5	5
7–12 m	7	5	5
1–3 y	7	5	5
4–10 y	—	5	5
Males			
10–50 y	—	5	5
51–70 y	10	10	10
>70 y	10	15	15
Females			
10–50 y	—	5	5
51–70 y	10	10	10
>70 y	10	15	15
Pregnant	10	5	5
Lactating	10	5	5

FAO, Food and Agriculture, Organization; WHO, World Health Organization.
Sources: Department of Health, 1991; Institute of Medicine, 1997; FAO/WHO, 2001.

equivalent of 20 to 50 μg per day, with possible benefits of preventing some cancer (Section 3.3.6), hypertension, and the progression of osteoarthritis.

3.6.1 Toxicity of Vitamin D

Intoxication with vitamin D causes weakness, nausea, loss of appetite, headache, abdominal pains, cramps, and diarrhea. More seriously, it also causes hypercalcemia, with plasma concentrations of calcium between 2.75 to 4.5 mmol per L, compared with the normal range of 2.2 to 2.5 mmol per L. At plasma concentrations of calcium above 3.75 mmol per L, vascular smooth muscle may contract abnormally, leading to hypertension and hypertensive encephalopathy. Hypercalciuria may also result in the precipitation of calcium phosphate in the renal tubules and hence the development of urinary calculi. Hypercalcemia can also result in calcinosis – the calcification of soft tissues, including kidneys, heart, lungs, and blood vessels. This is assumed to be the result of increased calcium uptake into tissues in response to excessive plasma concentrations of the vitamin and its metabolites.

Some children are sensitive to hypercalcemia and calcinosis as a result of vitamin D intakes as low as 45 μg per day (Chesney, 1990; Holick, 1990). There is thus a narrow margin between amounts of vitamin D adequate to ensure that rickets is prevented throughout the community and the level at which vulnerable infants will develop hypercalcemia. This became a significant problem in Britain in the 1950s. Widespread fortification of infant foods had resulted in eradication of rickets, but, by 1955, 200 cases of hypercalcemia had been reported. The amounts of vitamin D added to infant foods was reduced; as a result, rickets reappeared. The problem is to identify those children at risk of deficiency, who therefore require additional supplements, without putting those with a low threshold for intoxication at risk of hypercalcemia and calcinosis.

The U.S./Canadian report (Institute of Medicine, 1997) quotes a no adverse effect level of 60 μg per day, leading to a tolerable upper level of intake of 50 μg per day (and 25 μg per day for infants). The toxic threshold for adults has not been established, but reports of hypercalcemia in adults have involved intakes in excess of 1,000 μg per day. There is no evidence of adverse effects at plasma concentrations of calcidiol lower than 140 nmol per L, which requires an intake in excess of 250 μg per day, suggesting that the currently accepted no adverse effect level is lower than necessary (Vieth, 1999).

Hypercalcemia persists for many months after the cessation of excessive intakes of vitamin D, because of the accumulation of the vitamin in adipose tissue and its slow release into the circulation. The introduction of calcitriol and 1α-hydroxycalcidiol for the treatment of such conditions as hypoparathyroidism, renal osteodystrophy, hypophosphatemic osteomalacia, and vitamin D-dependent rickets has meant that hypercalcemia is less of a problem than when high doses of vitamin D were used in the treatment of these conditions. Because calcitriol has a short half-life in the circulation, the resultant hypercalcemia is of shorter duration than after cholecalciferol, and adjustment of the dose is easier.

3.6.2 Pharmacological Uses of Vitamin D

Multiple sclerosis is less common among people living at high altitude, where UV exposure is greater. Patients with multiple sclerosis have poor vitamin D status and low bone density, although this could be a result of the disease rather than a cause. Calcitriol prevents the development of experimental autoimmune encephalomyelitis in mice, a widely accepted model of multiple sclerosis, and it has been suggested that vitamin D supplements may protect genetically susceptible people from developing the disease (Hayes et al., 1997; Hayes, 2000).

A number of studies have shown that there is a north-south gradient in the incidence of type I diabetes mellitus and that children who are given vitamin

D supplements are less at risk of developing the disease. It is not known how vitamin D protects against the development of diabetes, but it may be by modulation of the differentiation of lymphocytes involved in the autoimmune destruction of pancreatic β-islet cells. The protective dose is above current reference intakes and indeed may be above the tolerable upper intake of 25 μg per day for infants (Harris, 2002).

The calcitriol molecule shows considerable conformational flexibility, and different conformations are required for binding to the plasma vitamin D binding protein (Section 3.2.7), nuclear receptors (Section 3.3.1), and cell surface receptors (Section 3.3.2; Norman et al., 1996; 2001a; 2001b). Because of the roles of vitamin D in regulating cell proliferation and differentiation (Section 3.3.6), there is considerable interest in the development of analogs of calcitriol that have little or no hypercalcemic action, for the treatment of psoriasis and some cancers. Such compounds include calcipotriol (1α,27-dihydroxycholecalciferol), 19-*nor*-calcidiol, doxercalciferol, 22-oxacalcitriol, and alfacacidiol (Brown, 1998, 2001; Guyton et al., 2001).

FURTHER READING

Beckerman P and Silver J (1999) Vitamin D and the parathyroid. *American Journal of Medical Science* **317**, 363–9.

Boyan BD, Sylvia VL, Dean DD, and Schwartz Z (2001) 24,25-(OH)(2)D(3) regulates cartilage and bone via autocrine and endocrine mechanisms. *Steroids* **66**, 363–74.

Bronner F and Pansu D (1999) Nutritional aspects of calcium absorption. *Journal of Nutrition* **129**, 9–12.

Brown AJ (2001) Therapeutic uses of vitamin D analogues. *American Journal of Kidney Disease* **38**, S3–S19.

Brown AJ, Dusso A, and Slatopolsky E (1999) Vitamin D. *American Journal of Physiology* **277**, F157–75.

Casteels K, Bouillon R, Waer M, and Mathieu C (1995) Immunomodulatory effects of 1,25-dihydroxyvitamin D_3. *Current Opinion in Nephrology and Hypertension* **4**, 313–18.

Chatterjee M (2001) Vitamin D and genomic stability. *Mutation Research* **475**, 69–87.

Christakos S, Raval-Pandya M, Wernyj RP, and Yang W (1996) Genomic mechanisms involved in the pleiotropic actions of 1,25-dihydroxyvitamin D_3. *Biochemical Journal* **316**, 361–71.

DeLuca HF and Zierold C (1998) Mechanisms and functions of vitamin D. *Nutrition Reviews* **56**, S4–S10; discussion S54–S75.

Gennari C (2001) Calcium and vitamin D nutrition and bone disease of the elderly. *Public Health Nutrition* **4**, 547–59.

Haussler MR, Haussler CA, Jurutka PW, Thompson PD, Hsieh JC, Remus LS, Selznick SH, and Whitfield GK (1997) The vitamin D hormone and its nuclear receptor: molecular actions and disease states. *Journal of Endocrinology* **154**(Suppl.) S57–S73.

Holick MF (1996) Vitamin D and bone health. *Journal of Nutrition* **126,** 1159S–64S.

Jones G, Strugnell SA, and DeLuca HF (1998) Current understanding of the molecular actions of vitamin D. *Physiological Reviews* **78,** 1193–1231.

Mawer EB and Davies M (2001) Vitamin D nutrition and bone disease in adults. *Reviews of Endocrine and Metabolic Disorders* **2,** 153–64.

Narvaez CJ, Zinser G, and Welsh J (2001) Functions of 1alpha,25-dihydroxyvitamin D(3) in mammary gland: from normal development to breast cancer. *Steroids* **66,** 301–8.

Norman AW, Henry HL, Bishop JE, Song XD, Bula C, and Okamura WH (2001) Different shapes of the steroid hormone 1alpha,25(OH)(2)-vitamin D(3) act as agonists for two different receptors in the vitamin D endocrine system to mediate genomic and rapid responses. *Steroids* **66,** 147–58.

Norman AW, Ishizuka S, and Okamura WH (2001) Ligands for the vitamin D endocrine system: different shapes function as agonists and antagonists for genomic and rapid response receptors or as a ligand for the plasma vitamin D binding protein. *Journal of Steroid Biochemistry and Molecular Biology* **76,** 49–59.

Omdahl JL, Bobrovnikova EA, Choe S, Dwivedi PP, and May BK (2001) Overview of regulatory cytochrome P450 enzymes of the vitamin D pathway. *Steroids* **66,** 381–9.

Omdahl JL, Morris HA, and May BK (2002) Hydroxylase enzymes of the vitamin D pathway: expression, function, and regulation. *Annual Reviews of Nutrition* **22,** 139–66.

Pike JW (1994) Vitamin D receptors and the mechanism of action of 1,25-dihydroxyvitamin D$_3$. In *Vitamin Receptors: Vitamins as Ligands in Cell Communication*, K Dakshinamurti (ed.), pp. 59–77. Cambridge, UK: Cambridge University Press.

Ross TK, Darwish HM, and DeLuca HF (1994) Molecular biology of vitamin D action. *Vitamins and Hormones* **49,** 281–326.

St-Arnaud R (1999) Targeted inactivation of vitamin D hydroxylases in mice. *Bone* **25,** 127–9.

Various authors (1997) Symposium on nutritional aspects of bone. *Proceedings of the Nutrition Society* **56,** 903–87.

Webb AR and Holick MF (1988) The role of sunlight in the cutaneous production of vitamin D$_3$. *Annual Reviews of Nutrition* **8,** 375–99.

White P and Cooke N (2000) The multifunctional properties and characteristics of vitamin D-binding protein. *Trends in Endocrinology and Metabolism* **11,** 320–7.

Wikvall K (2001) Cytochrome P450 enzymes in the bioactivation of vitamin D to its hormonal form (review). *International Journal of Molecular Medicine* **7,** 201–9.

Wood RJ and Fleet JC (1998) The genetics of osteoporosis: vitamin D receptor polymorphisms. *Annual Reviews of Nutrition* **18,** 233–58.

Yamada S, Yamamoto K, Masuno H, and Choi M (2001) Three-dimensional structure-function relationship of vitamin D and vitamin D receptor model. *Steroids* **66,** 177–87.

References cited in the text are listed in the Bibliography.

FOUR

Vitamin E: Tocopherols and Tocotrienols

For a long time, it was considered that, unlike the other vitamins, vitamin E had no specific functions; rather it was the major lipid-soluble, radical-trapping antioxidant in membranes. Many of its functions can be met by synthetic antioxidants; however, some of the effects of vitamin E deficiency in experimental animals, including testicular atrophy and necrotizing myopathy, do not respond to synthetic antioxidants. The antioxidant roles of vitamin E and the trace element selenium are closely related and, to a great extent, either can compensate for a deficiency of the other. The sulfur amino acids (methionine and cysteine) also have a vitamin E-sparing effect.

More recent studies have shown that vitamin E also has roles in cell signaling, by inhibition or inactivation of protein kinase C, and in modulation of gene expression, inhibition of cell proliferation, and platelet aggregation. These effects are specific for α-tocopherol and are independent of the antioxidant properties of the vitamin.

Deficiency of vitamin E is well established in experimental animals, resulting in reproductive failure, necrotizing myopathy, liver and kidney damage, and neurological abnormalities. In human beings, deficiency is less well defined, and it was only in 1983 that vitamin E was conclusively demonstrated to be essential. Deficiency is a problem only in premature infants with low birth weight and in patients with abnormalities of lipid absorption or congenital lack of β-lipoproteins – abetalipoproteinemia – or a genetic defect in the α-tocopherol transfer protein. In adults, lipid malabsorption only results in signs of vitamin E deficiency after many years.

4.1 VITAMIN E VITAMERS AND UNITS OF ACTIVITY

As shown in Figure 4.1, there are eight vitamers of vitamin E; the tocopherols and the tocotrienols differ in that the tocopherols have a saturated side chain,

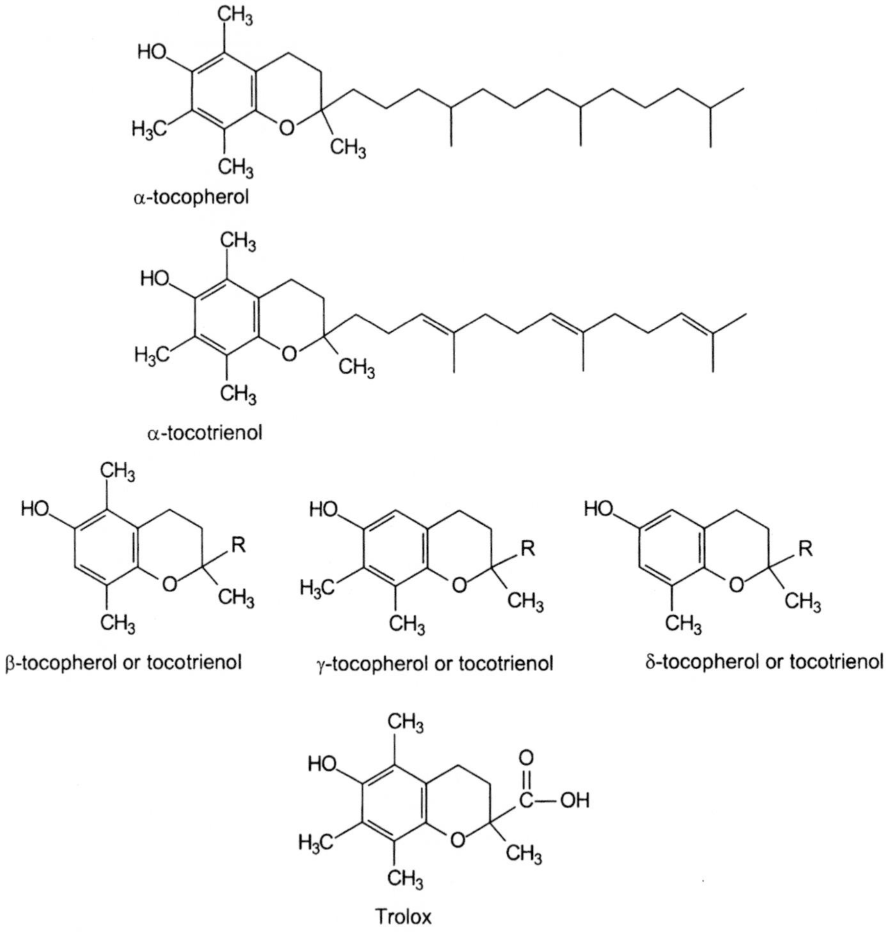

Figure 4.1. Vitamin E vitamers, tocopherols and tocotrienols, and the synthetic water-soluble vitamin E analog, Trolox. Relative molecular masses (M_r): α-tocopherol, 430.7 (acetate 488.8, succinate 546.8); β-tocopherol, 419.7; γ-tocopherol, 416.7; δ-tocopherol, 402.7; α-tocotrienol, 424.7; β-tocotrienol, 410.7; γ-tocotrienol, 410.7; δ-tocotrienol, 396.7; and Trolox, 250.3.

whereas the tocotrienols have an unsaturated side chain. The different to-copherols and tocotrienols (α, β, γ, and δ) differ in the methylation of the chromanol ring. Tocotrienols occur in foods as both the free alcohols and also as esters; tocopherols occur naturally as the free alcohols, but acetate and suc-cinate esters are used in pharmaceutical preparations because of their greater stability against oxidation. Trolox is a synthetic water-soluble compound with vitamin E activity.

Table 4.1 Relative Biological Activity of the Vitamin E Vitamers

	Fetal Resorption Bioassay		Binding to α-Tocopherol Transfer Protein
	IU/mg	Relative Activity	
D-α-tocopherol (*RRR*)	1.49	1.0	1.0
D-β-tocopherol (*RRR*)	0.75	0.50	0.38
D-γ-tocopherol (*RRR*)	0.15	0.10	0.09
D-δ-tocopherol (*RRR*)	0.05	0.03	0.02
D-α-tocotrienol	0.75	0.50	0.12
D-β-tocotrienol	0.08	0.05	—
D-γ-tocotrienol	—	—	—
D-δ-tocotrienol	—	—	—
L-α-tocopherol (*SRR*)	0.46	0.31	0.11
RRS-α-tocopherol	1.34	0.90	—
SRS-α-tocopherol	0.55	0.37	—
RSS-α-tocopherol	1.09	0.73	—
SSR-α-tocopherol	0.31	0.21	—
RSR-α-tocopherol	0.85	0.57	—
SSS-α-tocopherol	1.10	0.60	—
RRR-α-tocopheryl acetate	1.36	0.91	—
RRR-α-tocopheryl acid succinate	1.21	0.81	—
All-*rac*-α-tocopherol	1.10	0.74	0.02
All-*rac*-α-tocopheryl acetate	1.00	0.67	—
All-*rac*-α-tocopheryl acid succinate	0.89	0.60	—

Based on biological assay in vitamin E-deficient rats, the vitamers have widely varying biological activity. The original international unit (iu) of vitamin E potency was equated with the activity of 1 mg of (synthetic) DL-α-tocopherol acetate; on this basis, pure D-α-tocopherol (*RRR*-α-tocopherol, the most potent vitamer) is 1.49 iu per mg. The precise mixture of stereoisomers in this original standard is unknown, and the different stereoisomers have very different biological activities, so that different preparations may differ considerably.

It is now usual to express the vitamin E content of foods in terms of milligram-equivalents of (*RRR*)-α-tocopherol, based on their biological activities. In Table 4.1, the biological activity is shown in iu per milligram and relative to D-α-tocopherol on a molar basis. For the major vitamers present in foods, total α-tocopherol equivalent is calculated as the sum of mg α-tocopherol +

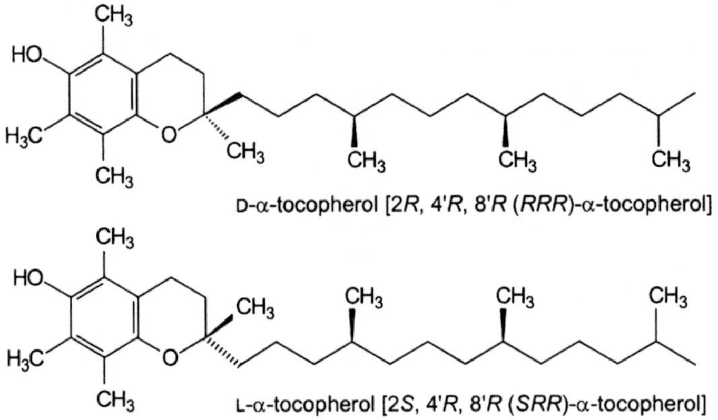

D-α-tocopherol [2R, 4'R, 8'R (RRR)-α-tocopherol]

L-α-tocopherol [2S, 4'R, 8'R (SRR)-α-tocopherol]

Figure 4.2. Stereochemistry of α-tocopherol.

$0.4 \times$ mg β-tocopherol $+ 0.3 \times$ mg γ-tocopherol $+ 0.01 \times$ mg δ-tocopherol $+$ $0.3 \times$ mg α-tocotrienol $+ 0.05 \times$ mg β-tocotrienol $+ 0.01 \times$ mg γ-tocotrienol (Holland et al., 1991).

As shown in Figure 4.2, the tocopherols have three asymmetric centers. The naturally occurring compound is D-α-tocopherol, in which all three asymmetric centers have the R-configuration [2R, 4'R, 8'R, or all-R (RRR)-α-tocopherol]. Chemical synthesis yields a mixture of the eight possible stereoisomers (all-rac-α-tocopherol); as shown in Table 4.1, the stereoisomers all have different biological activity in the rat biological assay, and the all-rac mixture has a relative biological activity of 0.74 × the activity of RRR-α-tocopherol. Conventionally, this factor of 0.74 has been used to calculate the nutritional contribution of synthetic all-rac-α-tocopherol. The most important determinant of biological activity is the chirality of C-2; the four isomers with the 2R configuration are all more active than the corresponding 2S isomers. Indeed, the 2S isomers are not maintained in plasma and do not bind to the α-tocopherol transfer protein in liver (Section 4.2).

The U.S./Canadian Dietary Reference intakes report (Institute of Medicine, 2000) departed from tradition by considering only the contribution of the 2R isomers to vitamin E intake, and proposed an equivalence of 0.45 iu per mg for synthetic all-rac-α-tocopherol, although in consideration of upper tolerable levels of intake (Section 4.6.1), they considered the contribution of all isomers equally. However, although the 2S isomers have a shorter half-life than RRR-α-tocopherol in the circulation, and hence a lower apparent biological availability, they are active in animal biological assays (Hoppe and Krennrich, 2000).

Tocopherols and tocotrienols are important constituents of chloroplast membranes in green plants, and are also found in large amounts in seeds. The aromatic ring arises from homogentisic acid, which is both a precursor of aromatic amino acids and also an intermediate in the catabolism of tyrosine. The side chain is formed by the addition of geranyl-geranyl pyrophosphate to yield initially δ-tocotrienol. Successive methylation of the ring results in the formation of γ-, β-, and α-tocotrienols; the tocopherols are formed by reduction of the side chain. Tocotrienols can be considered to be end-products of the hydroxymethylglutaryl coenzyme A (HMG CoA) reductase pathway that also leads to the synthesis of cholesterol, and they down-regulate HMG CoA reductase and cholesterol synthesis.

4.2 METABOLISM OF VITAMIN E

The absorption of vitamin E is relatively poor – only some 20% to 40% of a test dose is normally absorbed from the small intestine, in mixed lipid micelles with other dietary lipids. This absorption is enhanced by medium-chain triglycerides and inhibited by polyunsaturated fatty acids, possibly because of chemical interactions between tocopherols and polyunsaturated fatty acids or their peroxidation products in the intestinal lumen. Esters are hydrolyzed in the intestinal lumen by pancreatic esterase and also by intracellular esterases in the mucosal cells.

In intestinal mucosal cells, all vitamers of vitamin E are incorporated into chylomicrons, and tissues take up some vitamin E from chylomicrons. Most, however, goes to the liver in chylomicron remnants. α-Tocopherol, which binds to the liver α-tocopherol transfer protein, is then exported in very low-density lipoprotein (VLDL) and is available for tissue uptake (Traber and Arai, 1999; Stocker and Azzi, 2000). Later, it appears in low-density lipoprotein (LDL) and high-density lipoprotein, as a result of metabolism of VLDL in the circulation. The other vitamers, which do not bind well to the α-tocopherol transfer protein, are not incorporated into VLDL, but are metabolized in the liver and excreted. This explains the lower biological potency of the other vitamers, despite similar, or higher, in vitro antioxidant activity.

The affinities of α-tocopherol transfer protein for the other vitamers (relative to RRR-α-tocopherol $= 1$, and based on competition with RRR-α-tocopherol) are β-tocopherol, 0.38; γ-tocopherol, 0.09; δ-tocopherol, 0.02; SRR-α-tocopherol, 0.11; and α-tocotrienol, 0.09. As a result, whereas the half-life of α-tocopherol in the circulation is 48 hours, that of β- and γ-tocopherol (and the other vitamers) is only of the order of 13 to 15 hours. In patients with ataxia and vitamin E deficiency caused by a genetic lack of α-tocopherol transfer

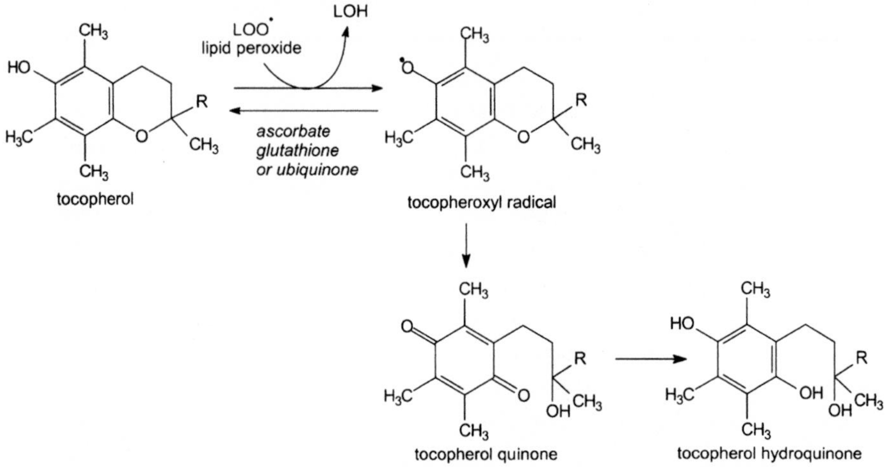

Figure 4.3. Reaction of tocopherol with lipid peroxides; the tocopheroxyl radical can be reduced to tocopherol or undergo irreversible onward oxidation to tocopherol quinone.

protein (Section 4.4.2), the half-life of plasma α-tocopherol is of the order of 13 hours (Hosomi et al., 1997).

Because vitamin E is transported in lipoproteins secreted by the liver, the plasma concentration depends to a great extent on total plasma lipids. Erythrocytes may also be important in transport, because there is a relatively large amount of the vitamin in erythrocyte membranes, and this is in rapid equilibrium with plasma vitamin E. There are two mechanisms for tissue uptake of the vitamin. Lipoprotein lipase releases the vitamin by hydrolyzing the triacylglycerol in chylomicrons and VLDL, whereas separately there is receptor-mediated uptake of LDL-bound vitamin E. Studies in knockout mice suggest that the main mechanism for tissue uptake of vitamin E from plasma lipoproteins is by way of the class B scavenger receptor (Mardones et al., 2002).

Retention within tissues depends on intracellular binding proteins which, like the liver α-tocopherol transfer protein, have the highest affinity for *RRR-α*-tocopherol. The retention of α-tocopherol in tissues varies. In the lungs the vitamin has a half-life of 7.6 days, in liver 9.8 days, in skin 23.4 days, in brain 29.4 days, and in the spinal cord 76.3 days (Ingold et al., 1987).

As shown in Figure 4.3, tocopherol can undergo reversible oxidation to an epoxide, followed by ring cleavage to yield a quinone, which is reduced to the hydroquinone and conjugated with glucuronic acid, for excretion in the bile, which is the major route of excretion. The side chain of the quinone and hydroquinone may be oxidized by β-oxidation, and small amounts of these oxidation products (carboxyethyl-hydroxychromans) and their conjugates are excreted in the urine. This is generally a minor route of metabolism, accounting for only about 1% of a test dose of labeled α-tocopherol, although larger amounts of the oxidation products of the other vitamers are excreted in urine. There may also be significant excretion of the vitamin by the skin. After the administration of chylomicron-incorporated [^{3}H]α-tocopherol to rats, there is not only a significant accumulation and retention of radioactivity in the skin, but also on the outer surface and in the fur (Shiratori, 1974).

4.3 METABOLIC FUNCTIONS OF VITAMIN E

The best-established function of vitamin E is as a lipid-soluble antioxidant in plasma lipoproteins and cell membranes. Many of the antioxidant actions are unspecific, and a number of synthetic antioxidants have a vitamin E-sparing effect. There is considerable overlap between the antioxidant roles of vitamin E and selenium (Section 4.3.2).

A number of studies have shown that α-tocopherol has a role in modulation of gene expression and regulation of cell proliferation (Section 4.3.3), suggesting that the potential beneficial effects of vitamin E against heart disease and cancer (Section 4.6.2) may not be because of its antioxidant action.

There is some evidence that tocopherols have a specific function in cell membranes; the phytyl side chain of *RRR-α*-tocopherol can interact closely with the methylene-interrupted *cis*-double bonds of arachidonic acid and other long-chain polyunsaturated fatty acids in membranes, both stabilizing membrane structure and also protecting the fatty acids from oxidative damage. The membrane phospholipids of fibroblasts grown in culture contain more arachidonate, and less linoleate, when grown in the presence of adequate amounts of α-tocopherol. This seems to be a specific effect on phospholipid synthesis and is not reflected in the fatty acid content of neutral lipids (Giasuddin and Diplock, 1981). α-Tocopherol forms a 1:1 complex with lysophospholipids and free fatty acids formed by the action of membrane phospholipases, thus negating the detergent-like action that might otherwise disrupt the membrane (Wang and Quinn, 1999, 2000).

It was noted in Section 4.1 that the tocotrienols can be considered to be derivatives of mevalonate, the product of HMG CoA reductase, which is the key regulatory enzyme of cholesterol synthesis. Dietary tocotrienols have a cholesterol-lowering effect; they act by reducing the activity of HMG CoA reductase. The main effect is posttranslational; tocotrienols cause an increased rate of catabolism of the enzyme protein (Parker et al., 1993; Theriault et al., 1999). Tocotrienols also induce cell cycle arrest in tumor cell lines in culture and induce apoptosis (Yu et al., 1999).

As well as being an end-product of the oxidation of α-tocopherol, α-tocopherol quinone (see Figure 4.3) is the cofactor for the mitochondrial fatty acid desaturation/elongation pathway, and it has been suggested that the severe neurological degeneration in patients with a genetic lack of α-tocopherol transfer protein or abetalipoproteinemia (Section 4.4.2) is caused by failure of synthesis of long-chain polyunsaturated fatty acids (Infante, 1999).

The main metabolite of γ-tocopherol is 2,7,8-trimethyl-2-(β-carboxyethyl)-6-hydroxychroman (γ-CEHC), which is excreted in the urine. It has potentially physiologically significant natriuretic activity, whereas the corresponding metabolite of α-tocopherol, which is formed in increasing amounts as intake increases, is inactive (Jiang et al., 2001).

4.3.1 Antioxidant Functions of Vitamin E

Vitamin E functions as a lipid antioxidant both in vitro and in vivo; a number of synthetic antioxidants will prevent or cure most of the signs of vitamin E deficiency in experimental animals. Polyunsaturated fatty acids undergo oxidative attack by hydroxyl radicals and superoxide to yield alkylperoxyl (alkyldioxyl) radicals, which perpetuate a chain reaction in the lipid – with potentially disastrous consequences for cells. Similar oxidative radical damage can occur to proteins (especially in a lipid environment) and nucleic acids.

Phenolic compounds can break such chain reactions by trapping the radicals, with the formation of stable nonradical products from the oxidized lipid and phenoxyl radicals that are relatively unreactive because they are stabilized by resonance. The phenoxyl radical may either react with a further alkylperoxyl radical to yield nonradical products, or it may be reduced back to the starting phenol by reaction with a water-soluble reducing agent.

Vitamin E is one of the most active radical-trapping, chain-breaking antioxidant phenols that has been investigated, and is the major lipid-soluble antioxidant in tissues (Burton and Ingold, 1984). As shown in Figure 4.4, the α- and β-tocopheroxyl radicals have three resonance forms compared with two for the γ- and δ-radicals, and are therefore more stable and have a greater antioxidant activity. In simple solution, α-tocopherol and α-tocotrienol are equipotent as antioxidants. In vitro, when it is incorporated into liposomes or microsome preparations, α-tocotrienol has greater antioxidant activity. This is probably because the unsaturated side chain causes more membrane perturbation, and tocotrienol is able to distribute more uniformly through the

resonance forms of the α- and β-tocopherol and tocotrienol radicals

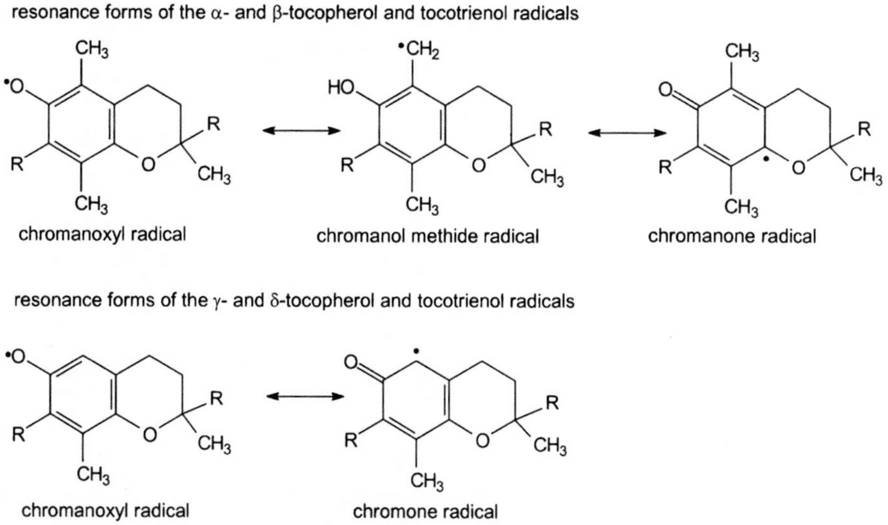

chromanoxyl radical   chromanol methide radical   chromanone radical

resonance forms of the γ- and δ-tocopherol and tocotrienol radicals

chromanoxyl radical   chromone radical

Figure 4.4. Resonance forms of the vitamin E radicals.

membrane, with a greater chance of interacting with lipid peroxides. However, the poor retention of tocotrienols in tissues means that in vivo tocotrienols have lower antioxidant activity than tocopherols (Packer et al., 2001).

Tocopherol can act catalytically as a chain-breaking lipophilic antioxidant in membranes and plasma lipoproteins, because the tocopheroxyl radical formed by reaction of α-tocopherol with a lipid peroxide radical can be reduced to tocopherol in four main ways:

1. By reaction with ascorbate to yield the monodehydroascorbate radical, which in turn can either be reduced to ascorbate or can undergo dismutation to yield dehydroascorbate and ascorbate (Section 13.4.7.1). In vitro, the formation of the tocopheroxyl radical can be demonstrated by the appearance of its characteristic absorbance peak, which normally has a decay time of 3 msec; in the presence of ascorbate, the tocopheroxyl peak has a decay time of 10 μsec, and its disappearance is accompanied by the appearance of the monodehydroascorbate peak. There is an integral membrane oxidoreductase that uses ascorbate as the preferred electron donor, linked either directly to reduction of tocopheroxyl radical or via an electron transport chain involving ubiquinone (see no. 4 below; May, 1999).

2. By reaction with glutathione, catalyzed by a membrane-specific isoenzyme of glutathione peroxidase, which is a selenoenzyme. Thus, in

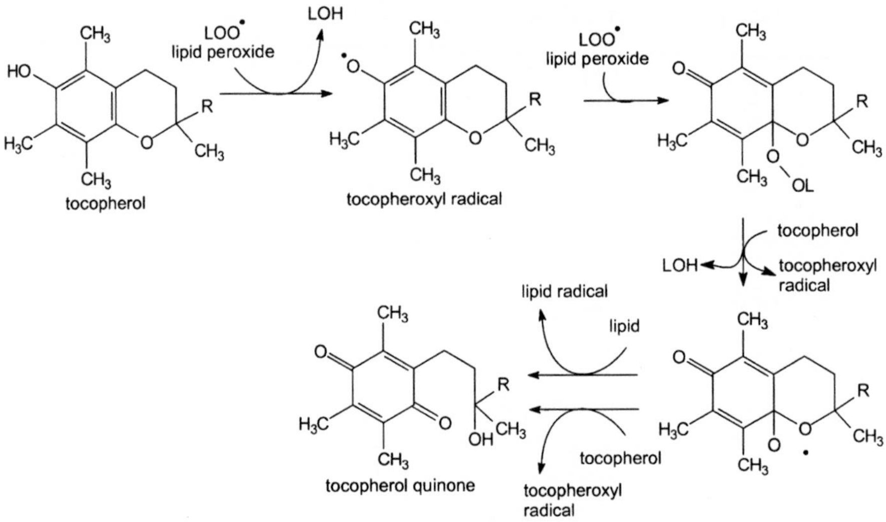

Figure 4.5. Role of vitamin E as a chain-perpetuating prooxidant.

addition to its role in removing products of lipid peroxidation (Section 4.3.2), selenium has a direct role in the recycling of tocopherol.

3. By reaction with other lipid-soluble antioxidants in the membrane or lipoprotein, including ubiquinone (Section 14.6), which is present in large amounts in all membranes as part of an electron transport chain, not just the mitochondrial inner membrane (Thomas et al., 1995; Crane and Navas, 1997; Thomas and Stocker, 2000; Villalba and Navas, 2000).

4. In mitochondria by reaction with the electron transport chain linked to the oxidation of NADH, succinate, or reduced cytochrome c (Maguire et al., 1989).

4.3.1.1 **Prooxidant Actions of Vitamin E** Most of the studies of the anti-oxidant activity of vitamin E have used relatively strong oxidants as the source of oxygen radicals to produce lipid peroxides in lipoproteins or liposomes in vitro. Studies of lipoproteins treated in vitro with low concentrations of sources of the perhydroxyl radical suggest that vitamin E may have a prooxidant action. Over 8 hours, there was 10-fold more formation of cholesterol ester hydroper-oxide (an index of lipid peroxidation) in native LDL than in vitamin E-depleted LDL (Bowry et al., 1992). This is perhaps unsurprising; vitamin E and other radical-trapping antioxidants are effective because they form stable radicals that persist long enough to undergo reaction to nonradical products. It is there-fore to be expected that they are also capable of perpetuating the radical chain reaction deeper into lipoproteins or membranes, as shown in Figure 4.5,

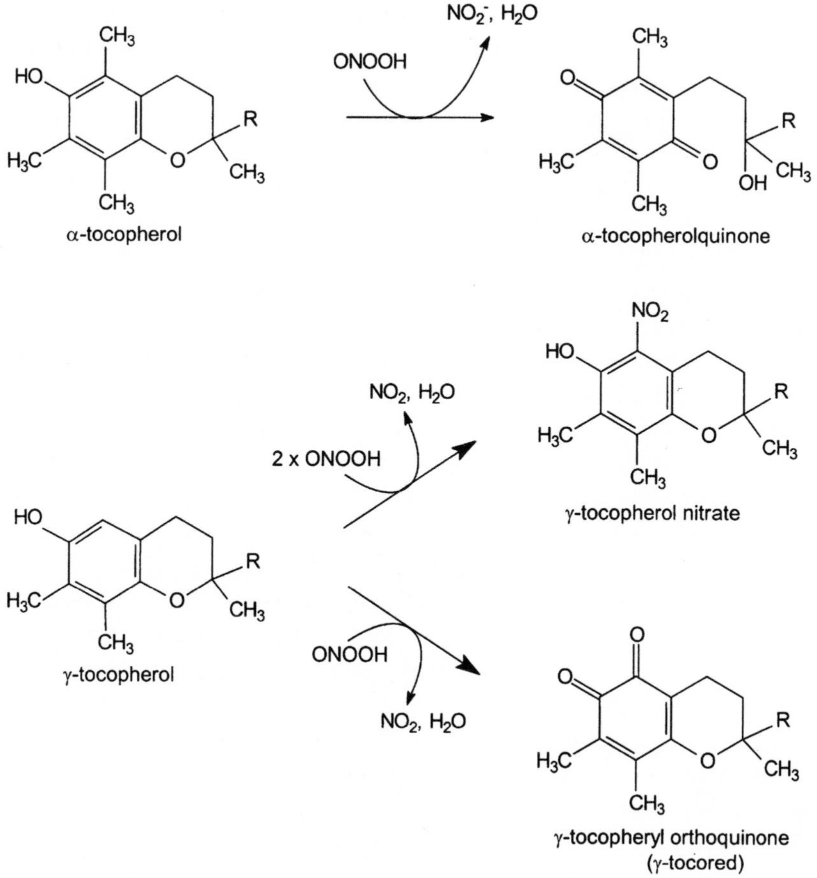

Figure 4.6. Reactions of α- and γ-tocopherol with peroxynitrite.

therefore causing increased oxidative damage to lipids, especially in the absence of coantioxidants such as ascorbate or ubiquinone (Upston et al., 1999; Carr et al., 2000).

4.3.1.2 Reaction of Tocopherol with Peroxynitrite Nitric oxide can react with superoxide to yield peroxynitrite, which, although not a radical, can react with proteins and lipids. Both α- and γ-tocopherol react with peroxynitrite; γ-tocopherol is more active and undergoes nitration as shown in Figure 4.6; it is significantly more active in preventing lipid peroxidation by peroxynitrite than is α-tocopherol, which does not undergo nitration (Jiang et al., 2001).

4.3.2 Nutritional Interactions Between Selenium and Vitamin E

The physiological role of vitamin E as an antioxidant and scavenger of free radicals explains the apparently complex nutritional interactions between vitamin E and selenium. Selenium is required, as the selenium analog of cysteine, selenocysteine, in the catalytic site of glutathione peroxidase. As noted previously, the membrane-specific isoenzyme of glutathione peroxidase catalyzes the reduction of the tocopheroxyl radical back to tocopherol. In addition, glutathione peroxidase reduces hydrogen peroxide and so lowers the amount of peroxide available for the generation of radicals, whereas vitamin E is involved in removing the products of attack by these radicals on lipids.

Thus, in vitamin E deficiency, selenium has a beneficial effect in lowering the concentrations of alkylperoxyl radicals, and conversely, in selenium deficiency, vitamin E has a protective effect in reducing the radicals. When selenium is adequate, but vitamin E is deficient, tissues with low activity of glutathione peroxidase [e.g., the central nervous system and (rat) placenta] are especially susceptible to lipid peroxidation, whereas tissues with high activity of glutathione peroxidase are not. Conversely, with adequate vitamin E and inadequate selenium, membrane lipid peroxidation will be inhibited, but tissues with high peroxide production and low catalase activity will still be at risk from peroxidative damage, especially to sulfhydryl proteins.

There is one selenocysteine residue per monomer unit of glutathione peroxidase, at the catalytic site, and this is incorporated during ribosomal protein synthesis, not as a postsynthetic modification. Tissues contain a selenocysteine tRNA (tRNA^{sec}) and the UGA codon in mRNA is read in a context-dependent manner as coding for selenocysteine. Under normal conditions, UGA is the *opal* terminator codon. However, in the presence of a stem-loop in the 3'-untranslated region of mRNA, with the selenocysteine insertion sequence AUG(N)$_m$AAA(N)$_n$UGR, UGA codes for the incorporation of selenocysteine. If selenocysteine tRNA is uncharged (i.e., in the absence of selenocysteine), the UGA codon is read as a terminator codon in the normal way, and translation of the mRNA ceases. For this reason, there is no abnormal selenium-deficient variant of the enzyme produced in selenium deficiency (Chambers et al., 1986). Selenocysteine is formed in a pyridoxal phosphate-dependent reaction between tRNA^{sec} charged with serine and selenophosphate. The incorporation of selenium into serine-tRNA^{sec} occurs cotranslationally, and the activity of the enzyme is controlled by the stem-loop sequence in the 3'-untranslated region of mRNA (Bock et al., 1991; Stadtman, 1996).

Selenocysteine is also incorporated in the same way into a number of other enzymes, including thyroxine deiodinase, which catalyzes the formation of

active triiodothyronine from thyroxine in target tissues, and two selenoproteins of unknown function in plasma and muscle. In addition, a number of
T-cell–associated genes (CD4, CD8, and HLA-DR) have as many as 10 UGA
codons in the open reading frame, as well as a potential stem-loop containing the selenocysteine insertion sequence, suggesting a role of selenium in
T-cell–mediated immune function (Taylor, 1995).

There may also be an effect of selenium deficiency on vitamin E nutrition.
Selenium deficiency causes a specific pancreatic atrophy, which is unresponsive to vitamin E supplements. In turn, this leads to impaired secretion of lipase, and hence impaired absorption of dietary lipids in general that will affect
the absorption of vitamin E (Thompson and Scott, 1970).

4.3.3 Functions of Vitamin E in Cell Signaling

Both in vitro and in vivo α-tocopherol inhibits platelet aggregation in response
to agonists such as arachidonic acid and phorbol ester; the effect seems to be
mediated by a protein kinase C mechanism. In cultured cells, α-tocopherol
lowers protein kinase C activity, apparently by activating diacylglycerol kinase and phosphoprotein phosphatase, and hence initiating the dephosphorylation and inactivation of protein kinase C (Steiner, 1999; Freedman and
Keaney, 2001). Inhibition of protein kinase C is also the mechanism by which
α-tocopherol inhibits vascular smooth muscle proliferation, a factor in throm-
bus formation and platelet aggregation. Neither β-tocopherol nor Trolox has
any effect on protein kinase C, platelet aggregation, or vascular smooth muscle
proliferation (Azzi et al., 2000, 2001, 2002).

There are a number of actions of α-tocopherol on monocytes and macrophages that suggest it may have a protective role against atherogenesis
quite distinct from its antioxidant actions. In monocytes, it reduces formation of reactive oxygen species, cell adhesion to the endothelium, and release of interleukins and tumor necrosis factor. The effect on cell adhesion
is from inhibition of protein kinase C. α-Tocopherol inhibits the assembly of
the respiratory burst NADPH oxidase that is responsible for oxygen radical
generation (Section 7.3.6). The effect on cytokine release is from inhibition
of 5-lipoxygenase. In macrophages, it reduces release of interleukin-1 and
the activity of the scavenger LDL receptor, thus reducing the accumulation
of cholesterol and transformation into foam cells (Jialal et al., 2001; Devaraj
et al., 2002).

α-Tocopherol modulates transcription of a number of genes. It induces
the liver α-tocopherol transfer protein and α-tropomyosin in smooth muscle, whereas it represses the age-related increase in synthesis of collagenase

(metalloproteinase 1) in fibroblasts in culture, and the expression of the scavenger receptors for oxidized LDL in macrophages and smooth muscle. With the exception of the induction of α-tocopherol transfer protein, in which β-tocopherol is also active, these responses are specific for α-tocopherol. It is likely that these transcriptional actions are mediated by binding to the α-tocopherol–associated protein, a hydrophobic-ligand protein that specifically binds α-tocopherol and is expressed in all tissues; however, as yet, no response element for this protein has been identified on any of the proposed target genes (Azzi et al., 2000, 2001, 2002).

In experimental animals, vitamin E deficiency depresses immune system function, with reduced mitogenesis of B and T lymphocytes, reduced phagocytosis and chemotaxis, and reduced production of antibodies and interleukin-2. This suggests a signaling role in the immune system (Moriguchi and Muraga, 2000).

4.4 VITAMIN E DEFICIENCY

Vitamin E deficiency in experimental animals was first described by Evans and Bishop in 1922, when it was discovered to be essential for fertility. It was not until 1983 that vitamin E was demonstrated to be a dietary essential for human beings, when Muller and coworkers (1983) described the devastating neurological damage from lack of vitamin E in patients with hereditary abetalipoproteinemia.

4.4.1 Vitamin E Deficiency in Experimental Animals

Vitamin E deficiency in experimental animals results in a number of different conditions, with considerable differences between different species in their susceptibility to different signs of deficiency. As shown in Table 4.2, some of the lesions can be prevented or cured by the administration of synthetic antioxidants, and others respond to supplements of selenium.

Vitamin E-deficient female animals suffer death and resorption of the fetuses. This was the basis of the original biological assay of the vitamin; female rats were maintained for 2 to 3 months on a vitamin E-free diet and then mated. Impregnation and implantation proceed normally; but, if they are not provided with vitamin E, the fetuses die and are resorbed. Five days after mating, the animals were killed, and the number of surviving fetuses gave an index of the biological activity of the test compound, relative to standard doses of α-tocopherol. Synthetic antioxidants can replace vitamin E for this function, but selenium cannot.

Table 4.2 Responses of Signs of Vitamin E or Selenium Deficiency to Vitamin E, Selenium, and Synthetic Antioxidants in Experimental Animals

	Vitamin E	Selenium	Synthetic Antioxidants
Fetal resorption	+	−	+
Testicular atrophy	+	+	−
Necrotizing myopathy/white muscle disease	Variable	+	−
CNS necrosis	+	−	+
Exudative diathesis	+	+	+
Erythrocyte hemolysis	+	+	+
Liver necrosis	Variable	+	+
Kidney necrosis	+	−	+

CNS, central nervous system.

In male animals, deficiency results in testicular atrophy, with degeneration of the germinal epithelium of the seminiferous tubules. This lesion responds to vitamin E or selenium, but not to synthetic antioxidants.

Vitamin E deficiency results in the development of necrotizing myopathy, sometimes including cardiac muscle. This has been called nutritional muscular dystrophy, an unfortunate term, because deficiency of the vitamin is not a factor in the etiology of human muscular dystrophies, and supplements of the vitamin have no beneficial effect. The myopathy responds to selenium, but not to synthetic antioxidants.

The nervous system is also affected in deficiency, with the development of central nervous system necrosis (nutritional encephalomacia), a condition that can be exacerbated by feeding a diet especially rich in polyunsaturated fatty acids. There is also axonal dystrophy in animals maintained for prolonged periods of time on vitamin E-deficient diets. Synthetic antioxidants, but not selenium, can prevent these changes. The neuropathy begins from axonal membrane injury, and then develops as a distal and dying-back type of axonopathy.

Vitamin E-deficient animals show exudative diathesis, in which there is leakage of blood plasma from capillaries into subcutaneous tissues, apparently the result of abnormal permeability of capillary blood vessels. There is an accumulation of (usually green-colored) fluid under the skin. This responds to synthetic antioxidants or selenium.

There is also increased erythrocyte hemolysis, which responds to synthetic antioxidants or selenium. The sensitivity of erythrocytes to chemically induced hemolysis can be used both as a biological assay of vitamin E in experimental animals and also as an index of vitamin E nutritional status (Section 4.5).

Deficient animals may also show necrosis of the liver (which responds to selenium and synthetic antioxidants) and kidney (which responds to synthetic antioxidants but not selenium).

Most of these effects of vitamin E deficiency can be attributed to membrane damage. In deficiency, there is an accumulation of lysophosphatidylcholine in membranes, which is cytolytic. The accumulation of lysophosphatidylcholine is a result of increased activity of phospholipase A. It is not clear whether α-tocopherol inhibits phospholipase A; whether there is increased phospholipase activity because of increased peroxidation of polyunsaturated fatty acids in phospholipids, and hence an attempt at membrane lipid repair; or whether the physicochemical effects of α-tocopherol on membrane organization and fluidity prevent the cytolytic actions of lysophosphatidylcholine (Douglas et al., 1986; Erin et al., 1986).

Lipid peroxidation is increased in vitamin E deficiency, and subsequent catabolism of the peroxides results in the formation of malondialdehyde and other aldehydes. These can form Schiff bases with amino groups of proteins, free amino acids, and nucleic acids. The resultant fluorescent pigments are called ceroid pigments, lipopigments, or lipofuscin, and accumulate in increased amounts in the liver and other tissues of deficient animals (Manwaring and Csallany, 1988).

The formation of the ceroid pigments seems to represent a mechanism for detoxication of the products of lipid peroxidation and, although a useful indicator of peroxidation, and hence of vitamin E deficiency, seems to have no adverse effects under most conditions. The exception is in neuronal ceroid lipofuschinosis, which can result in physiological disturbance and nerve loss. Malondialdehyde attack on nucleic acids may be a factor in the induction of carcinogenesis in response to some oxidative radical-generating carcinogens. The formation of ceroid pigments can be prevented by synthetic antioxidants and possibly also selenium.

Vitamin E deficiency is also associated with impaired mitochondrial oxidative metabolism and impaired activity of microsomal cytochrome P_{450}-dependent mixed-function oxidases, and hence the metabolism of xenobiotics. There is no evidence that vitamin E has any specific role in electron transport in mitochondria or microsomes. Again, changes in membrane lipids and oxidative damage presumably account for the observed metabolic abnormalities.

4.4.2 Human Vitamin E Deficiency

Vitamin E deficiency is not a problem, even among people living on relatively poor diets. In depletion studies, very low intakes of vitamin E must be maintained for many months before there is any significant fall in circulating α-tocopherol, because there are relatively large tissue reserves of the vitamin.

Deficiency develops in patients with severe fat malabsorption, cystic fibrosis, chronic cholestatic hepatobiliary disease, and in two rare groups of patients with genetic diseases:

1. Patients with congenital abetalipoproteinemia, who are unable to synthesize VLDL. This was the first condition reported that confirmed the essentiality of vitamin E in human beings (Muller et al., 1983). The patients have undetectably low plasma levels of α-tocopherol and develop devastating ataxic neuropathy and pigmentary retinopathy. The administration of massive oral supplements of vitamin E (of the order of 100 mg of α-tocopherol acetate per kg of body weight per day) halts the progression of the neuropathy in children diagnosed relatively late and can prevent the development of neuropathy altogether if therapy is started early enough (Muller, 1986).
2. Patients who lack the hepatic tocopherol transfer protein (Section 4.2) and suffer from what has been called AVED (ataxia with vitamin E deficiency) are unable to export α-tocopherol from the liver in VLDL.

In both groups of patients, the only source of vitamin E for peripheral tissues will be recently ingested vitamin E in chylomicrons. They develop cerebellar ataxia, axonal degeneration of sensory neurons, skeletal myopathy, and pigmented retinopathy similar to those seen in experimental animals.

In premature infants, whose reserves of the vitamin are inadequate, vitamin E deficiency causes a shortened half-life of erythrocytes, which can progress to increased intravascular hemolysis, and hence hemolytic anemia. In infants treated with hyperbaric oxygen, there is a risk of damage to the retina (retrolental fibroplasia), and vitamin E supplements may be protective, although this is not firmly established (Phelps, 1987).

4.5 ASSESSMENT OF VITAMIN E NUTRITIONAL STATUS

The most commonly used index of vitamin E nutritional status is the plasma concentration of α-tocopherol; because it is transported in plasma lipoproteins, it is best expressed per mole of cholesterol or per milligram of total plasma lipids (Horwitt et al., 1972; Winbauer et al., 1999). The reference range is

Table 4.3 Indices of Vitamin E Nutritional Status

	Deficient	Low	Acceptable	Desirable
Plasma tocopherol, μmol/L	<12	12–16	>16	>30
Plasma tocopherol, μmol/g plasma total lipid	<1.1	1.1–1.86	>1.86	>3.4
Plasma tocopherol, mmol/mol cholesterol	<2.2	2.2–2.25	>2.25	>4.0
Erythrocyte fragility	—	—	<0.05	—
Ratio of hemolysis by H_2O_2:H_2O				

Sources: From data reported by Horwitt et al., 1972; Sauberlich et al., 1974; Morrissey and Sheehy, 1999.

12 to 37 μmol per L; ranges associated with inadequate and desirable status are shown in Table 4.3; an optimum concentration for protection against cardiovascular disease and cancer is >30 μmol per L (Morrissey and Sheehy, 1999).

Erythrocytes are incapable of de novo lipid synthesis, so peroxidative damage resulting from oxygen stress has a serious effect: shortening red cell life and possibly precipitating hemolytic anemia in vitamin E deficiency. This has been exploited as a method of assessing status by measuring the hemolysis induced by either hydrogen peroxide or dialuric acid. This gives a means of assessing the functional adequacy of vitamin E intake, albeit one that will be masked by adequate selenium intake and may be affected by other, unrelated factors. Plasma concentrations of α-tocopherol greater than 14 μmol per L are adequate to prevent significant hemolysis; less than 2.2 mmol per mol of cholesterol or 1.1 μmol per g of total plasma lipid is associated with increased susceptibility of erythrocytes to induced hemolysis in vitro.

Overall antioxidant status, as opposed to specifically vitamin E status, can be assessed by a variety of measures of lipid peroxidation, including measurement of:

1. Plasma total thiobarbituric acid-reacting substances (TBARS);
2. F_2 isoprostanes, the isomers of prostaglandin F_2 that are formed by radical-catalyzed peroxidation of arachidonic acid in membranes.
3. The exhalation of pentane arising from the catabolism of the products of peroxidation of ω6 polyunsaturated fatty acids or ethane arising from ω3 polyunsaturated fatty acids. Intravenous infusion of a lipid mixture rich in linoleic acid stresses antioxidant capacity and results in increased

breath pentane; this is more marked in subjects with low vitamin E status, and the administration of vitamin E reduces the exhalation of pentane (Tappel and Dillard, 1981; Lemoyne et al., 1988; Van Gossum et al., 1988).

4.6 REQUIREMENTS AND REFERENCE INTAKES

Based on the plasma concentration of α-tocopherol to prevent significant hemolysis in vitro (14 to 16 μmol per L), the U.S./Canadian estimated average requirement is 12 mg per day, giving a Recommended Dietary Amount (RDA) of 15 mg per day (Institute of Medicine, 2000) – a 50% increase on the previous RDA (National Research Council, 1989). This increase arose partly as a result of considering only the $2R$ isomers in dietary intake (Section 4.1). Average intakes are of the order of 8 to 12 mg of α-tocopherol equivalent per day; it would be difficult meet this reference intake without significant changes in diet or use of supplements.

Horwitt (2001) has criticized this high reference intake, noting that it was based on reinterpretation of the same data as had been provided to the committee in 1998, which set a lower RDA. Data were from studies that he had performed (Horwitt, 1960), and he notes that to provide a diet rich in polyunsaturated fatty acids and low in vitamin E, the oils had been oxidized to remove vitamin E, and therefore contained large amounts of oxidized lipids that would increase apparent vitamin E requirements.

Early reports suggested that vitamin E requirements increase with the intake of polyunsaturated fatty acids. Neither the United Kingdom (Department of Health, 1991) nor the European Union (Scientific Committee for Food, 1993) set reference intakes for vitamin E, but both suggested that an acceptable intake was 0.4 mg of α-tocopherol equivalent per g of dietary polyunsaturated fatty acid (PUFA). This should be readily achievable from PUFA-rich foods, which are also rich sources of vitamin E, but PUFA supplements may not contain adequate vitamin E. There is little evidence to support the figure of 0.4 mg of α-tocopherol equivalent per g of dietary PUFA, and indeed the need for vitamin E (and other antioxidants) depends more on the degree of unsaturation of fatty acids than the total amount, and dietary PUFAs undergo chain elongation and further desaturation in the body. Valk and Hornstra (2000) suggested an intake of 0.6 mg of α-tocopherol per g of linoleic acid, and noted that there are no data on the requirements for people consuming long-chain PUFA.

Gey (1995) has shown that there is an inverse relationship between plasma α-tocopherol and risk of ischemic heart disease over a range of 2.5 to 4.0 mmol per mol of cholesterol, and has suggested an optimum or desirable plasma concentration > 4 mmol of α-tocopherol per mol of cholesterol (>3.4 μmol per g of total plasma lipid). This would require an intake of 17 to 40 mg of α-tocopherol equivalents per day.

4.6.1 Upper Levels of Intake

Vitamin E seems to have very low toxicity, and habitual intake of supplements of 200 to 600 mg per day (compared with an average dietary intake of 8 to 12 mg) seems to be without untoward effect; there are no consistent reports of adverse effects up to 3,200 mg per day, suggesting that an acceptable daily intake is in the very wide range between 0.15 to 2 mg per kg of body weight; the European Health Food Manufacturers' Association suggests an upper limit of 800 mg per day from supplements (Shrimpton, 1997).

Very high intakes may antagonize vitamin K and hence potentiate anti-coagulant therapy. This is probably the result of inhibition of the vitamin K quinone reductase, but α-tocopheryl quinone may compete with vitamin K hydroquinone and hence inhibit carboxylation of glutamate in target proteins (Section 5.3.1).

The U.S./Canadian tolerable upper level is set at 1,000 mg per day, based on reports of prolonged prothrombin time in people receiving anticoagulants and consuming 1,100 to 2,100 mg of vitamin E per day. It is noteworthy that although the report specifically excluded the 2S isomers of synthetic α-tocopherol from calculations of nutritional requirements, this tolerable upper level includes all forms of the vitamin, regardless of their tissue retention and biological activity (Institute of Medicine, 2000).

4.6.2 Pharmacological Uses of Vitamin E

Although vitamin E deficiency causes infertility in experimental animals (Section 4.4.1), there is no evidence that deficiency has any similar effects on human fertility. It is a considerable leap of logic from the effects of gross depletion in experimental animals to the popular, and unfounded, claims for vitamin E in enhancing human fertility and virility.

There are no established pharmacological uses of vitamin E except for the protection of preterm infants exposed to high partial pressure oxygen, who may develop the retinopathy of prematurity (retrolental fibroplasia); however, even here, the protective effect of vitamin E is controversial, and it is not routinely recommended (Phelps, 1987). There is some evidence that age-related macular degeneration is also associated with oxidative damage and that high circulating concentrations of vitamin E may be protective, although there is no evidence from intervention studies (Beatty et al., 2000).

Animal studies show some protective effects of tocopherol supplements against a variety of radical-generating chemical toxicants, and it has been assumed that vitamin E may similarly be protective against a variety of degenerative diseases that are associated with radical damage, including cancer, cardiovascular disease, cataracts, and neurodegenerative diseases.

4.6.2.1 **Vitamin E and Cancer**   The α-tocopherol β-carotene study was designed to test the efficacy of supplements against lung cancer; vitamin E had no effect, although as discussed in Section 2.5.2.3, carotene supplements were associated with increased mortality from lung cancer (Alpha-Tocopherol Beta-Carotene Cancer Prevention Study Group, 1994). However, there was a 32% lower incidence of, and 41% lower mortality from, prostate cancer in those people taking the vitamin E supplements (Heinonen et al., 1998). There is no clear evidence from other intervention trials that vitamin E reduces cancer risk.

4.6.2.2 **Vitamin E and Cardiovascular Disease**   Epidemiologically, there is a significant negative association between vitamin E status and coronary heart disease (Gey, 1995). This is generally assumed to be because of the antioxidant properties of the vitamin (Section 4.3.1), but the effects of α-tocopherol in regulating vascular smooth muscle proliferation and platelet aggregation (Section 4.3.3) and of tocotrienols on cholesterol synthesis (Section 4.1) may also be important (Traber, 2001). The results of intervention studies have been disappointing. In the Cambridge Heart AntiOxidant Study, there was a significant reduction in nonfatal myocardial infarctions among those taking vitamin E supplements, but a (nonsignificant) increase in fatal infarctions (Stephens et al., 1996). Other large intervention trials have shown no beneficial effect of vitamin E in coronary heart disease (Meydani, 2000; Kaul et al., 2001; Pruthi et al., 2001).

4.6.2.3 **Vitamin E and Cataracts**   There is good evidence that cataracts are the result of oxidative damage to α-crystallin in the lens of the eyes, and therefore high intakes of antioxidants might be expected to be beneficial. Of 10 controlled trials, 5 showed a protective effect of vitamin E supplements, and 5 showed no effect (Institute of Medicine, 2000).

4.6.2.4 **Vitamin E and Neurodegenerative Diseases**   There is evidence that Parkinson's disease is a result of oxidative radical damage, and there have been some suggestions that Alzheimer's disease may be as well (Sun and Chen, 1998; Grundman, 2000). The neurological damage associated with Down's syndrome (trisomy 21) may also be related to radical damage; the gene for superoxide dismutase is on chromosome 21, and overexpression in mice is associated with increased lipid peroxidation in the brain, apparently as a result of increased formation of hydrogen peroxide. However, controlled trials show no beneficial effects of vitamin E supplementation in any of these conditions (Shoulson, 1998; Tabet et al., 2000).

FURTHER READING

Asplund K (2002) Antioxidant vitamins in the prevention of cardiovascular disease: a systematic review. *Journal of Internal Medicine* **251**, 372–92.

Behne D and Kyriakopoulos A (2001) Mammalian selenium-containing proteins. *Annual Reviews of Nutrition* **21**, 453–73.

Brigelius-Flohe R and Traber MG (1999) Vitamin E: function and metabolism. *FASEB Journal* **13**, 1145–55.

Carr AC, Zhu BZ, and Frei B (2000) Potential antiatherogenic mechanisms of ascorbate (vitamin C) and alpha-tocopherol (vitamin E). *Circulation Research* **87**, 349–54.

Diplock AT (2001) Antioxidants, nutrition and health. In *Food and Nutritional Supplements: Their Role in Health and Disease*, JK Ransley, JK Donnelly, and NW Read (eds.), pp. 65–80. Berlin: Springer.

Frei B, Keaney JF Jr, Retsky KL, and Chen K (1996) Vitamins C and E and LDL oxidation. *Vitamins and Hormones* **52**, 1–34.

Herrera E and Barbas C (2001) Vitamin E: action, metabolism and perspectives. *Journal of Physiology and Biochemistry* **57**, 43–56.

Kamal-Eldin A and Appelqvist LA (1996) The chemistry and antioxidant properties of tocopherols and tocotrienols. *Lipids* **31**, 671–701.

Parks E and Traber MG (2000) Mechanisms of vitamin E regulation: research over the past decade and focus on the future. *Antioxidants and Redox Signaling* **2**, 405–12.

Rayman M (2000) The importance of selenium to human health. *Lancet* **356**, 233–41.

Ricciarelli R, Zingg JM, and Azzi A (2001) Vitamin E: protective role of a Janus molecule. *FASEB Journal* **15**, 2314–25.

Schorah CJ (1995) Micronutrients, antioxidants and risk of cancer. *Biblio Nutritio et Dieta*, 92–107.

Traber MG and Arai H (1999) Molecular mechanisms of vitamin E transport. *Annual Reviews of Nutrition* **19**, 343–55.

Traber MG and Sies H (1996) Vitamin E in humans: demand and delivery. *Annual Reviews of Nutrition* **16**, 321–47.

Various authors (2001) Symposium proceedings: molecular mechanisms of protective effects of vitamin E in atherosclerosis. *Journal of Nutrition* **131** (Suppl.).

References cited in the text are listed in the Bibliography.

FIVE

Vitamin K

Vitamin K was discovered in 1935 as a result of studies of a hemorrhagic disease of chickens fed on solvent-extracted fat-free diets and cattle fed on silage made from spoiled sweet clover. The problem in the chickens was a lack of the vitamin in the diet, whereas in the cattle it was from the presence of dicoumarol, an antimetabolite of the vitamin. It soon became apparent that vitamin K was required for the synthesis of several of the proteins required for blood clotting, but it was not until 1974 that Stenflo and coworkers (1974) elucidated the mechanism of action of the vitamin. A new amino acid, γ-carboxyglutamate (Gla) was found to be present in the vitamin K-dependent proteins, but absent from the abnormal precursors that circulate in deficiency. γ-Carboxyglutamate is chemically unstable and undergoes spontaneous decarboxylation to glutamate when proteins are hydrolyzed in acid for amino acid analysis.

Since then, a number other proteins that undergo the same vitamin K-dependent posttranslational modification, carboxylation of glutamate residues to γ-carboxyglutamate, have been discovered, including osteocalcin and the matrix Gla protein in bone, nephrocalcin in the kidneys, and the product of the growth arrest-specific gene Gas6, which is involved in both the regulation of differentiation and development in the nervous system, and control of apoptosis in other tissues. All of these γ-carboxyglutamate–containing proteins bind calcium, which causes a conformational change so that they interact with membrane phospholipids.

The impairment of blood clotting in response to antimetabolites of vitamin K has been exploited in two ways: the development of clinically useful anticoagulant drugs for patients at risk of thrombosis and, at higher doses, as rodenticides.

In green plants, vitamin K (phylloquinone) functions as a secondary electron acceptor in photosystem I, and in bacteria a variety of menaquinones
(which also have vitamin K activity) have a role in the plasma membrane in
electron transport, where they serve the same role as ubiquinone (Section 14.6)
in mitochondrial electron transport. There is no evidence that vitamin K has
any role in electron transport in animals.

5.1 VITAMIN K VITAMERS

As shown in Figure 5.1, compounds with vitamin K activity have a 2-methyl-1,4-
naphthoquinone ring. There are two naturally occurring vitamers: phylloquinone (from plants) has a phytyl side chain, whereas the menaquinones (from
bacteria) have a polyisoprenyl side chain, with up to 15 isoprenyl units (most
commonly 6 to 10), shown by menaquinone-n. Bacteria also form a variety of

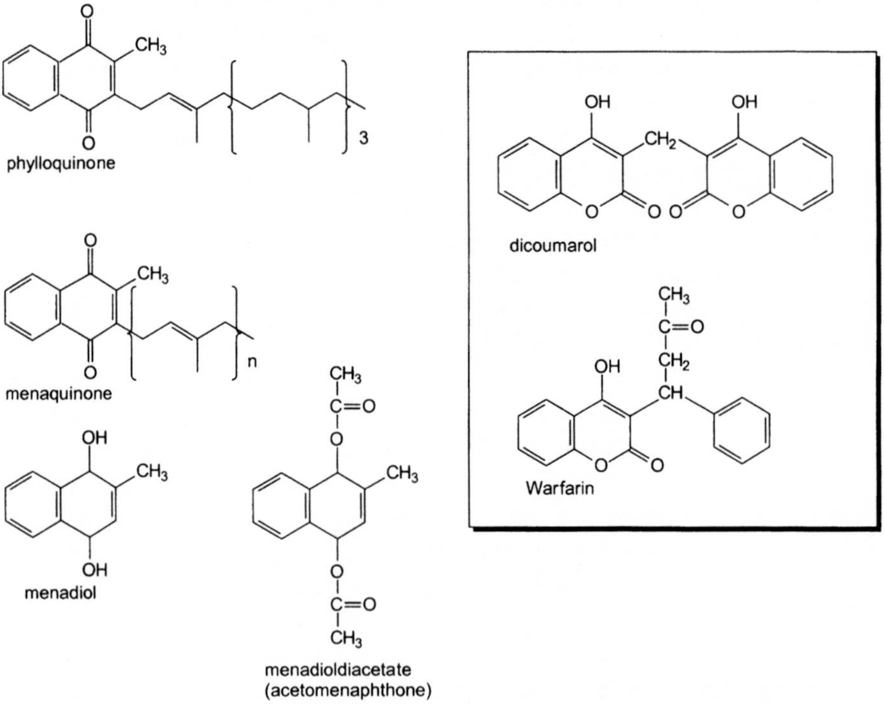

Figure 5.1. Vitamin K vitamers and the vitamin K antagonists dicoumarol and warfarin. Relative molecular masses (M_r): phylloquinone, 450.7; menaquinone-4, 447.4;
menaquinone-5, 512.8; menaquinone-6, 580.0; menaquinone-7, 649.0; menaquinone-
8, 717.1; menadione, 172.2; menadiol diacetate, 258.3; dicoumarol, 336.3; and warfarin
308.3.

menaquinones with differing degrees of saturation of the side chain, variations in the positions of the double bonds, and sometimes additional methylation of the naphthoquinone ring. The pattern of menaquinones synthesized is useful in the taxonomy and identification of bacteria.

Phylloquinone is vitamin K_1, menaquinones are vitamin K_2, and the synthetic compounds menadione and menadiol are vitamin K_3. In addition to menadione itself, menadiol diacetate (acetomenaphthone) is used in pharmaceutical preparations, and two water-soluble derivatives, menadione sodium bisulfite and menadiol sodium phosphate, have been used for administration of the vitamin by injection and in patients with malabsorption syndromes that would impair the absorption of menadione, phylloquinone, and menaquinones, which are lipid soluble.

Menaquinones are synthesized by intestinal bacteria, but it is unclear how much they contribute to vitamin K nutrition, because they are extremely hydrophobic, and will only be absorbed from regions of the gastrointestinal tract where bile salts are present – mainly the terminal ileum. However, prolonged use of antibiotics can lead to vitamin K deficiency and the development of vitamin K-responsive hypoprothrombinemia (Section 5.4), as can dietary deprivation of phylloquinone. In vitro, menaquinones 2 to 6 have the same activity as phylloquinone as coenzyme for the solubilized liver microsomal vitamin K-dependent carboxylase (Section 5.3.1), whereas menaquinones with a side chain longer than seven have lower activity (Suttie, 1995). In extrahepatic tissues, the principal active vitamer is menaquinone-4 (Thijssen and Drittij-Reijnders, 1996; Thijssen et al., 1996).

5.2 METABOLISM OF VITAMIN K

Phylloquinone is absorbed in the proximal small intestine, by an energy-dependent mechanism, and is incorporated into chylomicrons. Estrogens increase phylloquinone absorption in both male and female animals, and male animals are more susceptible to dietary vitamin K deprivation than females (Jolly et al., 1977). Even after an overnight fast, about half the plasma vitamin K is present in chylomicron remnants, and only a quarter in low-density lipoprotein. The plasma concentration of phylloquinone is associated with genetic variants of apoprotein E, which determines the binding of chylomicron remnants to the liver lipoprotein receptor (Kohlmeier et al., 1996).

Extrahepatic tissues take up phylloquinone from chylomicrons and very low-density lipoprotein, and synthesize menaquinone-4, which is the principal vitamer in tissues other than the liver. Some menaquinone-4 is also absorbed into the portal system from the colon.

Menaquinones are absorbed mainly from the terminal ileum, where bile salts are present, into the hepatic portal vein. Little of the menaquinones formed by colonic bacteria can be absorbed, because they remain tightly bound to bacterial cell membranes in the absence of bile salts. About 90% of the total liver content of vitamin K is menaquinones 7 to 13, and the hepatic pool of phylloquinone turns over considerably faster than that of menaquinones. Sixty percent to 70% of the daily intake of phylloquinone is excreted, mainly as conjugates in the bile, and the half-life of a tracer dose of phylloquinone is only about 17 hours.

After a tracer dose of radioactive phylloquinone, the label is rapidly accumulated in the liver, then lost from the body with turnover time of 1.5 days. This suggests that there is rapid turnover and little storage of vitamin K. However, there may be considerable enterohepatic recirculation of the conjugates excreted in the bile (Shearer et al., 1996; Olson et al., 2002). About 10% of the total liver vitamin K is normally present as the epoxide, which is formed by the vitamin K-dependent carboxylase and normally reduced back to the active vitamin (Section 5.3.1).

Menadione is mainly absorbed by way of the portal system, although some is also absorbed into the lymphatic system. In the liver, menadione is alkylated to menaquinone-4 by the addition of geranyl-geranyl pyrophosphate, rather than a stepwise formation of the polyisoprenyl side chain; liver microsomal preparations will catalyze the formation of menaquinone-2 (from geranyl pyrophosphate), menaquinone-3 (from farnesyl pyrophosphate), and menaquinone-4 at the same rate, depending on which isoprenyl pyrophosphate is provided (Dialameh et al., 1970). The vitamin K antagonist warfarin (see Figure 5.1) inhibits the alkylation of menadione, although this is not its major mode of action (Section 5.3.1). Menadione, which is not alkylated, is rapidly metabolized, largely by reduction to menadiol, followed by the formation of the glucuronide, which is excreted in the bile, and sulfate and phosphate conjugates, which circulate in the bloodstream and are excreted in both bile and urine. The metabolism of menadione is rapid, so that only a small proportion is converted to (biologically active) menaquinone-4.

About 20% of an oral dose of [^{3}H]phylloquinone is excreted in the feces unchanged, suggesting that 80% is absorbed. A further 35% to 40% of the radioactivity is recovered in the feces as a variety of conjugates of polar metabolites formed by ω-methyl oxidation of the side chain, followed by successive β-oxidation, as well as reduction of the quinone ring to the quinol, to provide the site for conjugation with glucuronic acid, phosphate, or sulfate. About 75% of the conjugates are excreted in the bile and the remainder in the urine. In the intestinal lumen, the glucuronides are largely hydrolyzed by bacterial glucuronidase, and the carboxyl group may be esterified by bacterial enzymes, so that the fecal metabolites are more lipophilic than those initially excreted in the bile. After the administration of the vitamin K antagonist warfarin, there is an increase in the urinary excretion of metabolites of phylloquinone epoxide and a compensatory decrease in fecal excretion (Olson, 1984).

5.2.1 Bacterial Biosynthesis of Menaquinones
The quinone ring is derived from isochorismic acid, formed by isomerization of chorismic acid, an intermediate in the shikimic acid pathway for synthesis of the aromatic amino acids. The first intermediate unique to menaquinone formation is o-succinyl benzoate, which is formed by a thiamin pyrophosphate-dependent condensation between 2-oxoglutarate and chorismic acid. The reaction catalyzed by o-succinylbenzoate synthetase is a complex one, involving initially the formation of the succinic semialdehyde-thiamin diphosphate complex by decarboxylation of 2-oxoglutarate, then addition of the succinyl moiety to isochorismate, followed by removal of the pyruvoyl side chain and the hydroxyl group of isochorismate.

o-Succinyl benzoate forms a coenzyme A thioester, which then undergoes dehydration and ring closure to naphthohydroquinone-3-carboxylate. This reacts with polyprenyl pyrophosphate, with the loss of carbon dioxide. The immediate product of the enzymic reaction, desmethylmenaquinol, is oxidized to desmethylmenaquinone, which is methylated to yield menaquinone (Meganathan, 2001)

5.3 THE METABOLIC FUNCTIONS OF VITAMIN K
The main metabolic function of vitamin K is as the coenzyme in the carboxylation of protein-incorporated glutamate residues to yield γ-carboxyglutamate – a unique type of carboxylation reaction, clearly distinct from the biotin-dependent carboxylation reactions (Section 11.2.1).

Four vitamin K-dependent proteins involved in blood coagulation, prothrombin and Factors VII, IX, and X, were discovered early in the investigations of the vitamin, as a result of the hemorrhagic disease caused by deficiency. The function of γ-carboxyglutamate in these proteins is to chelate calcium and induce a conformational change that permits binding of the proteins to membrane phospholipids. Unlike the high-affinity calcium binding proteins, such as calmodulin and calbindin (Section 3.3.3.1), the vitamin K-dependent calcium binding proteins have a relatively low affinity for calcium, with values of K_{diss} in the millimolar range.

In addition to blood clotting, γ-carboxyglutamate–containing proteins are found in

1. bone (osteocalcin and bone matrix Gla protein);
2. the kidney cortex (nephrocalcin); hydroxyapatite and calcium oxalate containing urinary stones;
3. atherosclerotic plaque – this protein is sometimes called atherocalcin, but is probably the same as the bone matrix Gla protein (Section 5.5.3.3);
4. the intermembrane space of mitochondria, where they may have a role in the mitochondrial accumulation of calcium; and
5. the central nervous system and probably other tissues (Gas6).

A number of proteins involved in cell signaling and as cell surface receptors also contain γ-carboxyglutamate (Nelsestuen et al., 2000).

A specific vitamin K binding protein has been identified in the nucleus in osteoblasts, suggesting that the vitamin may also have direct nuclear actions (Hoshi et al., 1999). Phylloquinone, but not menaquinones, down-regulates osteoclastic bone resorption by inducing apoptosis in osteoclasts (Kameda et al., 1996).

The vitamin also activates serine palmitoyltransferase, the first enzyme of phosphosphingolipid synthesis, and in bacteria it can, together with inorganic phosphate, replace part of the ATP requirement of galactocerebroside sulfo-transferase (Tsaioun, 1999). In animals, lipid sulfatides are decreased in vitamin K deficiency and increased with higher intakes (Sundaram et al., 1996).

5.3.1 The Vitamin K-Dependent Carboxylase

The vitamin K-dependent carboxylase is an integral membrane protein. Most of the proteins that are carboxylated are extracellular proteins, and the major activity of the carboxylase is at the luminal face of the rough endoplasmic reticulum. However, there is also significant carboxylase activity in mitochondria.

As shown in Figure 5.2, the initial reaction is oxidation of vitamin K hydroquinone to the epoxide, linked to γ-deprotonation of the glutamate residue to yield a carbanion, catalyzed by vitamin K epoxidase.

Menaquinones 2 to 7 have essentially the same activity as phylloquinone when the reduced hydroquinones are incubated with either isolated preparations of epoxidase or intact microsomes. If intact microsomes are incubated with the quinones, menaquinone-3 has higher activity than phylloquinone, which in turn has higher activity than menaquinone-4 or higher homologs, suggesting that quinone reductase (see below) has greater specificity for the length of the side chain than does the epoxidase. The glutamate

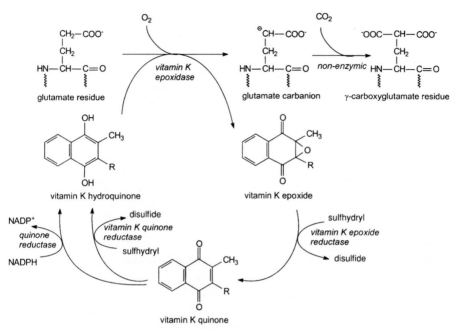

Figure 5.2. Reaction of the vitamin K-dependent carboxylase (vitamin K epoxidase) and recycling of vitamin K epoxide to the hydroquinone. Vitamin K epoxidase, EC 1.14.99.20; warfarin-sensitive epoxide/quinone reductase, EC 1.1.4.1; and warfarin-insensitive quinone reductase, EC 1.1.4.2.

carbanion formed by the epoxidase reaction reacts with carbon dioxide to form γ-carboxyglutamate. At saturating concentrations of carbon dioxide, there is equimolar formation of γ-carboxyglutamate and vitamin K epoxide.

The double bond at C-3 of the side chain is essential to activity and has *trans*-configuration; *cis*-phylloquinone has no biological activity. This double bond seems to be involved in the deprotonation of the glutamate substrate.

Target proteins for carboxylation have the recognition sequence X-Phe-X-aa-aa-aa-aa-Ala, where X is an aliphatic hydrophobic amino acid, and 10 to 12 glutamate residues in the first 40 amino acids of the amino terminal region. The peptide substrate binds to the amino terminal region of the carboxylase, whereas the carboxyl terminal region has the epoxidase activity. The epoxidase activity is regulated by the binding of the peptide substrate and in the absence of the peptide, epoxidation does not occur (Furie and Furie, 1997; Sugiura et al., 1997). The enzyme catalyzes multiple carboxylation of glutamate groups, each associated with epoxidation of vitamin K hydroquinone, during a single peptide binding event; no partially carboxylated peptide is released from the enzyme to be rebound and undergo further carboxylation, although in vitamin K deficiency partially carboxylated peptides are released (Morris et al., 1995).

Vitamin K epoxide is reduced to the quinone in a reaction involving oxidation of a dithiol to the disulfide, catalyzed by epoxide reductase. This enzyme has no activity toward menadione epoxide or the epoxides of a variety of xenobiotics, but is specific for alkylated vitamin K epoxides. Like other dithiol-linked flavoprotein reductases, this enzyme initially undergoes an internal dithiol–disulfide reaction, followed by reaction with the dithiol substrate. The initial step in the reaction is an attack on the epoxide ring by an enzyme-bound sulfhydryl group, with intermediate formation of a thioether adduct. Thiol reagents, such as iodoacetamide and N-ethylmaleimide, inhibit the enzyme irreversibly. The addition of vitamin K epoxide protects the enzyme against this inactivation. The physiological dithiol substrate has not been uniquevocally identified, but is assumed to be thioredoxin.

Vitamin K quinone is reduced to the active hydroquinone substrate for the epoxidase reaction by either a dithiol-linked reductase that is almost certainly the same enzyme as the epoxide reductase or NADPH-dependent quinone reductase. Like the epoxide reductase, the dithiol-linked reductase is inhibited by warfarin. In warfarin-resistant rats, there is a warfarin-insensitive epoxide reductase, which also has quinone reductase activity (Hildebrandt et al., 1984; Gardill and Suttie, 1990).

The NADPH-dependent reduction of vitamin K quinone to the hydroquinone is not inhibited by warfarin. In the presence of adequate amounts of vitamin K, the carboxylation of glutamate residues can proceed normally, despite the presence of warfarin, with the stoichiometric formation of vitamin K epoxide that cannot be reutilized. Small amounts of vitamin K epoxide, and hydroxides formed by its reduction by other enzymes, are normally found in plasma. In warfarin-treated animals and patients, there is a significant increase in the plasma concentration of both. There is also an increase in the urinary excretion of the products of side-chain oxidation of the epoxide and hydroxides.

Prothrombin normally contains 10 γ-carboxyglutamate residues in the amino terminal region. In the presence of high concentrations of warfarin, a completely uncarboxylated precursor, preprothrombin, is released into the circulation. Before the nature of this precursor protein was known, it was called *protein induced by vitamin K absence or antagonism* (PIVKA), a term that is sometimes still used.

At lower doses of anticoagulant, a variety of partially carboxylated preprothrombins are formed. Sequencing of these proteins shows that the carboxylation is not random, but proceeds in an orderly fashion from the amino terminal of the substrate. In preprothrombin containing 80% of the normal amount of γ-carboxyglutamate, it is the last two glutamate residues before the carboxy terminal that are not carboxylated; in 60% undercarboxylated preprothrombin, it is the last four glutamate residues that are not carboxylated (Liska and Suttie, 1988).

In some patients with combined deficiency of vitamin K-dependent coagulation factors, the deficiency can be partially corrected by high doses of vitamin K, suggesting that the defect is in the affinity of the carboxylase for its coenzyme (Mutucumarana et al., 2000).

5.3.2 Vitamin K-Dependent Proteins in Blood Clotting

The formation of blood clots is the result of the conversion of the soluble protein fibrinogen into fibrin, an insoluble network of fibers. This is achieved by specific proteolysis of fibrinogen at two arginine–glycine junctions, removing two pairs of small peptides (fibrinopeptides), catalyzed by thrombin. The resultant fibrin monomer aggregates into the insoluble fibrin polymer, which undergoes further covalent cross-linkage, catalyzed by a transamidase, the so-called fibrin-stabilizing factor or Factor XIII. Fibrin-stabilizing factor is normally present as an inactive dimer that is activated by thrombin. By the formation of an insoluble clot, bleeding is rapidly stopped.

Thrombin, which catalyzes the proteolysis of fibrinogen, circulates as an inactive precursor, prothrombin, which in turn is activated by partial proteolysis to remove a peptide sequence that masks the catalytic site. There are two distinct pathways leading to the activation of prothrombin to thrombin (see Figure 5.3):

1. The extrinsic pathway, which is initiated by thromboplastin released from injured tissues and the protease proconvertin (Factor VII).
2. The intrinsic pathway, which is initiated by the activation of Factor XII as a result of adsorption onto collagen, platelet membranes, or (under laboratory conditions) glass.

Factor X can also be activated by kallikrein – in turn prekallikrein is activated to kallikrein by activated Factor XII, thus prolonging the initial contact activation of Factor XII.

The intrinsic pathway is involved in the clotting of blood in glass tubes and in the undesirable intravascular clotting that results in thrombosis. Control of the clotting mechanism is thus central to hemostasis to avoid both hemorrhage and thrombosis.

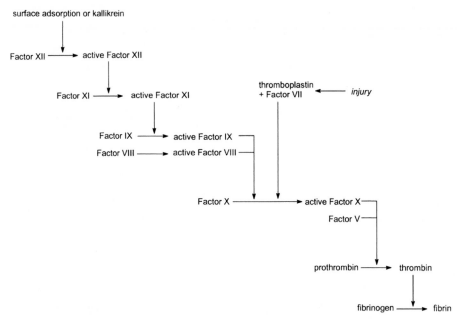

Figure 5.3. Intrinsic and extrinsic blood clotting cascades. Factor I, fibrinogen; Factor II, prothrombin (vitamin K-dependent); Factor III, thromboplastin; Factor V, proaccelerin; Factor VII, proconvertin (vitamin K-dependent); Factor VIII, antihemophilic factor; Factor IX, Christmas factor (vitamin K-dependent); Factor X, Stuart factor (vitamin K-dependent); Factor XI, plasma thromboplastin; Factor XII, Hageman factor; Factor XIII, fibrin-stabilizing factor; and Factor XIV, protein C (vitamin K-dependent). What was at one time called Factor IV is calcium; no factor has been assigned number VI.

As shown in Figure 5.3, both the intrinsic and extrinsic pathways for the activation of prothrombin and hence initiation of blood clotting involve a number of intermediate factors. The nomenclature of the factors is based on the history of their discovery, which was largely as a result of studies in patients with various congenital clotting defects. Most of the clotting factors are serine proteases, which circulate as inactive zymogens. Each factor is activated by partial proteolysis and then in turn activates the next factor – a cascade that results in considerable amplification of the original stimulus.

The cascade is not a simple linear one. The concerted action of activated Factors VIII and IX is required in the intrinsic pathway for the activation of Factor X. The rate of prothrombin activation by activated Factor X alone is inadequate to meet physiological needs; an additional protein, proaccelerin or Factor V, is also required. In addition to prothrombin, Factors VII, IX, and X contain γ-carboxyglutamate and hence are vitamin K-dependent, as are three further proteins: proteins C, S, and Z, which are anticoagulants. Protein C is a protease that hydrolyzes activated Factor V and activates clot lysis. It circulates as a zymogen and is activated by thrombin. Protein Z serves as the cofactor for the inhibition of activated Factor X by a protease inhibitor; in turn, the protease inhibitor–protein Z complex is hydrolyzed by activated Factors X and XI (Furie and Furie, 1988; Broze, 2001).

The initiation of the clotting process occurs on phospholipid surfaces, and the γ-carboxyglutamate residues in the various vitamin K-dependent clotting factors are essential for the calcium-dependent binding of the proteins to phospholipid. In the absence of calcium, the γ-carboxyglutamate residues are exposed, and the Gla domain has no affinity for lipid surfaces. After binding calcium ions, the protein undergoes a conformational change, so that the Gla residues are internalized and hydrophobic amino acids are exposed. These penetrate into phospholipid membranes and bind the protein to the cell surface.

5.3.3 Osteocalcin and Matrix Gla Protein

Treatment with warfarin or other anticoagulants during pregnancy can lead to bone abnormalities in the fetus, the so-called *fetal warfarin syndrome*, which is because of impaired synthesis of osteocalcin – a small calcium binding protein containing three γ-carboxyglutamate residues found in bone matrix and dentine. It also contains a hydroxyproline residue, and thus undergoes both vitamin K- and vitamin C-dependent posttranslational modifications (Section 13.4.3). It is the most abundant of the noncollagen proteins of bone matrix, accounting for 1% to 2% of total bone protein, or 15% of noncollagen bone protein. Osteocalcin synthesis is induced by physiological concentrations of calcitriol, and the release of osteocalcin into the circulation provides a sensitive marker of vitamin D action and metabolic bone disease (Section 3.5).

Osteocalcin is synthesized in the osteoblasts as a precursor protein that then undergoes γ-carboxylation of glutamate residues and cleavage of a peptide extension before secretion into the extracellular space, where it binds to hydroxyapatite. Osteocalcin binds to hydroxyapatite in bone and modifies the crystallization of calcium phosphates, retarding the conversion of brushite [$CaHPO_4.2H_2O$] to hydroxyapatite [$Ca_{10}(PO_4)_6(OH)_2$] and inhibiting the mineralization of bone. Osteocalcin knockout mice have a higher bone mineral density than normal. The absence of osteocalcin leads to increased bone formation without impairing resorption (Ducy et al., 1996).

The matrix Gla protein, which was originally described in bone, where it binds to both organic and mineral components, is found in a variety of tissues. It acts to prevent mineralization of cartilage and other connective tissues. Matrix Gla protein knockout mice develop a variety of abnormalities within 2 weeks of birth and die within 2 months from rupture of the aorta, associated with extensive calcification of blood vessels (Luo et al., 1997). Matrix Gla protein (atherocalcin) is also found in atherosclerotic plaque.

Nephrocalcin in the kidneys has considerable homology with matrix Gla protein. It is probably involved in renal reabsorption of calcium, but also acts to solubilize calcium salts in the urine. It is found in calcium oxalate renal stones.

5.3.4 Vitamin K-Dependent Proteins in Cell Signaling – Gas6

The fetal warfarin syndrome involves neurological as well as bone abnormalities. Vitamin K-dependent carboxylase is expressed in different brain regions at different times during embryological development.

The product of the growth arrest-specific gene 6 (Gas6) is a γ-carboxyglutamate containing growth factor that is important in the regulation of growth and development. The γ-carboxyglutamate region of Gas6 is required for binding to phosphatidylserine in cell membranes before interacting with a receptor tyrosine kinase, leading to the induction of mitogen-activated protein kinase. Phosphatidylserine is normally deep in the membrane phospholipid bilayer, but it is exposed at the cell surface in senescent red blood cells and apoptotic cells, suggesting that Gas6 has a role in the recognition of cells that are to undergo phagocytosis, and hence regulation of apoptosis and cell survival (Saxena et al., 2001).

5.4 VITAMIN K DEFICIENCY

Vitamin K deficiency results in prolonged prothrombin time, and eventually hemorrhagic disease, as a result of the impairment of synthesis of the vitamin K-dependent blood clotting proteins. Although osteocalcin synthesis is similarly impaired, the effects on blood clotting predominate, and effects of vitamin K deficiency on bone mineralization can only be demonstrated in experimental animals if they are transfused with preformed blood clotting factors. Otherwise, they suffer fatal hemorrhage before there is any detectable effect on osteocalcin and bone metabolism. However, there is some evidence that undercarboxylated osteocalcin is formed in subjects with marginal intakes of vitamin K with no evidence of impaired blood clotting (Binkley et al., 2000)

The coumarin anticoagulants act as vitamin K antimetabolites, inhibiting vitamin K quinone reductase and epoxide reductase, and hence cause functional vitamin K deficiency. As discussed in Section 5.3.1, the inhibition of vitamin K quinone reductase does not prevent the formation of the hydroquinone for epoxidase activity, because a warfarin-insensitive enzyme also catalyzes the reaction. Provision of high intakes of vitamin K will overcome the inhibition of glutamate carboxylation caused by anticoagulants by permitting more or less stoichiometric utilization of the vitamin, with excretion of metabolites of the epoxide. This may cause problems with patients receiving anticoagulant therapy who take supplements of vitamin K. High dietary intakes of vitamin K, rather than supplements, may also affect anticoagulant action. Although daily consumption of 250 g of a vitamin K-rich vegetable, such as spinach or broccoli, is required to affect prothrombin time, some of the vegetable oils in common use have a high vitamin K content (Karlson et al., 1986; Bolton-Smith et al., 2000).

The response of deficient animals to repletion with vitamin K, or the administration of the vitamin after anticoagulants, is more rapid than might be expected. There is a considerable intracellular accumulation of preprothrombin, which is immediately available for carboxylation when vitamin K becomes available.

5.4.1 Vitamin K Deficiency Bleeding in Infancy

Newborn infants present a special problem with respect to vitamin K. They have low plasma levels of prothrombin and the other vitamin K-dependent clotting factors (about 30% to 60% of the adult concentrations, depending on gestational age). To a great extent, this is the result of the relatively late development of liver glutamate carboxylase, but they are also short of vitamin K, as a result of the placental barrier that limits fetal uptake of the vitamin. This is probably a way of regulating the activity of Gas6 and other vitamin K-dependent proteins in development and differentiation (Section 5.3.4; Israels et al., 1997).

Over the first 6 weeks of postnatal life, the plasma concentrations of clotting factors gradually rise to the adult level; in the meantime, they are at risk of potentially fatal hemorrhage that was formerly called hemorrhagic disease of the newborn and is now known as vitamin K deficiency bleeding in infancy. It is usual to give all newborn infants prophylactic vitamin K, either orally or by intramuscular injection (Sutor et al., 1999). At one time, menadione was used, but, because of the association between menadione and childhood leukemia (Section 5.6.1), phylloquinone is preferred.

5.5 ASSESSMENT OF VITAMIN K NUTRITIONAL STATUS

The usual method of assessing vitamin K nutritional status, or monitoring the efficacy of anticoagulant therapy, is a functional test of blood clotting, and hence the ability to synthesize the vitamin K-dependent clotting factors.

The standard assay measures the time taken for the formation of a fibrin clot in citrated plasma after the addition of calcium ions and thromboplastin to activate the extrinsic clotting system – the prothrombin time. The normal prothrombin time is 11 to 13 seconds; greater than 25 seconds is associated with severe bleeding.

Measurement of plasma concentrations of preprothrombin permits a more sensitive means of detecting marginally inadequate vitamin K status than simple determination of prothrombin time. Preprothrombin is not activated by thromboplastin, although it is a substrate for the protease from snake venom, which does not require phospholipid binding of the substrate for activity. When so activated, descarboxy-thrombin will catalyze clot formation from fibrinogen. This provides a means of determining the relative amounts of prothrombin and preprothrombin in blood samples. If snake venom protease is used instead of thromboplastin, the prothrombin time will be shorter, depending on how much preprothrombin is present. In normal subjects, the ratio of prothrombin time using thromboplastin and that using snake venom protease is >0.6, whereas in vitamin K-deficient or anticoagulant-treated subjects, it is lower.

Preprothrombin can be determined immunologically, either using antiprothrombin antibodies, after adsorption of the γ-carboxylated protein onto barium carbonate or using anti-preprothrombin antibodies that do not cross-react with prothrombin. Circulating concentrations of preprothrombin in vitamin K deficiency are of the order of 150 to 1,500 nmol per L. If elevated preprothrombin is because of vitamin K deficiency, then it will fall on administration of the vitamin, whereas if it is the result of liver disease, then vitamin K supplements will have no effect.

In deficiency, there is also undercarboxylated osteocalcin in the circulation, and this provides a more sensitive index of marginal status; it is detectable, and responds to supplements of vitamin K, in subjects with normal clotting time and no circulating preprothrombin (Sokoll and Sadowski, 1996; Binkley et al., 2000).

Measurement of the plasma concentration of phylloquinone gives some information about status, but reflects not only intake but also plasma triacylglycerol, because most is carried in chylomicrons and chylomicron remnants. The plasma concentration of phylloquinone is higher in older subjects, but the phylloquinone:triacylglycerol ratio is lower than in younger people (Booth and Suttie, 1998).

The urinary excretion of γ-carboxyglutamate, as both the free amino acid and in small peptides, also reflects functional vitamin K status, because γ-carboxyglutamate released by the catabolism of proteins is neither reutilized nor metabolized. The normal range of γ-carboxyglutamate excretion is 0.2 to 0.6 μmol per mol of creatinine in adults. Children excrete more, presumably reflecting greater turnover of osteocalcin. In patients receiving anticoagulants, the urinary excretion of γ-carboxyglutamate falls to half as the prothrombin time increases two- to three-fold (Suttie et al., 1988).

5.6 VITAMIN K REQUIREMENTS AND REFERENCE INTAKES

The determination of vitamin K requirements is complicated by the intestinal bacterial synthesis of menaquinones and the extent to which these are absorbed and utilized (Section 5.1). Prolonged use of antibiotics leads to impaired blood clotting, but simple dietary restriction of vitamin K results in prolonged prothrombin time and increased circulating preprothrombin; so it is apparent that bacterial synthesis is inadequate to meet requirements in the absence of a dietary intake of phylloquinone. Preprothrombin is elevated at intakes between 40 to 60 μg per day, but not at intakes above 80 μg per day (Suttie et al., 1988).

The total body pool of vitamin K is 150 to 200 nmol (70 to 100 μg); the half-life of phylloquinone is 17 hours, suggesting a requirement for replacement of 50 to 70 μg per day. In subjects treated with neomycin to sterilize the gastrointestinal tract, a daily intake of 0.4 g of phylloquinone is adequate to maintain normal blood clotting. On the basis of preventive studies (as opposed to earlier curative studies in deficient patients), Olson (1987b) suggested a mean daily requirement of 0.4 μg per kg of body weight, and a Recommended Daily Amount based on 0.56 μg per kg of body weight. Depletion/repletion studies suggest a requirement of 0.5 to 1 mg of phylloquinone (Suttie et al., 1988).

Most reference intakes are based on 0.5 to 1 μg phylloquinone per kg of body weight (see Table 5.1). However, the U.S./Canadian Adequate Intake (Institute of Medicine, 2001) is 120 μg for men and 90 μg for women, based on observed intakes. There is some evidence that average intakes may, in fact, be inadequate to permit full carboxylation of osteocalcin; Binkley and coworkers (2000) showed that giving supplements to normal healthy subjects reduced the circulating concentration of undercarboxylated osteocalcin.

5.6.1 Upper Levels of Intake

Even large intakes of phylloquinone have no apparent toxic effects, although they may be dangerous in patients receiving anticoagulant therapy (Section 5.4).

Menadione and its water-soluble derivatives are potentially toxic in excess and have been reported to cause hemolytic anemia, hyperbilirubinemia, central nervous system toxicity, and methemoglobinemia in the newborn.

Table 5.1 Reference Intakes of Vitamin K
(μg/day)

Age	U.S./Canada 2001	FAO 2001
0–6 m	2	5
7–12 m	2.5	10
1–3 y	30	15
4–16 y	55	20
7–8 y	55	25
Males		
9–10 y	60	25
11–13 y	60	35–55
14–18 y	75	35–55
>19 y	120	65
Females		
9–10 y	60	25
11–13 y	60	35–55
14–18 y	75	35–55
>19 y	90	55
Pregnant	90	55
Lactating	90	55

FAO, Food and Agriculture Organization; WHO, World Health Organization.
Sources: Institute of Medicine, 2001; FAO/WHO, 2001.

Menadione prophylaxis against vitamin K deficiency bleeding in infancy (Section 5.4.1) has been associated with increased incidence of childhood leukemia and other cancers.

5.6.2 Pharmacological Uses of Vitamin K

The pharmacological uses of vitamin K are in the treatment of (rare) vitamin K deficiency, prophylaxis in newborn infants (Section 5.4.1), and as an antidote to overcome anticoagulant toxicity. Because of its cytoxic effects, menadione has been used in cancer chemotherapy. It also potentiates the analgesic actions of salicylates and opiates, and has been used to enhance pain relief in cancer patients.

It has been suggested that supplements may be beneficial in the healing of bone fractures. There is some evidence that patients with osteoporosis (Section 3.4.3) have low circulating concentrations of vitamin K, and some evidence that supplements of vitamin K may retard the progression of osteoporosis (Vermeer et al., 1995, 1996; Feskanich et al., 1999).

FURTHER READING

Berkner KL (2000) The vitamin K-dependent carboxylase. *Journal of Nutrition* **130,** 1877–80.

Cadenas E, Hochstein P, and Ernster L (1992) Pro- and antioxidant functions of quinones and quinone reductases in mammalian cells. *Advances in Enzymology and Related Areas of Molecular Biology* **65,** 97–146.

Dowd P, Ham SW, Naganathan S, and Hershline R (1995) The mechanism of action of vitamin K. *Annual Reviews of Nutrition* **15,** 419–40.

Furie B and Furie BC (1988) The molecular basis of blood coagulation. *Cell* **53,** 505–18.

Lian JB, Stein GS, Stein JL, and van Wijnen AJ (1999) Regulated expression of the bone-specific osteocalcin gene by vitamins and hormones. *Vitamins and Hormones* **55,** 443–509.

Meganathan R (2001) Biosynthesis of menaquinone (vitamin K_2) and ubiquinone (coenzyme Q): a perspective on enzymatic mechanisms. *Vitamins and Hormones* **61,** 173–218.

Nelsestuen GL, Shah AM, and Harvey SB (2000) Vitamin K-dependent proteins. *Vitamins and Hormones* **58,** 355–89.

Olson RE (1984) The function and metabolism of vitamin K. *Annual Reviews of Nutrition* **4,** 281–337.

Price PA (1988) Role of vitamin-K-dependent proteins in bone metabolism. *Annual Reviews of Nutrition* **8,** 565–83.

Shearer MJ (1997) The roles of vitamins D and K in bone health and osteoporosis prevention. *Proceedings of the Nutrition Society* **56,** 915–937.

Shearer MJ, Bach A, and Kohlmeier M (1996) Chemistry, nutritional sources, tissue distribution and metabolism of vitamin K with special reference to bone health. *Journal of Nutrition* **126,** 1181S–6S.

Suttie JW (1988) Vitamin K-dependent carboxylation of glutamyl residues in proteins. *Biofactors* **1,** 55–60.

Suttie JW (1995) The importance of menaquinones in human nutrition. *Annual Reviews of Nutrition* **15,** 399–417.

Suttie JW and Jackson CM (1977) Prothrombin structure, activation, and biosynthesis. *Physiological Reviews* **57,** 1–70.

Vermeer C, Jie KS, and Knapen MH (1995) Role of vitamin K in bone metabolism. *Annual Reviews of Nutrition* **15,** 1–22.

Weber P (2001) Vitamin K and bone health. *Nutrition* **17,** 880–7.

References cited in the text are listed in the Bibliography.

SIX

Vitamin B₁ – Thiamin

The peripheral nervous system disease, beriberi, caused by thiamin deficiency, has been known sporadically for nearly 1,300 years; it became a major problem of public health in the Far East in the nineteenth century with the introduction of the steam-powered rice mill, which resulted in more widespread consumption of highly milled (polished) rice. Thiamin was discovered as the factor in the discarded polishings that protected against the disease.

Although now largely eradicated, beriberi remains a problem in some parts of the world among people whose diet is especially high in carbohydrates. A different condition, affecting the central rather than peripheral nervous system, the Wernicke–Korsakoff syndrome, is also due to thiamin deficiency. It occurs in developed countries, especially among alcoholics and narcotic addicts.

Thiamin was the first of the vitamins to be demonstrated to have a clearly defined metabolic function as a coenzyme; indeed, the studies of Peters' group in the 1920s and 1930s laid the foundations not only of nutritional biochemistry, but also of modern metabolic biochemistry and neurochemistry. Despite this, the mechanism by which thiamin deficiency results in central and peripheral nervous system lesions remains unclear; in addition to its established coenzyme role, thiamin regulates the activity of a chloride transporter in nerve cells.

6.1 THIAMIN VITAMERS AND ANTAGONISTS

As shown in Figure 6.1, thiamin consists of pyrimidine and thiazole rings, linked by a methylene bridge; the alcohol group of the side chain can be esterified with one, two, or three phosphates, yielding thiamin monophosphate, thiamin diphosphate (also known as thiamin pyrophosphate, the metabolically active coenzyme), and thiamin triphosphate. The vitamin was originally named *aneurine*, the antineuritic vitamin, because of its function in preventing or

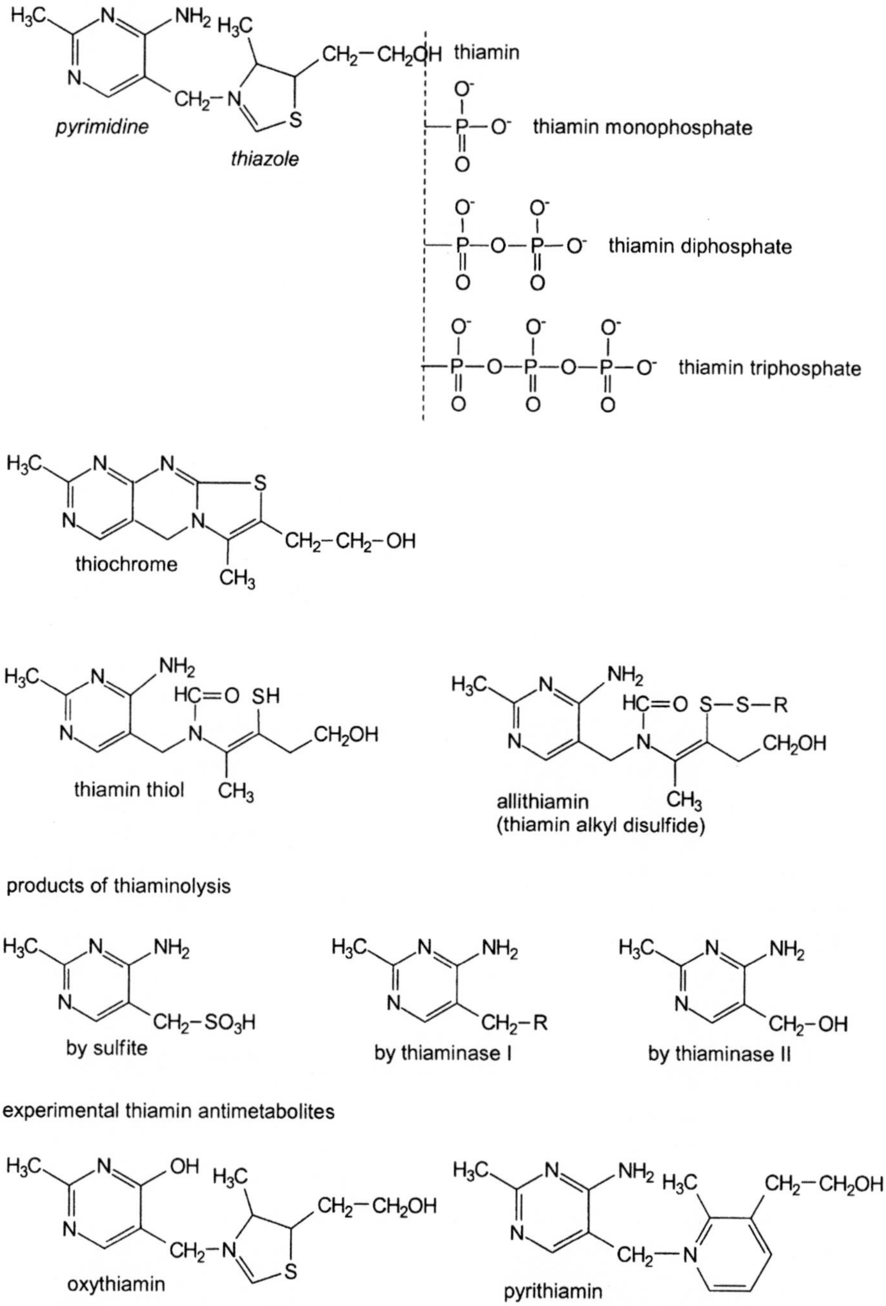

Figure 6.1. Thiamin and thiamin analogs, products of thiaminolysis, and experimental antimetabolites. Relative molecular masses (M_r): thiamin, 266.4 (chloridehydrochloride, 337.3); thiamin monophosphate, 345.3; thiamin diphosphate, 425.3; thiamin triphosphate, 505.3; thiochrome, 262.3; thiamin thiol, 282.4 (oxidizes to thiamin disulfide, 562.7); oxythiamin, 301.8; and pyrithiamin, 420.2.

curing polyneuritis in deficient animals. When its chemistry was discovered, it was called thiamine, because of the presence of both sulfur and an amino group in the molecule. The final -e has been dropped; the amino group is not involved in the metabolic role of the vitamin.

The free base is unstable, and two derivatives of thiamin are commonly used in food enrichment and pharmaceutical preparations: thiamin chloride hydrochloride (generally known simply as thiamin hydrochloride) and thiamin mononitrate. The mononitrate is less hygroscopic than the chloride hydrochloride and is the preferred form for food enrichment. There is considerable difficulty in interpreting much of the literature on thiamin requirements, because many authors quote mg of thiamin, without specifying whether it was as the free base, the chloride hydrochloride, or the equivalent amount of free base. Because the M_r of free thiamin is 266.4, and that of the chloride hydrochloride is 337.3, this confusion can result in errors of the order of 26%.

Oxidative cleavage of the thiazole ring occurs in alkaline solution, forming a reactive sulfhydryl group (thiamin thiol) that can react with other thiols, forming thiamin alkyl disulfides – allithiamins. A number of allithiamins occur in plants (especially members of the genus *Allium*). They are biologically active; on reductive cleavage of the disulfide bridge, they spontaneously dehydrate to yield thiamin. However, they do not undergo alkali-catalyzed ring closure to thiochrome, which is the basis of the most commonly used method for determining thiamin chemically, so may be overlooked in chemical analysis.

Synthetic allithiamin derivatives, such as thiamin propyl and tetrahydrofurfuryl disulfides, have been used for the prevention and treatment of thiamin deficiency. Because they are lipid soluble, and are not subject to the normal control of thiamin absorption by saturation of the intestinal transport system, they have potential benefits in the treatment of thiamin-deficient alcoholics, whose absorption of thiamin is impaired.

Thiamin is labile to sulfite, which cleaves the methylene bridge. The reaction is slow at acid pH, but rapid above pH 6. Sulfite treatment of dried fruit and other foods results in more or less complete loss of thiamin.

Two analogs of thiamin, oxythiamin and pyrithiamin, are potent antimetabolites and have been widely used to induce thiamin deficiency in experimental animals. The mechanisms of action of pyrithiamin, oxythiamin, and thiaminases found in foods are discussed in Section 6.4.7.

6.2 METABOLISM OF THIAMIN

Dietary thiamin phosphates are hydrolyzed by intestinal phosphatases, and the resultant free thiamin is absorbed by active transport in the duodenum and proximal jejunum, with little absorption in the rest of the small intestine.

Thiamin active transport is sodium-independent and requires an outwardly directed proton gradient (i.e., it is dependent on a proton-pumping ATPase). Antimetabolites, such as pyrithiamin, compete with thiamin for active transport. There is a similar proton-dependent thiamin uptake system in renal tubules (Rindi and Laforenza, 2000; Dudeja et al., 2001).

The transport system is saturated at relatively low concentrations of thiamin (about 2 μmol per L), thus limiting the amount of thiamin that can be absorbed. As a result, increasing test doses of thiamin from 2.5 to 20 mg have only a negligible effect on the plasma concentration of thiamin or on urinary excretion. By contrast, the absorption of lipid-soluble allithiamin derivatives is not apparently saturable, and they can be used to achieve high blood concentrations of thiamin.

Some thiamin is phosphorylated to thiamin monophosphate in the intestinal mucosa, although this is not essential for uptake, and isolated membrane vesicles will accumulate free thiamin against a concentration gradient. Thiamin does not accumulate in the mucosal cells; there is sodium-dependent active transport across the basolateral membrane, so that the mucosal concentration of thiamin is lower than that in the serosal fluid (Rindi et al., 1984; Rindi and Laforenza, 2000; Dudeja et al., 2001).

The absorption of thiamin is impaired in alcoholics, leading to thiamin deficiency (Section 6.4.4). In vitro, tissue preparations show normal uptake of the vitamin into the mucosal cells in the presence of ethanol, but impaired transport to the serosal compartment. The sodium-potassium–dependent ATPase of the basolateral membrane responsible for the active efflux of thiamin into the serosal fluid is inhibited by ethanol (Hoyumpa et al., 1977).

Both free thiamin and thiamin monophosphate circulate in plasma; about 60% of the total is the monophosphate. Under normal conditions, most is bound to albumin; when the albumin binding capacity is saturated, the excess is rapidly filtered at the glomerulus and excreted in the urine. Although a significant amount of newly absorbed thiamin is phosphorylated in the liver, all tissues can take up both thiamin and thiamin monophosphate, and are able to phosphorylate them to thiamin diphosphate and thiamin triphosphate. In most tissues, it is free thiamin that is the immediate precursor of thiamin diphosphate, which is formed by a pyrophosphokinase; both the β- and γ-phosphates of ATP are incorporated. Thiamin monophosphate arises mainly as a result of sequential hydrolysis of thiamin triphosphate and thiamin diphosphate.

Both thiamin monophosphate and free thiamin are found in cerebrospinal fluid. Uptake of thiamin monophosphate into cells in the central nervous system involves extracellular hydrolysis to free thiamin, probably catalyzed by membrane-bound phosphatases (Reggiani et al., 1984; Rindi et al., 1984; Patrini et al., 1988).

Genetic defects of the tissue thiamin transport protein and thiamin pyrophosphokinase cause megaloblastic anemia, presumably as a result of impaired synthesis of pentoses for DNA synthesis from low activity of transketolase (Section 6.3.2). In many cases, this megaloblastic anemia is thiamin-responsive, suggesting that the defect must be because of low affinity of either the transport protein or thiamin pyrophosphokinase for its substrate (Neufeld et al., 2001).

Two percent to 3% of the thiamin in nervous tissue is present as the triphosphate, which also occurs in significant amounts in skeletal muscle, especially in fast-twitch muscle fibers. In the nervous system, the triphosphate is found exclusively in the membrane fraction; muscle thiamin triphosphate is mainly cytosolic. There are two pathways for formation of thiamin triphosphate from the diphosphate:

1. Phosphorylation by ATP, catalyzed by thiamin diphosphate kinase, which acts only on protein-bound thiamin diphosphate; and
2. Phosphorylation by ADP, catalyzed by adenylate kinase – this enzyme is especially important in the rapid synthesis and turnover of thiamin triphosphate in slow-twitch white muscle fibers.

In both muscle and the central nervous system, there is an active thiamin triphosphatase, so that tissue concentrations of thiamin triphosphate are strictly regulated (Nishino et al., 1983; Miyoshi et al., 1990; Lakaye et al., 2002).

Thiamin that is not bound to plasma proteins is rapidly filtered at the glomerulus. Diuresis increases the excretion of the vitamin, and patients who are treated with diuretics are potentially at risk of thiamin deficiency. Some of the diuretics used in the treatment of hypertension may also inhibit cardiac (and other tissue) uptake of thiamin, thus further impairing thiamin status, which may be a factor in the etiology of heart failure (Suter and Vetter, 2000).

A small amount of thiamin is excreted in the urine unchanged, accounting for about 3% of a test dose, together with small amounts of thiamin monophosphate and thiamin diphosphate. As discussed in Section 6.5.1, this can be used to assess thiamin nutritional status. One of the major excretory products is thiochrome; cyclization to thiochrome is the basis of the normal method of determining thiamin; so, most reports of thiamin excretion are actually of thiamin plus thiochrome. In addition, small amounts of thiamin disulfide, formed by the oxidation of thiamin thiol, are also excreted.

Sweat may contain up to 30 to 60 nmol of thiamin per L. In very hot conditions, this may represent a significant loss of the vitamin.

The biological half-life of thiamin is 10 to 20 days, and deficiency signs can develop rapidly during depletion.

6.2.1 Biosynthesis of Thiamin

There are differences in the pathways of thiamin biosynthesis between prokaryotes and eukaryotes, and also between organisms that are aerobes and facultative anaerobes. Some organisms are completely autotrophic for thiamin, whereas others require the presence of either preformed pyrimidine or thiazole in the culture medium, and indeed some require both.

The pyrimidine moiety is not synthesized by the usual pathway for pyrimidine synthesis, but arises from amino-imidazole ribonucleotide, an intermediate in purine synthesis. Bacterial mutants with a single point mutation in the purine biosynthetic pathway show requirements for both thiamin and preformed purines. High concentrations of adenosine inhibit the synthesis of the hydroxymethylpyrimidine moiety of thiamin. The amino-imidazole ribonucleotide undergoes ring cleavage and the insertion of a two-carbon unit to yield the six-membered pyrimidine ring and a hydroxymethyl group, which becomes the methylene bridge of thiamin.

The thiazole ring is synthesized from a pentulose or deoxypentulose 5-phosphate and either glycine or tryptophan, depending on the organism. Incorporation of sulfur leads to formation of hydroxymethyl thiazole, which is then phosphorylated. The sulfur comes from cysteine and is incorporated by formation of a thiocarboxylate at the carboxyl terminal of the enzyme, unlike biotin synthesis (Section 11.1.1), where an iron-sulfur cluster at the active site of the enzyme is the donor.

The final step in the synthesis of thiamin involves condensation between the hydroxymethylpyrimidine diphosphate and hydroxyethylthiazole monophosphate to yield thiamin monophosphate, which can then be dephosphorylated to free thiamin or phosphorylated further to the diphosphate and triphosphate (Hohmann and Meacock, 1998; Begley et al., 1999).

6.3 METABOLIC FUNCTIONS OF THIAMIN

The studies of Peters in the 1920s and 1930s (Peters, 1963) established the coenzyme role of thiamin in the oxidative decarboxylation of pyruvate. Thiamin diphosphate is the coenzyme for three multienzyme complexes in mammalian mitochondria that are involved in the oxidative decarboxylation of oxoacids: pyruvate dehydrogenase and 2-oxoglutarate dehydrogenase in central

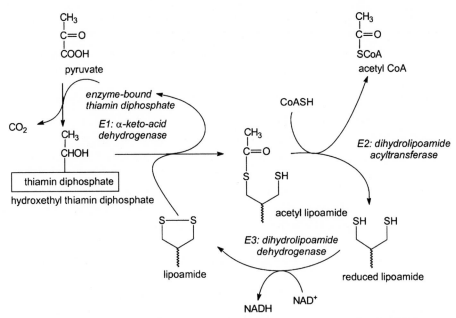

Figure 6.2. Reaction of the pyruvate dehydrogenase (EC 1.2.4.1) complex. CoASH, free coenzyme A.

energy-yielding metabolic pathways, and the branched-chain oxo-acid dehydrogenase in the catabolism of leucine, isoleucine, and valine.

Later studies established the coenzyme role of thiamin diphosphate in transketolase in the pentose phosphate pathway. More recent studies have shown that thiamin triphosphate acts to regulate a chloride channel in nerve tissue.

6.3.1 Thiamin Diphosphate in the Oxidative Decarboxylation of Oxo-acids

The reaction of the pyruvate dehydrogenase complex is shown in Figure 6.2; the reactions of the 2-oxoglutarate and branched-chain oxo-acid dehydrogenase complexes follow the same sequence, and the multienzyme complexes are similar.

Thiamin binds the oxo-acid substrate, decarboxylating it to an active aldehyde intermediate. This is then transferred to enzyme-bound lipoamide, reducing the disulfide bridge of the lipoamide and forming a thioester. The resultant acyl group is transferred to coenzyme A (CoA), and the dithiol lipoamide is reoxidized by NAD^+.

The multienzyme complexes are self-assembling and will reassemble to an active complex after resolution of the individual enzymes. The core enzyme of the complex is the dihydrolipoyl acyltransferase (E2); the oxo-acid dehydrogenase (E1) and dihydrolipoyl dehydrogenase (E3) subunits form noncovalent bonds to this central catalytic unit.

The pyruvate dehydrogenase E1 subunit is a tetramer of two classes of subunits. The α-subunit catalyzes the decarboxylation of pyruvate to hydroxyethylthiamine diphosphate and the β-subunit catalyzes the reductive acylation of the lipoamide prosthetic group of the acetyltransferase subunit. In the complete multienzyme complex, there are 20 (kidney) or 30 (heart) dehydrogenase tetramers, 60 acetyltransferase subunits, 10 (kidney) or 12 (heart) dihydrolipoyl dehydrogenase subunits, and 5 each of the kinase and phosphatase regulatory subunits.

6.3.1.1 Regulation of Pyruvate Dehydrogenase Activity

Pyruvate dehydrogenase is the key enzyme that commits pyruvate (and hence the products of carbohydrate metabolism) to complete oxidation (via the tricarboxylic acid cycle) or lipogenesis. It is subject to regulation by both product inhibition and a phosphorylation/dephosphorylation mechanism. Acetyl CoA and NADH are both inhibitors, competing with coenzyme A and NAD^+.

There are four isoenzymes of the kinase. Pyruvate dehydrogenase is inhibited by phosphorylation of three serine residues on the E1α subunit; this reduces the affinity for thiamin diphosphate 12-fold. Kinase 1 phosphorylates all three sites, whereas the other isoenzymes phosphorylate only two sites, and kinase 2 may catalyze only partial phosphorylation of one of these sites. Binding of thiamin diphosphate to E1α decreases the phosphorylation of the enzyme by the kinases. Pyruvate dehydrogenase is reactivated by dephosphorylation, catalyzed by two phosphatases, which have different activity on the enzyme depending on which kinase has catalyzed the phosphorylation. This means that regulation of pyruvate dehydrogenase activity will differ in different tissues, depending on the tissue-specific expression of the kinases and phosphatases (Kolobova et al., 2001; Korotchkina and Patel, 2001a, 2001b).

The kinases are inhibited by pyruvate and adenosine disphosphate (ADP), and the phosphatases are activated by calcium ions. There is normally a constant process of phosphorylation and dephosphorylation of the enzyme, so that it is very sensitive to changes in intracellular free calcium and the adenosine triphosphate (ATP):ADP ratio.

Pyruvate dehydrogenase kinases are induced by glucocorticoid hormones and long-chain fatty acids (acting via the peroxisome proliferation-activated receptor). This effect is antagonized by insulin, thus reducing pyruvate dehydrogenase activity in starvation and suggesting abnormalities of regulation of pyruvate metabolism in diabetes mellitus (Sugden et al., 2001a, 2001b; Huang et al., 2002).

In addition to its cofactor role, thiamin diphosphate, together with calcium or other divalent cations, activates pyruvate dehydrogenase by binding to a regulatory site and reducing the K_m for pyruvate (Czerniecki and Czygier, 2001).

6.3.1.2 Thiamin-Responsive Pyruvate Dehydrogenase Deficiency Genetic deficiency of pyruvate dehydrogenase E1α (which is on the X chromosome) leads to potentially fatal lactic acidosis, with psychomotor retardation, central nervous system damage, atrophy of muscle fibers and ataxia, and developmental delay. At least some cases respond to the administration of high doses (20 to 3,000 mg per day) of thiamin. In those cases where the enzyme has been studied, there is a considerable increase in the K_m of the enzyme for thiamin diphosphate. Female carriers of this X-linked disease are affected to a variable extent, depending on the X-chromosome inactivation pattern in different tissues (Robinson et al., 1996).

6.3.1.3 2-Oxoglutarate Dehydrogenase and the γ-Aminobutyric Acid (GABA) Shunt 2-Oxoglutarate dehydrogenase catalyzes the oxidative decarboxylation of 2-oxoglutarate to succinyl CoA in the citric acid cycle. There is considerably less impairment of citric acid cycle activity (and hence ATP formation) in thiamin deficiency than might be expected, and unlike pyruvate, 2-oxoglutarate does not accumulate in the brains of thiamin-deficient animals. There is a significant reduction in brain 2-oxoglutarate (Butterworth and Heroux, 1989).

As shown in Figure 6.3, the formation and catabolism of the neurotransmitter γ-aminobutyric acid (GABA) provides an alternative to 2-oxoglutarate dehydrogenase – the so-called GABA shunt. 2-Oxoglutarate is aminated to glutamate by the reaction of either glutamate dehydrogenase or a variety of transaminases; glutamate then undergoes decarboxylation to GABA. GABA is inactivated by transamination (in which 2-oxoglutarate is the amino acceptor), yielding succinic semialdehyde, which is oxidized to succinate, an intermediate in the tricarboxylic acid cycle.

Glutamate decarboxylase and GABA aminotransferase are found in regions of the central nervous system other than those in which GABA has a

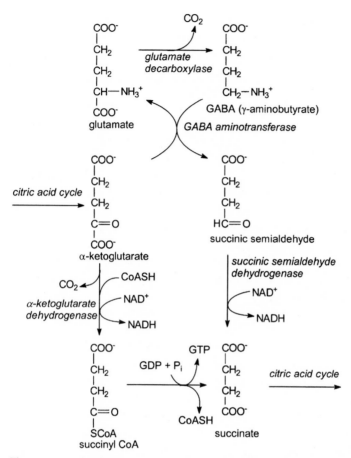

Figure 6.3. GABA shunt as an alternative to α-ketoglutarate dehydrogenase in the citric acid cycle. 2-Oxoglutarate dehydrogenase, EC 1.2.4.2; glutamate decarboxylase, EC 4.1.1.15; GABA aminotransferase, EC 2.6.1.19; and succinic semialdehyde dehydrogenase, EC1.2.1.16.

neurotransmitter role, and also in nonneuronal tissues, including the liver and kidneys. In a number of studies, there is evidence of a decrease in the total brain concentration of GABA in thiamin deficiency, but studies with [^{14}C]glutamate show that there is a considerable increase in the rate of GABA turnover, and suggest that the GABA shunt may be a significant alternative to 2-oxoglutarate dehydrogenase in energy-yielding metabolism in thiamin deficiency, permitting continued tricarboxylic acid cycle activity despite the impairment of 2-oxoglutarate dehydrogenase (Page et al., 1989).

6.3.1.4 Branched-Chain Oxo-acid Decarboxylase and Maple Syrup Urine Disease The third oxo-acid dehydrogenase catalyzes the oxidative decarboxylation of the branched-chain oxo-acids that arise from the transamination of the branched-chain amino acids, leucine, isoleucine, and valine. It has a similar subunit composition to pyruvate and 2-oxoglutarate dehydrogenases, and the E3 subunit (dihydrolipoyl dehydrogenase) is the same protein as in the other two multienzyme complexes. Genetic lack of this enzyme causes maple syrup urine disease, so-called because the branched-chain oxo-acids that are excreted in the urine have a smell reminiscent of maple syrup.

Like the other thiamin diphosphate-dependent dehydrogenases, branched-chain oxo-acid dehydrogenase is regulated by phosphorylation and dephosphorylation, and the proportion of the enzyme in the active (dephosphorylated) state is high in the liver, low in skeletal muscle, and intermediate in the kidneys and heart. In most tissues, the activity of the dehydrogenase is considerably lower than that of branched-chain amino acid aminotransferase, and regulation of the dehydrogenase is therefore important for control of branched-chain amino acid metabolism, and overall amino acid and nitrogen metabolism (Lombardo et al., 1999; Harris et al., 2001; Obayashi et al., 2001).

Various mutations affecting either the E1 or the E2 subunit of the dehydrogenase are involved in different forms of maple syrup urine disease. Acute infantile disease is caused by near complete lack of activity of the enzyme. The intermittent form of the disease is associated with marginally adequate residual activity of the enzyme that is able to cope with the branched-chain oxo-acids arising from the metabolism of modest amounts of branched-chain amino acids, but not relatively large amounts.

Some patients with maple syrup urine disease respond to high doses (10 to 1,000 mg per day) of thiamin; in some patients, the defect is clearly in the E1α subunit, which has a K_m for thiamin diphosphate 16-fold higher than normal. But in other thiamin-responsive cases, the mutation is in the E2 subunit, suggesting that assembly of the active multienzyme complex affects the affinity of the thiamin diphosphate binding site of the E1α subunit.

In vitro, thiamin diphosphate inhibits the kinase that phosphorylates and inactivates branched-chain oxo-acid dehydrogenase, and might be expected to increase the activity of the enzyme in tissues, thus offering an alternative mechanism for thiamin-responsive maple syrup urine disease. However, this seems not to be relevant in vivo, possibly because tissue concentrations of thiamin diphosphate do not rise high enough to affect the activity of the kinase. In thiamin-deficient animals, there is an increase in the total liver content of branched-chain oxo-acid dehydrogenase (suggesting induction of the enzyme) and an increase in the proportion in the dephosphorylated active form. Feeding animals with very high levels of thiamin results in a decrease in the total liver content of the dehydrogenase and a decrease in the proportion in the dephosphorylated state (Blair et al., 1999).

6.3.2 Transketolase

Transketolase is involved in the pentose phosphate pathway, which is the major pathway of carbohydrate metabolism in some tissues and a significant alternative to glycolysis in all tissues. The main importance of the pentose phosphate pathway is in the production of NADPH for use in biosynthetic reactions (and especially lipogenesis) and the de novo synthesis of ribose for nucleotide synthesis.

As shown in Figure 6.4, transketolase catalyzes the transfer of a two-carbon unit from a donor ketose onto an acceptor aldose sugar. The donor ketose forms a transient intermediate with thiamin diphosphate, which then undergoes cleavage to release an aldose two carbons smaller than the ketose substrate, leaving enzyme-bound dihydroxyethyl thiamin diphosphate. This reacts with an acceptor aldose to form a ketose two carbons larger.

Although the entry of glucose 6-phosphate into the pentose phosphate pathway is controlled by the need for NADPH and pentose sugars, transketolase has a high control strength (0.74) in the nonoxidative part of the pathway, and a proportion of the enzyme is normally present as the (inactive) apoenzyme. High intakes of thiamin, leading to more-or-less complete saturation of transketolase with its cofactor, might therefore disturb regulation of pentose metabolism (Berthon et al., 1992).

6.3.3 The Neuronal Function of Thiamin Triphosphate

Early studies showed that the development of neurological abnormalities in thiamin deficiency did not follow the same time course as the impairment of pyruvate and 2-oxoglutarate dehydrogenase or transketolase activities. The brain regions in which metabolic disturbances are most marked were not those that are vulnerable to anatomical lesions. These studies suggested a function for thiamin in the nervous system other than its coenzyme role.

Thiamin triphosphate is formed in brain and skeletal muscle by phosphorylation of thiamin diphosphate (Section 6.2), and its concentration is very precisely controlled, because there is also an active thiamin triphosphatase (Lakaye et al., 2002). In nervous tissue thiamin triphosphate is localized

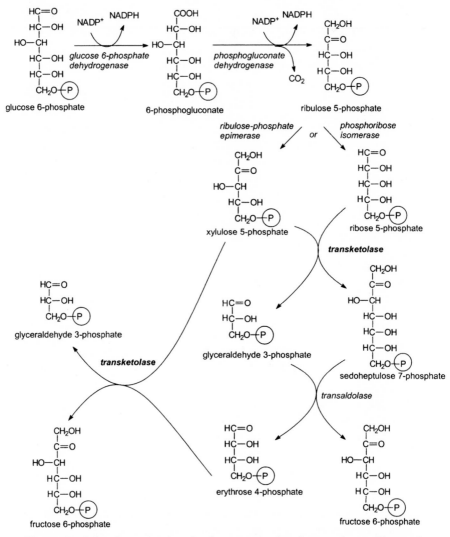

Figure 6.4. Role of transketolase in the pentose phosphate pathway. Glucose 6-phosphate dehydrogenase, EC 1.1.1.49; phosphogluconate dehydrogenase, EC 1.1.1.44; ribulose-phosphate epimerase, EC 5.1.3.1; phosphoribose isomerase, EC 5.3.1.6; transketolase, EC 2.2.1.1; and transaldolase, EC 2.2.1.2.

more-or-less completely in the membrane fraction, whereas in muscle it is mainly cytosolic.

Early studies showed that thiamin triphosphate had a role in electrical conduction in nerve cells; more recent studies have shown that it activates a chloride channel in the nerve membrane, acting as a phosphate donor (Bettendorff et al., 1994; Bettendorff, 1996). It also acts as a phosphate donor to other membrane proteins in nerve and synaptosome preparations (Nghiem et al., 2000).

6.4 THIAMIN DEFICIENCY

Thiamin deficiency can result in three distinct syndromes: a chronic peripheral neuritis, beriberi, which may or may not be associated with heart failure and edema; acute pernicious (fulminating) beriberi (shoshin beriberi), in which heart failure and metabolic abnormalities predominate, with little evidence of peripheral neuritis; and Wernicke's encephalopathy with Korsakoff's psychosis, a thiamin-responsive condition associated especially with alcoholism and narcotic abuse.

In general, a relatively acute deficiency is involved in the central nervous system lesions of the Wernicke–Korsakoff syndrome, and a high-energy intake, as in alcoholics, is also a predisposing factor. Dry beriberi is associated with a more prolonged, and presumably less severe, deficiency, with a generally low food intake, whereas higher carbohydrate intake and physical activity predispose to wet beriberi.

In experimental animals, thiamin deficiency is associated with severe anorexia. One of the problems in interpreting the literature on thiamin deficiency is distinguishing between effects of thiamin deficiency per se and effects of general lack of food and inanition. Even more than with other vitamins, studies of thiamin deficiency require strict pair-feeding of control animals with those receiving the deficient diet. The mechanism of anorexia is unclear. Its development shows a clear correlation with the loss of transketolase activity in the intestinal mucosa, but not the loss of pyruvate or 2-oxoglutarate dehydrogenase activity. Animals treated with oxythiamin, which does not cross the blood–brain barrier, and therefore has little effect on central nervous system metabolism, show anorexia. This suggests that the effect is on the intestinal mucosa rather than the central nervous system. In addition, it is possible that the changes in GABA and 5-hydroxytryptamine turnover in thiamin deficiency (Section 6.4.6) may be involved in the etiology of anorexia, because potentiation of GABA and 5-hydroxytryptamine activity is part of the action of a number of clinically used appetite suppressants.

6.4.1 Dry Beriberi

Chronic deficiency of thiamin, especially associated with a high carbohydrate diet, results in beriberi, which is a symmetrical ascending peripheral neuritis. Initially, the patient complains of weakness, stiffness, and cramps in the legs, and is unable to walk for more than a short distance. There may be numbness of the dorsum of the feet and ankles, and vibration sense may be diminished.

As the disease progresses, the ankle jerk reflex is lost, and the muscular weakness spreads upward, involving first the extensor muscles of the foot, then the muscles of the calf, and finally the extensors and flexors of the thigh. At this stage, there is pronounced toe and foot drop – the patient is unable to keep either the toe or the whole foot extended off the ground. When the arms are affected, there is a similar inability to keep the hand extended – wrist drop.

The affected muscles become tender, numb, and hyperesthetic. The hyperesthesia extends in the form of a band around the limb, the so-called stocking and glove distribution, and is followed by anesthesia. There is deep muscle pain and, in the terminal stages, when the patient is bedridden, even slight pressure (as from bed clothes), causes considerable pain.

In thiamin-deficient rats, electron microscopy of the sciatic and plantar nerves shows distally pronounced axonal degeneration, with an increase in the number of mitochondria and proliferation of vesicular elements of the endoplasmic reticulum. This is followed by disintegration of neurotubules and neurofilaments, and finally axonal shrinkage and myelin disruption (Pawlik et al., 1977).

6.4.2 Wet Beriberi

The heart may also be affected in beriberi, with dilatation of arterioles, rapid blood flow, and increased pulse rate and pressure, and increased jugular venous pressure leading to right-sided heart failure and edema (so-called wet beriberi).

The signs of chronic heart failure may be seen without peripheral neuritis. The arteriolar dilatation, and possibly also the edema, probably results from high circulating concentrations of lactate and pyruvate, a result of impaired activity of pyruvate dehydrogenase.

Together with the fall in pyruvate dehydrogenase, there is a fall in the concentration of ATP in the heart, although the ATP:ADP ratio in most tissues is not affected by thiamin deficiency (McCandless et al., 1970).

6.4.3 Acute Pernicious (Fulminating) Beriberi – Shoshin Beriberi

Heart failure without increased cardiac output, and no peripheral edema, may also occur acutely, associated with severe lactic acidosis. This was a common presentation of deficiency in Japan, where it was called shoshin (= acute) beriberi; in the 1920s, nearly 26,000 deaths a year were recorded.

With improved knowledge of the cause, and improved nutritional status, the disease has become more-or-less unknown, although it occurs among alcoholics, when the lactic acidosis may be life-threatening, without clear signs of heart failure. There have been a number of case reports among patients receiving total parenteral nutrition, when it may occur as early as 4 days after the start of parenteral nutrition in patients with initially low thiamin status (Campbell, 1984; Kitamura et al., 1996).

Acute infantile beriberi in infants breast-fed by deficient mothers may involve high-output cardiac failure, as in shoshin beriberi, as well as signs of central nervous system involvement similar to those seen in Wernicke's encephalopathy (Section 6.4.4).

6.4.4 The Wernicke–Korsakoff Syndrome

Although the classical signs of beriberi are of peripheral neuritis, most of the biochemical studies (Peters, 1963) were performed on the central nervous system of pigeons, because they show central nervous system abnormalities in thiamin deficiency and signs of peripheral neuritis. Although peripheral neuritis and acute cardiac beriberi and lactic acidosis occur in thiamin deficiency associated with alcohol abuse, the more usual presentation is as the Wernicke–Korsakoff syndrome caused by central nervous system lesions. There is some evidence that thiamin deficiency alone is not sufficient to cause the Wernicke–Korsakoff syndrome, but that alcohol is also a necessary factor (Homewood and Bond, 1999). However, although alcohol is neurotoxic and causes neuronal damage in the cerebral cortex, there is little evidence to support a separate classification of alcoholic dementia; most, if not all, of the organic brain damage associated with alcohol abuse can be considered to be from thiamin deficiency (Joyce, 1994).

Initially, there is a confused state, Korsakoff's psychosis, that is characterized by confabulation and loss of recent memory, although memory for past events may be unimpaired. Later, clear neurological signs develop – Wernicke's encephalopathy. This is characterized by nystagmus and extraocular palsy. Postmortem examination shows hemorrhagic lesions in the thalamus, pontine tegmentum, and mammillary body, with severe damage to astrocytes, neuronal dendrites, and myelin sheaths.

Wernicke's encephalopathy may be more common than is believed on clinical grounds. Harper (1979) reported that 1.7% of all postmortem examinations in Western Australia over a 4-year period showed clear anatomical evidence of the disease, yet only 13% of the patients had been diagnosed as suffering from the condition. Other studies have similarly shown that only 10% to 20% of cases confirmed by postmortem examination had been diagnosed on clinical grounds (Zubaran et al., 1997). There appears to have been a reduction in the prevalence of the Wernicke–Korsakoff syndrome in Australia after mandatory enrichment of flour with thiamin (Ma and Truswell, 1995).

The irreversible brain lesions are associated with decreased activity of 2-oxoglutarate dehydrogenase and increased activity of the GABA shunt (Section 6.3.1.3), leading to localized lactic acidosis and excitotoxic levels of glutamate, as well as localized increased permeability of the blood–brain barrier, evidence of radical activity, and inflammatory responses to radical action. In addition, the activity of kynurenine aminotransferase (Section 9.5.4) in glial cells is regulated by the availability of its oxo-acid substrates, pyruvate and 2-oxoglutarate, so that increased accumulation of these two metabolites will result in increased synthesis and release into synapses of kynurenic acid, which is an antagonist of both the N-methyl-D-aspartate glutamate receptor and acetylcholine receptors (Heroux and Butterworth, 1995; Langlais, 1995; Leong and Butterworth, 1996; McEntee, 1997; Hazell et al., 1998; Calingasan and Gibson, 2000).

A number of studies have suggested that there may be genetic polymorphism of transketolase (Section 6.3.2) and that some variants may be associated with increased susceptibility to the Wernicke–Korsakoff syndrome. Blass and Gibson (1977) showed that transketolase in cultured fibroblasts from patients with Wernicke–Korsakoff syndrome had a K_m for thiamin diphosphate 12-fold higher than that from control subjects. This difference persisted through serial passage in culture, suggesting it was a genetic rather than environmental effect. Nixon and coworkers (1984) demonstrated different patterns of multiple bands of transketolase on isoelectric focusing; 39 or their 42 patients with Wernicke–Korsakoff syndrome showed the same unusual pattern that was only seen in 8 of 36 control subjects. Wang and coworkers (1997) expressed human transketolase in *Escherichia coli* and showed that formation of the normal enzyme required a cytosolic factor derived from human cells that was absent in cultured cells from a patient with Wernicke–Korsakoff syndrome, in which the enzyme showed enhanced sensitivity to thiamin deficiency. However, from reviews of a number of studies, there is little evidence to support the hypothesis that susceptibility to the Wernicke–Korsakoff syndrome is a genetic defect (Blansjaar et al., 1991; Schenk et al., 1998).

6.4.5 Effects of Thiamin Deficiency on Carbohydrate Metabolism

The role of thiamin diphosphate in pyruvate dehydrogenase means that, in deficiency, there is impaired conversion of pyruvate to acetyl CoA, and hence impaired entry of pyruvate into the citric acid cycle. Especially in subjects on a relatively high carbohydrate diet, this results in increased plasma concentrations of lactate and pyruvate, which may lead to life-threatening lactic acidosis.

The increase in plasma lactate and pyruvate after a test dose of glucose was used historically as a means of assessing thiamin nutritional status (Section 6.5).

In addition to the potential to maintain citric acid cycle activity by way of the GABA shunt (Section 6.3.1.3), the activity of 2-oxoglutarate dehydrogenase is less impaired in thiamin deficiency than the activities of pyruvate dehydrogenase and transketolase. Synthesis of the apoenzymes of pyruvate dehydrogenase and transketolase is reduced, whereas there is no change in the expression of 2-oxoglutarate dehydrogenase, suggesting a potential role for thiamin or a metabolite in regulation of the expression of genes for thiamin dependent enzymes (Pekovich et al., 1996, 1998).

6.4.6 Effects of Thiamin Deficiency on Neurotransmitters
As noted in Section 6.3.1.3, brain GABA falls in thiamin deficiency, but there is increased flux through the GABA shunt. The changes in the cerebellum occur early, and asymptomatic animals are more sensitive than normal to the GABA antagonist picrotoxin. Brain concentrations of glutamate and aspartate are also reduced in thiamin deficiency, as are several other neurotransmitters.

6.4.6.1 **Acetylcholine**   One effect of the impaired activity of pyruvate dehydrogenase in thiamin deficiency is a reduction in the brain content of acetyl CoA, and a reduction in both the pool size and turnover of acetylcholine. This is reflected in functional impairment; within 1 day of the initiation of treatment with pyrithiamin, animals show impaired performance in a tight-rope test. Repletion with thiamin, or the administration of either the directly acting muscarinic cholinergic agonist arecoline or the centrally acting acetyl cholinesterase inhibitor physostigmine, rapidly restores normal performance (Barclay et al., 1981).

6.4.6.2 **5-Hydroxytryptamine**   There is no change in the concentration of 5-hydroxytryptamine (serotonin) in the brains of thiamin-deficient rats, but there is an increase in its metabolite, 5-hydroxy-indoleacetic acid, and an increase in the accumulation of 5-hydroxytryptamine after the administration of monoamine oxidase inhibitors, suggesting an increased rate of 5-hydroxytryptamine turnover. Pyrithiamin treatment leads to signs of increased serotoninergic activity, especially changes in sleep patterns, which are normalized by the administration of thiamin (Plaitakis et al., 1981; Crespi and Jouvet, 1982). There is no obvious metabolic role of thiamin in the synthesis or catabolism of 5-hydroxytryptamine, and it is likely that the changes are secondary to changes in GABA turnover, and hence the activity of GABA neurons that are organizationally superior to some serotoninergic tracts.

6.4.7 Thiaminases and Thiamin Antagonists

Thiaminolytic enzymes are found in a variety of microorganisms and foods, and a number of thermostable compounds present in foods (especially polyphenols) cause oxidative cleavage of thiamin, as does sulfite, which is widely used in food processing. The products of thiamin cleavage by sulfite and thiaminases are shown in Figure 6.1.

In people whose thiamin intake is marginal, colonization of the gastrointestinal tract with thiaminolytic microorganisms may be a factor in the development of beriberi. The thiaminases present in raw fish can result in so-called Chastek paralysis of foxes and mink, as a result of destruction of thiamin, and may be important in parts of the world where much of the apparent thiamin intake is from fish that is eaten raw or fermented. The polyphenols and thiaminase in bracken fern can cause thiamin deficiency (blind staggers) in horses, and tannic acid in tea and betel nut have been associated with human thiamin deficiency.

There are two classes of thiaminase. Thiaminase I catalyzes a base exchange reaction between the thiazole moiety of thiamin and a variety of bases, commonly primary, secondary, or tertiary amines, but also nicotinamide and other pyridine derivatives, and sometimes proline and sulfhydryl compounds. Thiaminase I is relatively widespread in a variety of microorganisms, plants, and fish. In addition to depleting thiamin, the products of base exchange catalyzed by thiaminase I are structural analogs of the vitamin and may have antagonistic effects (Edwin and Jackman, 1970). Similarly, the neurotoxic effects of the antibiotic metronidazole, which is a thiazole, may be from its activity as a substrate for thiaminase I, forming thiamin antimetabolites (Alston and Abeles, 1987).

Thiaminase II catalyzes a simple hydrolysis, releasing thiazole and methoxypyrimidine, which has some antivitamin B$_6$ antimetabolic activity. It is relatively rare and is restricted to a small number of microorganisms.

The destruction of thiamin by polyphenols is not a stoichiometric reaction, and reducing compounds such as ascorbate and cysteine inhibit the reaction. In alkaline conditions, the thiazole ring of thiamin undergoes a reversible cleavage to the thiol. Thiamin thiol can react with a variety of thiol or disulfide compounds to form alkyl thiamin derivatives (allithiamins), some of which have biological activity. However, the thiol can also undergo oxidation catalyzed by polyphenols, resulting in the formation of thiamin disulfide, which has no biological activity.

Experimentally, two analogs of thiamin, pyrithiamin and oxythiamin (see Figure 6.1), are used to induce thiamin deficiency. Both are inhibitors of, and substrates for, thiamin pyrophosphokinase. Pyrithiamin also competes with thiamin for the blood–brain barrier uptake mechanism and is accumulated in the central nervous system by metabolic trapping. Oxythiamin does not cross the blood–brain barrier and has little or no effect on the central nervous system. Oxythiamin diphosphate is a potent inhibitor of thiamin diphosphate-dependent enzymes, whereas pyrithiamin diphosphate is a poor inhibitor. In general, oxythiamin acts as a peripheral thiamin antagonist, while pyrithiamin depletes the vitamin.

6.5 ASSESSMENT OF THIAMIN NUTRITIONAL STATUS

The impairment of pyruvate dehydrogenase in thiamin deficiency (Section 6.4.5) results in a considerable increase in the plasma concentrations of lactate and pyruvate. This has been exploited as a means of assessing thiamin nutritional status by measuring changes in the plasma concentrations of lactate, pyruvate, and glucose after an oral dose of glucose and mild exercise. This is not specific for thiamin deficiency; a variety of other conditions can also result in metabolic acidosis. Although it may be useful in depletion/repletion studies, it is used little nowadays in screening or assessment of nutritional status, and a number of more sensitive and specific tests of thiamin status are available (as shown in Table 6.1).

6.5.1 Urinary Excretion of Thiamin and Thiochrome

Although there are a number of urinary metabolites of thiamin, a significant amount of the vitamin is excreted unchanged or as thiochrome, especially if intake is adequate, and therefore the urinary excretion can provide useful information on nutritional status. Excretion decreases proportionally with intake in adequately nourished subjects; but, at low intakes, there is a threshold below which further reduction in intake has little effect on excretion.

The excretion of a test dose of thiamin has also been used as an index of status; after a parenteral dose of 5 mg (19 μmol) of thiamin, adequately nourished subjects excrete more than 300 nmol of the vitamin over 4 hours, whereas deficient subjects excrete less than 75 nmol.

6.5.2 Blood Concentration of Thiamin

In experimental animals and in depletion studies, measurement of the concentration of thiamin in plasma or whole blood provides an indication of the progression of deficiency. The normal method is by the formation of thiochrome, which is fluorescent; only free thiamin, and not the phosphates, undergoes

Table 6.1 Indices of Thiamin Nutritional Status

	Adequate	Marginal	Deficient
Intake			
mmol/1,000 kcal	>1.1	0.75–1.1	<0.75
mmol/MJ	>0.27	0.18–0.27	<0.18
mg/1,000 kcal	>0.3	0.2–0.29	< 0.2
μg/MJ	>72	48–72	<48
Urinary excretion			
mmol/mol creatinine	>28	11–27	<11
mg/g creatinine	>66	27–65	<27
nmol/24 h	>375	150–375	<150
μg/24 h	>100	40–99	<40
Urinary excretion over 4 h after a 19 nmol (5 mg) parenteral dose			
nmol	>300	75–300	<75
μg	>80	20–79	<20
Transketolase activation coefficient			
	<1.15	1.15–1.24	>1.25
Erythrocyte thiamin diphosphate			
nmol/L	>150	120–150	<120
μg/L	>64	50–64	<50

Sources: From data reported by Brin, 1964; Sauberlich et al., 1974; Finglass, 1993.

oxidation to thiochrome, so measurement before and after reaction of the sample with alkaline phosphatase permits determination of free and total thiamin.

Erythrocytes and leukocytes contain mainly thiamin diphosphate, whereas plasma contains free thiamin and thiamin monophosphate. The concentration of thiamin diphosphate in erythrocytes is normally between 110 and 330 nmol per L of packed cells. The total thiamin concentration in erythrocytes is about 4- to 5-fold higher than in plasma and that in leukocytes is 10-fold higher again.

Whole blood total thiamin below 150 nmol per L is considered to indicate deficiency. However, the changes observed in depletion studies are small. Even in patients with frank beriberi, the total thiamin concentration in erythrocytes is only 20% lower than normal; whole blood thiamin is not a sensitive index of status.

6.5.3 Erythrocyte Transketolase Activation
Activation of apotransketolase in erythrocyte lysate by thiamin diphosphate added in vitro has become the most widely used and accepted index of thiamin nutritional status. Apotransketolase is unstable both in vivo and in vitro; therefore, problems may arise in the interpretation of results, especially if samples have been stored for any appreciable time. An activation coefficient >1.25 is

indicative of deficiency, and <1.15 is considered to reflect adequate thiamin nutrition.

6.6 THIAMIN REQUIREMENTS AND REFERENCE INTAKES

It is apparent from the central role of thiamin in carbohydrate metabolism that the requirement will depend on carbohydrate intake to a considerable extent. In practice, requirements are calculated on the basis of total energy intake, assuming that the average diet provides 40% of energy from fat. For diets that are lower in fat, and hence higher in carbohydrate and protein, thiamin requirements will be somewhat higher.

On the basis of depletion/repletion studies, an intake of 0.2 mg per 1,000 kcal is required to maintain normal urinary excretion, but an intake of 0.3 mg per 1,000 kcal is required for a normal transketolase activation coefficient. At low levels of energy intake, there will be a requirement for metabolism of endogenous substrates and to maintain nervous system thiamin triphosphate.

Reference intakes (see Table 6.2) are based on 0.5 mg per 1,000 kcal (0.12 mg per MJ) for adults consuming more than 2,000 kcal per day, with the proviso that even in fasting there is a requirement for 0.8 mg of thiamin per day to permit the metabolism of endogenous energy-yielding substrates.

6.6.1 Upper Levels of Thiamin Intake

There is no evidence of any toxic effect of high intakes of thiamin, although high parenteral doses have been reported to cause respiratory depression in animals and anaphylactic shock in human beings. Hypersensitivity and contact dermatitis have been reported in pharmaceutical workers handling thiamin. As noted in Section 6.2, absorption of dietary thiamin is limited, and no more than about 2.5 mg (10 μmol) can be absorbed from a single dose; free thiamin is rapidly filtered by the kidneys and excreted.

6.6.2 Pharmacological Uses of Thiamin

Apart from children with thiamin-responsive maple syrup urine disease (Section 6.3.1.4) and thiamin-responsive megaloblastic anemia (Section 6.2), there are no established pharmacological uses of thiamin other than the treatment of deficiency. Because of the neurological involvement in thiamin deficiency, the vitamin has been used in nerve tonics, although there is no evidence that it has any effect except in cases of deficiency.

Studies in thiamin-deficient animals revealed the presence of Alzheimer-like amyloid plaques in the brain. Although there is no evidence of similar plaque formation in the brains of patients with the Wernicke–Korsakoff syndrome, this has led to trials of thiamin for treatment of Alzheimer's disease

Table 6.2 Reference Intakes of Thiamin (mg/day)

Age	U.K. 1991	EU 1993	U.S./Canada 1998	FAO 2001
0–6 m	0.2	—	0.2	0.2
7–9 m	0.2	0.3	0.3	0.3
10–12 m	0.3	0.3	0.3	0.3
1–3 y	0.5	0.5	0.5	0.5
4–6 y	0.7	0.7	0.5	0.6
7–8 y	0.7	0.8	0.5	0.9
Males				
9–10 y	0.7	0.8	0.9	0.9
11–13 y	0.9	1.0	0.9	1.2
14–15 y	0.9	1.0	1.2	1.2
16–18 y	1.1	1.2	1.2	1.2
19–30 y	1.0	1.1	1.2	1.2
31–50 y	1.0	1.1	1.2	1.2
>50 y	0.9	1.1	1.2	1.2
Females				
9–10 y	0.7	0.8	0.9	0.9
11–13 y	0.7	0.9	0.9	1.1
14–15 y	0.7	0.9	1.0	1.1
16–18 y	0.8	0.9	1.1	1.1
19–30 y	0.8	0.9	1.1	1.1
31–50 y	0.8	0.9	1.1	1.1
>50 y	0.8	0.9	1.1	1.1
Pregnant	0.9	1.0	1.4	1.4
Lactating	0.9	1.1	1.4	1.5

EU, European Union; FAO, Food and Agriculture Organization; WHO, World Health Organization.
Sources: Department of Health, 1991; Scientific Committee for Food, 1993; Institute of Medicine, 1998; FAO/WHO, 2001.

(Calingasan et al., 1996). Whereas some studies have shown beneficial effects, a systematic review has concluded that there is no evidence of beneficial effects of thiamin supplementation in Alzheimer's disease (Rodriguez-Martin et al., 2001).

FURTHER READING

Bettendorff L (1996) A non-cofactor role of thiamine derivatives in excitable cells? *Archives of Physiology and Biochemistry* **104,** 745–51.

Butterworth RF (1982) Neurotransmitter function in thiamine deficiency. *Neurochemistry International* **4,** 449–65.

Kopelman MD (1995) The Korsakoff syndrome. *British Journal of Psychiatry* **166,** 154–73.

Kril JJ (1996) Neuropathology of thiamine deficiency disorders. *Metabolic Brain Diseases* **11,** 9–17.

Peters R (1963) *Biochemical Lesions and Lethal Synthesis.* Oxford: Pergamon Press.

Reuker JB, Girard DE, and Cooney TG (1985) Wernicke's encephalopathy. *New England Journal of Medicine* **312,** 1035–8.

Schellenberger A (1998) Sixty years of thiamin diphosphate biochemistry. *Biochimica et Biophysica Acta* **1385,** 177–86.

Zubaran C, Fernandes JG, and Rodnight R (1997) Wernicke-Korsakoff syndrome. *Postgraduate Medical Journal* **73,** 27–31.

References cited in the text are listed in the Bibliography.

SEVEN

Vitamin B$_2$ – Riboflavin

Riboflavin has a central role as a redox coenzyme in energy-yielding metabolism and a more recently discovered role as the prosthetic group of the cryptochromes in the eye – the blue-sensitive pigments that are responsible for day-length sensitivity and the setting of circadian rhythms.

Dietary deficiency is relatively widespread, yet is apparently never fatal; there is not even a clearly characteristic riboflavin deficiency disease. In addition to intestinal bacterial synthesis of the vitamin, there is very efficient conservation and reutilization of riboflavin in tissues. Flavin coenzymes are tightly enzyme bound, in some cases covalently, and control of tissue flavins is largely at the level of synthesis and catabolism of flavin-dependent enzymes.

Reoxidation of reduced flavin coenzymes is the major source of oxygen radicals in the body, and riboflavin is also capable of generating reactive oxygen species nonenzymically. As protection against this, there is very strict control over the body content of riboflavin. Absorption is limited, and any in excess of requirements is rapidly excreted.

In bacteria, flavin adenine dinucleotide (FAD) is the prosthetic group of the photolyases that catalyze reductive repair of light-induced pyrimidine dimers in DNA. Riboflavin is the light-emitting molecule in some bioluminescent fungi and bacteria, and is the precursor for synthesis of the dimethylbenzimidazole ring of vitamin B$_{12}$ (Section 10.7.3).

7.1 RIBOFLAVIN AND THE FLAVIN COENZYMES

As shown in Figure 7.1, riboflavin consists of a tricyclic dimethyl-isoalloxazine ring conjugated to the sugar alcohol ribitol. The metabolically active coenzymes are riboflavin 5′-phosphate and flavin adenine dinucleotide (FAD). In some enzymes the prosthetic group is riboflavin, bound covalently at the catalytic site.

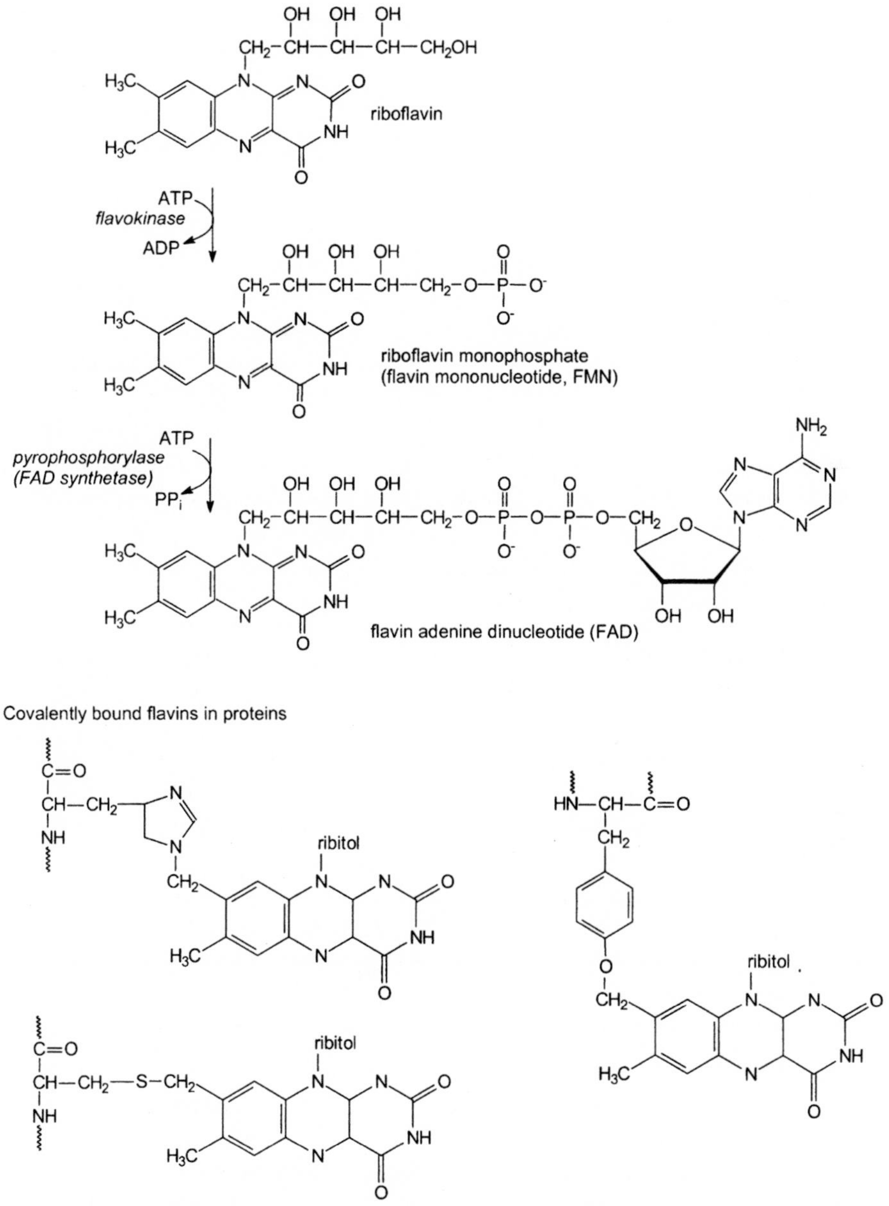

Figure 7.1. Riboflavin, the flavin coenzymes and covalently bound flavins in proteins. Relative molecular masses (M_r): riboflavin, 376.4; riboflavin phosphate, 456.6; and FAD, 785.6.

The ribityl moiety is not linked to the isoalloxazine ring by a glycosidic linkage, and it is not strictly correct to call FAD a dinucleotide. Nevertheless, this trivial name is accepted, as indeed is the even less correct term flavin mononucleotide for riboflavin phosphate.

Riboflavin phosphate and FAD may be either covalently or noncovalently bound at the catalytic sites of enzymes. Even in those enzymes in which the binding is not covalent, the flavin is tightly bound; in many cases, the flavin has a role in maintaining or determining the conformation of the enzyme protein. In some cases, the flavin is incorporated into the nascent polypeptide chain, while it is still attached to the ribosome. However, in others a flavin-free apoenzyme is synthesized and accumulates in riboflavin deficiency (Section 7.5.2).

Covalent binding of the flavin coenzymes is normally through the 8-α-methyl group. 8-Hydroxymethyl-riboflavin is formed by microsomal mixed-function oxidases (Section 7.2.5), but it is not known whether or not this is a precursor of covalently bound flavin coenzymes. A variety of amino acid residues may be involved in covalent binding of flavin coenzymes to enzymes, including the following:

1. flavin 8-α-carbon linkage to imidazole N-3 of a histidine residue (e.g., in succinate, sarcosine, and dimethylglycine dehydrogenases in mammals and bacterial 6-hydroxynicotine oxidase);
2. flavin 8-α-carbon linkage to imidazole N-1 of a histidine residue (e.g., in bacterial thiamin dehydrogenase and mammalian gulonolactone oxidase in those species for which ascorbate is not a vitamin) (Section 13.3.4);
3. flavin 8-α-carbon thioether linkage to a cysteine residue (e.g., in mono-amine oxidase) – the 8-ethyl analog of riboflavin is incorporated into monoamine oxidase, although the resultant holo-enzyme analog is catalytically inactive;
4. flavin 8-α-carbon thio-hemiacetal linkage to a cysteine residue (e.g., in bacterial cytochrome c$_{552}$);
5. flavin 8-α-carbon O-tyrosyl ether linkage (e.g., in bacterial p-cresol methyl hydroxylase); and
6. linkage from carbon-6 of the flavin to a cysteine residue (e.g., in bacterial trimethylamine dehydrogenase).

Although the ribitol moiety is not involved in the redox function of the flavin coenzymes, both the stereochemistry and nature of the sugar alcohol are important. Although some riboflavin analogs have partial vitamin action, most are inactive or have antivitamin activity, although they may be active in microbiological assays. The galactitol (dulcitol) analog, galactoflavin, has been widely used as a means of inducing riboflavin deficiency in animal and human studies.

Photolysis of riboflavin leads to the formation of lumiflavin in alkaline solution and lumichrome in acidic or neutral solution (see Figure 7.2). Because lumiflavin is chloroform extractable, photolysis in alkaline solution, followed by chloroform extraction and fluorimetric determination, is the basis of commonly used chemical methods of assaying riboflavin. The photolysis proceeds by way of intermediate formation of cytotoxic riboflavin radicals, and the addition of riboflavin and exposure to light has been suggested as a means of inactivating viruses and bacteria in blood products (Goodrich, 2000).

Exposure of milk in clear glass bottles to sunlight or fluorescent light (with a peak wavelength of 400 to 550 nm) can result in the loss of significant amounts of riboflavin as a result of photolysis. This is potentially nutritionally important, because on average, in Western diets, 25% to 30% of riboflavin comes from milk. The resultant lumiflavin and lumichrome catalyze the oxidation of vitamin C, so that even relatively brief exposure to light, causing little loss of riboflavin, can lead to a considerable loss of vitamin C. This is nutritionally unimportant, because milk is normally an insignificant source of vitamin C. Lumiflavin and lumichrome also catalyze oxidation of lipids (to lipid peroxides) and methionine (to methional), resulting in the development of an unpleasant flavor – the so-called sunlight flavor.

Photolysis of riboflavin occurs in vivo during phototherapy for neonatal hyperbilirubinemia (Section 7.4.4). There is no evidence that normal exposure to sunlight results in significant photolysis of riboflavin, although it is possible that some of the lumichromes found in urine may arise in this way.

7.2 THE METABOLISM OF RIBOFLAVIN

7.2.1 Absorption, Tissue Uptake, and Coenzyme Synthesis

Apart from milk and eggs, which contain relatively large amounts of free riboflavin bound to specific binding proteins, most of the vitamin in foods is as flavin coenzymes bound to enzymes, with about 60% to 90% as FAD.

FAD and riboflavin phosphate in foods are hydrolyzed in the intestinal lumen by nucleotide diphosphatase and a variety of nonspecific phosphatases to yield free riboflavin, which is absorbed in the upper small intestines by a sodium-dependent saturable mechanism; the peak plasma concentration is related to the dose only up to about 15 to 20 mg (40 to 50 μmol). Thereafter,

Table 7.1 Tissue Flavins in the Rat

		% Present as		
	Total μmol/kg	Riboflavin	Riboflavin P	FAD
Plasma	0.064	65	7	28
Liver	58.0	3	23	74
Kidney	63.2	4	41	55
Muscle	4.1	3	12	85

there is little or no absorption of higher single doses of riboflavin (Zempleni et al., 1996).

Intestinal bacteria synthesize riboflavin, and fecal losses of the vitamin may be five- to six-fold higher than intake. It is possible that bacterial synthesis makes a significant contribution to riboflavin intake, because there is carrier-mediated uptake of riboflavin into colonocytes in culture. The activity of the carrier is increased in riboflavin deficiency and decreased when the cells are cultured in the presence of high concentrations of riboflavin. The same carrier mechanism seems to be involved in tissue uptake of riboflavin (Said et al., 2000).

Much of the absorbed riboflavin is phosphorylated in the intestinal mucosa by flavokinase and enters the bloodstream as riboflavin phosphate; this metabolic trapping is essential for concentrative uptake of riboflavin into enterocytes (Gastaldi et al., 2000). Parenterally administered free riboflavin is also largely phosphorylated in the intestinal mucosa. It is not clear whether this is the result of enterohepatic recycling of the vitamin or simply uptake of free riboflavin into the intestinal mucosa from the bloodstream.

About 7% of dietary riboflavin is covalently bound to proteins (mainly as riboflavin-8-α-histidine or riboflavin-8-α-cysteine). The riboflavin–amino acid complexes released by proteolysis are not biologically available; although they are absorbed from the gastrointestinal tract, they are excreted in the urine (Chia et al., 1978).

As shown in Table 7.1, the total riboflavin concentration in plasma is very much lower than in most tissues. About 50% of plasma riboflavin is free riboflavin, which is the main transport form, with 44% as FAD and the remainder as riboflavin phosphate. The vitamin is largely protein bound in plasma; free riboflavin binds to both albumin and α- and β-globulins, and both riboflavin and the coenzymes also bind to immunoglobulins. The products of photolysis of riboflavin bind to albumin with considerably higher affinity than riboflavin itself; this albumin binding may represent a mechanism to prevent tissue uptake of these potential antimetabolites, or it may be an artifact of the exposure of samples to light during analysis.

Most tissues contain very little free riboflavin and, except in the kidneys, where 30% is as riboflavin phosphate, more than 80% is FAD, almost all bound to enzymes. Isolated hepatocytes (and presumably other tissues) show saturable concentrative uptake of riboflavin. The K_m of the uptake process is the same as that of flavokinase, and uptake is inhibited by inhibitors of flavokinase, suggesting that tissue uptake is the result of carrier-mediated diffusion, followed by metabolic trapping as riboflavin phosphate, then onward metabolism to FAD, catalyzed by FAD pyrophosphorylase. FAD is a potent inhibitor of the pyrophosphorylase and acts to limit its own synthesis. FAD, which is not protein bound is rapidly hydrolyzed to riboflavin phosphate by nucleotide pyrophosphatase; unbound riboflavin phosphate is similarly rapidly hydrolyzed to riboflavin by nonspecific phosphatases (Aw et al., 1983; Yamada et al., 1990).

FAD is cleaved by an FAD-adenosine monophosphate (AMP) lyase in liver to yield AMP and riboflavin 4′,5′-cyclic phosphate; it is not known whether this has any coenzyme or cell signaling function, but it is a substrate for phosphodiesterase and has also been identified in small amounts in yeast (Fraiz et al., 1998; Cabezas et al., 2001).

7.2.2 Riboflavin Binding Protein

There is a specific plasma riboflavin binding protein that is induced by estrogens; in female animals, its concentration varies through the estrous cycle. The same protein is also synthesized in the testes and is found on the acrosomal surface of spermatozoa. In females, it acts to transport the vitamin across the placenta, which is impermeable to free riboflavin or the coenzymes. The protein is essential for fetal uptake of riboflavin, and immunoneutralization of the protein causes a considerable decrease in the uptake of riboflavin by the fetus, leading to death of the fetus and termination of the pregnancy, with no apparent effect on maternal riboflavin metabolism. It has been suggested that active or passive immunization against the binding protein may provide both male and female contraception (Krishnamurthy et al., 1984; Adiga, 1994; Adiga et al., 1997).

In pregnant women, there is a progressive increase in the erythrocyte glutathione reductase activation coefficient (an index of functional riboflavin nutritional status; Section 7.5.2), which resolves on parturition despite the daily secretion of 200 to 400 μg (0.5 to 1 μmol) of riboflavin into milk. This suggests that the estrogen-induced riboflavin binding protein can sequester the vitamin for fetal uptake at the expense of causing functional deficiency in the mother.

In laying hens, induction of this riboflavin protein results in a 100-fold increase in plasma riboflavin, compared with males or nonlaying females. In mutant chickens lacking the protein, the adult has massive urinary loss of riboflavin. The embryo develops normally for about 10 days, then develops severe hypoglycemia associated with a reduction in medium-chain acyl coenzyme A (CoA) dehydrogenase to 20% of normal activity and the accumulation of intermediates of fatty acid oxidation (White, 1996).

The riboflavin binding protein that occurs in eggs has been exploited for the radio-ligand binding assay of riboflavin. Because binding to the protein quenches the native fluorescence of riboflavin, it can be exploited for a direct titrimetric fluorescence assay of the vitamin in urine and other biological samples (Kodentsova et al., 1995).

7.2.3 Riboflavin Homeostasis

There is no evidence of any significant storage of riboflavin; in addition to the limited absorption, any surplus intake is excreted rapidly; thus, once metabolic requirements have been met, urinary excretion of riboflavin and its metabolites reflects intake until intestinal absorption is saturated. In depleted animals, the maximum growth response is achieved with intakes that give about 75% saturation of tissues, and the intake to achieve tissue saturation is that at which there is quantitative urinary excretion of the vitamin.

Equally, there is very efficient conservation of tissue riboflavin in deficiency. There is only a four-fold difference between the minimum concentration of flavins in the liver in deficiency and the level at which saturation occurs. In the central nervous system, there is only a 35% difference between deficiency and saturation.

Control over tissue concentrations of riboflavin coenzymes seems to be largely by control of the activity of flavokinase, and the synthesis and catabolism of flavin-dependent enzymes. Almost all the vitamin in tissues is enzyme bound, and free riboflavin phosphate and FAD are rapidly hydrolyzed to riboflavin. If this is not rephosphorylated, it rapidly diffuses out of tissues and is excreted.

In deficiency, virtually the only loss of riboflavin from tissues will be the small amount that is covalently bound to enzymes. The 8α-linkage is not cleaved by mammalian enzymes and 8α-derivatives of riboflavin are not substrates for flavokinase and cannot be reutilized.

7.2.4 The Effect of Thyroid Hormones on Riboflavin Metabolism

The activities of a variety of flavin-dependent enzymes are depressed in hypothyroidism. They are increased by the administration of thyroxine or triiodothyronine, as a result of increased synthesis of riboflavin phosphate and

FAD, leading to increased saturation of enzyme proteins with coenzymes. This increases the stability of the enzymes against proteolysis and increases their activity in tissues (Rivlin and Langdon, 1966).

Tissue concentrations of flavin coenzymes in hypothyroid animals may be as low as in those fed a riboflavin-deficient diet. In hypothyroid patients, erythrocyte glutathione reductase (EGR) activity may be as low, and its activation by FAD added in vitro (Section 7.5.2) as high, as in riboflavin-deficient subjects. Tissue concentrations of flavin coenzymes and EGR are normalized by the administration of thyroid hormones, with no increase in riboflavin intake (Cimino et al., 1987).

The administration of thyroid hormones to hypothyroid animals results in a rapid increase in flavokinase activity as a result of the activation of an inactive precursor protein; as flavokinase activity increases, there is a parallel decrease in the tissue content of an apparently inactive riboflavin binding protein (Lee and McCormick, 1985).

Hyperthyroidism is not associated with elevated tissue concentrations of flavin coenzymes, despite increased activity of flavokinase. Again, this demonstrates the importance of the enzyme binding of flavin coenzymes and the rapid hydrolysis of unbound FAD and riboflavin phosphate in the regulation of tissue concentrations of the vitamin.

Riboflavin may also be involved in the metabolism of thyroid hormones. In the presence of oxygen, riboflavin phosphate catalyzes a photolytic deiodination of thyroxine. The lower tissue concentration of riboflavin phosphate in hypothyroidism may thus serve to protect such thyroid hormone as is available against catabolism and prolong its action.

7.2.5 Catabolism and Excretion of Riboflavin

Riboflavin and riboflavin phosphate that are not bound to plasma proteins are filtered at the glomerulus; the phosphate is generally dephosphorylated in the bladder. Renal tubular reabsorption of riboflavin is saturated at normal plasma concentrations, and there is also active tubular secretion of the vitamin, so that urinary clearance of riboflavin can be two- to three-fold greater than the glomerular filtration rate.

Under normal conditions, about 25% of the urinary excretion of riboflavin is as the unchanged vitamin, with a small amount as a variety of glycosides of riboflavin and its metabolites. Riboflavin-8-α-histidine and riboflavin-8-α-cysteine arising from the catabolism of enzymes in which the coenzyme is covalently bound are excreted unchanged.

Liver cytochrome P_{450}-linked mixed-function oxidases result in the production of 7- and 8-hydroxymethylriboflavin, both of which are substrates for

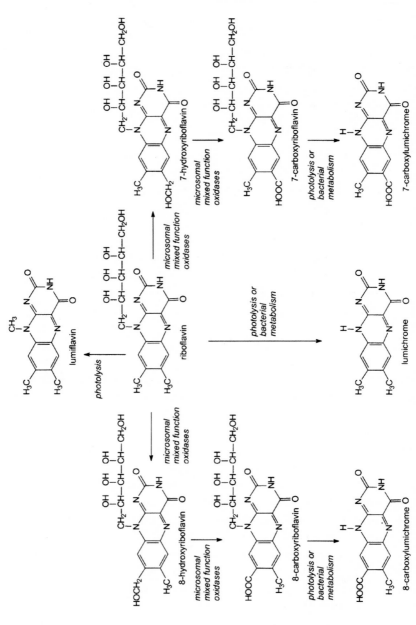

Figure 7.2. Products of riboflavin metabolism. Relative molecular masses (M_r): riboflavin, 376.4; lumiflavin, 256.3; 8- and 7-carboxylumichrome, 296.2; and lumichrome, 242.2.

Table 7.2 Urinary Excretion of
Riboflavin Metabolites

	% of Total
Riboflavin	60–70
7-Hydroxymethyl riboflavin	10–15
8-α-Sulfonyl riboflavin	5–10
8-Hydroxymethyl riboflavin	4–7
Riboflavin peptide esters	5
10-Hydroxyethyl riboflavin	1–3

flavokinase. There is some evidence that 8-hydroxymethylriboflavin may have biological activity; it is not known whether or not it is involved in the formation of 8-α-amino acid covalent links in proteins. Significant amounts of both of these hydroxylated derivatives and their onward oxidation products (7- and 8-carboxyriboflavin) are excreted in the urine (see Figure 7.2 and Table 7.2).

Intestinal bacterial cleavage of the ribityl side chain results in the formation of 10-hydroxyethylflavin (an oxidation product of lumiflavin), lumichrome, and 7- and 8-carboxy-lumichromes, which are also excreted in the urine. Some of the lumichromes detected in urine may result from photolysis of riboflavin in the circulation.

7.2.6 Biosynthesis of Riboflavin

A number of fungi have a failure of the normal regulation of riboflavin synthesis and are overproducers of the vitamin. Mutants of *Ashbya gossypii* may accumulate up to 150 μmol of riboflavin per gram of protein, compared with a normal content of 0.25 μmol per gram of protein. They can produce and excrete so much that riboflavin crystallizes in the culture medium. Such fungi are used for the commercial production of riboflavin by fermentation, as an alternative to chemical synthesis.

The precursors for riboflavin biosynthesis in plants and microorganisms are guanosine triphosphate and ribulose 5-phosphate. As shown in Figure 7.3, the first step is hydrolytic opening of the imidazole ring of GTP, with release of carbon-8 as formate, and concomitant release of pyrophosphate. This is the same as the first reaction in the synthesis of pterins (Section 10.2.4), but utilizes a different isoenzyme of GTP cyclohydrolase (Bacher et al., 2000, 2001).

In yeasts and fungi, opening of the imidazole ring is followed by reduction of the ribose side chain to ribitol, deamination, and dephosphorylation to yield amino-ribitylamino-pyrimidinedione; in bacteria, deamination occurs before

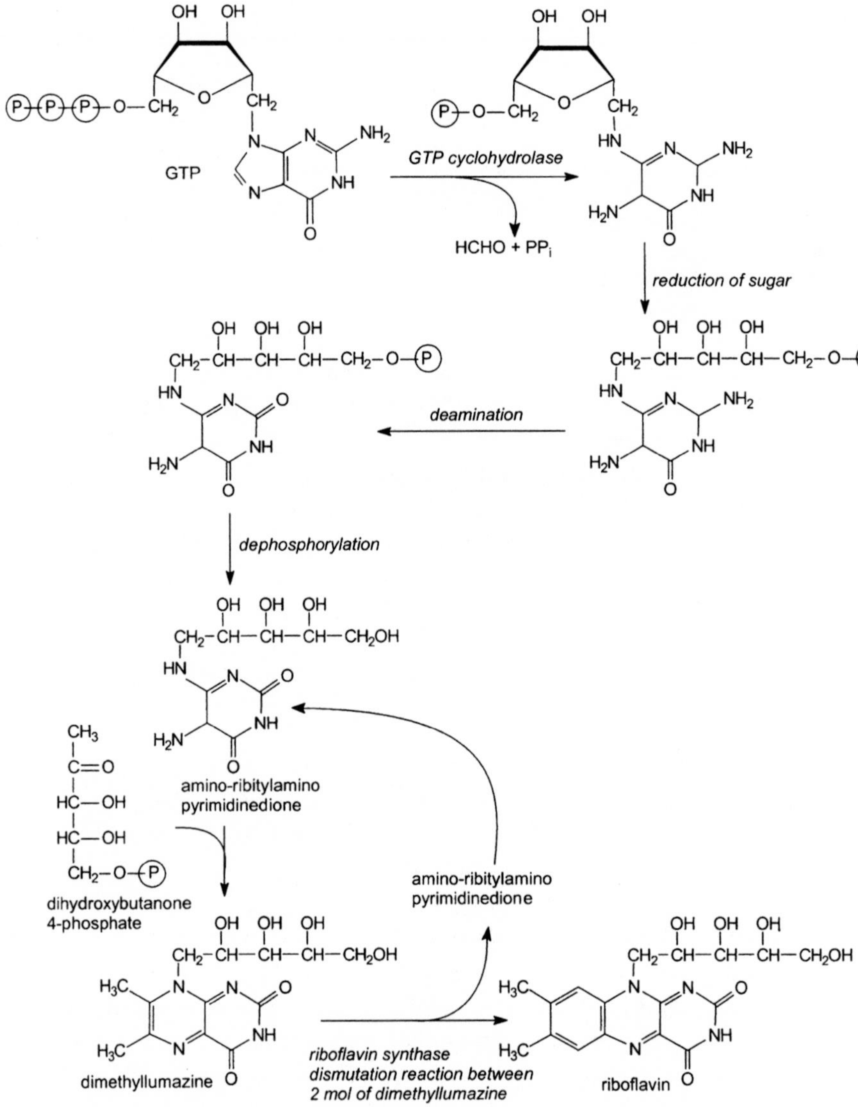

Figure 7.3. Biosynthesis of riboflavin in fungi; in bacteria, deamination precedes reduction of sugar. GTP cyclohydrolase, EC 3.5.4.25; and riboflavin synthase, EC 2.5.1.9.

reduction of the sugar. Amino-ribitylamino-pyrimidinedione condenses with dihydroxybutanone 4-phosphate, which is formed from ribulose 5-phosphate by an unusual reaction involving loss of carbon-4 via an intramolecular rearrangement. The product of the reaction catalyzed by lumazine synthase is dimethyllumazine.

The final step is a dismutation reaction between two molecules of dimethyl-lumazine, catalyzed by riboflavin synthase, yielding riboflavin and amino-ribitylamino-pyrimidinedione. This latter product can undergo reaction with dihydroxybutanone 4-phosphate to yield dimethyl-ribityllumazine.

7.3 METABOLIC FUNCTIONS OF RIBOFLAVIN

The metabolic function of the flavin coenzymes is as electron carriers in a wide variety of oxidation and reduction reactions central to all metabolic processes, including the mitochondrial electron transport chain. Unlike the nicotinamide nucleotide coenzymes (Section 8.4.1), which act as cosubstrates, leaving the catalytic site of the enzyme at the end of the reaction, the flavin coenzymes remain bound to the enzyme throughout the catalytic cycle.

FAD is the prosthetic group of the bacterial photolyase that reduces cyclobutane thymine dimers formed in DNA as a result of ultraviolet (UV) irradiation; closely homologous proteins in the human eye (the cryptochromes) are the blue-sensitive pigments that are responsible for day-length sensitivity and the setting of circadian rhythms.

7.3.1 The Flavin Coenzymes: FAD and Riboflavin Phosphate

The majority of flavoproteins have FAD as the prosthetic group rather than riboflavin phosphate. Some have both flavin coenzymes, and some have other prosthetic groups.

As shown in Figure 7.4, flavins can undergo a one-electron reduction to the semiquinone radical or a two-electron reduction to dihydroflavin. This means that flavins can act as intermediates between obligatory two-electron redox reactions involving nicotinamide nucleotides (Section 8.4.1) and obligatory one-electron reactions involving cytochromes, iron-sulfur proteins, and ubiquinone (Section 14.6).

In solution, the flavin semiquinone radical is highly unstable, undergoing rapid equilibration to a mixture of the oxidized and fully reduced flavins. It is stabilized by protein binding in enzymes.

The neutral flavin radical has an absorption maximum at 580 nm and hence a blue color; it is sometimes referred to as the blue radical. It can undergo either protonation at N-1 to yield a cation radical or deprotonation at N-5 to yield an anion radical, if the enzyme has appropriate proton donating or withdrawing amino acid residues at the catalytic site. Both protonation and deprotonation result in the same spectral shift to give an absorption maximum at 470 nm and hence a red color. Both the blue and red radicals are seen as intermediates in enzyme reactions, suggesting that some enzymes form the neutral radical, whereas others form one of the charged radicals.

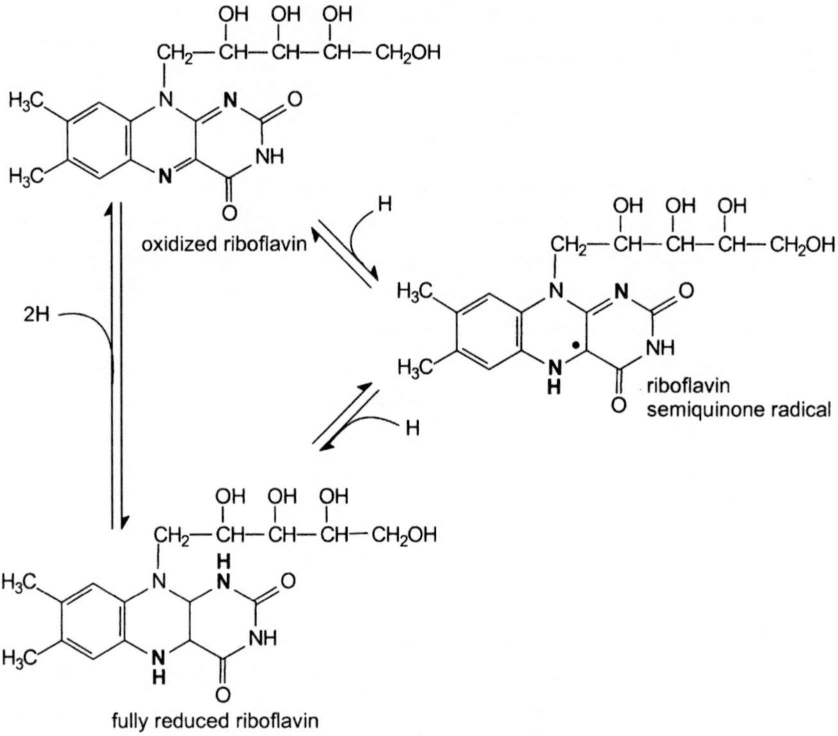

Figure 7.4. One- and two-electron redox reactions of riboflavin.

Dihydroflavin can be oxidized by reaction with a substrate, NAD(P)$^+$ or cytochromes in a variety of dehydrogenases, or can react with molecular oxygen in oxygenases and mixed-function oxidases.

7.3.2 Single-Electron–Transferring Flavoproteins
Flavoproteins catalyzing single-electron transfer provide the link between substrate oxidation catalyzed by dehydrogenases and the mitochondrial electron transport chain.

The simplest such single-electron–transferring flavoproteins are the flavodoxins of obligate anaerobic bacteria, which catalyze a single-electron–transfer reaction, cycling between dihydroflavin and the semiquinone radical. In all other organisms, the electron transport iron-sulfur flavoprotein undergoes a two-electron reduction at the expense of NADH; this may result in either the formation of dihydroflavin or reduction to the semiquinone radical of two molecules of flavin in the enzyme. The reduced enzyme then transfers electrons singly to cytochrome b.

The reduction of cytochrome P_{450} by NADPH involves a single enzyme, NADPH-cytochrome P_{450} reductase, which contains both FAD and riboflavin phosphate. The FAD undergoes a two-electron reduction at the expense of NADPH, then transfers electrons singly to the riboflavin phosphate, which in turn reduces cytochrome P_{450}. The semiquinone radicals of both FAD and riboflavin phosphate are intermediates in this reaction.

A distinct electron transfer flavoprotein (ETF) is the single-electron acceptor for a variety of flavoprotein dehydrogenases, including acyl CoA, glutaryl CoA, sarcosine, and dimethylglycine dehydrogenases. It then transfers the electrons to ETF-ubiquinone reductase, the iron-sulfur flavoprotein that reduces ubiquinone in the mitochondrial electron transport chain.

7.3.3 Two-Electron–Transferring Flavoprotein Dehydrogenases

The initial step of the two-electron–transferring reactions is the removal of a proton from the substrate, followed by the intermediate formation of an adduct between the substrate and prosthetic group at N-5 of the flavin. This undergoes cleavage to yield dihydroflavin and the oxidized product, which is commonly a carbon–carbon double bond. The reduced flavin is then reoxidized by reaction with an electron-transferring flavoprotein, as discussed above, or in some cases by reaction with nicotinamide nucleotide coenzymes.

The nicotinamide nucleotide independent flavoprotein dehydrogenases include the following:

1. Succinate dehydrogenase in the tricarboxylic acid cycle, which reacts directly with ubiquinone in the mitochondrial electron transport chain.
2. Acyl CoA dehydrogenases in fatty acid β-oxidation. These enzymes are especially sensitive to riboflavin depletion, and riboflavin deficiency is characterized by impaired fatty acid oxidation and organic aciduria (Section 7.4.1). These are also the enzymes affected in riboflavin-responsive organic acidurias.
3. Dimethylglycine and sarcosine dehydrogenases in the catabolism of choline (Section 14.2.1). In these reactions, a methyl group in the substrate is oxidized by FAD, then the intermediate adduct undergoes hydrolysis to release formaldehyde, which reacts with tetrahydrofolate to form 5,10-methylene tetrahydrofolate.

7.3.4 Nicotinamide Nucleotide Disulfide Oxidoreductases

Glutathione reductase, thioredoxin reductase, and lipoamide dehydrogenase are members of a group of flavoproteins that contain an active site disulfide as well as FAD. They catalyze the NAD(P)H dependent reduction of a disulfide

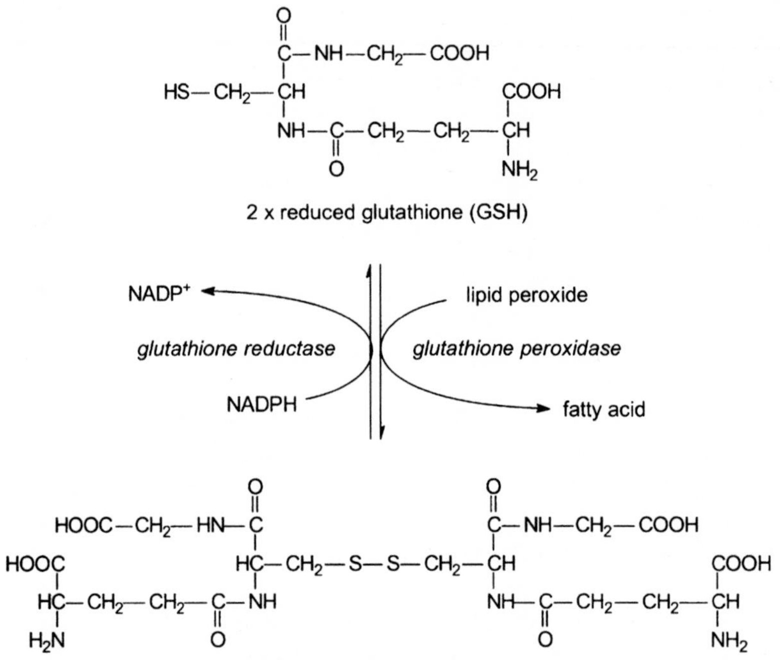

2 x reduced glutathione (GSH)

oxidized glutathione (GSSG)

Figure 7.5. Reaction of glutathione peroxidase (EC 1.11.1.19) and glutathione reductase (EC 1.6.4.2). Relative molecular masses (M_r): glutathione, 307.3; and oxidized glutathione, 612.6.

substrate to its dithiol form. The initial step in the reaction is reduction of the disulfide to yield a sulfhydryl group and a flavin-cysteine adduct, followed by release of the oxidized flavin to leave a second sulfhydryl group at the active site. It is these two disulfide groups that catalyze the reduction of the disulfide substrate. The reaction of glutathione reductase is shown in Figure 7.5, and that of lipoamide dehydrogenase in Figure 6.2.

7.3.5 Flavin Oxidases

Flavin oxidases include D- and L-amino acid oxidases, and some amine oxidases, although others are quinoproteins (Section 9.8.3). In these enzymes, the flavin is reduced by dehydrogenation of the substrate, by way of an intermediate substrate-flavin adduct, as occurs in the dehydrogenases (Section 7.3.3).

After the oxidized product has left the enzyme, the reduced flavin reacts with oxygen to form, initially, the flavin semiquinone radical and superoxide. These undergo the sequence of rapid reactions shown in Table 7.3, ultimately resulting in reoxidation of the flavin and formation of hydrogen peroxide.

Table 7.3 Reoxidation of Reduced Flavins in Flavoprotein Oxidases

The overall reaction is:
$$X-H_2 + flavin \rightarrow X + flavin\text{-}H_2, flavin\text{-}H_2 + O_2 \rightarrow flavin + H_2O_2$$
Fully reduced flavin-H_2 reacts with oxygen to form the flavin semiquinone radical and superoxide
$$flavin\text{-}H_2 + O_2 \rightarrow flavin\text{-}H^{\bullet} + {}^{\bullet}O_2{}^{-}$$
Flavin semiquinone and superoxide react to form flavin hydroperoxide
$$flavin\text{-}H^{\bullet} + {}^{\bullet}O_2{}^{-} \rightarrow flavin\text{-}HOOH$$
Flavin hydroperoxide slowly breaks down to yield flavin semiquinone and perhydroxyl
$$flavin\text{-}HOOH \rightarrow flavin\text{-}H^{\bullet} + {}^{\bullet}O_2H$$
Perhydroxyl decays to superoxide plus a proton
$${}^{\bullet}O_2H \rightarrow H^{+} + {}^{\bullet}O_2{}^{-}$$
In the presence of H^{+}, flavin semiquinone and superoxide yield peroxide and oxidized flavin
$$flavin\text{-}H^{\bullet} + H^{+} + {}^{\bullet}O_2{}^{-} \rightarrow flavin + H_2O_2$$

By their production of superoxide and perhydroxyl radicals and hydrogen peroxide, flavin oxidases make a significant contribution to the so-called oxidant stress of the body. Overall, some 3% to 5% of the daily consumption of about 30 mol of oxygen by an adult human being is converted to singlet oxygen, hydrogen peroxide, and the superoxide, perhydroxyl and hydroxyl radicals, rather than undergoing complete reduction to water in the electron transport chain. There is thus a total production of 1.5 mol of reactive oxygen species daily, potentially capable of causing damage to membrane lipids, proteins, and nucleic acids.

Paradoxically, although the oxidation of reduced flavins contributes significantly to oxidant stress, it may also have a protective role. There is a significant amount of reduced riboflavin bound to protein in hepatocyte (and presumably other cell) membranes. This undergoes oxidation when the cells are exposed to oxygen. The superoxide so produced may have a protective role in trapping the considerably more reactive and damaging hydroxyl radicals produced in other reactions (Nokubo et al., 1989). Nevertheless, the fact that reduced flavins react nonenzymically with oxygen to yield superoxide and perhydroxy radicals (the autoxidation of flavins) suggests that they are potentially toxic in excess. This may explain not only the limitation of the absorption of riboflavin from the intestinal tract, but also the active efflux of free riboflavin from the central nervous system and the active secretion of the vitamin in the renal tubule.

7.3.6 NADPH Oxidase, the Respiratory Burst Oxidase

NADPH oxidase was originally described in activated macrophages, whose function is to generate reactive oxygen species and halogen radicals as part of the cytotoxic action against phagocytosed microorganisms. It catalyzes transfer of electrons from NADPH onto cytochrome b$_{558}$, which then reduces oxygen to yield 2 mol of superoxide and two protons. Activation of the oxidase requires increased formation of NADPH and hence increased oxidation of glucose through the pentose phosphate pathway (see Figure 6.4), the so-called respiratory burst.

The oxidase is a cell membrane–multienzyme complex. It has a cell surface receptor linked to a G-protein that activates a phosphatidyl inositol cascade leading to assembly and activation of the oxidase complex. The receptor is activated by the following:

1. Complement fragment C$_{5a}$, which arises from the antibody–antigen reaction.
2. Peptides containing the sequence N-formyl-Met-Leu-Phe. These may be bacterial peptides or peptides arising from mitochondria of damaged tissue.
3. A variety of endogenous responses to infection and mediators of inflammatory reactions, including platelet activating factor, leukotriene β_4, and interleukin-8.

The NADPH binding site is on the cytosolic side of the membrane, whereas the superoxide release site is either extracellular or on the luminal side of the phagocytic vesicle. The enzyme acts as an ion pump, because it releases superoxide without an accompanying cation; protons remain inside the cell, resulting in considerable membrane depolarization (Babior, 1992; Chanock et al., 1994).

NADPH oxidase activity has also been demonstrated in a wide variety of cells other than macrophages, where it is not involved in cytotoxic action against engulfed microorganisms. It seems likely that reactive oxygen species produced by NADPH oxidase have a role in cell signaling, possibly as part of the mechanism of apoptosis.

7.3.7 Molybdenum-Containing Flavoprotein Hydroxylases

Xanthine oxidoreductase and aldehyde oxidase represent a distinct class of flavin-dependent oxidases. They both dehydrogenate and hydroxylate the substrate. However, unlike the mixed-function oxidases (Section 7.3.8), the oxygen introduced into the substrate by these enzymes is derived from water, and the role of molecular oxygen is in the reoxidation of the reduced flavin. Among other reactions, aldehyde oxidase is important in the oxidation of N^1-methyl nicotinamide to methyl pyridone carboxamide (Section 8.2.4) and pyridoxal to 4-pyridoxic acid (Section 9.2).

The initial step in the reaction is dehydrogenation of the substrate at the expense of the molybdenum, which is reduced from Mo^{VI} to Mo^{IV}. This is followed by attack by the persulfide group on the carbon atom of the substrate at which the hydroxyl group will be introduced. Hydrolysis of the persulfide–carbon bond then introduces a hydroxyl group. The reduced molybdopterin is reoxidized by the iron-sulfur groups, which in turn reduce the FAD. The reduced flavin then reacts with oxygen, eventually forming hydrogen peroxide.

Xanthine oxidoreductase contains two molecules of molybdenum as molybdopterin (Section 10.5), two molecules of FAD, eight non-heme iron-sulfide groups, and two persulfide (—S—S—) groups. It exists as two interconvertible forms:

1. The xanthine dehydrogenase form catalyzes the oxidation of xanthine to hypoxanthine at the expense of NAD^+.
2. The xanthine oxidase form cannot utilize NAD^+, but reduces oxygen to hydrogen peroxide.

The dehydrogenase form of the enzyme is converted to the oxidase form by reversible oxidation of cysteine to form a disulfide bridge. The redox potential of the dehydrogenase form of the enzyme is considerably lower than that of the oxidase form, because the protein confers greater stability on the neutral flavin semiquinone radical (Rajagopalan and Johnson, 1992; Kisker et al., 1997; Nishino and Okamoto, 2000).

7.3.8 Flavin Mixed-Function Oxidases (Hydroxylases)
The flavin-dependent mixed-function oxidases include amine N-oxidases and a variety of S-oxidases. They provide an alternative to cytochrome P_{450}-dependent enzymes in the metabolism of xenobiotics.

In most of these enzymes, the flavin is reduced to dihydroflavin by NADPH, although some also act as dehydrogenases, both oxidizing and hydroxylating the substrate so that it is the substrate that is the source of hydrogen to form the dihydroflavin. This then forms a hydroperoxide by reaction with oxygen. Rather than decaying to the flavin and perhydroxyl radicals as in the oxidases discussed in Section 7.3.5, the hydroperoxide is stabilized by the enzyme protein and is cleaved by the substrate, resulting in transfer of a hydroxyl group and leaving the flavin C-4 hydroxide, that breaks down to yield water and the oxidized flavin, via the sequence of reactions shown in Table 7.4, again adding to the overall radical burden in the body.

Table 7.4 Reoxidation of Reduced Flavins in Flavin Mixed-Function Oxidases

---

The overall reaction is:

$X-H_2 + O_2 \rightarrow X-OH + H_2O$ or $X + NADPH + O_2 \rightarrow X-OH + NADP^+ + H_2O$

Flavin is reduced by reaction with either substrate-H_2 or NADPH.

Fully reduced flavin-H_2 reacts with oxygen to form the flavin semiquinone radical and superoxide

$$\text{flavin-}H_2 + O_2 \rightarrow \text{flavin-}H^{\cdot} + {\cdot}O_2^{-}$$

Flavin semiquinone and superoxide react to form flavin hydroperoxide

$$\text{flavin-}H^{\cdot} + {\cdot}O_2^{-} \rightarrow \text{flavin-HOOH}$$

Flavin hydroperoxide reacts with substrate

$$\text{flavin-HOOH} + X \rightarrow \text{flavin-HOOH-X}$$

Intermediate complex breaks down to hydroxylated product + flavin hydroxide

$$\text{flavin-HOOH-X} \rightarrow \text{flavin-OH} + X\text{-OH}$$

Flavin hydroxide breaks down to regenerate fully oxidized flavin + H_2O

$$\text{flavin-OH} \rightarrow \text{flavin} + H_2O$$

7.3.9 The Role of Riboflavin in the Cryptochromes

One of the mechanisms of DNA repair in bacteria, acting to reduce cyclobutane dipyrimidines and pyrimidine-pyrimidone dimers formed by exposure
to UV light, is the blue light-activated photolyase. The primary light-trapping
pigment is 5,10-methylene tetrahydrofolate (Section 10.1), which then transfers the excitation energy of the trapped photon to FADH, which reduces the
substrate.

Cryptochromes in the human eye have a considerable sequence and structure homology with the photolyases, binding both methylene tetrahydrofolate
and FAD. They have the same DNA binding pocket as photolyase, although
they do not catalyze the reduction of DNA pyrimidine dimers. They are found
in the nucleus of cells of the inner layer of the retina, behind the rods and
cones involved in vision (Section 2.3.1), and absorb blue light, with maximum
absorbance at 420 nm.

The function of cryptochromes is in setting the circadian clock in response
to day-length. In response to excitation by light, there are changes in the expression in the retinal cells of genes known to regulate the circadian cycle,
possibly as a result of increased ubiquitination and hence catabolism of the
TIM protein (the product of the *timeless* gene). In addition, a nerve impulse is
generated along fibers of the optic nerve that innervate the suprachiasmatic
nucleus in the anterior hypothalamus, rather than the visual cortex. It is not
known how this nerve impulse is initiated in response to photoexcitation of
cryptochrome (Sancar, 2000; Lin et al., 2001).

7.4 **RIBOFLAVIN DEFICIENCY**

Riboflavin deficiency is relatively common, yet there is no clear deficiency disease and the condition never seems to be fatal. This presumably reflects the high degree of conservation of riboflavin in tissues (Section 7.2.3). There is only a relatively small difference between the concentration of flavins at which tissues are saturated and the lowest levels in prolonged depletion of experimental animals. In deficiency, most of the flavin coenzymes released by the catabolism of enzymes are reutilized.

Clinically, riboflavin deficiency is characterized by lesions of the margin of the lips (cheilosis) and corners of the mouth (angular stomatitis), a painful desquamation of the tongue, so that it is red, dry, and atrophic (so-called magenta tongue) and a sebhorroic dermatitis, with filiform excrescences, affecting especially the naso-labial folds, eyelids, and ears, with abnormalities of the skin around the vulva and anus and at the free border of the prepuce. The lesions of the mouth may respond to either riboflavin or vitamin B_6 in apparently riboflavin-deficient subjects (Lakshmi and Bamji, 1974).

There may also be conjunctivitis with vascularization of the cornea and opacity of the lens. This is the only lesion for which we know a possible biochemical basis – glutathione is important in maintaining the normal clarity of crystallin in the lens, and glutathione reductase is a flavoprotein that is particularly sensitive to riboflavin depletion.

7.4.1 Impairment of Lipid Metabolism in Riboflavin Deficiency

Riboflavin-deficient animals have a lower metabolic rate than controls, and require a 15% to 20% higher food intake to maintain body weight. There is increased accumulation of triglycerides in the liver, with an increase in liver weight as a proportion of body weight. There is no impairment of electron transport in the liver, although in brown adipose tissue both electron transport and the thermogenic response to adrenergic stimulation are impaired (Duerden and Bates, 1985).

The main effect of riboflavin deficiency is on lipid metabolism. In experimental animals on a riboflavin-free diet, feeding a high-fat diet leads to more marked impairment of growth, and a higher requirement for riboflavin to restore growth. There are also changes in the patterns of long-chain polyunsaturated fatty acids in membrane phospholipids.

Within a day of initiating a riboflavin-free diet in weanling rats, there is a 35% decrease in the oxidation of palmitoyl CoA. All three mitochondrial acyl CoA dehydrogenases are affected, although it is the short-chain acyl CoA dehydrogenase that is most severely impaired and that becomes the rate-limiting step of fatty acid oxidation. The accumulating short-chain fatty acyl CoA derivatives may undergo microsomal ω-oxidation and possibly peroxisomal β-oxidation of the resultant dicarboxylic acids. Although mitochondrial β-oxidation is impaired in riboflavin deficiency, the peroxisomes appear to be protected. As a result, a number of dicarboxylic acids (including adipic, suberic, sebacic, octenedioic, hexenedioic, and decendioic acids) are excreted in the urine. In addition, a number of conjugates of the substrates of impaired acyl CoA dehydrogenases are excreted, including butyryl-, isovaleryl-, 2-methylbutyl-, and isobutyl-glycine conjugates. There are a number of riboflavin responsive organic acidurias caused by impairment of one or another of the acyl CoA dehydrogenases (Goodman, 1981; Veitch et al., 1988).

In animals, the production of $^{14}CO_2$ from [^{14}C]palmitate or octanoate is not consistently affected by riboflavin deficiency, possibly as a result of increased activity of carnitine palmitoyl transferase, which is more a response to food deprivation than to riboflavin deficiency. However, the production of $^{14}CO_2$ from [^{14}C]adipic acid is significantly reduced, and responds rapidly (with some overshoot) to repletion with the vitamin. It has been suggested that the ability to metabolize a test dose of [^{13}C]adipic acid may provide a sensitive means of investigating riboflavin nutritional status in human beings (Bates, 1989, 1990).

7.4.2 Resistance to Malaria in Riboflavin Deficiency

A number of studies have noted that, in areas where malaria is endemic, riboflavin-deficient subjects are relatively resistant and have a lower parasite burden than adequately nourished subjects. Dietary deficiency of riboflavin, hypothyroidism, which induces functional riboflavin deficiency by lowering the synthesis of flavokinase (Section 7.2.4), or the administration of chlorpromazine, which inhibits flavokinase and can cause functional riboflavin deficiency (Section 7.4.4), all inhibit the growth of malarial parasites in experimental animals. However, although parasitemia is less in riboflavin deficiency, the course of the disease may be more severe (Dutta et al., 1985; Dutta, 1991; Akompong et al., 2000a, 2000b; Shankar, 2000).

The biochemical basis of this resistance to malaria in riboflavin deficiency is not known, but a number of mechanisms have been proposed, including the following:

1. The malarial parasites may have a particularly high requirement for riboflavin. A number of flavin analogs have antimalarial action.

2. The impairment of glutathione reductase activity may result in lower availability of glutathione in erythrocytes and hence a more oxidizing environment, which is hostile to the parasites.

3. As a result of impaired antioxidant activity in erythrocytes, there may be increased fragility of erythrocyte membranes or reduced membrane fluidity. As in sickle cell trait, which also protects against malaria, this may result in:

    (a) Exposure of the parasites to the host's immune system at a vulnerable stage in their development, resulting in the production of protective antibodies.

    (b) Release into the circulation of immature forms of the parasite that are not capable of either surviving outside the erythrocyte or infecting new cells.

In vitro, high concentrations of riboflavin also impair parasite growth, apparently as a result of increased reduction of methemoglobin. The parasites can only hydrolyze methemoglobin, not native hemoglobin.

7.4.3 Secondary Nutrient Deficiencies in Riboflavin Deficiency

Riboflavin deficiency is associated with hypochromic anemia as a result of secondary iron deficiency. The absorption of iron is impaired in riboflavin-deficient animals, with a greater proportion of a test dose retained in the intestinal mucosal cells bound to ferritin, and hence lost in the feces, rather than being absorbed. The mobilization of iron bound to ferritin, in either intestinal mucosal cells or the liver, for transfer to transferrin, requires oxidation of Fe^{2+} to Fe^{3+}, a reaction catalyzed by NAD-riboflavin phosphate oxidoreductase (Powers et al., 1991; Powers, 1995; Williams et al., 1995).

At least part of the impairment of iron absorption in riboflavin deficiency is a result of morphological changes in the intestinal mucosa, with hyperproliferation, an increased rate of enterocyte transit along the villi and a reduced number of (longer) villi and deeper crypts (Williams et al., 1996).

In addition to the role of flavoproteins in iron metabolism, it is possible that the anemia associated with riboflavin deficiency is a consequence of the impairment of vitamin B_6 metabolism in riboflavin deficiency. Pyridoxine oxidase is a flavoprotein and, like glutathione reductase, is very sensitive to riboflavin depletion (McCormick, 1989). Vitamin B_6 deficiency can result in hypochromic anemia as a result of impaired porphyrin synthesis. Although riboflavin depletion decreases the oxidation of dietary vitamin B_6 to pyridoxal (Section 9.2), it is not clear to what extent there is secondary vitamin B_6 deficiency in riboflavin deficiency. This is partly because vitamin B_6 nutritional status is commonly assessed by the metabolism of a test dose of tryptophan (Section 9.5.4), and kynurenine hydroxylase in the tryptophan oxidative pathway is a flavoprotein (Section 8.3.3.1); riboflavin deficiency can therefore disturb tryptophan metabolism quite separately from its effects on vitamin B$_6$ nutritional status. In riboflavin-deficient animals, despite a decrease in pyridoxine oxidase to 15% of the control activity, and an increase in the concentration of pyridoxine in tissues, there is no significant decrease in the tissue concentration of pyridoxal phosphate (Lakshmi and Bamji, 1974).

The disturbance of tryptophan metabolism in riboflavin deficiency, caused by impairment of kynurenine hydroxylase, can also result in reduced synthesis of NAD from tryptophan. This may therefore be a factor in the etiology of pellagra (Section 8.3.3.1).

In species for which ascorbate is not a vitamin, riboflavin deficiency can also lead to considerably reduced synthesis and low tissue concentrations of ascorbate, since gulonolactone oxidase, the key enzyme in ascorbate synthesis (Section 13.2), is a flavoprotein.

7.4.4 Iatrogenic Riboflavin Deficiency

The phenothiazines, such as chlorpromazine, used in the treatment of schizophrenia, the tricyclic antidepressant drugs such as imipramine and amitryptiline, antimalarials such as quinacrine, and the anticancer agent adriamycin are structural analogs of riboflavin (see Figure 7.6) and inhibit flavokinase. In experimental animals, administration of these drugs at doses equivalent to those used clinically results in an increase in the EGR activation coefficient (Section 7.5.2) and increased urinary excretion of riboflavin, with reduced tissue concentrations of riboflavin phosphate and FAD, despite feeding diets providing more riboflavin than is needed to meet requirements (Pinto et al., 1981).

Although there is no evidence that patients treated with these drugs for a prolonged period develop clinical signs of riboflavin deficiency, long-term use of chlorpromazine is associated with a reduction in metabolic rate.

Neonatal hyperbilirubinemia is normally treated by phototherapy. The peak wavelength for photolysis of bilirubin is 450 nm, the same as that for photolysis of riboflavin (Section 7.1). Infants undergoing phototherapy show biochemical evidence of riboflavin depletion, with a significant increase in the EGR activation coefficient. Provision of additional riboflavin to maintain plasma concentrations enhances the photolysis of bilirubin, apparently as a result of reactive oxygen radicals generated by the products of photolysis of riboflavin.

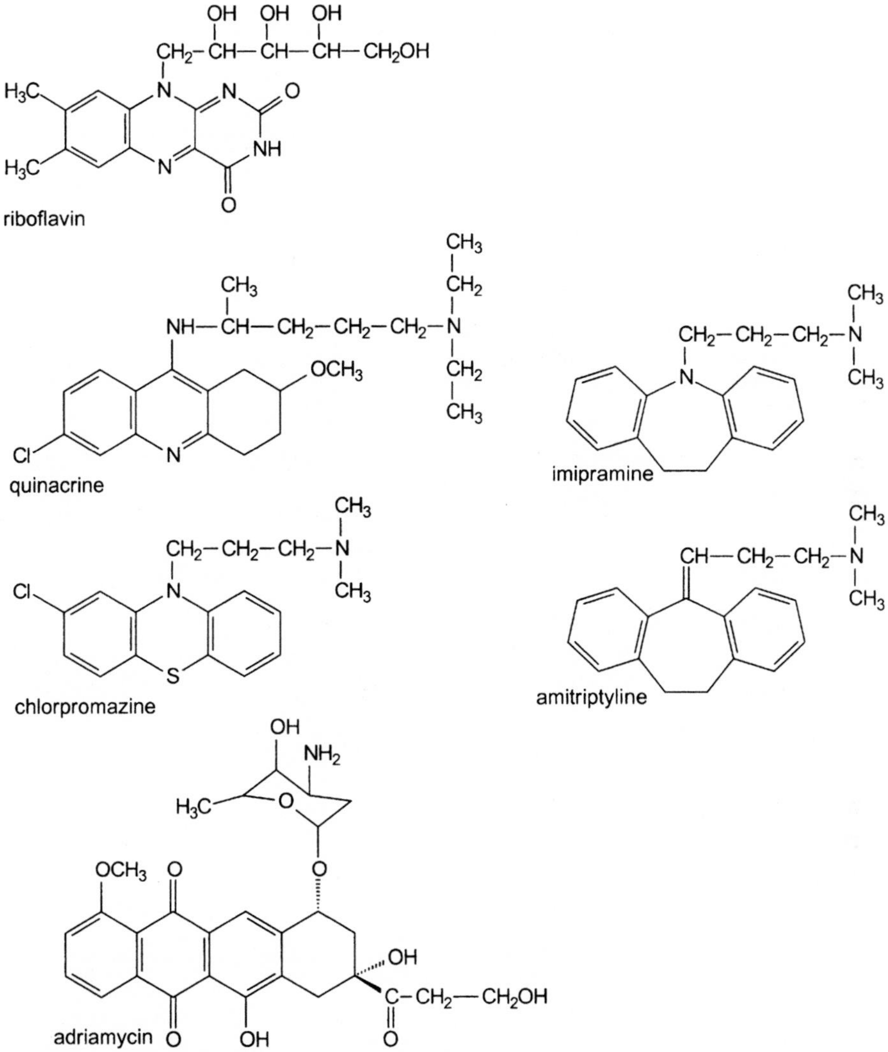

Figure 7.6. Drugs that are structural analogs of riboflavin and may cause deficiency. Relative molecular masses (M_r): riboflavin, 376.4; quinacrine, 472.9 (dihydrochloride); chlorpromazine, 318.9; imipramine, 280.4; amitryptyline, 277.4; and adriamycin (doxorubicin), 543.5.

However, even relatively low concentrations of riboflavin can cause damage to DNA under conditions of photolysis, with damage to deoxy-guanosine in isolated DNA, and activation of DNA repair mechanisms in cells in culture. It is therefore not common practice to use riboflavin supplements as an adjunct

Table 7.5 Indices of Riboflavin Nutritional Status

		Adequate	Marginal	Deficient
Urine riboflavin	μg/g creatinine	>80	27–80	<27
	mol/mol creatinine	>24	8–24	<8
	μg/24 h	>120	40–120	<40
	nmol/24 h	>300	100–300	<100
	mg over 4 h after 5 mg dose	>1.4	1.0–1.4	<1.0
	μmol over 4 h after 5 mg dose	>3.7	2.7–3.7	<2.7
Erythrocyte riboflavin	μg/g hemoglobin	>0.45	—	—
	nmol/g hemoglobin	>1.2	—	—
Glutathione reductase	Activation coefficient	<1.4	1.4–1.7	>1.7

Sources: From data reported by Sauberlich et al., 1974; Bates, 1993.

to phototherapy of neonatal hyperbilirubinemia (Speck et al., 1975; Gromisch et al., 1977).

7.5 ASSESSMENT OF RIBOFLAVIN NUTRITIONAL STATUS

Two methods of assessing riboflavin status are generally used: urinary excretion of the vitamin and its metabolites, and activation of EGR. Criteria of riboflavin adequacy are shown in Table 7.5.

7.5.1 Urinary Excretion of Riboflavin

Clinical signs of riboflavin deficiency are seen at intakes below about 1 mg per day. At intakes below about 1.1 mg per day, there is very little urinary excretion of riboflavin; thereafter, as intake increases, there is a sharp increase in excretion. Up to about 2.5 mg per day, there is a linear relationship between intake and excretion. At higher levels of intake, excretion increases sharply, reflecting active renal secretion of excessive vitamin (Section 7.2.5).

Riboflavin excretion is only correlated with intake in subjects who are maintaining nitrogen balance. In subjects in negative nitrogen balance, there may be more urinary excretion than would be expected, largely as a result of the catabolism of tissue flavoproteins and loss of their prosthetic groups. Higher intakes of protein than are required to maintain nitrogen balance do not affect the requirement for riboflavin or indices of riboflavin nutritional status, although, as might be expected, more riboflavin is retained in subjects in positive nitrogen balance, as a result of increased net synthesis of flavoproteins.

7.5.2 Erythrocyte Glutathione Reductase (EGR) Activation Coefficient

Glutathione reductase is especially sensitive to riboflavin depletion. In deficient animals, the activity of glutathione reductase responds earlier and more markedly than any other index of riboflavin status apart from liver concentrations of flavin coenzymes and the activity of hepatic flavokinase (Prentice and Bates, 1981a, 1981b). The activity of the enzyme in erythrocytes can therefore be used as an index of riboflavin status.

Interpretation of the results can be complicated by anemia, and it is more usual to use the activation of EGR by FAD added in vitro. An activation coefficient of 1.0 to 1.3 reflects adequate nutritional status; >1.7 indicates deficiency. EGR activation coefficient between 1.3 to 1.7 represents a marginal status, with no clinical signs of deficiency.

Like glutathione reductase, pyridoxine oxidase is sensitive to riboflavin depletion. In normal subjects and in experimental animals, the EGR and pyridoxine oxidase activation coefficients are correlated, and both reflect riboflavin nutritional status. In subjects with glucose 6-phosphate dehydrogenase deficiency, there is an apparent protection of EGR, so that even in riboflavin deficiency it does not lose its cofactor, and the EGR activation coefficient remains within the normal range. The mechanism of this protection is unknown. In such subjects, the erythrocyte pyridoxine oxidase activation coefficient gives a response that mirrors riboflavin nutritional status (Clements and Anderson, 1980).

7.6 RIBOFLAVIN REQUIREMENTS AND REFERENCE INTAKES

On the basis of depletion/repletion studies, the minimum adult requirement for riboflavin is 0.5 to 0.8 mg per day. In population studies, values of the EGR activation coefficient <1.3 are seen in subjects whose habitual intake of riboflavin is 1.2 to 1.5 mg per day. At intakes between 1.1 to 1.6 mg per day, urinary excretion rises sharply, suggesting that tissue reserves are saturated. On the basis of such studies, reference intakes (see Table 7.6) are in the range of 1.2 to 1.6 mg per day (Bates, 1987a, 1987b).

Because of the central role of flavin coenzymes in energy-yielding metabolism, reference intakes are sometimes calculated on the basis of energy intake: 0.6 to 0.8 mg per 1,000 kcal (0.14 to 0.19 mg per MJ). However, in view of the wide range of riboflavin-dependent reactions, in addition to energy-yielding metabolism, it is difficult to justify this basis for the calculation of requirements.

Table 7.6 Reference Intakes of Riboflavin (mg/day)

Age	U.K. 1991	EU 1993	U.S./Canada 1998	FAO 2001
0–6 m	0.4	—	0.3	0.3
7–12 m	0.4	0.4	0.4	0.4
1–3 y	0.6	0.8	0.5	0.5
4–6 y	0.8	1.0	0.6	0.6
7–8 y	1.0	1.2	0.6	0.9
Males				
9–10 y	1.0	1.2	0.9	0.9
11–13 y	1.2	1.4	0.9	1.3
>14 y	1.3	1.6	1.3	1.3
Females				
9–10 y	1.0	1.2	0.9	0.9
11–13 y	1.1	1.2	0.9	1.1
>14 y	1.1	1.3	1.1	1.1
Pregnant	1.4	1.6	1.4	1.4
Lactating	1.6	1.7	1.6	1.6

EU, European Union; FAO, Food and Agriculture Organization; WHO, World Health Organization. *Sources:* Department of Health, 1991; Scientific Committee for Food, 1993; Institute of Medicine, 1998; FAO/WHO, 2001.

7.7 PHARMACOLOGICAL USES OF RIBOFLAVIN

Because of its intense yellow color and low toxicity, riboflavin is widely used as a food color (E-101). It is also used in relatively high doses in the treatment of recessive familial methemoglobinemia and some organic acidurias.

Recessive familial methemoglobinemia is from lack of NADH-dependent cytochrome b$_5$ methemoglobin reductase, which is the major enzyme involved in reduction of methemoglobin. (The more common methemoglobinemias are caused by mutations in the hemoglobin gene and are commonly dominant conditions.) Reduced flavins will reduce methemoglobin nonenzymically, and doses of the vitamin of 20 to 40 mg per day result in a significant accumulation of reduced flavins in erythrocytes, as a result of the activity of NADH-flavin reductase (Yubisui et al., 1977).

Some congenital organic acidurias resulting from apparent deficiency of acyl CoA dehydrogenases are riboflavin responsive. The defects seem to result from impaired coenzyme binding to either the electron-transferring flavoprotein that transfers electrons from a variety of acyl CoA dehydrogenases into the electron transport chain, or electron-transferring flavoprotein-ubiquinone reductase. Administration of 100 mg of riboflavin per day or more seems to permit accumulation of sufficient flavin coenzymes to give useful activity of the affected enzyme, despite the limitation of riboflavin absorption (Christensen et al., 1984; Gregersen et al., 1986).

There is some evidence that riboflavin status affects the stability of the thermolabile variant of methylene tetrahydrofolate reductase (Section 10.3.2.1), and that supplements of riboflavin may lower plasma homocysteine (Section 10.3.4.2) in people who are homozygous for the variant enzyme (McNulty et al., 2002).

Because of its low solubility and limited absorption from the gastrointestinal tract, riboflavin has no significant or measurable toxicity by mouth. At extremely high parenteral doses (300 to 400 mg per kg of body weight), there may be crystallization of riboflavin in the kidney because of its low solubility.

FURTHER READING

Adiga PR (1994) Riboflavin carrier protein in reproduction. In *Vitamin Receptors: Vitamins as Ligands in Cell Communication*, K Dakshinamurti (ed.), pp. 137–76. Cambridge: Cambridge University Press.

Bates CJ (1987) Human requirements for riboflavin. *American Journal of Clinical Nutrition* **46,** 122–3.

Bates CJ (1987) Human riboflavin requirements, and metabolic consequences of deficiency in man and animals. *World Review of Nutrition and Dietetics* **50,** 215–65.

Ghisla S and Massey V (1989) Mechanisms of flavoprotein-catalyzed reactions. *European Journal of Biochemistry* **181,** 1–17.

Massey V (2000) The chemical and biological versatility of riboflavin. *Biochemical Society Transactions* **28,** 283–96.

McCormick DB (1989) Two interconnected B vitamins: riboflavin and pyridoxine. *Physiological Reviews* **69,** 1170–98.

Shankar AH (2000) Nutritional modulation of malaria morbidity and mortality. *Journal of Infectious Diseases* **182**(Suppl. 1), S37–S53.

References cited in the text are listed in the Bibliography.

EIGHT

Niacin

Niacin is unusual among the vitamins in that it was discovered as a chemical compound, nicotinic acid produced by the oxidation of nicotine, in 1867 – long before there was any suspicion that it might have a role in nutrition. Its metabolic function as part of what was then called coenzyme II [nicotinamide adenine dinucleotide phosphate (NADP)] was discovered in 1935, again before its nutritional significance was known.

It is not strictly correct to regard niacin as a vitamin. Its metabolic role is as the precursor of the nicotinamide moiety of the nicotinamide nucleotide coenzymes, nicotinamide adenine dinucleotide (NAD) and NADP, and this can also be synthesized in vivo from the essential amino acid tryptophan. At least in developed countries, average intakes of protein provide more than enough tryptophan to meet requirements for NAD synthesis without any need for preformed niacin. It is only when tryptophan metabolism is disturbed, or intake of the amino acid is inadequate, that niacin becomes a dietary essential.

The nicotinamide nucleotide coenzymes function as electron carriers in a wide variety of redox reactions. In addition, NAD is the precursor of adenine dinucleotide phosphate (ADP)-ribose for ADP-ribosylation and poly(ADP-ribosylation) of proteins and cADP-ribose and nicotinic acid adenine dinucleotide phosphate (NAADP). They act as second messengers and stimulate increases in intracellular calcium concentrations.

Pellagra was first described as *mal de la rosa* in Asturias in central Spain by Casal in 1735. He observed that the condition was apparently related to diet and was distinct from scurvy and other then known causes of superficially similar dermatitis. The name pellagra was coined by the Italian physician Frapolli in 1771 to describe the most striking feature of the disease: the roughened, sunburn-like appearance of the skin. Pellagra became common in Europe when maize was introduced from the New World as a convenient high-yielding dietary staple. By the nineteenth century, it was widespread throughout southern Europe and north Africa. The disease was unknown in southern Africa until the outbreak of rinderpest in 1897, which led to widespread death of cattle and a major change in the dietary habits of the Bantu.

From being a meat- and milk-eating community, they became, and have remained, largely maize eaters, and pellagra continued to be a major problem of public health nutrition in South Africa throughout much of the twentieth century. The other region where pellagra was a major problem at the beginning of the twentieth century was the southern part of the United States. The social and economic upheaval of the American Civil War led to a poor maize-based diet for large sections of the population, and it was not until the entry of the United States into the second World War that increasing employment and a rise in the general standard of living solved the dietary problem. Although Casal had considered pellagra to be due to a dietary deficiency, investigations at the beginning of the twentieth century started from the assumption that, like other diseases, it was an infection. It was the pioneering studies of Goldberger and coworkers in the United States that showed that the condition was neither contagious nor infectious, but could be prevented or cured by dietary means.

After it had been established that pellagra was a nutritional deficiency disease, the next problem was to discover the missing nutrient. Additional dietary protein was shown to be beneficial, thus it was concluded that pellagra was because of a protein deficiency. This view, and later that it was more specifically from a deficiency of tryptophan, was held for some time. In 1938, Spies and coworkers showed that nicotinic acid would cure pellagra; thereafter it was gradually accepted that it was a niacin deficiency disease.

The role of additional dietary protein in curing pellagra was elucidated in 1947, when it was shown that the administration of tryptophan to human beings led to an increase in the urinary excretion of N^1-methyl nicotinamide, the major urinary metabolite of niacin. It is usual to regard pellagra as a niacin deficiency disease, and tryptophan as a substitute for niacin when the dietary intake of the vitamin is inadequate. However, this is not strictly correct, and pellagra should be regarded as being a deficiency of both tryptophan and niacin.

8.1 NIACIN VITAMERS AND NOMENCLATURE

The term niacin is the generic descriptor for the two compounds that have the biological action of the vitamin: nicotinic acid and nicotinamide (see

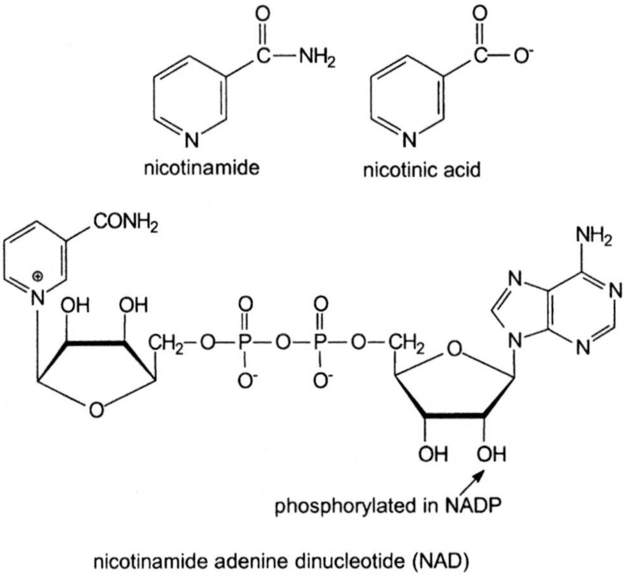

Figure 8.1. Niacin vitamers, nicotinamide and nicotinic acid, and the nicotinamide nucleotide coenzymes. Relative molecular masses (M_r): nicotinic acid, 123.1; nicotinamide, 122.1; NAD, 663.4; and NADP 743.4.

Figure 8.1). Although the amino acid tryptophan is quantitatively the major precursor of the nicotinamide ring of the coenzymes (Section 8.3), it is not considered to be a niacin vitamer.

Nicotinic acid was discovered and named as a product of the chemical oxidation of nicotine in 1867. When it was later discovered to be the pellagra-preventing vitamin, it was not assigned a number among the B vitamins because its chemistry was already known. Niacin is generally placed between vitamins B_2 and B_6, although it is incorrect to call it vitamin B_3, which was at one time assigned to pantothenic acid (Section 12.1).

There is confusion in the literature because of the North American usage of the name niacin to mean specifically nicotinic acid, whereas the amide is known as niacinamide. The name *niacin* was coined in the late 1940s when the role of deficiency in the etiology of pellagra was realized, and it was decided that dietary staples should be fortified with the vitamin. It was felt that nicotinic acid was not a suitable name for a substance that was to be added to foods, both because of its phonetic (and chemical) relationship to nicotine and because it is an acid.

Nicotinic acid and nicotinamide have equal biological activity. As discussed in Section 8.3, approximately 60 mg of tryptophan is equivalent to 1 mg of dietary preformed niacin. Requirements and intakes are calculated as mg niacin equivalents – the sum of preformed niacin + 1/60 × tryptophan.

8.2 NIACIN METABOLISM

8.2.1 Digestion and Absorption

Niacin is present in tissues, and therefore in foods, largely as the nicotinamide nucleotides. The postmortem hydrolysis of NAD(P) is extremely rapid in animal tissues, and it is likely that much of the niacin of meat (a major dietary source of the vitamin) is free nicotinamide.

Nicotinamide nucleotides present in the intestinal lumen are hydrolyzed to nicotinamide. A number of intestinal bacteria have high nicotinamide deamidase activity, and a significant proportion of dietary nicotinamide may be deamidated in the intestinal lumen. Both nicotinic acid and nicotinamide are absorbed from the small intestine by a sodium-dependent saturable process.

8.2.1.1 Unavailable Niacin in Cereals
Chemical analysis reveals niacin in cereals (largely in the bran), but this is biologically unavailable, because it is bound as niacytin – nicotinoyl esters to a variety of macromolecules ranging between M_r 1,500 to 17,000. In wheat bran, 60% is esterified to polysaccharides, and the remainder to polypeptides and glycopeptides (Mason et al., 1973). In calculation of niacin intakes, it is conventional to ignore the niacin content of cereals completely.

A small fraction of the niacin in niacytin may be biologically available as a result of hydrolysis by gastric acid. About 10% of the total is released as free nicotinic acid after extraction of maize or sorghum meal with 0.1 mol per L of hydrochloric acid, and Carter and Carpenter (1982) have shown that about 10% of the total niacin content of maize is biologically available to humans beings.

Treatment of cereals with alkali (e.g., soaking overnight in a calcium hydroxide solution, as is the traditional method for the preparation of tortillas in Mexico) and baking with alkaline baking powder, releases much of the nicotinic acid. This may explain why pellagra has always been rare in Mexico, despite the fact that maize is the dietary staple. Roasting of whole grain maize has a similar effect, because there is enough ammonia released from glutamine to form free nicotinamide by ammonolysis.

8.2.2 Synthesis of the Nicotinamide Nucleotide Coenzymes

As shown in Figure 8.2, the nicotinamide nucleotide coenzymes NAD and NADP can be synthesized from either of the niacin vitamers or from quinolinic

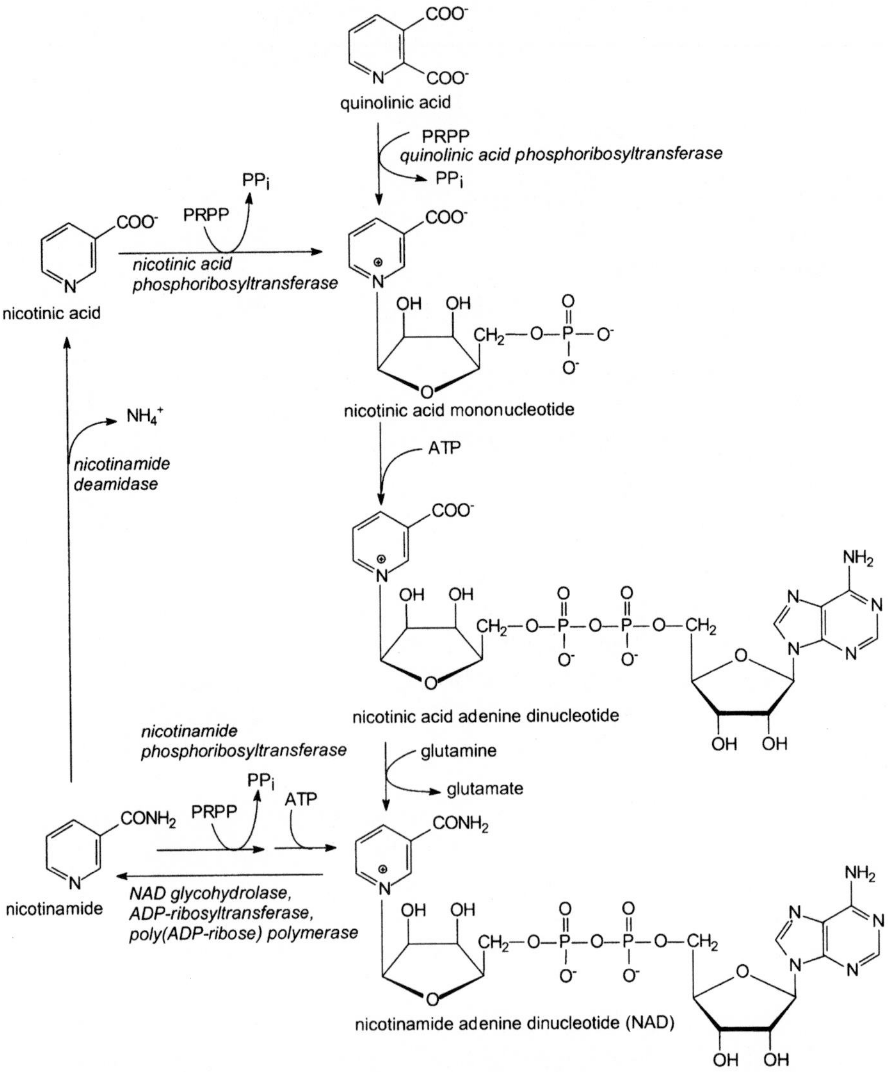

Figure 8.2. Synthesis of NAD from nicotinamide, nicotinic acid, and quinolinic acid. Quinolinate phosphoribosyltransferase, EC 2.4.2.19; nicotinic acid phosphoribosyltransferase, EC 2.4.2.11; nicotinamide phosphoribosyltransferase, EC 2.4.2.12; nicotinamide deamidase, EC 3.5.1.19; NAD glycohydrolase, EC 3.2.2.5; NAD pyrophosphatase, EC 3.6.1.22; ADP-ribosyltransferases, EC 2.4.2.31 and EC 2.4.2.36; and poly(ADP-ribose) polymerase, EC 2.4.2.30. PRPP, phosphoribosyl pyrophosphate.

acid, a metabolite of the amino acid tryptophan. Incorporation of nicotinic acid is catalyzed by nicotinate phosphoribosyltransferase to yield nicotinic acid mononucleotide, which is converted to nicotinic acid adenine
dinucleotide (desamido-NAD) by the action of nicotinic acid mononucleotide
pyrophosphorylase, and then amidated to NAD. Incorporation of nicotinamide may either be direct, catalyzed by nicotinamide phosphoribosyltransferase to yield nicotinamide mononucleotide, and then NAD by the action of
nicotinamide mononucleotide pyrophosphorylase, or indirect, by deamidation to nicotinic acid, catalyzed by nicotinamide deamidase. The dicarboxylic
acid intermediate in the quinolinate phosphoribosyltransferase reaction undergoes spontaneous decarboxylation to nicotinic acid mononucleotide.

In the liver, there is little utilization of preformed niacin for nucleotide
synthesis. Although isolated hepatocytes will take up both vitamers from the
incubation medium, they seem not to be used for NAD synthesis and cannot
prevent the fall in intracellular NAD(P), which occurs during incubation. The
enzymes for nicotinic acid and nicotinamide utilization are more or less saturated with their substrates at normal concentrations in the liver, and hence
are unlikely to be able to use additional niacin for nucleotide synthesis. By
contrast, incubation of isolated hepatocytes with tryptophan results in a considerable increase in the rate of synthesis of NAD(P) and accumulation of
nicotinamide and nicotinic acid in the incubation medium. Similarly, feeding experimental animals on diets providing high intakes of nicotinic acid or
nicotinamide has relatively little effect on the concentration of NAD(P) in the
liver, whereas high intakes of tryptophan lead to a considerable increase. It
thus seems likely that the major role of the liver is to synthesize NAD(P) from
tryptophan, followed by hydrolysis to release niacin for use by extrahepatic tissues (Bender et al., 1982; McCreanor and Bender, 1986; Bender and Olufunwa,
1988).

In most extrahepatic tissues, nicotinic acid is a better precursor of nucleotides than is nicotinamide. However, muscle, brain, and to a lesser extent
the testis are able to take up nicotinamide from the bloodstream effectively,
and apparently utilize it without prior deamidation (Gerber and Deroo, 1970).

8.2.3 Catabolism of NAD(P)
The nicotinamide nucleotide coenzymes are catabolized by four enzymes,
which act on the oxidized, but not the reduced, coenzymes:

1. NAD pyrophosphatase, which releases nicotinamide mononucleotide. This can either be hydrolyzed by NAD glycohydrolase to release

nicotinamide, or can be a substrate for nicotinamide mononucleotide pyrophosphorylase, to form NAD.

2. NAD glycohydrolase, which releases nicotinamide and ADP-ribose. As discussed in Section 8.4.4, this enzyme also catalyzes the synthesis of cADP-ribose and nicotinic acid ADP, which have roles in intracellular signaling.

3. ADP-ribosyltransferase, which catalyzes ADP-ribosylation of proteins, releasing nicotinamide (Section 8.4.2).

4. Poly(ADP-ribose) polymerase, which catalyzes poly-ADP-ribosylation of proteins, again releasing nicotinamide (Section 8.4.3).

The total NADase activity of tissues from these four enzymes is very high, and the total tissue content of nicotinamide nucleotides can be hydrolyzed within a few minutes. Two factors prevent this in vivo. Apart from NAD pyrophosphatase, the enzymes that catalyze the release of nicotinamide from NAD(P) are biosynthetic rather than catabolic, and their activity is highly regulated under normal conditions. Furthermore, the values of K_m of the enzymes are of the same order of magnitude as those of many of the NAD(P)-dependent enzymes in the cell, so that there is considerable competition for the nucleotides. Only that relatively small proportion of the nicotinamide nucleotide pool in the cell that is free at any one time will be immediately available for hydrolysis.

8.2.4 Urinary Excretion of Niacin Metabolites

Under normal conditions, there is little or no urinary excretion of either nicotinamide or nicotinic acid, because both vitamers are actively reabsorbed from the glomerular filtrate. It is only when the concentration is so high that the transport mechanism is saturated that there is any significant excretion.

As shown in Figure 8.3, the principal metabolites of nicotinamide are N^1-methyl nicotinamide and methyl pyridone carboxamides. N^1-Methyl nicotinamide is actively secreted into the urine by the proximal renal tubules. Nicotinamide N-methyltransferase is an S-adenosylmethionine–dependent enzyme that is present in most tissues. Very high intakes of nicotinamide may deplete tissue pools of one-carbon fragments – indeed, this was the basis for the use of nicotinamide in the treatment of schizophrenia (Section 8.8).

N^1-Methyl nicotinamide can also be oxidized to methyl pyridone-2-carboxamide and methyl pyridone-4-carboxamide. The extent to which this oxidation occurs, and the relative proportions of the two pyridones formed,

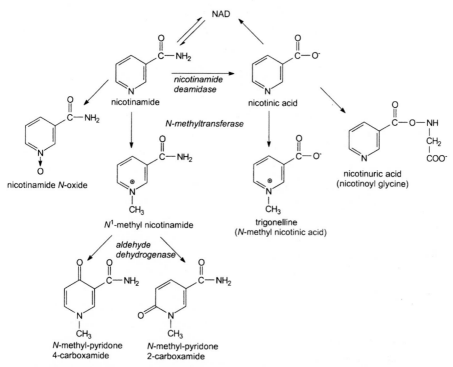

Figure 8.3. Metabolites of nicotinamide and nicotinic acid. Nicotinamide deamidase (nicotinamidase), EC 3.5.1.19; nicotinamide *N*-methyltransferase, EC 2.1.1.1; aldehyde dehydrogenase, EC 1.1.1.1. Relative molecular masses (M_r): nicotinamide, 123.1; nicotinic acid, 122.1; nicotinamide *N*-oxide, 139.1; N^1-methyl nicotinamide, 139.1; trigonelline, 137.1; nicotinuric acid, 179.2; and methyl pyridone carboxamides, 154.1.

not only varies from one species to another, but shows considerable variation between different strains of the same species. Aldehyde oxidase catalyzes the formation of both pyridones, and some additional 2-pyridone arises from the activity of xanthine oxidase. Aldehyde oxidase is activated by androgens, and male mice excrete 2 to 3 times more pyridone than do females (Felsted and Chaykin, 1967; Stanulovic and Chaykin, 1971a, 1971b).

Nicotinamide can also undergo oxidation to nicotinamide *N*-oxide. This is normally a minor metabolite in human beings, unless large amounts (about 200 mg) of nicotinamide are ingested. In the mouse, nicotinamide *N*-oxide is the major excretory product of niacin metabolism. At high levels of nicotinamide intake, some 6-hydroxynicotinamide may also be excreted.

Nicotinic acid can be conjugated with glycine to form nicotinuric acid (nicotinoyl-glycine), or may be methylated to trigonelline (N^1-methyl nicotinic acid). It is not clear to what extent urinary excretion of trigonelline reflects endogenous methylation of nicotinic acid, because there are significant amounts of trigonelline in foods that may be absorbed, but cannot be utilized as a source of niacin, and are excreted unchanged. Small amounts of 6-hydroxynicotinic acid may also be formed.

8.3 THE SYNTHESIS OF NICOTINAMIDE NUCLEOTIDES FROM TRYPTOPHAN

As shown in Figure 8.2, NAD(P) can be synthesized from the tryptophan metabolite quinolinic acid. The oxidative pathway of tryptophan metabolism is shown in Figure 8.4. Under normal conditions, almost all of the dietary intake of tryptophan, apart from the small amount that is used for net new protein synthesis, is metabolized by this pathway, and hence is potentially available for NAD synthesis. About 1% of tryptophan metabolism is by way of 5-hydroxylation and decarboxylation to 5-hydroxytryptamine (serotonin), which is excreted mainly as 5-hydroxyindoleacetic acid.

A number of studies have investigated the equivalence of dietary tryptophan and preformed niacin as precursors of the nicotinamide nucleotides, generally by determining the excretion of N^1-methyl nicotinamide and methyl pyridone carboxamide in response to test doses of the precursors, in subjects maintained on deficient diets.

The most extensive such study was that of Horwitt and coworkers (1956). They found that there was a considerable variation between subjects in the response to tryptophan and niacin, and suggested that in order to allow for individual variation, it should be assumed that 60 mg of tryptophan was equivalent to 1 mg of preformed niacin. This ratio has been generally accepted, and is the basis for expressing niacin requirements and intake in terms of niacin equivalents – the sum of preformed niacin and 1/60 of the tryptophan.

Changes in hormonal status may result in considerable changes in this ratio, between 7 to 30 mg of dietary tryptophan equivalent to 1 mg of preformed niacin in late pregnancy. The intake of tryptophan also affects the ratio. At low intakes, 1 mg of tryptophan may be equivalent to only 1/125 mg of preformed niacin (Nakagawa et al., 1969).

Tryptophan dioxygenase (Section 8.3.2) is only found in the liver; other tissues have an indoleamine dioxygenase, with lower specificity, that catalyzes the same reaction. However, the pathway for onward metabolism of kynurenine is found only in liver and mononuclear phagocytes, and induction of indoleamine dioxygenase by cytokines, such as interferon-γ, leads to increased circulating concentrations and urinary excretion of kynurenine, with little or

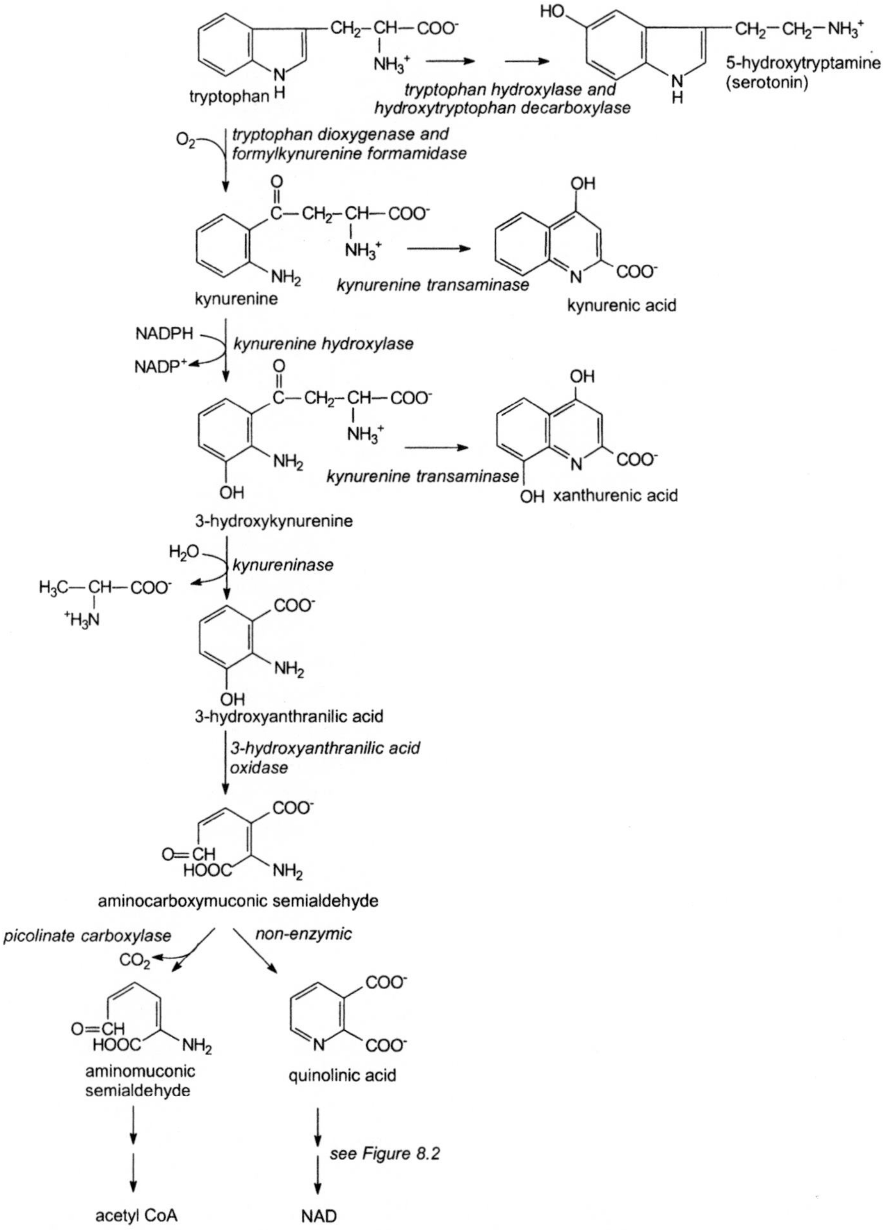

Figure 8.4. Pathways of tryptophan metabolism. Tryptophan dioxygenase, EC 1.13.11.11; formylkynurenine formamidase, EC 3.5.1.9; kynurenine hydroxylase, EC 1.14.13.9; kynureninase, EC 3.7.1.3; 3-hydroxyanthranilate oxidase, EC 1.10.3.5; picolinate carboxylase, EC 4.1.1.45; kynurenine oxoglutarate aminotransferase, EC 2.6.1.7; kynurenine glyoxylate aminotransferase, 2.6.1.63; tryptophan hydroxylase, EC 1.14.16.4; and 5-hydroxytryptophan decarboxylase, EC 4.1.1.26. Relative molecular masses (M_r): tryptophan, 204.2; serotonin, 176.2; kynurenine, 208.2; 3-hydroxykynurenine, 223.2; kynurenic acid, 189.2; xanthurenic acid, 205.2; and quinolinic acid 167.1. CoA, coenzyme A.

no formation of quinolinic acid and hence NAD(P). Induction of indoleamine dioxygenase may therefore be a factor in tryptophan depletion leading to the development of pellagra (Section 8.5).

8.3.1 Picolinate Carboxylase and Nonenzymic Cyclization to Quinolinic Acid

As shown in Figure 8.4, the synthesis of NAD from tryptophan involves the nonenzymic cyclization of aminocarboxymuconic semialdehyde to quinolinic acid. The alternative metabolic fate of aminocarboxymuconic semialdehyde is decarboxylation, catalyzed by picolinate carboxylase, leading into the oxidative branch of the pathway, and catabolism via acetyl coenzyme A. There is thus competition between an enzyme-catalyzed reaction that has hyperbolic, saturable kinetics, and a nonenzymic reaction that has linear, first-order kinetics.

The result of this is that at low rates of flux through the kynurenine pathway, which result in concentrations of aminocarboxymuconic semialdehyde below that at which picolinate carboxylase is saturated, most of the flux will be by way of the enzyme-catalyzed pathway, leading to oxidation. There will be little accumulation of aminocarboxymuconic semialdehyde to undergo nonenzymic cyclization. As the rate of formation of aminocarboxymuconic semialdehyde increases, and picolinate carboxylase nears saturation, there will be an increasing amount available to undergo the nonenzymic reaction and onward metabolism to NAD. Thus, there is not a simple stoichiometric relationship between tryptophan and niacin, and the equivalence of the two coenzyme precursors will vary as the amount of tryptophan to be metabolized and the rate of metabolism vary.

As might be expected, the synthesis of NAD from tryptophan is inversely related to the activity of picolinate carboxylase. Inhibition with pyrazinamide results in increased availability of aminocarboxymuconic semialdehyde, and hence increased NAD formation. Equally, activation of picolinate carboxylase results in reduced availability of aminocarboxymuconic semialdehyde for cyclization, and hence a reduced formation of NAD.

Cats, which have some 30- to 50-fold higher activity of picolinate carboxylase than other species, are entirely reliant on a dietary source of preformed niacin, and are not capable of any significant synthesis of NAD from tryptophan.

It is thus apparent that the utilization of tryptophan as a precursor for NAD synthesis depends on both the amount of tryptophan to be metabolized and also the rate of metabolic flux through the pathway. The activities of three enzymes (tryptophan dioxygenase, kynurenine hydroxylase, and kynureninase) may all affect the rate of formation of aminocarboxymuconic semialdehyde, as may the rate of tryptophan uptake into the liver.

8.3.2 Tryptophan Dioxygenase

The first enzyme of the pathway, tryptophan dioxygenase (also known as tryptophan oxygenase or tryptophan pyrrolase), is rate-limiting under normal conditions. In isolated hepatocytes, the control coefficient for flux through the pathway of tryptophan dioxygenase is 0.75 and that for tryptophan uptake into the cells is 0.25 (Salter et al., 1986).

Tryptophan dioxygenase has a short half-life (of the order of 2 hours) and is subject to regulation by three mechanisms: saturation with its heme cofactor, hormonal induction and feedback inhibition, and repression by NAD(P).

8.3.2.1 Saturation of Tryptophan Dioxygenase with Its Heme Cofactor

Unlike many other heme enzymes, the hematin of tryptophan dioxygenase behaves more like a dissociating cofactor than a tightly bound prosthetic group, and dissociation of the holo-enzyme can occur in the presence of protoporphyrin or mesoporphyrin. The holo-enzyme is more resistant to proteolysis than is the apoenzyme. In the presence of relatively large amounts of heme, both the activity of the enzyme and the total amount of immunoreactive tryptophan dioxygenase protein in the liver are increased. It has been suggested that induction of tryptophan dioxygenase apoenzyme may provide a metabolic sink for excess heme synthesis (Badawy and Evans, 1975; Badawy, 1977).

Tryptophan and a number of tryptophan analogs also enhance the stability of tryptophan dioxygenase by enhancing conjugation of the apoenzyme with hematin and stabilizing the holo-enzyme. The tryptophan analogs that promote heme conjugation are not substrates, nor do they compete with tryptophan at the catalytic site of the enzyme. They appear to bind to a domain on the enzyme protein that is distinct from the catalytic site. The conjugation promoting site of the apoenzyme has a relatively broad specificity, and a K_m for tryptophan of 26 μmol per L, whereas the catalytic site on the holo-enzyme binds only L-tryptophan and has a 10-fold higher K_m.

8.3.2.2 Induction of Tryptophan Dioxygenase by Glucocorticoid Hormones

The de novo synthesis of tryptophan dioxygenase is induced by glucocorticoid hormones (cortisol in human beings and corticosterone in the rat). This is true induction of new mRNA and protein synthesis; indeed, tryptophan dioxygenase was the first mammalian enzyme to be shown to be inducible.

In isolated hepatocytes, after maximum induction of tryptophan dioxygenase by glucocorticoids, the uptake of tryptophan into the cells has a control coefficient of 0.75, whereas the control coefficient of tryptophan dioxygenase falls to 0.25. Therefore, the induction of tryptophan dioxygenase has only a limited effect on tryptophan catabolism and NAD synthesis (Salter and Pogson, 1985; Salter et al., 1986). In isolated perfused liver, although cortisol leads to a several-fold increase in tryptophan dioxygenase activity, there is only a relatively small increase in the rate of clearance of tryptophan from the perfusion medium (Kim and Miller, 1969).

8.3.2.3 **Induction of Tryptophan Dioxygenase by Glucagon** Glucagon (mediated by cAMP) increases the synthesis of tryptophan dioxygenase after the administration of glucocorticoids, although it has little effect in unstimulated animals. The effect of glucagon appears to be the result of an increase in the rate of translation of mRNA rather than an increase in transcription and is antagonized by insulin.

8.3.2.4 **Repression and Inhibition of Tryptophan Dioxygenase by Nicotinamide Nucleotides** High concentrations of the nicotinamide nucleotide coenzymes, and especially NADPH, both inhibit preformed tryptophan oxygenase and also repress its synthesis. It is not clear how important this is in terms of physiological regulation, because concentrations of the nucleotides required to achieve significant inhibition are relatively high, but some of the effects of alcohol on tryptophan metabolism may be because of the increase in NADH and NADPH that follows the ingestion of relatively large amounts of alcohol (Badawy, 2002).

8.3.3 Kynurenine Hydroxylase and Kynureninase
The entry of tryptophan into the oxidative pathway is limited by tryptophan dioxygenase activity and the uptake of tryptophan; under basal conditions, these two processes have control coefficients of 0.75 and 0.25, respectively. Neither kynurenine hydroxylase nor kynureninase has a significant control coefficient in metabolic flux studies in hepatocytes isolated from normal animals (Salter and Pogson, 1985; Salter et al., 1986). However, the activity of both enzymes is only slightly higher than that of tryptophan dioxygenase under basal conditions, and increased tryptophan dioxygenase activity is accompanied by

increased accumulation and excretion of kynurenine, hydoxykynurenine, and their transamination products, kynurenic and xanthurenic acids (see Figures 8.4 and 9.4).

Even without induction of tryptophan dioxygenase, impairment of the activity of either enzyme may impair the onward metabolism of kynurenine and thus reduce the accumulation of aminocarboxymuconic semialdehyde and synthesis of NAD.

During the first half of the twentieth century, when 87,000 people died from pellagra in the United States, there was a two-fold excess of females over males. Reports of individual outbreaks of pellagra show a similar sex ratio. This may well be the result of inhibition of kynureninase, and impairment of the activity of kynurenine hydroxylase, by estrogen metabolites, and hence reduced synthesis of NAD from tryptophan (Bender and Totoe, 1984b).

8.3.3.1 Kynurenine Hydroxylase Kynurenine hydroxylase is an FAD-dependent mixed-function oxidase of the outer mitochondrial membrane, which uses NADPH as the reductant. The activity of kynurenine hydroxylase in the liver of riboflavin-deficient rats is only 30% to 50% of that in control animals, and deficient rats excrete abnormally large amounts of kynurenic and anthranilic acids after the administration of a loading dose of tryptophan, and, correspondingly lower amounts of quinolinate and niacin metabolites. Riboflavin deficiency may thus be a contributory factor in the etiology of pellagra when intakes of tryptophan and niacin are marginal (Section 8.5.1).

In a number of studies, sexually mature women show a higher ratio of urinary kynurenine:hydroxykynurenine than do children, postmenopausal women, or men, suggesting impairment of kynurenine hydroxylase activity by estrogens or their metabolites. In experimental animals, the administration of estrogens results in a reduction in kynurenine hydroxylase activity to about 30% of the control activity. The mechanism of this effect is unclear, because the addition of estrogens or their metabolites has no effect on the enzyme in vitro. It is possible that the effect is indirect, and due to the inhibition of kynureninase by estrogen conjugates, and hence an accumulation of hydroxykynurenine, and increased formation of kynurenic and xanthurenic acids. Kynurenine hydroxylase is inhibited by micromolar concentrations of xanthurenic acid (Bender and McCreanor, 1985).

8.3.3.2 Kynureninase Kynureninase is a pyridoxal phosphate (vitamin B_6)-dependent enzyme that catalyzes the hydrolysis of 3-hydroxykynurenine to 3-hydroxyanthranilic acid, releasing the side chain as alanine. Impairment of kynureninase activity in vitamin B_6 deficiency leads to accumulation of kynurenine and hydroxykynurenine, and their transamination products, kynurenic and xanthurenic acids. This is the basis of the tryptophan load test for vitamin B_6 nutritional status (Section 9.5.4). Vitamin B_6 deficiency, or inhibition of kynureninase by estrogen metabolites, would therefore be expected to reduce the rate of metabolic flux through the oxidative pathway and reduce the formation of quinolinic acid and NAD from tryptophan.

8.4 METABOLIC FUNCTIONS OF NIACIN

Nicotinamide is the reactive moiety of the nicotinamide nucleotide coenzymes NAD (nicotinamide adenine dinucleotide) and NADP (nicotinamide adenine dinucleotide phosphate), which are coenzymes (or more correctly cosubstrates) in a wide variety of oxidation and reduction reactions (Section 8.4.1). The notation NAD(P) is used to mean either NAD or NADP, without specifying the oxidation state.

NAD is the source of ADP-ribose for the modification of proteins by mono-ADP-ribosylation, catalyzed by ADP-ribosyltransferases (Section 8.4.2), and poly(ADP-ribosylation), catalyzed by poly(ADP-ribose) polymerase (Section 8.4.3). It is also the precursor of two second messengers that act to increase the intracellular concentration of calcium, cADP-ribose, and nicotinic acid adenine dinucleotide phosphate (Section 8.4.4).

Nicotinic acid has been tentatively identified as the organic component of the (as yet uncharacterized) chromium-containing glucose tolerance factor that enhances the interaction of insulin with cell surface receptors.

8.4.1 The Redox Function of NAD(P)

The nicotinamide coenzymes are involved as proton and electron carriers in a wide variety of oxidation and reduction reactions. Before their chemical structures were known, NAD and NADP were known as coenzymes I and II. Later, when the chemical nature of the pyridine ring of nicotinamide was discovered, they were called diphosphopyridine nucleotide (DPN = NAD) and triphosphopyridine nucleotide (TPN = NADP). The nicotinamide nucleotide coenzymes are sometimes referred to as the pyridine nucleotide coenzymes.

As shown in Figure 8.5, the oxidized coenzymes have a formal positive charge, and are represented as NAD^+ and $NADP^+$, whereas the reduced forms, carrying two electrons and one proton (and associated with an additional proton), are represented as NADH and NADPH. The two-electron reduction of $NAD(P)^+$ proceeds by way of a hydride (H^-) ion transfer to carbon-4 of the nicotinamide ring.

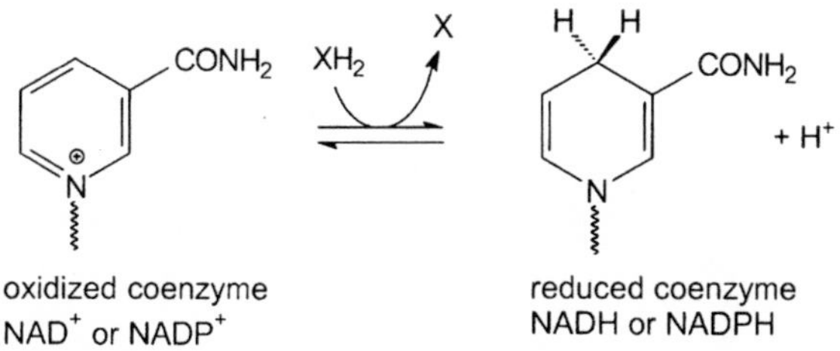

oxidized coenzyme
NAD⁺ or NADP⁺

reduced coenzyme
NADH or NADPH

Figure 8.5. Redox function of the nicotinamide nucleotide coenzymes.

In general, NAD^+ is involved as an electron acceptor in energy-yielding metabolism, and the resultant NADH is oxidized by the mitochondrial electron transport chain. The major coenzyme for reductive synthetic reactions is NADPH. An exception here is the pentose phosphate pathway (see Figure 6.4), which reduces $NADP^+$ to NADPH and is the source of about half the reductant for lipogenesis.

In the reduced coenzymes, the hydrogen atoms at carbon-4 of the nicotinamide ring lie above and below the plane of the ring. Isotope studies have shown that they are not equivalent, and enzymes specifically remove or add the pro-R hydrogen (above the plane of the ring) or the pro-S hydrogen (below the plane of the ring). The result of this is that, although NAD(P) acts as a cosubstrate, binding to and being released from the enzyme catalytic site during the reaction, rather than remaining enzyme bound, there can be considerable channeling between pairs of enzymes that use the opposite faces of the coenzyme and effectively sequester the coenzyme between them.

8.4.1.1 **Use of NAD(P) in Enzyme Assays** The reduced coenzymes have an absorption maximum at 340 nm, whereas the oxidized coenzymes do not. This is widely exploited to provide sensitive and specific methods for determining a variety of analytes using purified NAD(P)-linked enzymes and following the change in absorption at 340 nm as the coenzyme is either reduced or oxidized by the substrate.

8.4.2 **ADP-Ribosyltransferases**
ADP-ribosylation is a reversible modification of proteins, as shown in Figure 8.6, and there are specific hydrolases that remove the ADP-ribose from target proteins.

ADP-ribosyltransferases are enzymes of the cytosol, plasma membrane, and nuclear envelope that catalyze the transfer of ADP-ribose onto arginine,

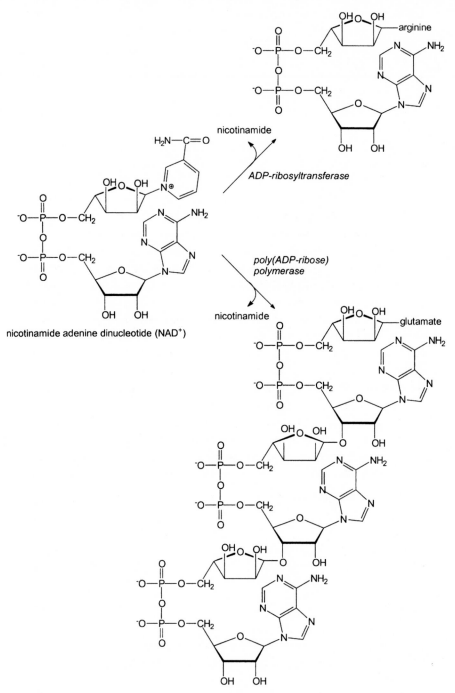

Figure 8.6. Reactions of ADP-ribosyltransferase (EC 2.4.2.31) and poly(ADP-ribose) polymerase (EC 2.4.2.30).

216

lysine, or asparagine residues in acceptor proteins to form N-glycosides. The plasma membrane ADP-ribosyltransferases are ecto-enzymes, anchored in the membrane by a glycosyl phosphoinositol tail, and have been implicated in cell adhesion and also in chemotaxis in lymphocytes. In addition to endogenous ADP-ribosyltransferases, a number of bacterial toxins, including diphtheria and cholera toxins, *Escherichia coli* enterotoxin LT and *Pseudomonas aeruginosa* exotoxin A also have ADP-ribosyltransferase activity.

In the absence of an acceptor protein, ADP-ribosyltransferase catalyzes the hydrolysis of NAD^+ to release nicotinamide and free ADP-ribose. The carboxy terminal region of the enzyme has NAD glycohydrolase activity, but does not catalyze the transfer of ADP-ribose onto target proteins.

The ribosomal elongation Factor II is the acceptor protein for the ADP-ribosyltransferase activity of diphtheria toxin and *P. aeruginosa* exotoxin A, as well as a mammalian cytosolic ADP-ribosyltransferase. ADP-ribosylation results in loss of activity. The uncontrolled action of the bacterial toxins causes the cessation of protein synthesis and hence cell death. The more regulated action of the endogenous ADP-ribosyltransferase is part of the normal regulation of protein synthesis.

A variety of guanine nucleotide binding proteins (G-proteins) involved with the regulation of adenylate cyclase activity and transducin in the retina (Section 2.3.1) are substrates for ADP-ribosylation. Cholera toxin and *E. coli* enterotoxin LT ADP-ribosylate, and hence activate, the stimulatory G-protein of adenylate cyclase, whereas pertussis toxin ADP-ribosylates, and inactivates the inhibitory G-protein of adenylate cyclase. The result of ADP-ribosylation by either mechanism is increased adenylate cyclase activity, and an increase in intracellular cAMP and the opening of membrane calcium channels. Again, there are endogenous ADP-ribosyltransferases that modify the same G-proteins, but in a controlled manner (Moss et al., 1997, 1999).

8.4.3 Poly(ADP-ribose) Polymerases

Poly(ADP-ribose) polymerases are a family of enzymes that catalyze transfer of multiple ADP-ribose units onto target proteins, as shown in Figure 8.6. They are DNA-binding proteins with a zinc-finger motif and require nicked DNA (with single- or double-strand breaks) for activity. They are present in the cell in high concentrations; about one molecule of enzyme for each kilobase of DNA (Hayaishi and Ueda, 1977; D'Amours et al., 1999).

Poly(ADP-ribose) polymerases catalyze three reactions:

1. Initial ADP-ribosylation of the γ-carboxyl group of glutamate residues (or the carboxyl group of a C-terminal lysine) in the target protein, forming an O-glycoside.

2. Elongation of the poly-(ADP-ribose) chain by reaction with the 2′-hydroxyl group of the nonreducing end of the growing chain, with up to 200 elongation events catalyzed per initiation event.
3. Introduction of branch points every 30 to 40 ADP-ribose units.

Poly(ADP-ribose) formed by the polymerase turns over rapidly in DNA-damaged cells, with a half-life of the order of a minute. Cells contain a glycohydrolase with both endo- and exoglycosidase activity, which acts initially to remove the complete branched poly(ADP)ribose from the protein, then, more slowly, hydrolyze it to oligomers and free ADP-ribose.

The best-established function of poly(ADP-ribose) polymerase is in repair of damaged DNA; it is activated by DNA strand breaks, and acts to clear histones and other nucleoproteins away from the DNA to permit access of the DNA repair enzymes. Both in vitro and in experimental animals, niacin deficiency leads to increased genomic instability, as the ability to repair damaged DNA is impaired, and may increase tumor risk. There is little information about genomic instability and cancer risk in human niacin deficiency (Hageman and Stierum, 2001).

Activation of poly(ADP-ribose) polymerase also provides a mechanism of cell death in response to severe DNA damage. In undamaged cells, the concentration of NAD is 400 to 500 μmol per L, with a half-life of the order of an hour. In response to DNA damage, it may fall by 80% within 5 to 15 minutes, resulting in significant impairment of oxidative metabolic activity, hence depletion of ATP and cell death.

The main target for poly(ADP-ribosylation) is poly(ADP-ribose) polymerase itself; the enzyme has up to 28 sites for automodification, and several thousand molecules of ADP-ribose may be added to each molecule of enzyme. Both while bound to the enzyme, and more importantly after release from the enzyme by glycohydrolase action, poly(ADP-ribose) has a considerably higher affinity for the basic amino acids in the α-helical tail of histones than does DNA, and serves to remove histones from DNA binding and hence unravel nucleosomes.

A variety of other nuclear proteins are also targets for poly(ADP-ribosylation), including histones, topo-isomerases, DNA ligases, and DNA-dependent RNA polymerase, suggesting that in addition to its role in DNA repair, poly(ADP-ribose) polymerase may be important in DNA replication and transcription. In DNA replication, it controls the progression of the replication fork, and in preadipocytes, there is a considerable increase in activity in response to the induction of differentiation (Simbulan-Rosenthal et al., 1996).

Poly(ADP-ribose) polymerase may also have a role in regulating the activity of nuclear-acting hormone receptors; it binds directly to retinoid X receptors (Section 2.3.2.1) and inhibits the transcriptional activity of retinoid X receptor-thyroid hormone dimers. It also acts as a transcription factor in pancreatic β-islet cells (Miyamoto et al., 1999).

The diabetogenic compounds streptozotocin and alloxan cause DNA strand breaks by radical generation, leading to necrosis of pancreatic β-islet cells as a result of NAD, and hence ATP, depletion. Release of cell contents from necrotic cells leads to the development of anti-β-cell antibodies and the autoimmune development of diabetes mellitus. In poly(ADP-ribose) polymerase knockout mice, these compounds are not diabetogenic, and inhibitors of poly(ADP-ribose) polymerase, such as nicotinamide, protect against streptozotocin- and alloxan-induced diabetes (Pieper et al., 1999). This has led to trials of nicotinamide as a protective agent against insulin-dependent diabetes mellitus in people at high risk (Section 8.8).

8.4.4 cADP-Ribose and Nicotinic Acid Adenine Dinucleotide Phosphate (NAADP)

Cells contain high activities of enzymes that were originally identified as NAD glycohydrolases, catalyzing the hydrolysis of $NAD(P)^+$ to nicotinamide and ADP-ribose (phosphate). As shown in Figure 8.7, these glycohydrolases also catalyze two additional reactions that lead to products that have a role in calcium release from intracellular stores, and act as second messengers (Dousa et al., 1996; Lee, 1996, 1999, 2000, 2001):

1. Cyclization of ADP-ribose arising from NAD to cADP-ribose.
2. Exchange of the nicotinamide moiety of NADP for nicotinic acid, forming NAADP. Early studies also showed that the enzyme can catalyze exchange of nicotinamide with a variety of other bases, including histamine. Formation of histamine adenine dinucleotide phosphate was thought to provide an alternative to amine oxidase for rapid inactivation of histamine.

The intracellular NAD glycohydrolase is now known as ADP-ribose cyclase; there is also a cell surface ectozyme, identical with the lymphocyte CD38 antigen. CD38 also occurs intracellularly, in endosomes. Both enzymes catalyze the formation of both cADP-ribose and NAADP, as well as the glycohydrolase reaction. The cyclase reaction predominates at neutral pH, and the nicotinamide/nicotinic acid exchange reaction at acid pH, suggesting that in cytosol the main product of the soluble enzyme is cADP-ribose. In endosomes,

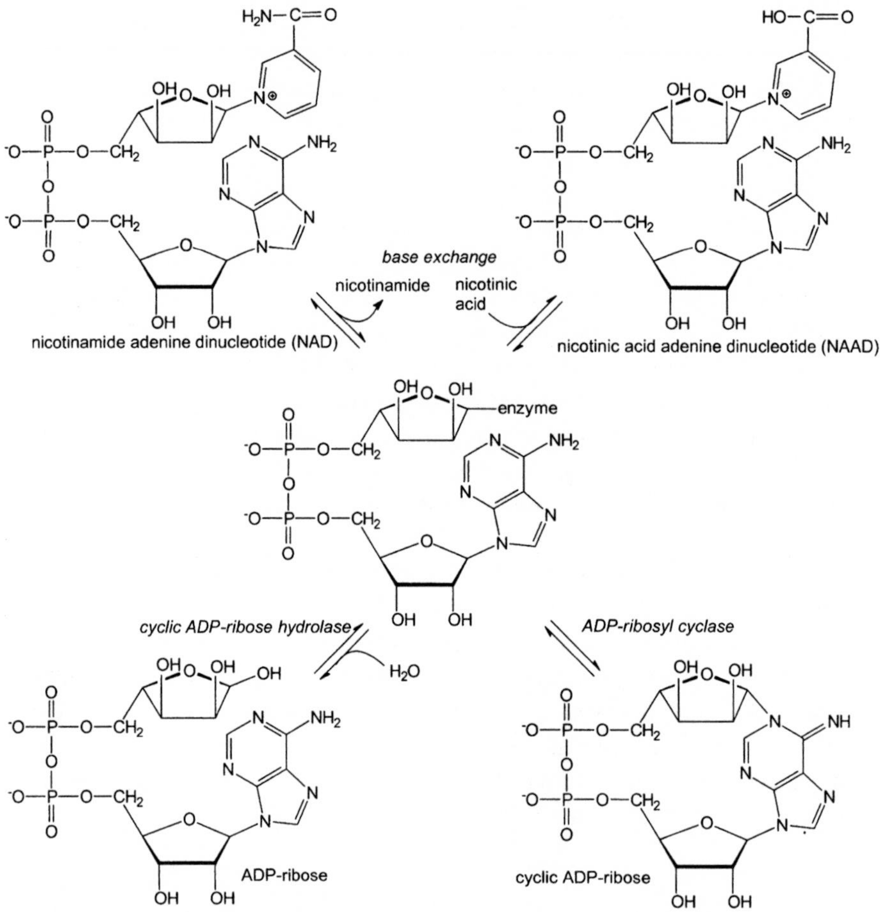

Figure 8.7. Reactions catalyzed by ADP ribose cyclase (NAD glycohydrolase, EC 3.2.2.5). When the substrate is NADP, the base exchange reaction leads to the formation of nicotinic acid adenine dinucleotide phosphate.

the main product of CD38 is NAADP (Lee, 1999). In sea urchin eggs, cyclase activity is increased by a cGMP-dependent kinase, whereas NAADP formation is activated by a cAMP-dependent kinase, although in mammalian cells cGMP has no effect on the enzymes (Wilson and Galione, 1998).

Both cADP-ribose and NAADP act to increase cytosolic calcium concentrations, releasing calcium from intracellular stores via a receptor distinct from that which responds to inositol trisphosphate (Section 14.4.1). The responses to cADP-ribose and NAADP are additive, and they seem to act on different intracellular calcium stores (Jacobson et al., 1995; Patel et al., 2001).

The role of cADP-ribose and NAADP in regulating cytosolic calcium may provide an alternative explanation to the serotonin hypothesis for the psychiatric and neurological signs of the niacin deficiency disease pellagra (Section 8.5; Petersen and Cancela, 1999).

cADP-ribose and NAADP act as second messengers in response to nitric oxide, acetyl choline, and cholecystokinin, and are therefore involved in responses to neurotransmitters and hormones (Cancela, 2001). Pancreatic β-islet cells express CD38, and cADP-ribose has been implicated in the calcium release that signals insulin secretion. In response to increased intracellular ATP, as a result of increased uptake and metabolism of glucose, there is inhibition of the NAD glycohydrolase activity of the enzyme, and increased formation of cADP-ribose. This acts synergistically with palmitate, formed as a result of increased glucose uptake and metabolism, to release calcium from intracellular stores (Okamoto, 1999a, 1999b).

All-*trans*-retinoic acid (Section 2.2.3.2) stimulates the synthesis of cADP-ribose in kidney cells in culture, apparently as a result of the induction of CD38 (Beers et al., 1995; Takahashi et al., 1995); in ovariectomized rats, estradiol induces cytosolic ADP-ribosyl cyclase in the uterus, but not in estrogen unresponsive tissues (Chini et al., 1997). If this induction of ADP-ribose cyclase by estrogens leads to significant depletion of nicotinamide nucleotides, it may provide an additional explanation for the 2:1 excess of females to males in the incidence of pellagra (Section 8.5).

8.5 PELLAGRA – A DISEASE OF TRYPTOPHAN AND NIACIN DEFICIENCY

Pellagra is characterized by a photosensitive dermatitis, like severe sunburn, typically with a butterfly-like pattern of distribution over the face, affecting all parts of the skin that are exposed to sunlight. Similar skin lesions may also occur in areas not exposed to sunlight, but subject to pressure, such as the knees, elbows, wrists, and ankles. Advanced pellagra is also accompanied by a dementia or depressive psychosis, and there may be diarrhea. Untreated pellagra is fatal.

Pellagra was a major problem of public health in the early part of the twentieth century and continued to be a problem until the 1980s in some parts of the world. It is now rare, although there were reports of outbreaks among refugees in Africa (Malfait et al., 1993), and occasional cases are reported in alcoholics in developed countries and among people being treated with isoniazid (Section 8.5.6) and some other drugs, and people with chronic gastrointestinal disease.

Despite our understanding of the biochemistry of niacin, we still cannot account for the characteristic photosensitive dermatitis in terms of the known metabolic lesions. There is no apparent relationship between reduced availability of tryptophan and niacin, and sensitivity of the skin to ultraviolet (UV) light. The only biochemical abnormalities that have been reported in the skin of pellagrins involve increased catabolism of the amino acid histidine leading to a reduction in the concentration of urocanic acid, a histidine metabolite that is the major UV-absorbing compound in normal dermis (see Figure 10.6).

The other characteristic feature of pellagra is the development of a depressive psychosis, superficially similar to schizophrenia and the organic psychoses, but clinically distinguishable by the sudden lucid phases that alternate with the most florid psychiatric signs. The mental symptoms may be the result of tryptophan depletion, and hence a lower availability of tryptophan for synthesis of the neurotransmitter serotonin (5-hydroxytryptophan). But the role of cADP-ribose and NAADP in controlling calcium release in response to neurotransmitters (Section 8.4.4) and impaired energy-yielding metabolism in the central nervous system as a result of depletion of NAD(P) may also be important.

The diarrhea associated with pellagra is caused by rectal inflammation; within 5 to 7 days of starting treatment with niacin, rectal histology is normalized and the diarrhea ceases (Segal et al., 1986).

8.5.1 Other Nutrient Deficiencies in the Etiology of Pellagra

Although the nutritional etiology of pellagra is well established, and additional tryptophan or niacin will prevent or cure the disease, there are a number of reports that suggest that additional factors may be involved. Carpenter and Lewin (1985) reexamined the diets associated with the development of pellagra in the United States during the early part of the twentieth century and showed that the total intake of tryptophan and niacin was apparently adequate, as judged by current knowledge of requirements. They suggested that deficiency of riboflavin (and hence impaired activity of kynurenine hydroxylase; Section 8.3.3.1) or vitamin B_6 (and hence impaired activity of kynureninase; Section 8.3.3.2) may have been a significant factor in the etiology of pellagra when intakes of tryptophan and niacin were only marginally adequate.

Iron deficiency may also be a factor, because impairment of heme synthesis will both reduce the activity of the enzyme and increase its susceptibility to proteolysis (Section 8.3.2; Oduho et al., 1994).

8.5.2 Possible Pellagragenic Toxins

Early studies suggested the presence of a toxin from some samples of maize that was pellagragenic in experimental animals. Schoental (1983) suggested that mycotoxins resulting from fungal spoilage of maize and other grain stored under damp conditions may have been responsible for some outbreaks of pellagra. Certainly, the only known outbreak of the disease in Mexico can be traced to a consignment of maize that was shipped under damp conditions and had a significant fungal overgrowth. As discussed in Section 8.4.2 and Section 8.4.3, a number of bacterial, fungal, and environmental toxins activate ADP-ribosyltransferase or poly(ADP-ribose) polymerase, and it is possible that chronic exposure to such toxins will deplete tissue NAD(P) and be a contributory factor in the development of pellagra when intakes of tryptophan and niacin are marginal.

8.5.3 The Pellagragenic Effect of Excess Dietary Leucine

Pellagra was a major problem in parts of India where jowar (*Sorghum vulgare*) is the dietary staple, despite the fact that the tryptophan content of sorghum proteins is higher than that of maize, the cereal traditionally associated with endemic pellagra. The intake of tryptophan and niacin was as great among people whose dietary staple was jowar as that of rice eaters, yet pellagra was common among jowar eaters and not in rice-eating communities, suggesting that the relatively high content of leucine in the proteins of jowar might be a contributory factor.

A number of studies have demonstrated a pellagragenic effect of excess dietary leucine in experimental animals and in human beings (Gopalan and Rao, 1975), although other studies failed to show any effect (Manson and Carpenter, 1978a, 1978b). Magboul and Bender (1983) showed that, when rats were fed diets that were only marginally adequate with respect to tryptophan and niacin, the addition of 1.5% leucine led to a significant depletion of liver and blood nicotinamide nucleotides. This effect was only apparent when the niacin content of the diet was such that it provided less than half the minimum requirement and was most marked when the diets provided virtually no preformed niacin, suggesting that the effect of leucine is on the metabolism of tryptophan and not on the utilization of niacin.

Studies with [^{14}C]tryptophan in animals and isolated hepatocytes show that leucine does inhibit the synthesis of NAD from tryptophan, inhibiting metabolism at the level of kynurenine hydroxylase and kynureninase, causing the accumulation of intermediates. In isolated hepatocytes, the more important effect seems to be at the level of uptake of tryptophan into the cells; leucine and tryptophan share a common transport mechanism with the other large neutral amino acids (Bender, 1983, 1989a; Salter et al., 1985).

It is likely that leucine is only a factor in the etiology of pellagra when the dietary intakes of both tryptophan and niacin are extremely low – a condition that may occur when sorghum is the dietary staple, especially during food shortage.

8.5.4 Inborn Errors of Tryptophan Metabolism
A number of inborn errors of metabolism of the tryptophan oxidative pathway (see Figure 8.4) have been reported, all of which result in the development of pellagra that responds to high doses of niacin. These conditions include vitamin B_6-responsive xanthurenic aciduria, caused by a defect of kynureninase (Section 9.4.3); hydroxykynureninuria, apparently caused by a defect of kynureninase; tryptophanuria, apparently caused by tryptophan dioxygenase deficiency; a hereditary pellagra-like condition, apparently caused by an increase in activity of picolinate carboxylase; and Hartnup disease.

Hartnup disease is a rare genetic condition in which there is a defect of the membrane transport mechanism for tryptophan and other large neutral amino acids. The result is that the intestinal absorption of free tryptophan is impaired, although dipeptide absorption is normal. There is a considerable urinary loss of tryptophan (and other amino acids) as a result of the failure of the normal reabsorption mechanism in the renal tubules – renal aminoaciduria. In addition to neurological signs that can be attributed to a deficit of tryptophan for the synthesis of serotonin in the central nervous system, the patients show clinical signs of pellagra, which respond to the administration of niacin.

8.5.5 Carcinoid Syndrome
Carcinoid is a tumor of the enterochromaffin cells that normally synthesize 5-hydroxytrytophan and 5-hydroxytryptamine. The carcinoid syndrome is seen when there are significant metastases of the primary tumor in the liver. It is characterized by increased gastrointestinal motility and diarrhea, as well as by regular periodic flushing. These symptoms can be attributed to systemic release of large amounts of serotonin and can be controlled with inhibitors of tryptophan hydroxylase, such as p-chlorophenylalanine. The synthesis of 5-hydroxytryptamine in advanced carcinoid syndrome may be so great that as much as 60% of the body's tryptophan metabolism proceeds by this pathway, compared with about 1% under normal conditions. A significant number of

patients with advanced carcinoid syndrome develop clinical signs of pellagra, because of this diversion of tryptophan away from the oxidative pathway of metabolism.

8.5.6 Drug-Induced Pellagra

The antituberculosis drug isoniazid (*iso*-nicotinic acid hydrazide) can cause pellagra by forming a biologically inactive complex with pyridoxal phosphate, the metabolically active form of vitamin B_6, and hence reducing the activity of kynureninase (Section 8.3.3.2). This isoniazid-induced pellagra responds to the administration of niacin supplements. However, isoniazid may also cause peripheral neuropathy, which responds to vitamin B_6 and not niacin, and therefore it became usual to give vitamin B_6 supplements together with isoniazid.

During the 1960s, the doses of isoniazid used in the treatment of tuberculosis were considerably reduced, as a result of the introduction of other effective antimycobacterial agents in therapeutic cocktails, and it was no longer thought necessary to give patients supplements of vitamin B_6. There have been some reports of the development of pellagra in patients treated with relatively low doses of isoniazid; most patients were of Indian origin, and it is likely that they were genetically slow acetylators of isoniazid, so that an apparently low dose of the drug was, in fact, high. Up to 60% of Indians are slow acetylators of isoniazid.

Two further drugs are also capable of forming hydrazones with pyridoxal phosphate: the anti-Parkinsonian drugs Benserazide and Carbidopa. Both drugs inhibit the oxidative metabolism of tryptophan and cause reduced excretion of N^1-methylnicotinamide (Bender, 1980). Although no case of clinical pellagra has been unequivocally reported, there have been a number of cases of pellagra-like conditions among patients receiving Benserazide or Carbidopa, and Parkinsonian patients treated with these drugs excrete less N^1-methyl nicotinamide than control patients with similarly severe disease, but receiving different medication.

8.6 ASSESSMENT OF NIACIN NUTRITIONAL STATUS

Although the nicotinamide nucleotide coenzymes function in a large number of oxidation and reduction reactions, this cannot be exploited as a means of assessing the state of the body's niacin reserves, because the coenzymes are not firmly attached to their apoenzymes, as are coenzymes derived from thiamin (Section 6.5.3), riboflavin (Section 7.5.3), and vitamin B_6 (Section 9.5.3), but act as cosubstrates of the reactions, binding to and leaving the enzyme as the reaction proceeds. No specific metabolic lesions associated with NAD(P) depletion have been identified.

The two methods of assessing niacin nutritional status are measurement of blood nicotinamide nucleotides and the urinary excretion of niacin metabolites, neither of which is wholly satisfactory.

8.6.1 Tissue and Whole Blood Concentrations of Nicotinamide Nucleotides

Measurement of liver and other tissue concentrations of NAD(P) gives a precise estimate of niacin nutritional status and seems to be the most sensitive indicator in experimental animals. Measurement of the whole blood concentration of NAD(P) may serve the same purpose; there is a good correlation between blood and liver concentrations of nicotinamide nucleotides in experimental animals. The sensitivity of the method is such that reproducible determinations can be carried out on finger-prick samples of 200 μL of blood (Bender et al., 1982).

Erythrocyte NAD falls during niacin depletion and rises during repletion in human subjects. Erythrocyte NADP is unaffected, and the ratio of NAD:NADP provides a useful index of niacin nutritional status, with a ratio <1.0 indicating deficiency (Fu et al., 1989).

8.6.2 Urinary Excretion of N^1-Methyl Nicotinamide and Methyl Pyridone Carboxamide

The most widely used method for assessing niacin nutritional status is measurement of the urinary excretion of niacin metabolites. Table 8.1 shows the excretion of N^1-methyl nicotinamide and methyl pyridone carboxamide in niacin adequacy and deficiency.

The excretion of methyl pyridone carboxamide is more severely reduced in marginal niacin deficiency than is that of N^1-methyl nicotinamide. The excretion of methyl pyridone carboxamide decreases rapidly in subjects fed on a niacin-deficient diet, and virtually ceases several weeks before the appearance of clinical signs of deficiency; by contrast, a number of studies have shown continuing excretion of N^1-methyl nicotinamide even in pellagrins. A better estimate of niacin nutritional status can be obtained by determining the ratio of urinary methyl pyridone carboxamide:N^1-methyl nicotinamide, which is relatively constant, despite the administration of loading doses of tryptophan or niacin to adequately nourished subjects (between 1.3 to 4.0), and a ratio of less than 1.0 indicates depletion of niacin reserves (de Lange and Joubert, 1964; Dillon et al., 1992).

Table 8.1 Indices of Niacin Nutritional Status

	Elevated	Adequate	Marginal	Deficient
N^1-Methyl nicotinamide				
μmol/24 h	>48	17–47	5.8–17	<5.8
mg/g creatinine	>4.4	1.6–4.3	0.5–1.6	<0.5
mmol/mol creatinine	>4.0	1.3–3.9	0.4–1.3	<0.4
Methyl pyridone carboxamide				
μmol/24 h	—	>18.9	6.4–18.9	<6.4
mg/g creatinine	—	>4.0	2.0–3.9	<2.0
mmol/mol creatinine	—	>4.4	0.44–4.3	<0.44
Ratio, methyl pyridone carboxamide:N^1-methyl nicotinamide				
	—	1.3–4.0	1.0–1.3	<1.0
Ratio, erythrocyte NAD:NADP				
	—	>1.0	—	<1.0

Sources: From data reported by de Lange and Joubert, 1964; Kelsay, 1969; Gontzea et al., 1976; Fu et al., 1989.

8.7 NIACIN REQUIREMENTS AND REFERENCE INTAKES

In view of the central role of the nicotinamide nucleotides in energy-yielding metabolism, and the fact that, at least in theory, the nicotinamide released by ADP-ribosyltransferase (Section 8.4.2) and poly(ADP-ribose) polymerase (Section 8.4.3) is available to be reutilized for nucleotide synthesis (although this may not occur when these enzymes are significantly activated), niacin requirements are conventionally calculated on the basis of energy expenditure.

The depletion/repletion studies of Horwitt et al. (1956) and others have suggested, on the basis of restoration of urinary excretion of N^1-methyl nicotinamide, that the average niacin requirement is 5.5 mg per 1,000 kcal (1.3 mg per MJ). Allowing for individual variation, reference intakes (see Table 8.2) are set at 6.6 mg niacin equivalents (preformed niacin + 1/60 of the dietary tryptophan) per 1,000 kcal (1.6 mg per MJ). Even when energy intakes are very low, it must be assumed that energy expenditure will not fall below 2,000 kcal, and this is the basis for the calculation of reference intakes for subjects with low energy intakes.

There is probably little or no requirement for any preformed niacin in the diet, because it is likely that average intakes of protein (at least in developed countries) will provide enough tryptophan to meet requirements (Section 8.3). Assuming that the average diet provides some 15% of energy from protein, and this protein meets the reference pattern for essential amino acids and provides 14 g of tryptophan per kg of dietary protein, this implies an intake of

Table 8.2 Reference Intakes of Niacin (mg/day)

Age	U.K. 1991	EU 1993	U.S./Canada 1998	FAO 2001
0–6 m	3	—	2	2
7–9 m	4	5	4	4
10–12 m	5	5	4	4
1–3 y	8	9	6	6
4–6 y	11	11	8	8
7–8 y	12	13	8	12
Males				
9–10 y	12	13	12	12
11–13 y	15	15	12	16
14–15 y	15	15	16	16
16–18 y	18	18	16	16
19–50 y	17	18	16	16
>50 y	16	18	16	16
Females				
9–10 y	12	13	12	12
11–13 y	12	14	12	16
14–15 y	14	14	14	16
16–18 y	14	14	14	16
19–50 y	13	14	14	14
>50 y	12	14	14	14
Pregnant	12	14	18	18
Lactating	16	16	17	17

EU, European Union; FAO, Food and Agriculture Organization; WHO, World Health Organization.

Sources: Department of Health, 1991; Scientific Committee for Food, 1993; Institute of Medicine 1998; FAO/WHO, 2001.

37.5 g of protein = 525 mg of tryptophan per 1,000 kcal. Assuming that 60 mg of tryptophan is equivalent to 1 mg of dietary niacin, this suggests that an average diet provides 8.75 mg niacin equivalents per 1,000 kcal from tryptophan alone – more than the reference intake.

8.7.1 Upper Levels of Niacin Intake

Nicotinic acid (but not nicotinamide) causes a marked vasodilatation, with flushing, burning, and itching of the skin. Very large single doses of nicotinic acid may cause sufficient vasodilatation to lead to hypotension; after the administration of 1 to 3 g of nicotinic acid daily for several days, the effect wears off to a considerable extent. A number of nicotinoyl esters have been developed to permit sensitive patients to benefit from the hypolipidemic effect of nicotinic acid without the vasodilatation. The tolerable upper limit is 35 mg per day for adults (Institute of Medicine, 1998).

At intake in excess of 1 g of niacin per day, there is evidence of toxicity, with changes in liver function tests, carbohydrate tolerance, and uric acid metabolism that are reversible on withdrawal of niacin (Parsons, 1961a, 1961b). Baggenstoss and coworkers (1967) reported changes in liver ultra-structure in patients receiving high doses of niacin, including dilatation of the endoplasmic reticulum with formation of vesicles and sacs, and a diminution in the parallel arrays of rough endoplasmic reticulum, with fewer ribosomes on the outer surface. There was also elongation of the mitochondria, with bud-like projections and crystalloid inclusions. The mechanism of niacin hepatotox-icity is not known. Sustained release preparations are associated with more severe liver damage than crystalline preparations, presumably because they permit more prolonged maintenance of high concentrations of the vitamin, whereas after an acute high dose there is normally considerable excretion of unchanged nicotinic acid and nicotinamide, as the renal threshold is exceeded.

The European Health Food Manufacturers' Federation restricts over-the-counter supplements to 500 mg per day (Shrimpton, 1997). Where niacin is being used to treat clinically significant hyperlipidemia, and in trials for the prevention of type I diabetes mellitus, a tentative upper limit has been set at 3 g per day (Knip et al., 2000).

8.8 PHARMACOLOGICAL USES OF NIACIN

Nicotinic acid is used clinically in large doses (of the order of 1 to 3 g per day) as a hypolipidemic agent. It reduces both triglycerides and total cholesterol by about 20%, acting as an inhibitor of cholesterol synthesis. It has a more marked effect on cholesterol in low-density and very low-density lipoproteins, and increases high-density lipoprotein cholesterol; nicotinamide is ineffective (Brown, 1995; Capuzzi et al., 2000). It also inhibits the release of nonesterified fatty acids from adipose tissue by inhibiting the adenylate cyclase that is acti-vated in response to hormonal stimulation, thus decreasing synthesis of very low-density lipoprotein in the liver, and stimulates the 5-lipoxygenase path-way, leading to increased formation of prostaglandin E_2, thromboxane B_2, and leukotriene E_4 from arachidonic acid (Saareks et al., 1999). There is, however, some evidence that doses of about 1 g of nicotinic acid per day may lead to elevation of plasma homocysteine (Section 10.3.4.2), which would reduce the benefits to be expected from its hypolipidemic action (Garg et al., 1999).

In experimental animals, nicotinamide protects against the destruction of pancreatic β-islet cells caused by diabetogenic agents, such as alloxan and streptozotocin. Type I diabetes mellitus is caused by autoimmune destruction of β-cells, and autoantibodies against β-cell proteins can be detected in the circulation several years before the clinical onset of diabetes. It has been suggested that nicotinamide may delay the development of diabetes in susceptible subjects (Gale, 1996a, 1996b). Nicotinamide may act by either inhibiting ADP-ribosyltransferase (Section 8.4.2) and poly(ADP-ribose) polymerase (Section 8.4.3) or by inhibiting the induction of cell surface antigens by interferon and tumor necrosis factor (Kolb and Burkart, 1999; Kim et al., 2002). Although nicotinamide has no effect once diabetes has developed (Vidal et al., 2000), preliminary studies in people at risk of developing type I diabetes are promising. A prospective study involving first-degree relatives of patients with type I diabetes, who are being screened for autoantibodies and randomized to receive either nicotinamide or placebo, the European Nicotinamide Diabetes Intervention Trial (ENDIT) is expected to report in 2004 (Schatz and Bingley, 2001).

Gram doses of nicotinamide have been used in so-called orthomolecular psychiatry as a treatment for schizophrenia, originally because of the similarities between schizophrenia and the depressive psychosis of pellagra. The underlying rationale for this use is that such high doses of niacin may deplete methyl donors, and at least one of the theories of the biochemical basis of schizophrenia was that the condition is caused by inappropriate methylation of neurotransmitter metabolites to yield psychotogenic compounds (Hoffer et al., 1957). There is no independent confirmation of the efficacy of nicotinamide in the treatment of schizophrenia.

FURTHER READING

Bender DA (1983) Biochemistry of tryptophan in health and disease. *Molecular Aspects of Medicine* **6,** 101–97.

Bender DA (1996) Tryptophan and niacin nutrition – is there a problem? *Advances in Experimental Medicine and Biology* **398,** 565–9.

Bender DA and Bender AE (1986) Niacin and tryptophan metabolism: the biochemical basis of niacin requirements and recommendations. *Nutrition Abstracts and Reviews (Series A)* **56,** 695–719.

D'Amours D, Desnoyers S, D'Silva I, and Poirier GG (1999) Poly(ADP-ribosyl)ation reactions in the regulation of nuclear functions. *Biochemical Journal* **342,** 249–68.

Lee HC (1999) A unified mechanism of enzymatic synthesis of two calcium messengers: cyclic ADP-ribose and NAADP. *Biological Chemistry* **380,** 785–93.

Lee HC (2001) Physiological functions of cyclic ADP-ribose and NAADP as calcium messengers. *Annual Reviews of Pharmacology and Toxicology* **41,** 317–45.

Magni G, Amici A, Emanuelli M, Raffaelli N, and Ruggieri S (1999) Enzymology of NAD$^+$ synthesis. *Advances in Enzymology and Related Areas of Molecular Biology* **73**, 135–82, xi.

Roe DA (1973) *A Plague of Corn: The Social History of Pellagra*. Ithaca, NY: Cornell University Press.

Shall S (1995) ADP-ribosylation reactions. *Biochimie* **77**, 313–18.

Ueda K and Hayaishi O (1985) ADP-ribosylation. *Annual Reviews of Biochemistry* **54**, 73–100.

Ziegler M (2000) New functions of a long-known molecule. Emerging roles of NAD in cellular signaling. *European Journal of Biochemistry* **267**, 1550–64.

References cited in the text are listed in the Bibliography.

NINE

Vitamin B$_6$

Vitamin B$_6$ has a central role in the metabolism of amino acids: in transaminase reactions (and hence the interconversion and catabolism of amino acids and the synthesis of nonessential amino acids), in decarboxylation to yield biologically active amines, and in a variety of elimination and replacement reactions. It is also the cofactor for glycogen phosphorylase and a variety of other enzymes. In addition, pyridoxal phosphate, the metabolically active vitamer, has a role in the modulation of steroid hormone action and the regulation of gene expression.

The vitamin is widely distributed in foods, and clinical deficiency is virtually unknown, apart from an outbreak during the 1950s, which resulted from overheating of infant milk formula.

Marginal inadequacy, affecting amino acid metabolism and possibly also steroid hormone responsiveness, may be relatively common. A number of vitamin B$_6$ dependency syndromes have been reported – inborn errors of metabolism in which the defect is in the coenzyme binding site of the affected enzyme.

Estrogens cause abnormalities of tryptophan metabolism that resemble those seen in vitamin B$_6$ deficiency, and the vitamin is widely used to treat the side effects of estrogen administration and estrogen-associated symptoms of the premenstrual syndrome, although there is little evidence of its efficacy. High doses of the vitamin, of the order of 100 times requirements, cause peripheral sensory neuropathy.

In a number of enzymes that catalyze reactions that might be assumed to be pyridoxal phosphate-dependent, pyruvate provides the reactive carbonyl group (Section 9.8.1). Other enzymes have reactive carbonyl groups provided by a variety of quinones. One of these quinones, pyrrolidone quinolinequinone, may be a dietary essential, although no mammalian enzymes have been demonstrated to use it; the other catalytic quinones are covalently bound to the enzyme and are formed by postsynthetic modification of amino acid residues in a precursor protein, so are unlikely to be nutritionally relevant (Section 9.8.3).

9.1 VITAMIN B$_6$ VITAMERS AND NOMENCLATURE

The generic descriptor vitamin B$_6$ includes six vitamers: the alcohol pyridoxine, the aldehyde pyridoxal, the amine pyridoxamine, and their 5′-phosphates. There is some confusion in the older literature, because at one time pyridoxine was used as a generic descriptor, with pyridoxol as the specific name for the alcohol. As shown in Figure 9.1, the vitamers are metabolically interconvertible and, as far as is known, they have equal biological activity.

A significant proportion (in some cases up to 75%) of the vitamin B$_6$ in plant foods is present as glycosides, mainly to the 5′-hydroxyl group, although 4′-glycosides also occur. There may be some hydrolysis of glycosides in the

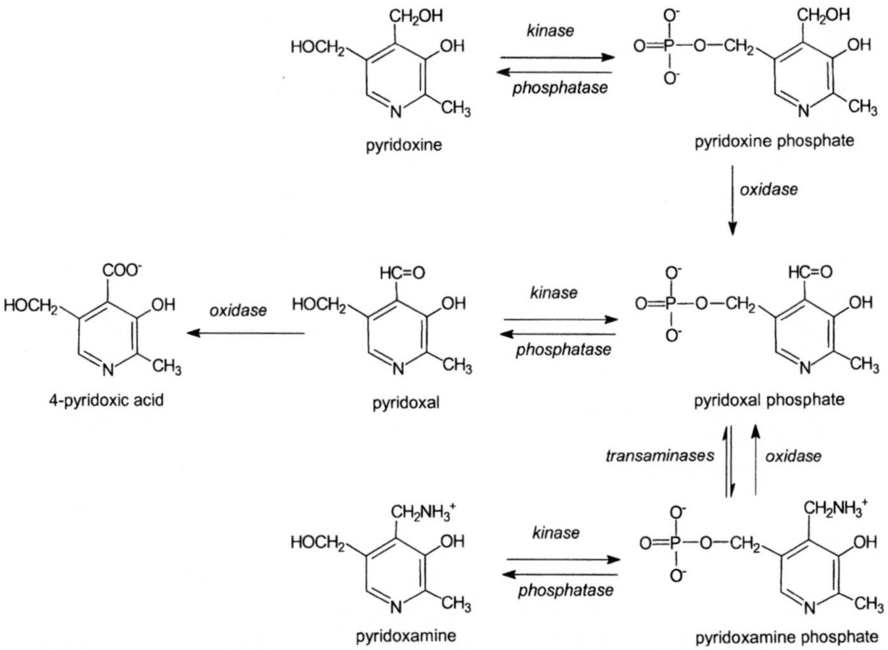

Figure 9.1. Interconversion of the vitamin B$_6$ vitamers. Pyridoxal kinase, EC 2.7.1.38; pyridoxine oxidase, EC 1.1.1.65; pyridoxamine phosphate oxidase, EC 1.4.3.5; and pyridoxal oxidase, EC 1.1.3.12. Relative molecular masses (M_r): pyridoxine, 168.3 (hydrochloride, 205.6); pyridoxal, 167.2; pyridoxamine, 168.3 (dihydrochloride, 241.1); pyridoxal phosphate, 247.1; pyridoxamine phosphate, 248.2; and 4-pyridoxic acid, 183.2.

gastrointestinal tract, and about half the vitamin present in foods as glyco-sides may be biologically available. However, pyridoxine glycosides are also absorbed intact (and excreted unchanged in the urine), and may compete for intestinal absorption and tissue uptake with the vitamin, thus having antivi-tamin activity (Gregory, 1998).

A proportion of the vitamin B$_6$ in foods may be biologically unavailable after heating, as a result of the formation of (phospho)pyridoxyllysine by re-duction of the aldimine (Schiff base) by which pyridoxal and the phosphate are bound to the ε-amino groups of lysine residues in proteins. A proportion of this pyridoxyllysine may be useable, because it is a substrate for pyridoxamine phosphate oxidase to form pyridoxal and pyridoxal phosphate. However, it is also a vitamin B$_6$ antimetabolite, and even at relatively low concentrations can accelerate the development of deficiency in experimental animals maintained on vitamin B$_6$-deficient diets (Gregory, 1980a, 1980b).

9.2 METABOLISM OF VITAMIN B$_6$

The phosphorylated vitamers are dephosphorylated by membrane-bound al-kaline phosphatase in the intestinal mucosa; pyridoxal, pyridoxamine, and pyridoxine are all absorbed rapidly by carrier-mediated diffusion. Intesti-nal mucosal cells have pyridoxine kinase and pyridoxine phosphate oxidase (see Figure 9.1), so that there is net accumulation of pyridoxal phosphate by metabolic trapping. Much of the ingested pyridoxine is released into the portal circulation as pyridoxal, after dephosphorylation at the serosal surface.

Tissue uptake of vitamin B$_6$ is again by carrier-mediated diffusion of pyri-doxal (and other unphosphorylated vitamers), followed by metabolic trapping by phosphorylation. Circulating pyridoxal and pyridoxamine phosphates are hydrolyzed by extracellular alkaline phosphatase. All tissues have pyridoxine kinase activity, but pyridoxine phosphate oxidase is found mainly in the liver, kidney, and brain.

Pyridoxine phosphate oxidase is a flavoprotein, and activation of the ery-throcyte apoenzyme by riboflavin 5'-phosphate in vitro can be used as an index of riboflavin nutritional status (Section 7.4.3). However, even in riboflavin de-ficiency, there is sufficient residual activity of pyridoxine phosphate oxidase to permit normal metabolism of vitamin B$_6$ (Lakshmi and Bamji, 1974). Pyridox-ine phosphate oxidase is inhibited by its product, pyridoxal phosphate, which binds a specific lysine residue in the enzyme. In the brain, the K$_i$ of pyridoxal phosphate is of the order of 2 μmol per L – the same as the brain concen-tration of free and loosely bound pyridoxal phosphate, suggesting that this inhibition may be a physiologically important mechanism in the control of tissue pyridoxal phosphate (Choi et al., 1987).

Pyridoxine is rapidly converted to pyridoxal phosphate in the liver and other tissues. Pyridoxal phosphate does not cross cell membranes, and efflux of the vitamin from most tissues is as pyridoxal. Pyridoxal phosphate is exported from the liver bound to albumin by formation of a Schiff base to lysine (Zhang et al., 1999). Much of the free pyridoxal phosphate in the liver (i.e., that which is not protein bound) is hydrolyzed to pyridoxal, which is also exported, and circulates bound to both albumin and hemoglobin in erythrocytes.

Extrahepatic tissues take up both pyridoxal and pyridoxal phosphate from the plasma. Pyridoxal phosphate is hydrolyzed to pyridoxal, which can cross cell membranes, by extracellular alkaline phosphatase, which is then trapped intracellularly by phosphorylation.

In subjects with hypophosphatasia, the rare genetic lack of extracellular alkaline phosphatase, plasma concentrations of pyridoxal phosphate are very much higher than normal (up to 4 μmol per L, compared with a normal range of about 100 nmol per L), and intracellular concentrations of pyridoxal phosphate are lower than normal (Narisawa et al., 2001).

Tissue concentrations of pyridoxal phosphate are controlled by the balance between phosphorylation and dephosphorylation. The activity of phosphatases acting on pyridoxal phosphate is greater than that of the kinase in most tissues, although this may be an artifact of determining alkaline phosphatase activity at its pH optimum rather than at a more physiological pH, when the two activities are approximately equal. This means that pyridoxal phosphate that is not bound to enzymes is readily dephosphorylated.

Free pyridoxal either leaves the cells or is oxidized to 4-pyridoxic acid by aldehyde dehydrogenase (which is present in all tissues) and also by hepatic and renal aldehyde oxidases. 4-Pyridoxic acid is actively secreted by the renal tubules, so measurement of the plasma concentration provides an index of renal function (Coburn et al., 2002). There is some evidence that oxidation to 4-pyridoxic acid increases with increasing age; in elderly people, the plasma concentration of pyridoxal phosphate is lower, and that of 4-pyridoxic acid higher, than in younger subjects even when there is no evidence of impaired renal function (Bates et al., 1999b). Small amounts of pyridoxal and pyridoxamine are also excreted in the urine, although much of the active vitamin B₆ that is filtered in the glomerulus is reabsorbed in the kidney tubules.

Although pyridoxine is taken up and phosphorylated by muscle (and other tissues), the resultant pyridoxine phosphate is not oxidized to pyridoxal phosphate. It has been suggested that the neurotoxicity of high intakes of pyridoxine (Section 9.9.6.4) may be caused by the uptake and trapping of pyridoxine, and hence competition with pyridoxal, resulting in depletion of tissue pyridoxal phosphate and a deficiency of the metabolically active form of the vitamin.

9.2.1 Muscle Pyridoxal Phosphate

Some 80% of the body's total vitamin B$_6$ is as pyridoxal phosphate in muscle, and some 80% of this is associated with glycogen phosphorylase. This does not seem to function as a reserve of the vitamin and is not released from the muscle in deficiency.

Muscle pyridoxal phosphate is released into the circulation (as pyridoxal) in starvation as muscle glycogen reserves are exhausted and there is less requirement for glycogen phosphorylase activity. Under these conditions, it is potentially available for redistribution to other tissues, especially the liver and kidneys, to meet the increased requirement for gluconeogenesis from amino acids (Black et al., 1978). However, during both starvation and prolonged bed rest, there is a considerable increase in urinary excretion of 4-pyridoxic acid, suggesting that much of the vitamin B$_6$ released as a result of depletion of muscle glycogen and atrophy of muscle is not redistributed, but rather is catabolized and excreted (Coburn et al., 1995).

The normal muscle concentration of pyridoxal phosphate is of the order of 10 nmol per g; in patients with McArdle's disease (glycogen storage disease from congenital lack of glycogen phosphorylase), the muscle content of pyridoxal phosphate is reduced to one-fifth of this. There is some evidence that patients with McArdle's disease show signs of vitamin B$_6$ deficiency, suggesting that the muscle pool of the vitamin is important in maintenance of vitamin B$_6$ homeostasis (Beynon et al., 1995).

9.2.2 Biosynthesis of Vitamin B$_6$

In most bacteria, pyridoxal phosphate is synthesized by condensation between 4-hydroxythreonine and the pentose sugar 1-deoxy-xylulose phosphate, formed by condensation between pyruvate and glyceraldehyde 3-phosphate. The details of the pathway are not known. In plants, fungi, and some bacteria, the precursor is glutamine; again, the pathway is unknown (Drewke and Leistner, 2001; Gupta et al., 2001).

9.3 METABOLIC FUNCTIONS OF VITAMIN B$_6$

The metabolically active vitamer is pyridoxal phosphate, which is involved in many reactions of amino acid metabolism, where the carbonyl group is the reactive moiety; in glycogen phosphorylase, where it is the phosphate group that is important in catalysis; and in the recycling of steroid hormone receptors from tight nuclear binding, where again it is the carbonyl group that is important. In vitro, pyridoxal and the phosphate will catalyze many of the reactions of pyridoxal phosphate (PLP)-dependent enzymes in protein-free solution; investigation of the crystal structures of the enzymes has shown how

Table 9.1 Pyridoxal Phosphate-Catalyzed Enzyme Reactions of Amino Acids

	Bond Cleaved	Reaction
Reactions at the α-carbon		
Racemization	αC—H	DL-isomerization
Aminotransfer	αC—NH$_2$	Amino acid → keto-acid + PMP
Decarboxylation	αC—COOH	Amino acid → primary amine + CO$_2$
Loss of side chain	αC—βC	Threonine aldolase
		Serine hydroxymethyltransferase
Reactions at the β-carbon		
Elimination	βC—R and αC—H	Tryptophanase
		Serine dehydratase
		Aspartate β-decarboxylase
Replacement	βC—R and αC—H	Tryptophan synthetase
Reactions at the γ-carbon		
Elimination	γC—R and αC—H	Homocysteine desulfhydrylase
Replacement	γC—R and αC—H	Cystathionine γ-synthetase

PMP, pyridoxamine phosphate.

reactive groups in the active sites of the enzymes provide specificity for both the substrates bound and the reaction catalyzed.

Pyridoxal phosphate reacts with ε-amino groups of lysine residues in a wide variety of proteins, forming an aldimine (Schiff base). Before the development of site-directed mutagenesis as a way of investigating the role of individual amino acids in substrate binding and catalytic mechanisms, loss of catalytic activity after reaction with pyridoxal was used to provide evidence of the involvement of a lysine residue. The aldimine can be reduced chemically to the stable covalent aldamine (e.g., with sodium borohydride), thus providing a means of labeling this reactive lysine residue for sequencing studies.

Pyridoxal phosphate has a clear role in lipid metabolism as the coenzyme for the decarboxylation of phosphatidylserine, leading to the formation of phosphatidylethanolamine, and then phosphatidylcholine (Section 14.2.1), and membrane lipids from vitamin B$_6$-deficient animals are low in phosphatidylcholine (She et al., 1995). It also has a role, less well defined, in the metabolism of polyunsaturated fatty acids; vitamin B$_6$ deficiency results in reduced activity of Δ^6 desaturase and impairs the synthesis of eicosapentanoic and docosahexanoic acids (Tsuge et al., 2000).

9.3.1 Pyridoxal Phosphate in Amino Acid Metabolism

The various reactions of pyridoxal phosphate in amino acid metabolism shown in Table 9.1 all depend on the same chemical principle – the ability to stabilize amino acid carbanions and to labilize bonds about the α-carbon, by

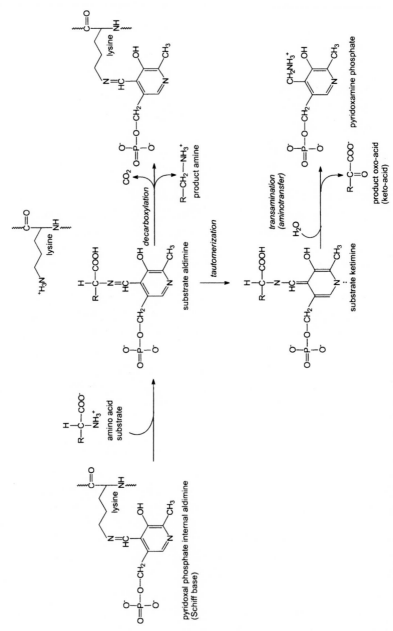

Figure 9.2. Reactions of pyridoxal phosphate-dependent enzymes with amino acids.

reaction of the α-amino group of the substrate with the carbonyl group of the coenzyme.

The ring nitrogen of pyridoxal phosphate exerts a strong electron withdrawing effect on the aldimine, and this leads to weakening of all three bonds about the α-carbon of the substrate. In nonenzymic reactions, all the possible pyridoxal-catalyzed reactions are observed – α-decarboxylation, aminotransfer, racemization and side-chain elimination, and replacement reactions. By contrast, enzymes show specificity for the reaction pathway followed; which bond is cleaved will depend on the orientation of the Schiff base relative to reactive groups of the catalytic site. As discussed in Section 9.3.1.5, reaction specificity is not complete, and a number of decarboxylases also undergo transamination.

In the absence of the amino acid substrate, pyridoxal phosphate is bound to enzymes by the formation of a Schiff base to the ε-amino group of a lysine residue at the active site. As shown in Figure 9.2, the first reaction between the substrate and the coenzyme is transfer of the aldimine linkage from the ε-amino group of the lysine residue to the α-amino group of the substrate.

Because pyridoxal phosphate is bound to lysine in this way, the resolution of holoenzymes to yield the apoenzyme is difficult unless this Schiff base can be reacted with a carbonyl-trapping reagent to give an adduct that can be removed by dialysis. The pyridoxamine phosphate form of aminotransferases (Section 9.3.1.3) can be resolved more readily, because the coenzyme is only held by ionic bonds to the 5'-phosphate, hydrophobic interactions with the 2'-methyl group, and hydrogen bonding to the heterocyclic nitrogen.

9.3.1.1 α-Decarboxylation of Amino Acids   If the electron-withdrawing effect of the heterocyclic nitrogen of pyridoxal phosphate is primarily centered on the α-carbon–carboxyl bond, the result is decarboxylation of the amino acid aldimine and release of CO_2. The resultant carbanion is then protonated, and the primary amine corresponding to the amino acid is displaced by the lysine residue at the active site, with reformation of the internal Schiff base.

A number of the products of the decarboxylation of amino acids shown in Table 9.2 are important as neurotransmitters and hormones, such as dopamine, noradrenaline, adrenaline, serotonin (5-hydroxytryptamine), histamine, and γ-aminobutyric acid (GABA), and as the diamines agmatine and putrescine and the polyamines spermidine and spermine, which are involved in the regulation of DNA metabolism. The decarboxylation of phosphatidylserine to phosphatidylethanolamine is important in phospholipid metabolism (Section 14.2.1).

Table 9.2 Amines Formed by Pyridoxal Phosphate-Dependent Decarboxylases

Amine	Parent Amino Acid	Enzyme	EC No.
Agmatine	Arginine	Arginine decarboxylase	4.1.1.19
Dopamine[a] (dihydroxyphenylethylamine)	DOPA (3,4-dihydroxyphenylalanine)	Aromatic amino acid decarboxylase	4.1.1.28
Phosphatidylethanolamine[b]	Phosphatidylserine	Phosphatidylserine decarboxylase	4.1.1.65
GABA	Glutamate	Glutamate decarboxylase	4.1.1.15
Histamine	Histidine	Histidine decarboxylase	4.1.1.22
Phenylethylamine	Phenylalanine	Bacterial phenylalanine decarboxylase	4.1.1.53
Putrescine[c]	Ornithine	Ornithine decarboxylase	4.1.1.17
Serotonin (5-hydroxytryptamine)	5-Hydroxytryptophan	Aromatic amino acid decarboxylase	4.1.1.28
Tryptamine	Tryptophan	Bacterial tryptophan decarboxylase	4.1.1.28
Tyramine	Tyrosine	Bacterial tyrosine decarboxylase	4.1.1.25

DOPA, dihydroxylphenylalanine; GABA, γ-aminobutyric acid.
[a] Dopamine is also the precursor for noradrenaline and adrenaline biosynthesis.
[b] Phosphatidylethanolamine is the precursor for choline synthesis, Section 14.2.1.
[c] Putrescine is the precursor for synthesis of spermine and spermidine by reaction with decarboxylated S-adenosyl-methionine.

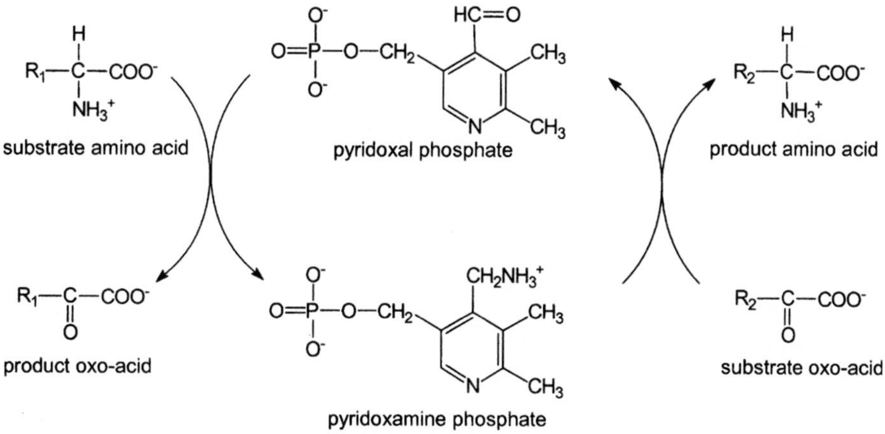

Figure 9.3. Transamination of amino acids.

9.3.1.2 Racemization of the Amino Acid Substrate

Deprotonation of the α-carbon of the amino acid leads to tautomerization of the Schiff base to the quinonoid ketimine, as shown in Figure 9.2. The simplest reaction that the ketimine can undergo is reprotonation at the now symmetrical α-carbon. This is not a stereospecific process; therefore, displacement of the substrate by the reactive lysine residue results in the racemic mixture of D- and L-amino acid.

Amino acid racemases have long been known to be important in bacterial metabolism, because several D-amino acids are required for the synthesis of cell wall mucopolysaccharides. D-Serine is found in relatively large amounts in mammalian brain, where it acts as an agonist of the N-methyl-D-aspartate (NMDA) glutamate receptor. Serine racemase has been purified from rat brain and cloned from human brain (Wolosker et al., 1999; De Miranda et al., 2000).

9.3.1.3 Transamination of Amino Acids (Aminotransferase Reactions)

Hydrolysis of the α-carbon–amino bond of the ketimine results in the release of the oxo-acid corresponding to the amino acid substrate, leaving pyridoxamine phosphate at the catalytic site of the enzyme. In this case, there is no reformation of the internal Schiff base to the reactive lysine residue. This is the half-reaction of transamination. The process is completed by reaction of pyridoxamine phosphate with a second oxo-acid substrate, forming an intermediate ketimine, then by the reverse of the reaction sequence shown in Figure 9.2, releasing the amino acid corresponding to this second substrate after displacement from the aldimine by the reactive lysine residue to reform the internal Schiff base. Transamination is thus a bisubstrate ping-pong reaction, with the amino acid substrate binding to the pyridoxal phosphate form of the enzyme and the oxo-acid substrate to the pyridoxamine phosphate form (as shown in Figure 9.3).

Table 9.3 Transamination Products of the Amino Acids

Amino Acid	Oxo-acid
Alanine	Pyruvate
Arginine	α-Oxo-γ-guanidoacetate
Aspartic acid	Oxaloacetate
Cysteine	β-Mercaptopyruvate
Glutamic acid	α-Oxoglutarate
Glutamine	α-Oxoglutaramic acid
Glycine	Glyoxylate
Histidine	Imidazolepyruvate
Isoleucine	α-Oxo-β-methylvalerate
Leucine	α-Oxo-isocaproate
Lysine[a]	α-Oxo-ε-aminocaproate $\rightarrow$ pipecolic acid
Methionine	S-methyl-β-thiol 1α-oxopropionate
Ornithine	Glutamic-γ-semialdehyde
Phenylalanine	Phenylpyruvate
Proline	γ-Hydroxypyruvate
Serine	Hydroxypyruvate
Threonine	α-Oxo-β-hydroxybutyrate
Tryptophan	Indolepyruvate
Tyrosine	p-Hydroxyphenylpyruvate
Valine	α-Oxo-isovalerate

[a] Lysine does not usually undergo transamination; if it does, the product, α-oxo-ε-aminocaproate, undergoes spontaneous dehydration and cyclization to pipecolic acid.

Transamination is of central importance in amino acid metabolism, providing pathways for the catabolism of all amino acids other than lysine (which does not undergo transamination), although pathways other than transamination may be more important for the catabolism of some amino acids. It also provides a pathway for the synthesis of those amino acids for which there is an alternative source of the oxo-acid (the nonessential amino acids). As can be seen from Table 9.3, many of the oxo-acids are common metabolic intermediates.

Many of the transaminase reactions are linked to the amination of 2-oxoglutarate to glutamate or glyoxylate to glycine, which are substrates for oxidative deamination, reforming the oxo-acids, and thus providing a pathway for net deamination of most amino acids.

9.3.1.4 **Steps in the Transaminase Reaction** Purified aspartate aminotransferase is capable of catalyzing the half-reaction of transamination (very slowly) in the crystal. This means that the conformational changes that occur during the reaction can be followed by X-ray diffraction crystallography.

The first stage in the reaction is a noncovalent (Michaelis-type) binding of the substrate to the enzyme. The substrate amino acid then attacks the internal Schiff base aldimine nucleophilically at the C=N bond. To acquire nucleophilicity, the substrate must be deprotonated; this is the role of the phenolic –OH group of the coenzyme, which will be partially ionized, and therefore carry some negative charge, when protein bound. This group then forms a dative bond to the imino nitrogen of the internal aldimine, increasing its electrophilicity and thus enhancing the nucleophilic attack by the substrate.

Once the substrate-coenzyme aldimine has been formed, it loses a proton from the α-carbon. The base that catalyzes this is the ε-amino group of the lysine residue that was involved in the internal aldimine. At this stage, the attachment of the coenzyme to the enzyme, through the ring nitrogen, the 2'-methyl group and the 5'-phosphate group, is especially important in maintaining a rigid geometry as the cofactor ring rotates about its C-2 to C-5 axis to bring the reacting region of the Schiff base into juxtaposition with the various groups at the catalytic site that catalyze the successive stages of the reaction.

The deprotonated aldimine is reprotonated at carbon-4 by reaction with a histamine residue to form the pyridoxamine phosphate ketimine. Hydrolysis of this complex yields the free oxoacid (oxaloacetate), leaving pyridoxamine phosphate at the catalytic site (Ivanov and Karpeisky, 1969).

9.3.1.5 Transamination Reactions of Other Pyridoxal Phosphate Enzymes
In addition to their main reactions, a number of pyridoxal phosphate-dependent enzymes also catalyze the half-reaction of transamination. Such enzymes include serine hydroxymethyltransferase (Section 10.3.1.1), several decarboxylases, and kynureninase (Section 8.3.3.2).

The result of this half-transaminase reaction is formation of pyridoxamine phosphate at the active site of the enzyme, and hence loss of activity. Pyridoxamine phosphate dissociates from the active site, so that if adequate pyridoxal phosphate is available the resultant apoenzyme can be reactivated.

The ratio of transamination:decarboxylation is relatively small – of the order of 1:10,000 for glutamate decarboxylase. Nevertheless, this is sufficient to result in significant loss of active enzyme, and Meister (1990) suggested that this may be a control mechanism rather than simply a lack of reaction specificity.

9.3.1.6 Transamination and Oxidative Deamination Catalyzed by Dihydroxyphenylalanine (DOPA) Decarboxylase
DOPA decarboxylase catalyzes the decarboxylation of dihydroxyphenylalanine to yield dopamine (and hence the other catecholamine neurotransmitters; see Figure 13.4) and

5-hydroxytryptophan to yield serotonin (5-hydroxytryptamine). Early studies suggested that it was especially susceptible to self-inactivation as a result of catalyzing the half-reaction of transamination (Section 9.3.1.5). However, it only catalyzes half-transamination under anaerobic conditions (commonly used in vitro because of the ready polymerization of DOPA and dopamine to yield melanin in the presence of oxygen). Under aerobic conditions, the enzyme does not seem to catalyze half-transamination, but it does catalyze oxidative deamination of its aromatic amine products (Bertoldi and Borri Voltattorni, 2000, 2001).

9.3.1.7 Side-Chain Elimination and Replacement Reactions The third bond in the Schiff base aldimine that can be labilized by the electron-withdrawing effect of the ring nitrogen is that between the α-carbon and the side chain of the amino acid, resulting in a variety of α, β-elimination and β-replacement reactions, such as the reactions of serine dehydratase; serine hydroxymethyltransferase (Section 10.3.1.1); cysteine lyase, which cleaves cysteine to form alanine and H$_2$S; selenocysteine lyase, which cleaves selenocysteine to release H$_2$Se and alanine; and cystathionine β-synthetase (see Figure 10.9). The reaction may also involve β, γ-elimination, as in γ-cystathionase (see Figure 10.9).

Cystathionine β-synthetase contains heme as well as pyridoxal phosphate, but this seems to have a regulatory rather than catalytic role; the yeast enzyme does not contain heme (Jhee et al., 2000; Kabil et al., 2001). A common genetic polymorphism in human cystathionine β-synthetase (a 68-base-pair insertion, occurring in about 12% of the general population) is associated with a lower than normal increase in plasma homocysteine after a methionine load in patients with low vitamin B$_6$ status, suggesting that the variant enzyme may have higher affinity for its cofactor than the normal form – the reverse of the position in the vitamin B$_6$ responsive genetic diseases discussed in Section 9.4.3 (Tsai et al., 1999).

9.3.2 The Role of Pyridoxal Phosphate in Glycogen Phosphorylase
Glycogen phosphorylase catalyzes the sequential phosphorolysis of glycogen to release glucose 1-phosphate; it is thus the key enzyme in the utilization of tissue glycogen reserves.

Unlike other pyridoxal phosphate-dependent enzymes, in which it is the carbonyl group that is essential for catalysis, the internal Schiff base between pyridoxal phosphate and lysine in glycogen phosphorylase can be reduced with sodium borohydride without affecting catalytic activity. Thus, while pyridoxal phosphate is essential for phosphorylase activity, it does not act by the same kind of mechanism as in amino acid metabolism.

Studies on the reactivation of apoglycogen phosphorylase with a variety of analogs of pyridoxal phosphate have shown that the catalytic moiety is the 5′-phosphate group – only analogs with a reversibly protonatable dianion in this position have any activity. In the nonactivated form of phosphorylase b, the phosphate is monoprotonated ($-OPO_3H^-$); when the enzyme has been activated, either allosterically or by phosphorylation (phosphorylase a), it is dianionic ($-OPO_3{}^{2-}$). A glutamate residue in the active site acts as the proton acceptor or donor for this transition between the inactive and active forms of the cofactor.

The initial stage in the phosphorolysis of glycogen is protonation of the glycosidic oxygen of the polysaccharide by inorganic phosphate. The resultant oxycarbonium ion is stabilized by the inorganic phosphate. The role of pyridoxal phosphate is as a proton shuttle or buffer to stabilize the oxycarbonium–phosphate ion pair, permitting covalent binding of the phosphate to the oxycarbonium ion to form glucose 1-phosphate (Palm et al., 1990).

9.3.3 The Role of Pyridoxal Phosphate in Steroid Hormone Action and Gene Expression

Steroid hormones act by binding to, and activating, nuclear receptors that then bind to hormone response elements on DNA, increasing (or sometimes decreasing) the transcription of specific genes.

Early studies showed that pyridoxal phosphate reacts with a lysine residue in the steroid receptor protein and extracts the steroid–receptor complex from tight nuclear binding. The specificity for pyridoxal phosphate, but not pyridoxal, and the low physiological concentration at which the effect can be observed, led Cidlowski and Thanassi (1981) to propose that pyridoxal phosphate may have a physiological role in the action of steroid hormones. It acts to release the hormone–receptor complex from tight nuclear binding, resulting in release of the steroid from the nucleus, and frees or recycles receptors for further uptake of steroid.

In experimental animals, vitamin B$_6$ deficiency results in increased and prolonged nuclear uptake and retention of steroids and hormones in target tissues, and enhanced end-organ responsiveness to low doses of hormones (Bender, 1987). Allgood and Cidlowski (1992) prepared a variety of gene constructs of marker genes with response elements for estrogens, androgens, glucocorticoids, and progestagens and transfected these into various cell lines in culture. Incubation in a vitamin B$_6$-deficient medium (together with the addition of the antimetabolite 4-deoxypyridoxone) led to a two-fold increase in expression of the reporter gene, whereas addition of a high concentration of pyridoxine led to a halving of gene expression. It thus seems likely that pyridoxal phosphate acts to terminate the nuclear action of steroid hormones.

Later studies showed that the effect of pyridoxal phosphate is mediated by the nuclear transcription factor NF1; gene constructs lacking an NF1 binding site are insensitive to the effects of pyridoxal phosphate deficiency or excess. Pyridoxal phosphate does not only regulate the expression of hormone-inducible genes, it inactivates the tissue-specific transcription factor for albumin (by forming a Schiff base to an essential lysine residue), and the expression of a variety of housekeeping genes is increased in experimental vitamin B$_6$ deficiency (Oka et al., 1994; Tully et al., 1994; Oka, 2001). Proliferation of both steroid-dependent and steroid-independent cancer cells in culture is reduced by higher than normal concentrations of pyridoxal in the culture medium (Davis and Cowing, 2000).

9.4 VITAMIN B$_6$ DEFICIENCY

Gross clinical deficiency of vitamin B$_6$ is extremely rare. The vitamin is widely distributed in foods (although a significant proportion in plant foods may be biologically unavailable; Section 9.1), and intestinal flora synthesize relatively large amounts, at least some of which may be absorbed and hence available.

A variety of studies have shown that 10% to 20% of the population of developed countries have marginal or inadequate status, as assessed by erythrocyte transaminase activation coefficient (Section 9.5.36) or plasma pyridoxal phosphate (Section 9.5.1; Bender, 1989b). This may be sufficient to enhance the responsiveness of target tissues to steroid hormones (Section 9.3.3), and may be important in the induction and subsequent development of hormone-dependent cancer of the breast and prostate. Vitamin B$_6$ supplementation may be a useful adjunct to other therapy in these common cancers; certainly, there is evidence that poor vitamin B$_6$ nutritional status is associated with a poor prognosis in women with breast cancer.

In vitamin B$_6$-deficient experimental animals, there are skin lesions (e.g., acrodynia in the rat) and fissures or ulceration at the corners of the mouth and over the tongue, as well as a number of endocrine abnormalities; defects in the metabolism of tryptophan (Section 9.5.4), methionine (Section 9.5.5), and other amino acids; hypochromic microcytic anemia (the first step of heme biosynthesis is pyridoxal phosphate dependent); changes in leukocyte count and activity; a tendency to epileptiform convulsions; and peripheral nervous system damage resulting in ataxia and sensory neuropathy. There is also impairment of immune responses, as a result of reduced activity of serine hydroxymethyltransferase and hence reduced availability of one-carbon substituted folate for nucleic acid synthesis (Section 10.3.3). It has been suggested that the vitamin B_6 antagonist deoxypyridoxine may be a useful adjunct to immunosuppressive therapy (Trakatellis et al., 1997).

Much of our knowledge of human vitamin B_6 deficiency is derived from an outbreak in the early 1950s, which resulted from an infant milk preparation that had undergone severe heating in manufacture, leading to the formation of pyridoxyllysine by reaction between pyridoxal phosphate and the ε-amino groups of lysine in proteins. Pyridoxyllysine has little biological activity, and may also be an antimetabolite of vitamin B_6, thus exacerbating the deficiency. In addition to a number of metabolic abnormalities, many of the affected infants convulsed. They responded to the administration of vitamin B_6 supplements.

Investigation of the neurochemical basis of the convulsions seen in vitamin B_6 deficiency revealed the role of GABA as an inhibitory neurotransmitter; GABA is synthesized by decarboxylation of glutamate (see Figure 6.3). More recent studies have suggested that the accumulation of hydroxykynurenine as a result of impaired activity of kynureninase (see Figure 9.4) may be the critical factor precipitating convulsions, and GABA depletion may be a necessary but not sufficient condition for convulsions in vitamin B_6 deficiency (Guilarte and Wagner, 1987).

The excretion of oxalate is increased in vitamin B_6 deficiency because of depressed activity of alanine glyoxylate transaminase, leading to increased accumulation of glyoxylate and onward metabolism to oxalate catalyzed by lactate dehydrogenase. The administration of vitamin B_6 supplements has some beneficial effects in reducing oxalate excretion in idiopathic oxalate stone formers, by increasing the activity of glyoxylate transaminase and hence reducing the metabolic burden of oxalate. In primary hyperoxaluria, alanine glyoxylate transaminase either has very low activity or is mistargeted into mitochondria rather than peroxisomes. Again, there is an increased glyoxylate burden, leading to considerably increased synthesis and urinary excretion of oxalate. Some cases of primary hyperoxaluria are vitamin B_6 responsive (Section 9.4.3).

9.4.1 Enzyme Responses to Vitamin B₆ Deficiency

Some pyridoxal phosphate-dependent enzymes are normally fully saturated with cofactor and show the same activity on assay in vitro whether additional pyridoxal phosphate is present in the incubation medium or not. Examples of this class of enzymes include liver cysteine sulfinate decarboxylase (which is involved in the synthesis of taurine from cysteine; Section 14.5.1) and the brain and liver glutamate and aspartate aminotransferases.

Other enzymes appear not to be fully saturated with cofactor and show increased activity in vitro when additional pyridoxal phosphate is present

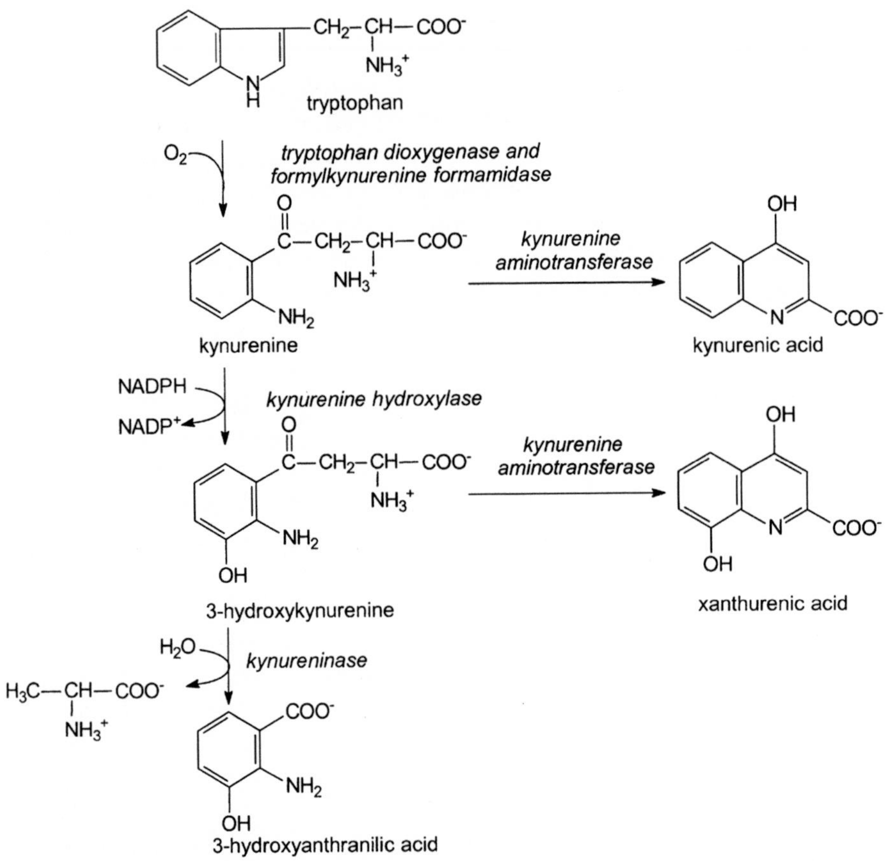

Figure 9.4. Tryptophan load test for vitamin B$_6$ status. Tryptophan dioxygenase, EC 1.13.11.11; formylkynurenine formamidase, EC 3.5.1.9; kynurenine hydroxylase, EC 1.14.13.9; kynureninase, EC 3.7.1.3; kynurenine oxoglutarate aminotransferase, EC 2.6.1.7; and kynurenine glyoxylate aminotransferase, 2.6.1.63. Relative molecular masses (M_r): tryptophan, 204.2; kynurenine, 208.2; 3-hydroxykynurenine, 223.2; kynurenic acid, 189.2; and xanthurenic acid, 205.2.

in the incubation medium; examples of these enzymes include brain glutamate decarboxylase, liver kynureninase and cystathionase, and aspartate aminotransferase from red blood cells. The activities of these enzymes thus vary with vitamin B$_6$ nutritional status. As discussed in Section 9.3.1.5, enzymes that undergo transamination, and hence mechanism-dependent inactivation, may show greater inactivation under conditions of vitamin B$_6$ deficiency and an exaggerated response to the addition of pyridoxal phosphate. When the intake of vitamin B$_6$ is increased to a relatively high level, these apoenzymes are activated in vivo, resulting in increased activity of enzymes that may well be rate controlling in metabolic pathways.

The rates of synthesis and catabolism of some pyridoxal phosphate-dependent enzymes are altered in deficiency. For example, within a few days of feeding a vitamin B$_6$-free diet to animals, there is a fall in the activity of cysteine sulfinate decarboxylase in liver; after 2 weeks, the amount of the enzyme protein has fallen to extremely low levels. It is likely that these enzymes are sacrificed to release pyridoxal phosphate for other, more essential enzymes. Other enzymes show the opposite response – apparent induction of the apoenzyme in vitamin B$_6$ deficiency, presumably in an attempt to trap as much of the available pyridoxal phosphate as possible. Sato and coworkers (1996) demonstrated increased catabolism of apocystathionase in vitamin B$_6$ deficiency, but no decrease in the amount of immunoreactive protein in the liver, as a result of increased transcription.

Katunuma and coworkers (1971) described a protease in the rat that hydrolyzes the apoenzymes of a number of pyridoxal phosphate-dependent enzymes; it has no effect on other proteins or the holoenzymes. Presumably, it attacks the conserved amino acid sequence around the active lysine residue to which the internal Schiff base is formed. The activity of the enzyme is increased some 10- to 20-fold in vitamin B$_6$ deficiency, suggesting that its function is to degrade those enzymes that lose their coenzyme more readily, and so make more pyridoxal phosphate available for use by other enzymes. There is also evidence that some pyridoxal phosphate-dependent apoenzymes are modified to become incapable of activation by pyridoxal phosphate, although retaining immunological cross-reactivity with the normal form of the enzyme in vitamin B$_6$ deficiency (Nagata and Okada, 1985).

As discussed in Section 9.3.3, pyridoxal phosphate is involved in the regulation of gene expression, terminating the responses to steroid hormones and inactivating some tissue-specific transcription factors. There is decreased synthesis of pancreatic digestive enzymes in vitamin B$_6$ deficiency, although the synthesis of other pancreatic proteins is unaffected (Dubick et al., 1995).

9.4.2 Drug-Induced Vitamin B$_6$ Deficiency

Vitamin B$_6$ deficiency may result from the prolonged administration of drugs that are carbonyl-trapping reagents, and hence can form biologically inactive adducts with pyridoxal and pyridoxal phosphate. Such compounds include penicillamine, the antituberculosis drug isoniazid, and the anti-Parkinsonian drugs Benserazide and Carbidopa. Such drug-induced vitamin B$_6$ deficiency frequently manifests as the tryptophan-niacin deficiency disease pellagra (Section 8.5.6). As discussed in Section 8.3, synthesis of the nicotinamide nucleotide coenzymes from tryptophan is pyridoxal phosphate-dependent; dietary intakes of preformed niacin seem to be inadequate to meet requirements without de novo synthesis from tryptophan, and endogenous synthesis

Table 9.4 Vitamin B$_6$-Responsive Inborn Errors of Metabolism

	Enzyme Affected	EC No.
Convulsions of the newborn	Unknown	—
Cystathioninuria	Cystathionase (see Figure 9.5)	4.4.1.1
Gyrate atrophy with ornithinuria	Ornithine-δ-aminotransferase	2.6.1.13
Homocystinuria	Cystathionine synthase (see Figure 9.5)	4.2.1.22
Primary hyperoxaluria, type I	Peroxisomal alanine-glyoxylate transaminase	2.6.1.44
Sideroblastic anemia	δ-Aminolevulinate synthase (↓ heme synthesis)	2.3.1.37
Xanthurenic aciduria	Kynureninase (see Figure 9.4)	3.7.1.3

of NAD from tryptophan seems to be more important than the utilization of dietary preformed niacin.

9.4.3 Vitamin B$_6$ Dependency Syndromes

A small number of patients show one or the other of the biochemical signs associated with vitamin B$_6$ deficiency despite apparently adequate status, and require high intakes of the vitamin to normalize the abnormal metabolic marker. These are genetic diseases and have been termed vitamin B$_6$ dependency syndromes.

As shown in Table 9.4, vitamin B$_6$ dependency has been reported in cases of type I primary hyperoxaluria, xanthurenic aciduria, homocystinuria, hypochromic sideroblastic anemia, gyrate atrophy with ornithinemia, and vitamin B$_6$ responsive infantile convulsions. In this last condition, the underlying defect has not been identified, but is almost certainly not impaired activity of glutamate decarboxylase.

The molecular basis of the other vitamin B$_6$ dependency syndromes is a severely reduced affinity of the defective enzyme for its cofactor, and the patients respond well to doses of 50 to 1,000 mg of vitamin B$_6$ per day. Apart from the affected enzyme, other biochemical indices of vitamin B$_6$ nutritional status are normal in these patients (Frimpter et al., 1969; Mudd, 1971).

9.5 THE ASSESSMENT OF VITAMIN B$_6$ NUTRITIONAL STATUS

As shown in Table 9.5, there are a number of indices of vitamin B$_6$ status available: plasma concentrations of the vitamin, urinary excretion of 4-pyridoxic acid, activation of erythrocyte aminotransferases by pyridoxal phosphate added in vitro, and the ability to metabolize test doses of tryptophan and methionine. None is wholly satisfactory; and where more than one index has been used in population studies, there is poor agreement between the different methods (Bender, 1989b; Bates et al., 1999a).

Table 9.5 Indices of Vitamin B$_6$ Nutritional Status

	Adequate Status
Plasma total vitamin B$_6$	>40 nmol (10 μg)/L
Plasma pyridoxal phosphate	>30 nmol (7.5 μg)/L
Erythrocyte alanine aminotransferase activation coefficient	<1.25
Erythrocyte aspartate aminotransferase activation coefficient	<1.80
Erythrocyte aspartate aminotransferase	>0.13 units (8.4 μkat)/L
Urine 4-pyridoxic acid	>3.0 μmol/24 h
	>1.3 mmol/mol creatinine
Urine total vitamin B$_6$	>0.5 μmol/24 h
	>0.2 mmol/mol creatinine
Urine xanthurenic acid after 2 g tryptophan load	<65 μmol/24 h increase
Urine cystathionine after 3 g methionine load	<350 μmol/24 h increase

Sources: From data reported by McChrisley et al., 1988; Leklem, 1990; Bitsch, 1993.

9.5.1 Plasma Concentrations of Vitamin B$_6$

Fasting plasma total vitamin B$_6$ (measured microbiologically), or more specifically pyridoxal phosphate, is widely used as an index of status. Conditions involving increased plasma activity of alkaline phosphatase may result in reduced plasma concentrations of pyridoxal phosphate, without affecting vitamin B$_6$ nutritional status or tissue concentrations of pyridoxal phosphate. There is a compensatory increase in the circulating concentration of pyridoxal, which, as discussed in Section 9.2, is the main form for extrahepatic uptake of vitamin B$_6$. Barnard and coworkers (1987) have shown that, despite the fall in plasma pyridoxal phosphate in pregnancy, which has been widely interpreted as indicating vitamin B$_6$ depletion or a greatly increased requirement for the vitamin, the plasma concentration of pyridoxal phosphate plus pyridoxal is unchanged. Plasma pyridoxal phosphate is also affected by acute phase responses (Bates et al., 1999a). This suggests that determination of plasma pyridoxal phosphate alone may not be a reliable index of vitamin B$_6$ nutritional status.

9.5.2 Urinary Excretion of Vitamin B$_6$ and 4-Pyridoxic Acid

Some biologically active vitamin B$_6$ is excreted in the urine, and a number of studies have assessed nutritional status by microbiological measurement of this excretion; it is difficult to interpret the results in terms of underlying nutritional status rather than as a reflection of recent intake, although the excretion does fall in deficiency (Sauberlich et al., 1972, 1974). A possibly important source of error here is that minor renal damage, resulting in albuminuria, will result in a considerable increase in urinary albumin-bound pyridoxal phosphate.

About half of the normal dietary intake of vitamin B$_6$ is excreted as 4-pyridoxic acid (see Figure 9.1). Urinary excretion of 4-pyridoxic acid will largely reflect recent intake of the vitamin rather than underlying nutritional status. More importantly, renal clearance of 4-pyridoxic acid is a marker of renal function, irrespective of vitamin B$_6$ status (Bates et al., 1999a; Coburn et al., 2002).

9.5.3 Coenzyme Saturation of Transaminases

A number of studies have measured the activation of plasma transaminases by pyridoxal phosphate added in vitro; however, it is difficult to interpret the results, because plasma transaminases arise largely accidentally, as a result of cell turnover, and the amount released will depend on tissue damage. Furthermore, there is a considerable amount of pyridoxal phosphate in plasma, largely associated with serum albumin, and the extent to which plasma transaminases are saturated will depend largely on the relative affinity of albumin and the enzyme concerned for the coenzyme, rather than reflecting the availability of pyridoxal phosphate for intracellular metabolism. Studies on erythrocyte transaminase activation coefficient are easier to interpret, because the extent to which the enzymes are saturated depends mainly on the availability of pyridoxal phosphate.

It seems likely that it is normal for a proportion of pyridoxal phosphate-dependent enzymes to be present as inactive apoenzyme, without coenzyme. This may be a mechanism of metabolic regulation. It is possible that increasing the intake of vitamin B$_6$, to ensure complete saturation of pyridoxal phosphate-dependent enzymes, may not be desirable.

9.5.4 The Tryptophan Load Test

The tryptophan load test for vitamin B$_6$ nutritional status (the ability to metabolize a test dose of tryptophan) is one of the oldest metabolic tests for functional vitamin nutritional status. It was developed as a result of observation of the excretion of an abnormal-colored compound, later identified as the tryptophan metabolite xanthurenic acid.

Apart from the relatively small amounts that are required for synthesis of the neurotransmitter serotonin (5-hydroxytryptamine), and for net new protein synthesis, essentially the whole of the dietary intake of tryptophan is metabolized by way of the oxidative pathway shown in Figures 8.4 and 9.4, which provides both a mechanism for total catabolism by way of acetyl coenzyme A and a pathway for synthesis of the nicotinamide nucleotide coenzymes (Section 8.3).

Under normal conditions, the rate-limiting enzyme of the pathway is tryptophan dioxygenase (Section 8.3.2), and there is little accumulation of intermediates. Kynurenine transaminase, the enzyme which catalyzes the transamination and ring closure of kynurenine to kynurenic acid, and of hydroxykynurenine to xanthurenic acid, has a high K_m relative to the normal steady-state concentrations of its substrates in the liver. Kynureninase and kynurenine hydroxylase have lower values of K_m, so that there is normally little accumulation of kynurenine or hydroxykynurenine.

Kynureninase is especially sensitive to vitamin B$_6$ depletion, because it undergoes self-inactivation as a result of catalyzing the half-reaction of transamination (Section 9.3.1.5). In vitamin B$_6$ deficiency, the activity of kynureninase is lower than that of tryptophan oxygenase, the normal rate-limiting enzyme of the pathway, and there is accumulation of both hydroxykynurenine and kynurenine, permitting greater metabolic flux than usual through kynurenine transaminase, resulting in increased formation of kynurenic and xanthurenic acids.

Xanthurenic and kynurenic acids, and kynurenine and hydroxykynurenine, are easy to measure in urine, so the tryptophan load test, the ability to metabolize a test dose of 2 to 5 g (150 to 380 μmol per kg of body weight) of tryptophan, was widely adopted as a convenient and sensitive index of vitamin B$_6$ nutritional status.

9.5.4.1 Artifacts in the Tryptophan Load Test Associated with Increased Tryptophan Dioxygenase Activity

As discussed in Section 8.3.2, tryptophan dioxygenase is subject to both induction by glucocorticoid hormones and also stabilization by tryptophan and heme, and increased activity may well result in a greater rate of entry of tryptophan into the pathway than the capacity of kynureninase or kynurenine hydroxylase, thus leading to increased formation of kynurenic and xanthurenic acids. Because of the problem of enhanced stability of tryptophan oxygenase in the presence of high concentrations of tryptophan, it was suggested that the test dose used should be no more than 150 mol per kg of body weight, or 2 g for adults – a level at which there is only a negligible increase in the rate of tryptophan oxidation.

In patients suffering from a wide variety of unrelated diseases, including Hodgkins' lymphoma, rheumatoid arthritis, schizophrenia, porphyria, renal tuberculosis and aplastic anemia, there is abnormal excretion of kynurenine metabolites after a test dose of tryptophan (Altman and Greengard, 1966; Coon and Nagler, 1969). It is unlikely that such disparate conditions would all be associated with vitamin B$_6$ deficiency. Liver biopsy shows elevated tryptophan dioxygenase activity, presumably because of increased glucocorticoid secretion as a result of the general stress of illness.

Induction of extrahepatic indoleamine dioxygenase (which catalyzes the same reaction as tryptophan dioxygenase, albeit by a different mechanism) by bacterial lipopolysaccharides and interferon-γ may result in the production of relatively large amounts of kynurenine and hydroxykynurenine in tissues that lack the enzymes for onward metabolism. Kidney has kynurenine transaminase activity, and therefore extrahepatic metabolism of tryptophan may result in significant excretion of kynurenic and xanthurenic acids, even when vitamin B$_6$ nutrition is adequate.

It is apparent that abnormally increased excretion of kynurenine metabolites after a test dose of tryptophan cannot necessarily be regarded as evidence of vitamin B$_6$ deficiency. This means that the tryptophan load test is unreliable as an index of status in epidemiological studies, although it is (probably) reliable in depletion/repletion studies to determine requirements.

9.5.4.2 Estrogens and Apparent Vitamin B$_6$ Nutritional Status   Rose (1966a, 1966b) was the first to report apparent vitamin B$_6$ deficiency in women taking combined progestagen–estrogen oral contraceptives. There was increased urinary excretion of xanthurenic acid after a tryptophan load, which was normalized by the administration of relatively high doses of vitamin B$_6$. A great many later reports have confirmed abnormal tryptophan metabolism among women taking the now obsolete high-dose oral contraceptives, and estrogens as menopausal hormone replacement therapy. Although they have been widely interpreted as evidence of estrogen-induced vitamin B$_6$ deficiency or depletion, when other indices of vitamin B$_6$ nutritional status have been measured, they have been unaffected by contraceptive use. This suggests an effect on tryptophan metabolism per se, rather than on vitamin B$_6$ nutritional status.

Modern low-dose estrogen oral contraceptives do not affect tryptophan metabolism, although they may cause increased plasma and erythrocyte concentrations of pyridoxal (Masse et al., 1996).

As discussed in Section 8.3.3, estrogen metabolites inhibit kynureninase and reduce the activity of kynurenine hydroxylase to such an extent that, even without induction of tryptophan dioxygenase (Section 9.5.4.1), the activity of these enzymes is lower than is needed for the rate of flux through the pathway, thus leading to increased formation of xanthurenic and kynurenic acids.

The gender difference in pellagra (Section 8.5) suggests that endogenous estrogens may have an effect on tryptophan metabolism similar to that of exogenous estrogens used as contraceptives. It implies that not only is the

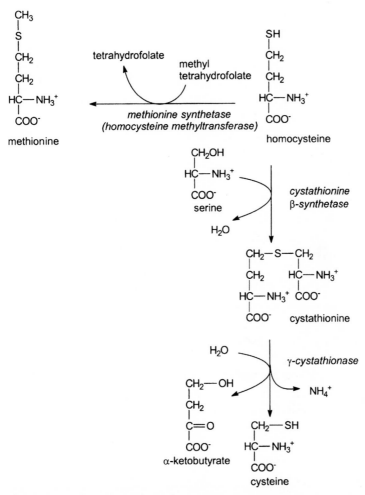

Figure 9.5. Methionine load test for vitamin B$_6$ status. Methionine synthetase, EC 2.1.1.13 (vitamin B$_{12}$-dependent); 2.1.1.5 (betaine as methyl donor); cystathionine synthetase, EC 4.2.1.22; and cystathionase, EC 4.4.1.1. Relative molecular masses (M_r): methionine, 149.2; homocysteine, 135.2; cystathionine, 222.3; and cysteine, 121.2.

tryptophan load test unreliable as an index of vitamin B$_6$ nutritional status in women taking estrogens, but also that it may be inappropriate for women in general (Bender, 1987).

9.5.5 The Methionine Load Test

The metabolism of methionine, shown in Figure 9.5, includes two pyridoxal phosphate-dependent steps: cystathionine synthetase and cystathionase. Cystathionine synthetase is little affected by vitamin B$_6$ deficiency, presumably because it has a high affinity for its cofactor, and possibly also a slow rate of turnover. However, cystathionase activity falls in vitamin B$_6$ deficiency, and there is an increase in the tissue content of inactive apoenzyme. The result of this is that in vitamin B$_6$ deficiency there is an increase in the urinary excretion of cystathionine, both after a loading dose of methionine and under basal conditions, suggesting that methionine metabolism provides a useful index of status.

As discussed in Section 10.3.4.2, the metabolic fate of homocysteine arising from methionine is determined not only by the activity of cystathionine synthetase and cystathionase, but also the rate at which it is remethylated to methionine (which is dependent on vitamin B$_{12}$ and folate status) and the requirement for cysteine.

It is apparent that abnormally increased excretion of homocysteine and cystathionine metabolites after a test dose of methionine cannot necessarily be regarded as evidence of vitamin B$_6$ deficiency. This means that, like the tryptophan load test, the methionine load test is unreliable as an index of status in epidemiological studies, although it is (probably) reliable in depletion/repletion studies to determine requirements.

9.6 VITAMIN B$_6$ REQUIREMENTS AND REFERENCE INTAKES

Vitamin B$_6$ requirements have been estimated both by isotopic tracer studies to determine turnover of the body pool (Section 9.6.1) and also by depletion/repletion studies using a variety of indices of status (Section 9.6.2). These studies have generally been conducted on young adults, and there is inadequate information to determine the requirements of elderly people, because apparent status assessed by a variety of indices declines with increasing age, despite intake as great as in younger people (Bates et al., 1999a). As discussed in Section 9.6.3, there is also inadequate information to estimate the requirements of infants.

9.6.1 Vitamin B$_6$ Requirements Estimated from Metabolic Turnover

There is a variety of estimates of the body pool of vitamin B$_6$. Short-term studies with isotopic tracers suggest a total body content of between 160 to 600 μmol (40 to 150 mg), with a half-life of 33 days, suggesting a minimum requirement for replacement in the wide range between 0.6 to 2.27 mg per day.

About 80% of the total body vitamin B$_6$ is in skeletal muscle glycogen phosphorylase, with a relatively slow turnover. Based on longer term tracer studies, Coburn (1990, 1996) has suggested a total body pool of 250 mg, or 15 nmol (3.7 μg) per g of body weight, with a loss of about 0.13% per day, hence a minimum requirement for replacement of 0.02 μmol (5 μg) per kg of body weight – 350 μg per day for a 70-kg adult.

This is considerably lower than the requirements estimated from depletion/repletion studies (Section 9.6.2) and may reflect dilution of the small pool associated with amino acid metabolism, which has a rapid turnover, by the larger and more stable pool associated with glycogen phosphorylase.

9.6.2 Vitamin B$_6$ Requirements Estimated from Depletion/Repletion Studies

Early studies of vitamin B$_6$ requirements used the development of abnormalities of tryptophan or methionine metabolism during depletion, and normalization during repletion with graded intakes of the vitamin. Although tryptophan and methionine load tests are unreliable as indices of vitamin B$_6$ status in epidemiological studies (Section 9.5.4 and Section 9.5.5), under the controlled conditions of depletion/repletion studies they do give a useful indication of the state of vitamin B$_6$ nutrition. More recent studies have used more sensitive indices of status, including the plasma concentration of pyridoxal phosphate, urinary excretion of 4-pyridoxic acid, and erythrocyte transaminase activation coefficient.

Because of the role of vitamin B$_6$ in amino acid metabolism, it is likely that protein intake will affect requirements. A number of studies have shown that adults maintained on vitamin B$_6$-deficient diets develop abnormalities of tryptophan and methionine metabolism faster, and their blood vitamin B$_6$ falls more rapidly, when their protein intake is relatively high (80 to 160 g per day in various studies) than on low protein intakes (30 to 50 g per day). Similarly, during repletion of deficient subjects, tryptophan and methionine metabolism are normalized faster at low than at high levels of protein intake (Miller and Linkswiler, 1967; Kelsay et al., 1968a, 1968b; Canham et al., 1969; Miller et al., 1985). However, Coburn (1994) noted that the requirement for growth in young animals was the same for carnivorous species, with a high protein intake, as for herbivorous species.

From such studies, the mean requirement for vitamin B$_6$ was estimated at 13 μg per g of dietary protein. Reference intakes (see Table 9.6) were based on 15 to 16 μg per g of dietary protein.

More recent depletion/repletion studies, using more sensitive indices of status in which subjects were repleted with either a constant intake of vitamin B$_6$ and varying amounts of protein, or a constant amount of protein and varying amounts of vitamin B$_6$, have shown average requirements of 15 to 16 μg per g of dietary protein, suggesting a reference intake of 18

Table 9.6 Reference Intakes of Vitamin B$_6$ (mg/day)

Age	U.K. 1991	EU 1993	U.S./Canada 1998	FAO 2001
0–6 m	0.2	—	0.1	0.1
7–9 m	0.3	0.4	0.3	0.3
10–12 m	0.4	0.4	0.3	0.3
1–3 y	0.7	0.7	0.5	0.5
4–6 y	0.9	0.9	0.6	0.6
7–8 y	1.0	1.1	0.6	1.0
Males				
9–10 y	1.0	1.1	1.0	1.0
11–13 y	1.2	1.3	1.0	1.3
14–15 y	1.2	1.3	1.3	1.3
16–18 y	1.5	1.5	1.3	1.3
19–30 y	1.4	1.5	1.3	1.3
31–50 y	1.4	1.5	1.3	1.3
51–70 y	1.4	1.5	1.7	1.7
>70 y	1.4	1.5	1.7	1.7
Females				
9–10 y	1.0	1.1	1.0	1.0
11–13 y	1.0	1.1	1.0	1.2
14–15 y	1.0	1.1	1.2	1.2
16–18 y	1.2	1.1	1.2	1.2
19–30 y	1.2	1.1	13	1.3
31–50 y	1.2	1.1	1.3	1.3
51–70 y	1.2	1.1	1.5	1.5
>70 y	1.2	1.1	1.5	1.5
Pregnant	1.2	1.3	1.9	1.9
Lactating	1.2	1.4	2.0	2.0

EU, European Union; FAO, Food and Agriculture Organization; WHO, World Health Organization.
Sources: Department of Health, 1991; Scientific Committee for Food, 1993; Institute of Medicine 1998; FAO/WHO, 2001.

to 20 μg per g protein (Kretsch et al., 1995; Hansen et al., 1996a, 1996b, 2001).

In 1998, the reference intake in the United States and Canada was reduced from the previous Recommended Daily Allowance of 2 mg per day for men and 1.6 mg per day for women (National Research Council, 1989) to 1.3 mg per day for both (Institute of Medicine, 1998). The report cites six studies that demonstrated that this level of intake would maintain a plasma concentration of pyridoxal phosphate at least 20 nmol per L although, as shown in Table 9.5, the more generally accepted criterion of adequacy is 30 nmol per L.

9.6.3 Vitamin B$_6$ Requirements of Infants

Estimation of the vitamin B$_6$ requirements of infants presents a problem, and there is a clear need for further research. Human milk, which must be assumed to be adequate for infant nutrition, provides only 2.5 to 3.5 μg of vitamin B$_6$ per g of protein – lower than the requirement for adults. Although their requirement for catabolism of amino acids may be lower than in adults (because they have net new protein synthesis), they must also increase their body content of the vitamin as they grow. Coburn (1994) noted that the requirement for growth in a number of animal species was less than that to maintain saturation of transaminases or minimum excretion of tryptophan metabolites after a test dose and was about 15 nmol per g of body weight gain across a range of species.

Based on the body content of 15 nmol (3.7 μg) of vitamin B$_6$ per g body weight, and the rate of weight gain, Coburn (1990) suggested that a minimum requirement for infants over the first 6 months of life is 100 μg (417 nmol) per day to establish tissue reserves, and an additional 20% to allow for turnover. Even if the mother receives daily supplements of 2.5 mg of vitamin B$_6$ throughout lactation, thus more than doubling her normal intake, the infant's intake ranges from 100 to 300 μg per day over the first 6 months of life. At 1 month, this is only 8.5 μg per g of protein, rising to 15 μg per g by 2 months (Borschel et al., 1986).

A first approximation to the vitamin B$_6$ needs of infants came from studies of those who convulsed as a result of gross deficiency caused by overheated infant milk formula. At intakes of 60 μg per day, there was an incidence of convulsions of 0.3%. Provision of 260 μg per day prevented or cured convulsions, but 300 μg per day was required to normalize tryptophan metabolism (Bessey et al., 1957). This is almost certainly a considerable overestimate of requirements, because pyridoxyllysine, formed by heating the vitamin with proteins, has antivitamin activity, and would therefore result in a higher apparent requirement.

9.6.4 Toxicity of Vitamin B$_6$

Animal studies have shown that vitamin B$_6$ is neurotoxic, causing peripheral neuropathy, with ataxia, muscle weakness, and loss of balance in dogs given 200 mg of pyridoxine per kg of body weight for 40 to 75 days, and the development of a swaying gait and ataxia within 9 days at a dose of 300 mg per kg body weight (Phillips et al., 1978; Krinke et al., 1980). At the lower dose of 50 mg per kg of body weight, there are no clinical signs of toxicity, but histologically there is loss of myelin in dorsal nerve roots. At higher doses, there is widespread neuronal damage, with loss of myelin and degeneration of sensory fibers in peripheral nerves, the dorsal columns of the spinal cord, and the descending tract of the trigeminal nerve. The clinical signs of toxicity after 200 to 300 mg of vitamin B$_6$ per kg of body weight regress within 3 months after withdrawal of these massive doses, but sensory nerve conduction velocity, which decreases during the development of the neuropathy, does not recover fully (Schaeppi and Krinke, 1982).

In 1983, sensory neuropathy was reported in seven patients who had been taking between 2,000 to 7,000 mg of pyridoxine/day for several months (Schaumburg et al., 1983). On withdrawal of the vitamin supplements, there was considerable recovery of neuronal function, although there was residual nerve damage in some patients.

There has been one report of the development, within 2 years, of sensory neuropathy in an infant with vitamin B$_6$-dependent seizures treated with 2,000 mg per day, but over the following 16 years, the neuropathy did not progress (McLachlan and Brown, 1995). However, most reports of patients with vitamin B$_6$ dependency diseases (Section 9.4.3) do not mention sensory neuropathy. One study has reported electrophysiological and neurological examination of 17 homocystinuric patients who had been treated with 200 to 500 mg of vitamin B$_6$ per day for 10 to 24 years; there was no evidence of neuropathy (Mpofu et al., 1991).

None of the studies in which there has been objective neurological examination has shown any evidence of sensory nerve damage at intakes of vitamin B$_6$ below 200 mg per day. Most have shown adverse effects only at considerably higher levels of intake.

Studies with cells in culture show a cytotoxic effect of vitamin B$_6$. This may be from the formation of cytotoxic products when the vitamin is subjected to ultraviolet irradiation and may not be relevant in vivo (Maeda et al., 2000).

9.6.4.1 Upper Levels of Vitamin B$_6$ Intake

Although there is no doubt that vitamin B$_6$ is neurotoxic in gross excess, there is considerable controversy over the way in which toxicological data have been translated into limits on the amounts that may be sold freely as nutritional supplements. This appears to have been achieved by the application of standard toxicology safety margins, and taking as the upper safe limit of intake 1% of the no adverse effect level. Whereas this is appropriate for setting limits on additives and contaminants, it can be argued that it is not appropriate as a basis for setting limits on a nutrient; for many nutrients, an upper limit of intake established in this way would be below the average requirement to prevent deficiency.

There is little evidence that intakes of up 200 to 500 mg of vitamin B$_6$ per day for prolonged periods are associated with any adverse effects. The U.S.

Food and Nutrition Board set a tolerable upper level for adults of 100 mg per day (Institute of Medicine, 1998); the European Union (Scientific Committee on Food, 2000) set 25 mg per day.

9.7 PHARMACOLOGICAL USES OF VITAMIN B$_6$

Supplements of vitamin B$_6$ ranging from 25 to 500 mg per day have been recommended for treatment of a variety of conditions in which there is an underlying physiological or biochemical mechanism to justify their use, although in most cases there is little evidence of efficacy.

Some reports have shown vitamin B$_6$ to be effective in suppression of lactation, although others have shown no difference from placebo. Because the vitamin suppresses the increase in prolactin induced by treatment with the dopamine receptor antagonist pimozide, and because lactation is also suppressed by the dopamine agonist bromocriptine, it has been suggested that it acts to stimulate dopaminergic activity in the hypothalamus. However, it is more likely that its action is by reduction in target tissue responsiveness to the steroid hormones that stimulate prolactin secretion. High doses of vitamin B$_6$ are also effective in controlling tardive dyskinesia induced by neuroleptic drugs (Lerner et al., 2001).

9.7.1 Vitamin B$_6$ and Hyperhomocysteinemia

The identification of hyperhomocysteinemia as an independent risk factor in atherosclerosis and coronary heart disease (Section 10.3.4.2) has led to suggestions that intakes of vitamin B$_6$ higher than are currently considered adequate to meet requirements may be desirable. Homocysteine is an intermediate in methionine metabolism and may undergo one of two metabolic fates, as shown in Figure 9.5: remethylation to methionine (a reaction that is dependent on vitamin B$_{12}$ and folic acid) or onward metabolism leading to the synthesis of cysteine (trans-sulfuration). Therefore, intakes of folate, vitamin B$_{12}$, and/or vitamin B$_6$ may affect homocysteine metabolism.

Epidemiological studies suggest that hyperhomocysteinemia is most significantly correlated with low folate status, but there is also a significant association with low vitamin B$_6$ status (Selhub et al., 1993). Trials of supplementation have shown that whereas folate supplements lower fasting homocysteine in moderately hyperhomocysteinemic subjects, supplements of 10 mg per day of vitamin B$_6$ have no effect, although supplements do reduce the peak plasma concentration of homocysteine after a test dose of methionine (Ubbink et al., 1994; Ubbink, 1997; Dierkes et al., 1998). This can probably be explained on the basis of the kinetics of the enzymes involved; the K_m of cystathionine synthetase is 10-fold higher than that of methionine synthetase. Under basal conditions, little homocysteine is metabolized by way of the transsulfuration pathway; it is only after a loading dose of methionine, when homocysteine rises to high levels, that the activity of cystathionine synthetase, rather than the concentration of its substrate, is limiting.

It thus seems unlikely that intakes of vitamin B$_6$ above amounts that are adequate to prevent metabolic signs of deficiency will be beneficial in lowering plasma concentrations of homocysteine (Homocysteine Lowering Trialists' Collaboration, 1998).

9.7.2 Vitamin B$_6$ and the Premenstrual Syndrome

Studies during the 1960s showing that vitamin B$_6$ supplements were effective in overcoming some of the side effects of (high-dose) oral contraceptives have led to the use of vitamin B$_6$ in treatment of the premenstrual syndrome – the condition of nervousness, irritability, emotional disturbance, headache, and/or depression suffered by many women for up to 10 days before menstruation.

Twelve placebo-controlled, double-blind trials of vitamin B$_6$ in the premenstrual syndrome were reviewed by Kleijnen et al. (1990); the evidence of beneficial effects was weak. In three of the studies, there was a significant beneficial effect of vitamin B$_6$ supplements of 100 to 500 mg per day. Five studies yielded ambiguous results and a further three reported the following: an improvement for 82% of subjects receiving 100 mg of vitamin B$_6$ per day, and 70% of those receiving placebo; a positive trend but no statistical significance using 200 mg per day; and disappointing and not clear results using 50 mg per day. The remaining four studies reported no beneficial effects of doses between 100 to 500 mg per day. A systematic review of 25 controlled trials concluded that only 10 met the criteria for inclusion, and concluded that overall the results were suggestive of beneficial effects, but did not provide evidence of efficacy (Wyatt et al., 1999).

Interestingly, one study, which reported no significant difference between vitamin B$_6$ (100 mg per day) and placebo, showed that whichever treatment was used second in a double-blind crossover trial was significantly better than the treatment used first (Hagen et al., 1985).

9.7.3 Impaired Glucose Tolerance

Impaired glucose tolerance is common in pregnancy and may be severe enough to be classified as diabetes mellitus (so-called gestational diabetes), which usually resolves on parturition. A number of studies have shown that supplements of 100 mg of vitamin B$_6$ per day result in improved glucose tolerance (Rose et al., 1975).

As discussed in Section 9.5.4.2, estrogen metabolites inhibit kynureninase, and they also lead to reduced activity of kynurenine hydroxylase. As a result, in pregnancy or in response to (high-dose) oral contraceptives, tissue concentrations of kynurenine, hydroxykynurenine, xanthurenic, and kynurenic acids are higher than normal.

The impairment of glucose tolerance associated with high plasma levels of estrogens may be caused by a high plasma concentration of xanthurenic acid, which forms a biologically inactive complex with insulin (Kotake et al., 1975). The improvement following high doses of vitamin B$_6$ could then be explained by reactivation of kynureninase that has been inactivated as a result of transamination (Section 9.3.1.5). However, animal studies have failed to demonstrate any effect of xanthurenic acid administration on glucose tolerance, and it has been suggested that the improvement in glucose tolerance in response to vitamin B$_6$ was because of increased formation of quinolinic acid as a result of relief of the impairment of kynureninase activity (Adams et al., 1976). Quinolinic acid is an inhibitor of phosphoenolpyruvate carboxykinase, one of the key enzymes of gluconeogenesis, and the administration of tryptophan (to increase synthesis of quinolinic acid) has also been reported to improve glucose tolerance.

9.7.4 Vitamin B$_6$ for Prevention of the Complications of Diabetes Mellitus

A number of studies have suggested that vitamin B$_6$ may be effective in preventing the adverse effects of poor glycemic control that lead to the development of the complications of diabetes mellitus (Jain and Lim, 2001). Many of these effects are mediated by nonenzymic glycation of proteins. Target proteins include the following:

1. hemoglobin, forming hemoglobin A$_{1c}$, which has a reduced capacity to transport oxygen;
2. α-crystallin in the lens of the eye, reducing its transparency and leading to the development of cataracts;
3. collagen in joints and connective tissue, resulting in the development of arthritis, retinopathy, and nephropathy; and
4. apolipoprotein B, resulting in impaired receptor-mediated clearance of low-density lipoproteins by the liver and increased macrophage uptake, and thus a factor in the development of atherosclerosis.

Administration of vitamin B$_6$ to genetically diabetic mice has been reported to reduce the thickening of the glomerular basement membrane (Hayakawa and Shibata, 1991). In men with non-insulin-dependent diabetes, supplements of 150 mg per day led to a significant reduction in glycated hemoglobin and improved oxygen transport capacity, although there was no change in glycemic control (Solomon and Cohen, 1989).

Pyridoxamine is a potent inhibitor of the irreversible rearrangement of the initial product of reversible glycation to the advanced glycation end-product. Such inhibitors of this reaction are collectively known as amadorins, and the name *pyridorin* has been coined for pyridoxamine used in this way (Khalifah et al., 1999). Pyridoxamine also inhibits protein modification caused by lipid peroxides; in both cases, it seems to act by trapping carbonyl compounds formed as intermediates (Onorato et al., 2000; Voziyan et al., 2002).

9.7.5 Vitamin B$_6$ for the Treatment of Depression

There is a great deal of evidence that deficiency of serotonin (5-hydroxytryptamine) is a factor in depressive illness, and many antidepressant drugs act to decrease its catabolism or enhance its interaction with receptors. A key enzyme involved in the synthesis of serotonin (and the catecholamines) is aromatic amino acid decarboxylase, which is pyridoxal phosphate-dependent. Therefore, it has been suggested that vitamin B$_6$ deficiency may result in reduced formation of the neurotransmitters and thus be a factor in the etiology of depression. Conversely, it has been suggested that supplements of vitamin B$_6$ may increase aromatic amino acid decarboxylase activity, and increase amine synthesis and have a mood-elevating or antidepressant effect. There is little evidence that vitamin B$_6$ deficiency affects the activity of aromatic amino acid decarboxylase. In patients with kidney failure, undergoing renal dialysis, the brain concentration of pyridoxal phosphate falls to about 50% of normal, with no effect on serotonin, catecholamines, or their metabolites (Perry et al., 1985).

In rats, high doses of vitamin B$_6$ (10 mg per kg of body weight) lead to decreased oxidative metabolism of tryptophan, presumably as a result of impaired responsiveness to glucocorticoid hormones, an increased plasma concentration of tryptophan, and increased uptake of tryptophan into the brain, leading to an increased rate of serotonin turnover (Bender and Totoe, 1984a). This suggests that vitamin B$_6$ supplements might be a useful adjunct to tryptophan for the treatment of depression.

9.7.6 Antihypertensive Actions of Vitamin B$_6$

Vitamin B$_6$ depletion leads to the development of hypertension in experimental animals, which is normalized within 24 hours by repletion with the vitamin.

Four mechanisms, which are not mutually exclusive, have been proposed to account for this (Dakshinamurti and Lal, 1992; Dakshinamurti, 2001):

1. Central effects on blood pressure regulation as a result of decreased synthesis of brain GABA and serotonin (5-hydroxytryptamine). Glutamate decarboxylase activity in the nervous system is especially sensitive to vitamin B_6 depletion, possibly as a result of mechanism-dependent inactivation by transamination. Although there is no evidence that aromatic amino acid decarboxylase activity is reduced in vitamin B_6 deficiency, there is reduced formation of serotonin in the central nervous system.

2. Increased sympathetic nervous system activity. There is evidence of elevated plasma concentrations of adrenaline and noradrenaline in vitamin B_6-deficient animals.

3. Increased uptake of calcium by arterial smooth muscle, leading to increased muscle tone, and hence increased circulatory resistance and blood pressure. This could reflect increased sensitivity of vascular smooth muscle to calcitriol (vitamin D) action in vitamin B_6 deficiency; the membrane calcium-binding protein is regulated by vitamin D, and vascular tissue has calcitriol receptors.

4. Increased end-organ responsiveness to glucocorticoids, mineralocorticoids, and aldosterone (Section 9.3.3). Oversecretion of (and presumably also enhanced sensitivity to) any of these hormones can result in hypertension. Vitamin B_6 supplementation would be expected to reduce end-organ sensitivity to these hormones, and thus might have a hypotensive action.

A number of studies suggest that supplements of vitamin B_6 may have a hypotensive action. Supplements of 300 mg of vitamin B_6 per kg of body weight per day attenuated the hypertensive response of rats treated with deoxycorticosterone acetate (Fregly and Cade, 1995). At a more realistic level of supplementation (five times the usual amount provided in the diet), vitamin B_6 prevented the development of hypertension in the Zucker (*fa/fa*) obese rat. Withdrawal of the vitamin supplement led to the development of hypertension (Lal et al., 1996).

9.8 OTHER CARBONYL CATALYSTS

A number of enzymes contain other carbonyl compounds that catalyze reactions in the same way as does pyridoxal phosphate or that catalyze redox reactions. Such compounds include pyruvate (Section 9.8.1); pyrroloquinoline quinone, which may be a dietary essential (Section 9.8.2); and a variety of other quinones that are not dietary essentials because they are formed by postsynthetic modification of precursor proteins (Section 9.8.3).

9.8.1 Pyruvoyl Enzymes

A number of enzymes that catalyze the same reactions as do pyridoxal phosphate-dependent enzymes contain a catalytic pyruvate residue at the amino terminal of the peptide chain. The catalytic mechanism is assumed to be the same as for pyridoxal phosphate-dependent enzymes, except that the proton donor is a glutamate residue rather than lysine.

The most studied enzyme is histidine decarboxylase from *Lactobacillus* 30a. There are pyruvate residues at the amino terminals of each of 5 of the 10 subunits in this enzyme. When the organism is grown on [¹⁴C]serine, the specific radioactivity of the pyruvate is the same as that of serine incorporated into the protein and much greater than that of free lactate or pyruvate in the culture medium. This suggests that pyruvate arises by postsynthetic modification of a serine residue.

A mutant strain of the organism has been isolated that produces an inactive precursor of histidine decarboxylase with only five separable subunits and no pyruvate residues. Prolonged incubation of this zymogen leads to apparently autocatalytic activation. Each subunit gives rise to two subunits, with the formation of an amino terminal pyruvate from the serine residue adjacent to the point of cleavage. The precursor protein has two adjacent serine residues, and undergoes an autocatalytic nonhydrolytic cleavage between these two serines. Oxygen from the side chain of the distal serine becomes the second oxygen of what now becomes the carboxy terminal of one peptide, leaving an imine at the amino terminal of the other. This imine undergoes hydrolysis to release ammonia and yield the amino terminal pyruvate residue.

Other pyruvate-containing enzymes include aspartate β-decarboxylase from *Escherichia coli*, the enzyme that catalyzes the formation of β-alanine for the synthesis of pantothenic acid (Section 12.2.4); proline reductase from *Clostridium sticklandii*; phosphatidylserine decarboxylase from *E. coli*; and phenylalanine aminotransferase from *Pseudomonas fluorescens*. Phosphopantetheinoyl cysteine decarboxylase, involved in the synthesis of coenzyme A (Section 12.2.1), and S-adenosylmethionine decarboxylase seem to be the only mammalian pyruvoyl enzymes (Snell, 1990).

9.8.2 Pyrroloquinoline Quinone (PQQ) and Tryptophan Tryptophylquinone (TTQ)

PQQ and TTQ (see Figure 9.6) are the cofactors for a number of dehydrogenases in gram-negative microorganisms. They can undergo reversible reduction to

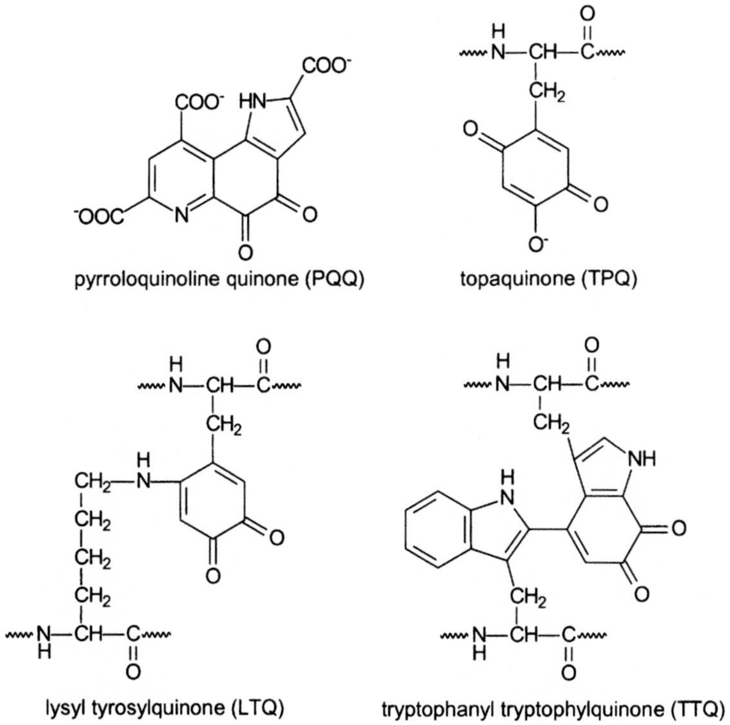

Figure 9.6. Quinone catalysts.

the semiquinone and quinol, and can therefore function as electron transport cofactors in redox reactions.

PQQ is present as a noncovalently bound coenzyme in bacterial enzymes, and organisms that are incapable of its de novo synthesis can import it from the culture medium. It is synthesized by reaction between glutamate and tyrosine residues in a small (24 amino acid) peptide that is coded for by one of the bacterial genes known to be required for PQQ synthesis (Stites et al., 2000b).

No mammalian enzymes have been shown to utilize PQQ or TTQ as a cofactor, although there is some evidence PPQ may be a dietary essential. Mice fed a defined diet completely devoid of PQQ show impaired growth, friable skin with hemorrhages, hunched posture, decreased fertility, and fewer mitochondria, which are less viable in vitro than normal. These abnormalities are corrected by providing 1 nmol (300 ng) of PQQ per g diet (Stites et al., 2000a). It is not known how much PQQ may normally be present in foods, nor how much may be synthesized by intestinal bacteria.

9.8.3 Quinone Catalysts in Mammalian Enzymes

Mammalian copper-dependent oxidases, including plasma amine oxidases (the semicarbazide-sensitive, chlorgyline-resistant, amine oxidases, not the flavin-dependent monoamine oxidases) and lysyl oxidase (which is involved in the cross-linking of collagen and elastin) have long been known to contain a reactive carbonyl group that is essential for activity. Although this was originally assumed to be pyridoxal phosphate, there is no evidence for its presence in these enzymes, and the apoenzymes cannot be reactivated with pyridoxal phosphate. In amine oxidases, the reactive quinone is topaquinone, whereas in lysyl oxidase it is lysyltopaquinone (see Figure 9.6).

Topaquinone and lysyltopaquinone are formed by postsynthetic modification of the precursor proteins of the active enzymes. A tyrosine residue in the enzyme undergoes autocatalytic oxidation in the presence of enzyme-bound copper and oxygen. Lysyltopaquinone in lysyl oxidase is believed to be synthesized by reaction between topaquinone and the ε-amino group of a lysine residue to form the cross-linked imino adduct. This means that it is highly improbable that either topaquinone or lysyltopaquinone is a dietary essential, because there is no way in which preformed quinone could be incorporated into the precursor protein or a hypothetical apoenzyme.

FURTHER READING

Anthony C (1996) Quinoprotein-catalysed reactions. *Biochemical Journal* **320,** 697–711.
Bender DA (1987) Oestrogens and vitamin B$_6$ – actions and interactions. *World Review of Nutrition and Dietetics* **51,** 140–88.
Bender DA (1989) Vitamin B$_6$ requirements and recommendations. *European Journal of Nutrition* **43,** 289–309.
Bender DA (1999) Non-nutritional uses of vitamin B$_6$. *British Journal of Nutrition* **81,** 7–20.
Coburn SP (1994) A critical review of minimal vitamin B$_6$ requirements for growth in various species with a proposed method of calculation. *Vitamins and Hormones* **48,** 259–300.
Coburn SP (1996) Modeling vitamin B$_6$ metabolism. *Advances in Food and Nutrition Research* **40,** 107–32.
Dakshinamurti K and Dakshinamurti S (2001) Blood pressure regulation and micronutrients. *Nutrition Research Reviews* **14,** 3–43.
Dakshinamurti K and Lal KJ (1992) Vitamins and hypertension. *World Review of Nutrition and Dietetics* **69,** 40–73.
Fasella P and Turano C (1970) Structure and catalytic role of the functional groups of aspartate aminotransferase. *Vitamins and Hormones* **28,** 157–94.
Hayashi H (1995) Pyridoxal enzymes: mechanistic diversity and uniformity. *Journal of Biochemistry (Tokyo)* **118,** 463–73.

Hayashi H, Wada H, Yoshimura T, Esaki N, and Soda K (1990) Recent topics in pyridoxal 5′-phosphate enzyme studies. *Annual Reviews of Biochemistry* **59,** 87–110.

Ink SL and Henderson LM (1984) Vitamin B₆ metabolism. *Annual Reviews of Nutrition* **4,** 455–70.

Ivanov VI and Karpeisky MY (1969) Dynamic three-dimensional model for enzymic transamination. *Advances in Enzymology and Related Areas of Molecular Biology* **32,** 21–53.

John RA (1995) Pyridoxal phosphate-dependent enzymes. *Biochimica et Biophysica Acta* **1248,** 81–96.

Kruger WD (2000) Vitamins and homocysteine metabolism. *Vitamins and Hormones* **60,** 333–52.

Martell AE (1982) Reaction pathways and mechanisms of pyridoxal catalysis. *Advances in Enzymology and Related Areas of Molecular Biology* **53,** 163–99.

McIntire WS (1998) Newly discovered redox cofactors: possible nutritional, medical, and pharmacological relevance to higher animals. *Annual Reviews of Nutrition* **18,** 145–77.

Mehta PK and Christen P (2000) The molecular evolution of pyridoxal-5′-phosphate-dependent enzymes. *Advances in Enzymology and Related Areas of Molecular Biology* **74,** 129–84.

Merrill AH Jr and Henderson JM (1987) Diseases associated with defects in vitamin B₆ metabolism or utilization. *Annual Reviews of Nutrition* **7,** 137–56.

Oka T (2001) Modulation of gene expression by vitamin B₆. *Nutrition Research Reviews* **14,** 257–65.

Selhub J (1999) Homocysteine metabolism. *Annual Reviews of Nutrition* **19,** 217–46.

Stites TE, Mitchell AE, and Rucker RB (2000) Physiological importance of quinoenzymes and the O-quinone family of cofactors. *Journal of Nutrition* **130,** 719–27.

References cited in the text are listed in the Bibliography.

TEN

Folate and Other Pterins and Vitamin B$_{12}$

Folic acid functions in the transfer of one-carbon fragments in a wide variety of biosynthetic and catabolic reactions; it is therefore metabolically closely related to vitamin B$_{12}$, which also functions in one-carbon transfer. Deficiency of either vitamin has similar clinical effects, and it seems likely that the main effects of vitamin B$_{12}$ deficiency are exerted by effects on folate metabolism.

The pterins include the redox cofactors biopterin and molybdopterin, as well as various insect pigments. Folic acid is a conjugated pterin, in which the pteridine ring is linked to p-aminobenzoyl-poly-γ-glutamate; it is this linkage that renders folate a dietary essential, because it is the ability to condense p-aminobenzoate to a pteridine, rather than to synthesize the pteridine nucleus itself, which has been lost by higher animals. Biopterin (Section 10.4) and molybdopterin (Section 10.5) are coenzymes in mixed-function oxidases; they are not vitamins, but can be synthesized in the body. Rare genetic defects of biopterin synthesis render it a dietary essential for affected individuals.

Although folate is widely distributed in foods, dietary deficiency is not uncommon, and a number of commonly used drugs can cause folate depletion. Marginal folate status is a factor in the development of neural tube defects and supplements of 400 μg per day periconceptually reduce the incidence of neural tube defects significantly. High intakes of folate lower the plasma concentration of homocysteine in people genetically at risk of hyperhomocysteinemia and may reduce the risk of cardiovascular disease, although as yet there is no evidence from intervention studies. There is also evidence that low folate status is associated with increased risk of colorectal and other cancers and that folate may be protective. Mandatory enrichment of cereal products with folic acid has been introduced in the United States and other countries, and considered in others.

By contrast, dietary deficiency of vitamin B_{12} is rare, except among strict vegetarians, despite the fact that the vitamin is found only in animal foods and some bacteria; rather, pernicious anemia caused by vitamin B_{12} lack is normally the result of a defect in the mechanism for intestinal absorption of the vitamin.

10.1 FOLATE VITAMERS AND DIETARY FOLATE EQUIVALENTS

As shown in Figure 10.1, folic acid consists of a pteridine linked at C-9 to *p*-aminobenzoic acid, forming pteroic acid. The carboxyl group of the *p*-aminobenzoic acid moiety is linked by a peptide bond to the amino group of glutamate, forming pteroyl monoglutamate.

In the folate coenzymes, the pteridine ring is fully reduced to tetrahydro-folate, although the oxidized form, dihydrofolate, is an important metabolic intermediate. In the reactions of thymidylate synthetase (Section 10.3.3) and methylene tetrahydrofolate reductase (Section 10.3.2.1), the pteridine ring has a redox role in the reaction. The folate coenzymes are conjugated with up to six additional glutamate residues, linked by γ-glutamyl peptide bonds.

Although the terms folic acid and folate are often used interchangeably, correctly *folic acid* refers to the oxidized compound, pteroyl monoglutamate, and the various tetrahydrofolate derivatives are collectively known as folates.

Figure 10.1 also shows the structures of the folate antagonist methotrex-ate (N^{10}-methyl aminopterin) and the pterin coenzymes tetrahydrobiopterin (Section 10.4) and molybdopterin (Section 10.5).

As shown in Figure 10.3, tetrahydrofolate can carry one-carbon fragments attached to N-5 (formyl, formimino, or methyl groups), N-10 (formyl), or bridg-ing N-5 to N-10 (methylene or methenyl groups).

5-Formyl-tetrahydrofolate is more stable to atmospheric oxidation than folic acid itself and is commonly used in pharmaceutical preparations; it is also known as folinic acid and the synthetic (racemic) compound as leucov-orin. Although the [6S, 6R] racemic mixture might be expected to have only 50% of the biological activity of the naturally occurring 6S isomer, between 10% to 40% of the 6R isomer is biologically active (Baggott et al., 2001).

10.1.1 Dietary Folate Equivalents

The folate in foods consists of a mixture of the different one-carbon substi-tuted derivatives, with varying numbers of conjugated glutamyl residues. The biological availability of these vitamers differs and is consistently lower than that of free folic acid (pteroyl monoglutamate), which is the compound that

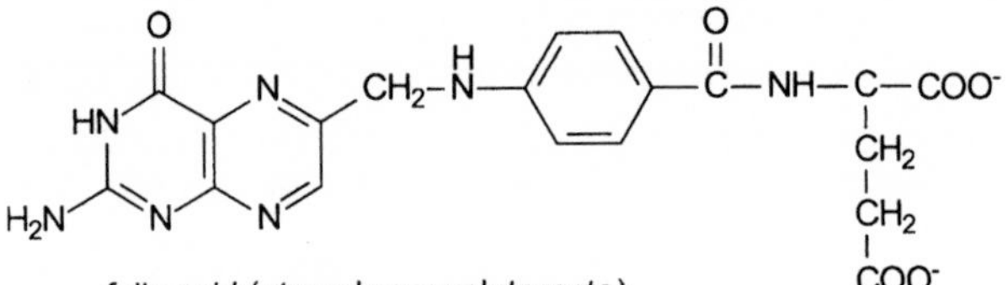

folic acid (pteroyl monoglutamate)

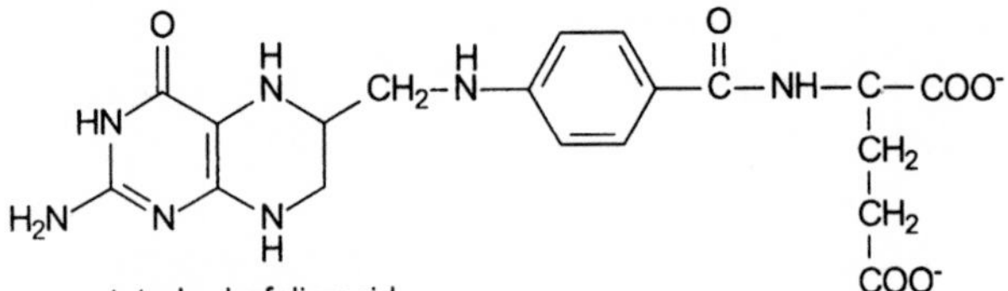

tetrahydrofolic acid

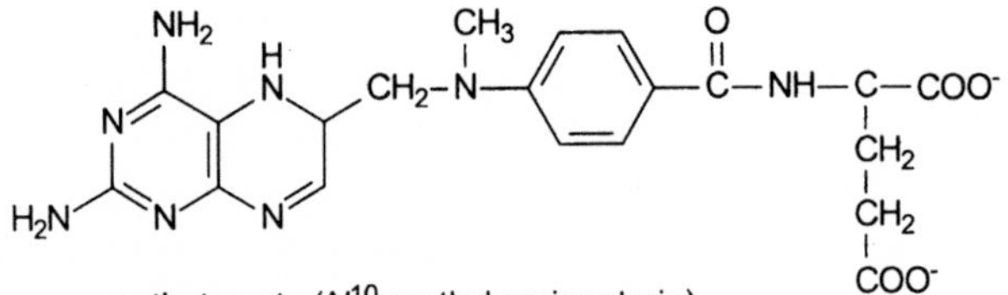

methotrexate (N^{10}-methyl aminopterin)

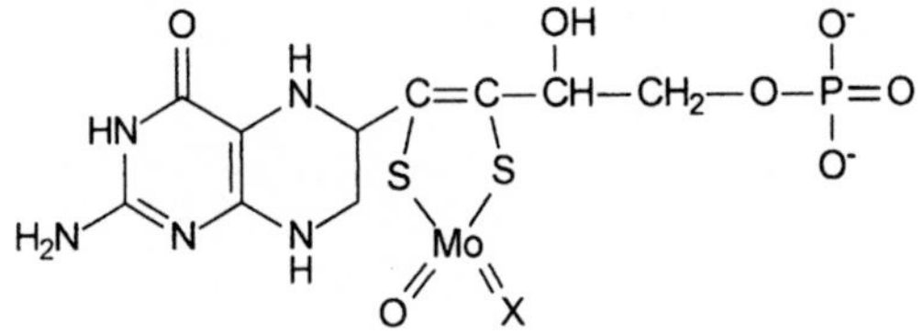

molybdopterin
(X = O in sulfite oxidase, S in aldehyde and xanthine oxidases)

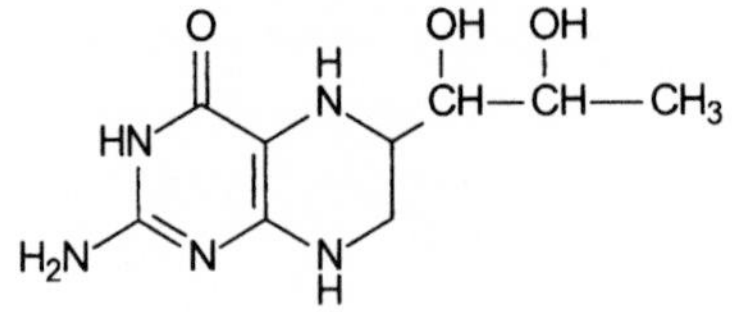

tetrahydrobiopterin

Figure 10.1. Folate vitamers, the folate antagonist methotrexate, molybdopterin, and tetrahydrobiopterin. Relative molecular masses (M_r): tetrahydrofolic acid, 445.4; methotrexate, 454.5; and tetrahydrobiopterin, 290.3.

has been used in depletion/repletion studies to determine folate requirements, and is the form used in food fortification. In order to permit calculation of folate intakes in terms of both naturally occurring mixed food folates and added folic acid, 1 μg of dietary folate equivalent has been defined as the sum of μg of food folate + 1.7 $\times$ μg of folic acid (Institute of Medicine, 1998).

10.2 METABOLISM OF FOLATES

Most of the dietary folate consists of polyglutamates; a variable amount may be substituted with various one-carbon fragments or be present as dihydrofolate derivatives. There is little information on either the distribution of folate vitamers in foods or their relative biological activities (Gregory, 2001). Unsubstituted reduced folates in foods are chemically unstable and readily undergo cleavage to p-aminobenzoic acid and the pteridine; between 50% to 75% of the folate in food may be lost during processing and storage (Scott, 1999).

The growth responses of the microorganisms used for bioassay are different for the different vitamers, and polyglutamates are used by bacteria only after enzymic hydrolysis with conjugase. Endogenous conjugase in foods may cause breakdown of polyglutamates during extraction and sample preparation, whereas autoclaving and the addition of preservatives will also hydrolyze some of the vitamers and may oxidize tetrahydrofolates, so that chromatographic analysis also may not reflect the true distribution of vitamers in foods.

10.2.1 Digestion and Absorption of Folates

Within the intestinal lumen, folate conjugates are hydrolyzed by glutamate carboxypeptidase (pteroylpolyglutamate hydrolase, also known as conjugase), a zinc-dependent enzyme of the pancreatic juice, bile, mucosal brush border, and lysosomes of enterocytes and other cells. In the rat, conjugase is mainly a pancreatic enzyme, acting in the intestinal lumen, whereas in human beings the conjugase of the mucosal brush border and enterocytes is more important.

Conjugase is a general poly-γ-glutamyl hydrolase, with a broad specificity for the pterin moiety. It acts randomly as both an exopeptidase removing γ-glutamyl groups sequentially and as an endopeptidase removing oligo-γ-glutamyl peptides, suggesting that there may be more than one enzyme. Similar endo- and exopeptidase conjugases are found in all tissues; the extent to which the conjugases in foods may contribute to the hydrolysis of folate polyglutamates is not known. Some foods contain conjugase inhibitors that will reduce the availability of conjugated folates.

Because conjugase is a zinc metallo-enzyme, zinc deficiency can impair the absorption of conjugated food folates, but not folate monoglutamate.

Conjugase responds rapidly to zinc depletion and repletion, and it has been suggested that the absorption of a test dose of folate polyglutamates may provide a sensitive index of zinc nutritional status (Canton and Cremin, 1990). The absorption of folate monoglutamates (from pharmaceutical preparations or foods) is not affected.

Free folate, released by conjugase action, is absorbed by a carrier-mediated mechanism in the jejunum. However, the folate in milk is mainly bound to a specific binding protein (which has been used in radioligand binding assays for folate); the protein–folate complex is absorbed intact, mainly in the ileum, by a mechanism that is distinct from the jejunal transport system for free folate. The biological availability of folate from milk, or of folate from diets to which milk has been added, is considerably greater than that of unbound folate, whereas that of folate from cereal foods, or of free folic acid taken with cereal foods, is lower.

Most of the dietary folate undergoes reduction and methylation within the intestinal mucosa and what enters the portal bloodstream is a largely 5-methyl-tetrahydrofolate. Single doses of more than about 200 μg of folic acid saturate the intestinal dihydrofolate reductase, so that free folic acid is absorbed and circulates in the bloodstream. It can be taken up by tissues, reduced to tetrahydrofolate, and utilized.

There is considerable enterohepatic circulation of folate, equivalent to about one-third of the dietary intake. Methyl-tetrahydrofolate is secreted in the bile, then reabsorbed in the jejunum together with food folates. In experimental animals, bile drainage for 6 hours results in a reduction of serum folate to 30% to 40% of normal (Steinberg et al., 1979). There is very little loss of folate; jejunal absorption is very efficient, and the fecal excretion of 450 nmol (200 μg) of folates per day largely represents synthesis by intestinal flora and does not reflect intake to any significant extent.

10.2.2 Tissue Uptake and Metabolism of Folate

Methyl-tetrahydrofolate from the intestinal mucosa circulates bound to albumin and is the main vitamer for uptake by extrahepatic tissues. Small amounts of other one-carbon substituted folates also circulate (about 10% to 15% of plasma folate is 10-formyl-tetrahydrofolate) and are also available for tissue uptake. There are two mechanisms for tissue uptake of folate:

1. The reduced folate transporter is a transmembrane protein with a high affinity for methyl-tetrahydrofolate and a low affinity for other vitamers. It is especially active in enterocytes and renal tubule epithelium, but is also found in other cells (Sirotnak and Tolner, 1999).

2. The folate receptor is a glycophosphatidyl inositol anchored cell surface protein with a broader specificity that permits uptake of folate by receptor-mediated endocytosis. At times of low folate requirement, the receptor is in intracellular vesicles, which migrate to the cell surface when the requirement for folate increases (Doucette and Stevens, 2001).

Demethylated tetrahydrofolate monoglutamate is released by extrahepatic tissues and is transported bound to a plasma folate binding protein similar to that in milk. It has a very low affinity for methyl-tetrahydrofolate and other one-carbon substituted derivatives. It functions mainly to return folate to the liver, where it is either conjugated for storage or methylated to 5-methyl-tetrahydrofolate that is secreted in the bile.

Red blood cells contain a several-hundred-fold higher concentration of folate than does plasma, incorporated during erythropoeisis rather than taken up from the circulation, as polyglutamates bound to hemoglobin. Folate binds to deoxyhemoglobin in competition with 2,3-bisphophoglycerate, but does not bind significantly to oxyhemoglobin. The binding affinity is low, but because of the high concentration of hemoglobin in erythrocytes, essentially all of the folate in cells from venous blood will be bound. The function of hemoglobin binding is not known, and it may not serve any physiological purpose, although, as discussed in Section 10.10.1, it may result in falsely low values for erythrocyte folate as an index of folate status.

10.2.2.1 Poly-γ-glutamylation of Folate Folate monoglutamates cross cell membranes readily, whereas polyglutamates do not; therefore, formation of conjugates permits intracellular accumulation of folate. Rapid formation of at least a diglutamate is essential for tissue retention of folate. Further elongation of the polyglutamate chain to form the metabolically active coenzymes can proceed in a more leisurely fashion.

A single enzyme, folate polyglutamate synthetase, catalyzes the formation of all the polyglutamates. At high concentrations of substrate, the diglutamate is the main or sole product; increasing amounts of tri-, tetra-, penta-, and hexaglutamates are formed as the concentration of tetrahydrofolate monoglutamate decreases. This is because, unlike other folate utilizing enzymes, the K_m of folate polyglutamate synthetase increases as the length of the polyglutamate chain increases, and short-chain polyglutamates are the preferred substrates. The rate of reaction with long-chain polyglutamates is considerably lower than with short-chain substrates.

Folate polyglutamate synthetase binds ATP, tetrahydrofolate-(oligo)-glutamate, then glutamate sequentially forming an intermediate folate polyglutamate phosphate. The immediate product is released to compete for enzyme binding with other intracellular folate conjugates, rather than undergoing sequential glutamyl transfer while remaining enzyme bound (Cichowicz and Shane, 1987).

The principal substrate for glutamylation is free tetrahydrofolate; one-carbon substituted folates are poor substrates. Because the main circulating folate, and the main form that is taken up into tissues, is methyl-tetrahydrofolate, demethylation by the action of methionine synthetase (Section 10.3.3) is essential for effective metabolic trapping of folate. In vitamin B_{12} deficiency, when methionine synthetase activity is impaired, there will be impairment of the retention of folate in tissues.

Under normal conditions, the predominant folates in liver are pentaglutamates, with small amounts of tetra- and hexaglutamates. The extent of polyglutamylation is controlled to a great extent by the availability of folate; in deficient animals, hexa- to octaglutamates predominate, whereas in supplemented animals, liver folate is mainly as the tri- to pentaglutamates (Cassady et al., 1980).

10.2.3 Catabolism and Excretion of Folate

There is very little urinary loss of folate, some 5 to 10 nmol of microbiologically active material per day. Not only is most folate in plasma bound to proteins (either folate binding protein for unsubstituted folate or albumin for methyltetrahydrofolate), and thus protected from glomerular filtration, but also the renal brush border has a high concentration of folate binding protein that acts to reabsorb any that is filtered.

Folate polyglutamate in cells that is not enzyme bound undergoes hydrolysis of the γ-glutamyl side chain, catalyzed by lysosomal conjugase, yielding folate monoglutamate, which then leaves the cell freely. Both carboxypeptidase G and ferritin catalyze hydrolysis of the C-9 to N-10 bond of folate monoglutamate to yield p-aminobenzoylglutamate (much of which is acetylated before excretion) and pterin, which is excreted either unchanged or as isoxanthopterin and other biologically inactive compounds. As a result of increased synthesis of ferritin in pregnancy, the catabolism of folate and excretion of p-aminobenzoylglutamate increase significantly, suggesting that the folate requirement in pregnancy may be considerably higher that would be expected on the basis of fetal requirements (Suh et al., 2001).

10.2.4 Biosynthesis of Pterins

As shown in Figure 10.2, the pteridine nucleus is synthesized from GTP, in a sequence of reactions catalyzed by a different isoenzyme of GTP cyclohydrolase

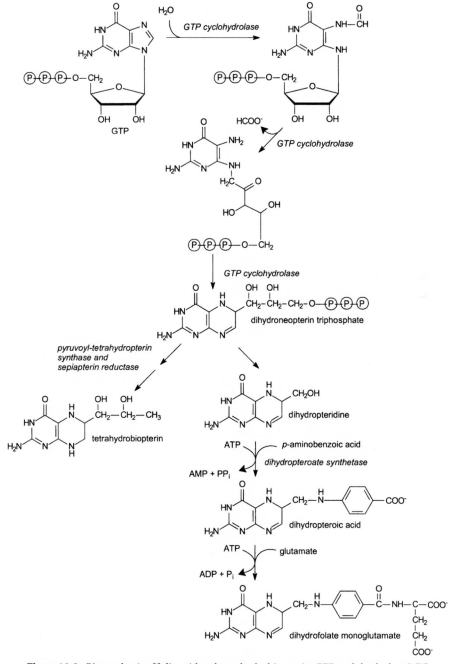

Figure 10.2. Biosynthesis of folic acid and tetrahydrobiopterin. GTP cyclohydrolase I, EC 3.5.4.16; dihydropteroate synthase, EC 2.5.1.15; pyruvoyl-tetrahydrobiopterin synthase, EC 4.6.1.10; and sepiapterin reductase, EC 1.1.1.153.

from that involved in riboflavin synthesis (Section 7.2.6). The reaction sequence involves loss of C-8 of guanine as formate, followed by rearrangement of the ribose moiety, condensation, and ring closure to yield dihydroneopterin triphosphate. Dephosphorylation and loss of the side chain leads to the formation of dihydropteridine, the immediate precursor of folate. In mammals, GTP cyclohydrolase is inhibited by unconjugated reduced pterins; folate, which is not an end-product of the mammalian enzyme, is not a significant inhibitor (Thony et al., 2000).

Mammals lack dihydropteroate synthetase, which catalyzes the condensation of dihydropteridine with p-aminobenzoic acid. The bacterial enzyme can utilize either p-aminobenzoate (yielding dihydropteroic acid) or p-aminobenzoyl-glutamate (yielding dihydrofolate directly). p-Aminobenzoyl-glutamate is not formed under normal conditions, although it is a product of mammalian catabolism of folate. The usual product of the reaction is dihydropteroic acid, followed by conjugation with glutamate. Dihydropteroate synthetase is inhibited by sulfonamides, which compete with p-aminobenzoate as substrate; this is the basis of their action as bacteriostatic agents. They deplete the organisms of pteridines by forming metabolically inactive sulfonamide analogs of dihydropteroate.

In both mammals and microorganisms, dihydrofolate is reduced to tetrahydrofolate by dihydrofolate reductase, which will act on free folate or various polyglutamate conjugates, although the affinity of the enzyme for its substrate falls as the length of the polyglutamate chain increases. In microorganisms, this enzyme is important for the de novo synthesis of tetrahydrofolate, whereas in mammals it is mainly required to reduce the dihydrofolate formed in the reaction of thymidylate synthetase (Section 10.3.3). As discussed in Section 10.3.3.1, dihydrofolate reductase is an important target for chemotherapy of cancer, bacterial infections, and malaria.

The formation of biopterin involves dephosphorylation and reduction of the side chain of dihydroneopterin triphosphate, followed by inversion of the conformation of the two hydroxyl groups, by way of intermediate oxidation to (symmetrical) oxo-groups, catalyzed by sepiapterin reductase.

Patients with a variety of cancers and some viral diseases excrete relatively large amounts of neopterin, formed by dephosphorylation and oxidation of dihydroneopterin triphosphate, an intermediate in biopterin synthesis. This reflects the induction of GTP cyclohydrolase by interferon-γ and tumor necrosis factor-α in response to the increased requirement for tetrahydrobiopterin for nitric oxide synthesis (Section 10.4.2). It is thus a marker of cell-mediated immune reactions and permits monitoring of disease progression (Werner et al., 1993, 1998; Berdowska and Zwirska-Korczala, 2001).

10.3 METABOLIC FUNCTIONS OF FOLATE

The metabolic role of folate is as a carrier of one-carbon fragments, both in catabolism and biosynthetic reactions. As shown in Figure 10.3, tetrahydrofolate can carry one-carbon fragments attached to N-5 (formyl, formimino, or methyl groups), N-10 (formyl), or bridging N-5 to N-10 (methylene or methenyl groups). The major sources of these one-carbon fragments, their major uses, and the interconversions of the substituted folates are shown in Figure 10.4.

The metabolically active forms of folate are all substituted tetrahydropteroyl polyglutamates. Whereas some folate-dependent enzymes will use the monoglutamate in vitro, most have a considerably lower K_m for polyglutamates, and some have higher V_{max}. In multienzyme complexes, in which the folate cofactor serves to transport one-carbon fragments from one catalytic site to another, the length of the polyglutamate tail may be especially important in anchoring the cofactor, but permitting considerable movement of the pteroyl moiety and in channeling intermediates between catalytic sites.

10.3.1 Sources of Substituted Folates

The major point of entry for one-carbon fragments into substituted folates is methylene-tetrahydrofolate, which is formed by the catabolism of glycine, serine, and choline.

10.3.1.1 Serine Hydroxymethyltransferase

Serine hydroxymethyltransferase is a pyridoxal phosphate-dependent aldolase that catalyzes the cleavage of serine to glycine and methylene-tetrahydrofolate (as shown in Figure 10.5). Serine is the major source of one-carbon substituted folates for biosynthetic reactions. At times of increased cell proliferation, the activities of serine hydroxymethyltransferase and the enzymes of the serine biosynthetic pathway are increased. The other product of the reaction, glycine, is also required in increased amounts under these conditions (for de novo synthesis of purines).

There are two isoenzymes of serine hydroxymethyltransferase; the cytosolic enzyme is involved in the provision of one-carbon fragments for biosynthetic reactions, whereas the mitochondrial enzyme is important in the fasting state as a source of serine for gluconeogenesis. The activity of the cytosolic enzyme is regulated by the state of folate substitution and the availability of folate, rather than by the state of serine metabolism. Methyl-tetrahydrofolate is a potent inhibitor, so when there is an adequate concentration of substituted folates from other sources, serine can be spared for energy-yielding metabolism or gluconeogenesis (Snell et al., 2000). The other reactions leading to the formation of one-carbon substituted tetrahydrofolate shown in Figure 10.4 are

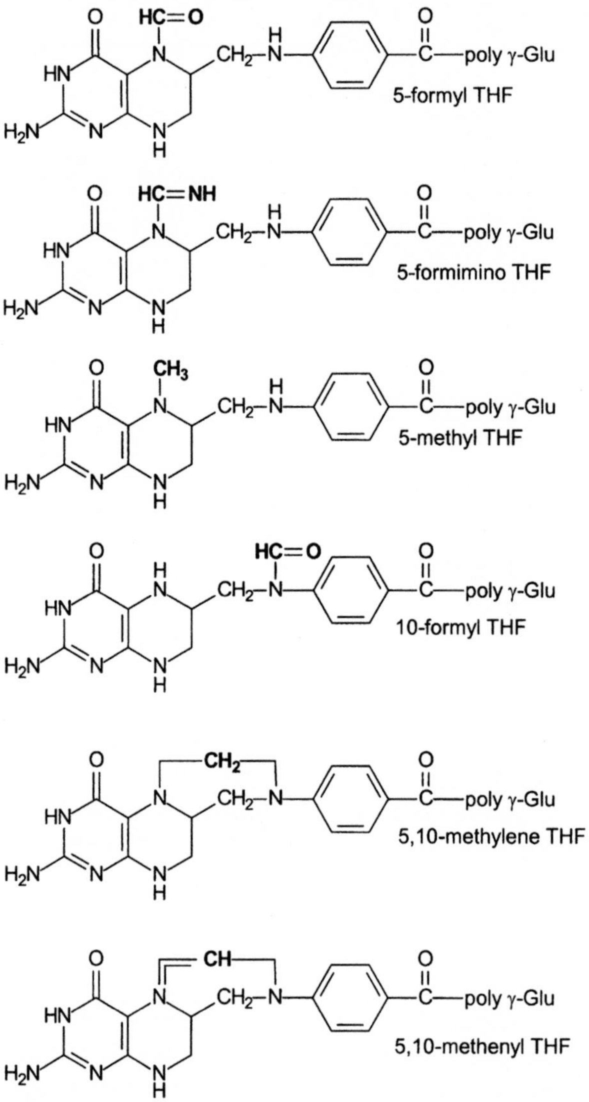

Figure 10.3. One-carbon substituted tetrahydrofolic acid derivatives. THF, tetrahydrofolate.

primarily catabolic reactions and are not subject to feedback inhibition by methyl-tetrahydrofolate.

When the glycine formed by serine hydroxymethyltransferase is not required for purine synthesis, it undergoes cleavage to carbon dioxide and ammonium, catalyzed by the glycine cleavage system. This is a multienzyme

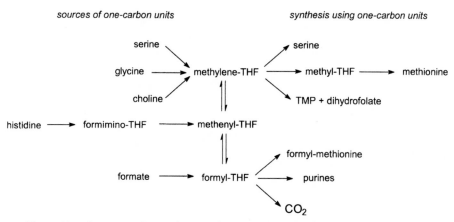

Figure 10.4. Sources and uses of one-carbon units bound to folate. THF, tetrahydrofolate.

complex with a number of similarities to the thiamin-dependent 2-oxo-acid dehydrogenases (Section 6.3.1), although it does not contain thiamin. It consists of the following:

1. a pyridoxal phosphate-dependent glycine decarboxylase;
2. a lipoamide-containing aminomethyltransferase, which acts to oxidize the one-carbon fragment to a methylene residue at the expense of reducing lipoamide to the disulfhydryl form;
3. a methylene-tetrahydrofolate synthesizing enzyme that transfers the one-carbon fragment onto tetrahydrofolate, releasing the nitrogen as ammonium; and
4. an NAD-dependent flavoprotein, dihydrolipoyl dehydrogenase, that oxidizes disulfhydryl lipoamide back to the disulfide.

10.3.1.2 **Histidine Catabolism** As shown in Figure 10.6, the catabolism of histidine leads to the formation of formiminoglutamate (FIGLU). The

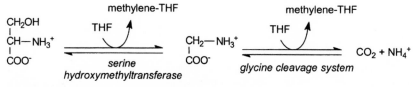

Figure 10.5. Reactions of serine hydroxymethyltransferase (EC 2.1.2.1) and the glycine cleavage system (EC 2.1.2.10). THF, tetrahydrofolate.

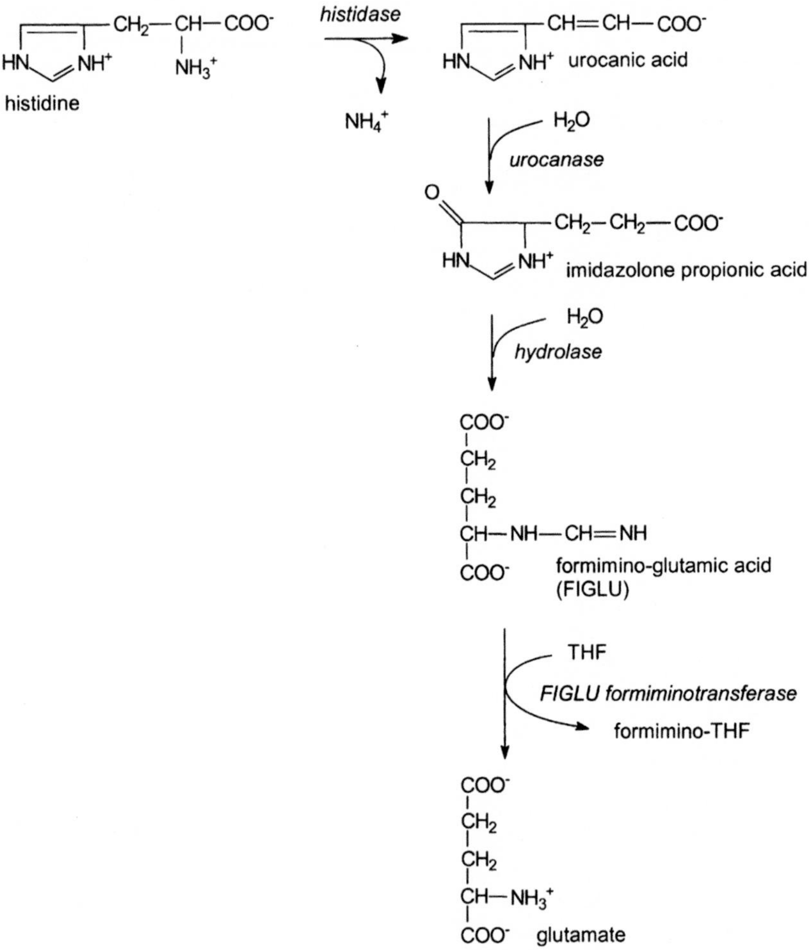

Figure 10.6. Catabolism of histidine – basis of the FIGLU test for folate status. Histidase, EC 4.3.1.3; urocanase, EC 4.2.1.49; FIGLU formiminotransferase, EC 2.1.2.5. THF, tetrahydrofolate.

formimino group is transferred onto tetrahydrofolate to form formimino-tetrahydrofolate, which is subsequently deaminated to form methenyl-tetrahydrofolate.

A single bifunctional enzyme catalyzes the FIGLU formiminotransferase and formiminofolate cyclodeaminase reactions, so there is little or no free formimino-tetrahydrofolate in tissues under normal conditions. The two catalytic sites are separate, and with tetrahydrofolate monoglutamate, there is release of the formimino derivative. However, when polyglutamates are used, there is channeling of the intermediate between the two sites, and no release of the formimino derivative (Mackenzie and Baugh, 1980; Paquin et al., 1985).

Although catabolism of histidine is not a major source of substituted folate, the reaction is of interest because it has been exploited as a means of assessing folate nutritional status. In folate deficiency, the activity of the formiminotransferase is impaired by lack of cofactor. After a loading dose of histidine, there is impaired oxidative metabolism of histidine and accumulation of FIGLU, which is excreted in the urine (Section 10.10.4).

10.3.1.3 **Other Sources of One-Carbon Substituted Folates** As shown in Figure 14.4, choline is oxidized to betaine (trimethylglycine), then the first methyl group is transferred directly to homocysteine, forming methionine. The resultant dimethylglycine is demethylated to methylglycine (sarcosine) by an iron-flavoprotein, dimethylglycine dehydrogenase, which oxidizes the methyl group to formaldehyde before transferring it to tetrahydrofolate to form methylene-tetrahydrofolate. The demethylation of sarcosine to glycine yields methylene-tetrahydrofolate in the same way.

Formylglutamate can transfer its formyl group directly onto tetrahydrofolate to yield 5-formyl-tetrahydrofolate. Formyl-glutamate is not a normal physiological intermediate, and the formation of 5-formyl-tetrahydrofolate is probably a side reaction of FIGLU formiminotransferase.

Free formate can react with tetrahydrofolate to form 10-formyl-tetrahydrofolate; the plasma concentration of formate rises in folate deficiency, and the ability to metabolize [14C]formate has been used as an index of folate depletion in experimental animals.

10.3.2 **Interconversion of Substituted Folates**
Methylene-, methenyl-, and 10-formyl-tetrahydrofolates are freely interconvertible. The two activities involved – methylene-tetrahydrofolate dehydrogenase and methenyl-tetrahydrofolate cyclohydrolase – form a trifunctional enzyme with 10-formyl-tetrahydrofolate synthetase (Paukert et al., 1976). This means that single-carbon fragments entering the folate pool in any form other than as methyl-tetrahydrofolate can be readily available for any of the biosynthetic reactions shown in Figure 10.4.

The conversion of 5-formyl-tetrahydrofolate to methenyl-tetrahydrofolate, catalyzed by 5-formyl-tetrahydrofolate cyclohydrolase, is important. Although 5-formyl-tetrahydrofolate is the most commonly used pharmaceutical preparation of the vitamin, a relatively large proportion of orally administered

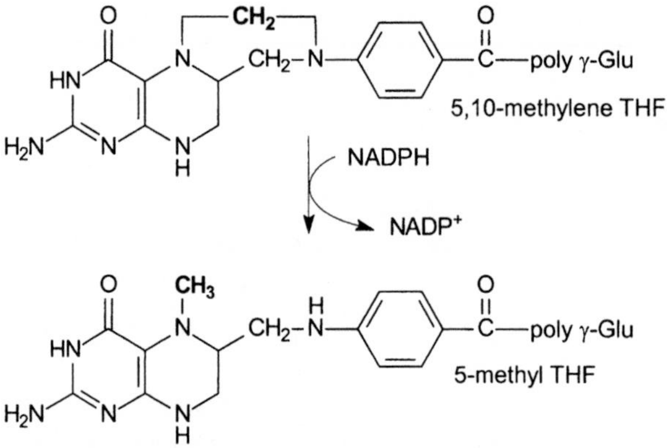

Figure 10.7. Reaction of methylene-tetrahydrofolate reductase (EC 1.7.99.5). THF, tetrahydrofolate.

5-formyl-tetrahydrofolate undergoes nonenzymic cyclization to methenyl-tetrahydrofolate in the acid conditions of the stomach.

10.3.2.1 Methylene-Tetrahydrofolate Reductase   The reduction of methylene-tetrahydrofolate to methyl-tetrahydrofolate, shown in Figure 10.7, is catalyzed by methylene-tetrahydrofolate reductase, a flavin adenine dinucleotide-dependent enzyme; during the reaction, the pteridine ring of the substrate is oxidized to dihydrofolate, then reduced to tetrahydrofolate by the flavin, which is reduced by nicotinamide adenine dinucleotide phosphate (NADPH; Matthews and Daubner, 1982). The reaction is irreversible under physiological conditions, and methyl-tetrahydrofolate – which is the main form of folate taken up into tissues (Section 10.2.2) – can only be utilized after demethylation catalyzed by methionine synthetase (Section 10.3.4).

Methylene-tetrahydrofolate reductase is inhibited by S-adenosylmethionine, which inhibits reduction of the flavin prosthetic group by NADPH. S-Adenosylhomocysteine overcomes this inhibition to some extent, as might be expected for an enzyme that is indirectly involved in the regulation of methionine and S-adenosylmethionine concentrations in the cell.

Kang and coworkers (1991) reported a variant of methylene-tetrahydrofolate reductase, in which cytosine[677] is replaced by thymidine, resulting in a change of alanine[226] to valine, in people who were hyperhomocysteinemic (Section 10.3.4.2). The variant enzyme is thermolabile (i.e., it is unstable to heating to about 40° to 45°C), and subjects who are homozygous for the thermolabile enzyme have about 50% of normal enzyme activity in tissues. Not only is the enzyme labile on moderate heating in vitro, but it is also unstable in vivo. There are considerable differences in the frequency of the variant gene in different population groups, ranging from 1.2% of Brazilian Amerindians, 3.1% of British South Asians, and 10% of British and Australian white people, with up to 30% of Italians and 35% of Japanese being homozygous for the thermolabile variant. Some studies have shown that the thermolabile variant was two- to three-fold more common among people with atherosclerosis and coronary heart disease than among disease-free people of the same ethnic origin.

Being homozygous for the thermolabile variant of methylene-tetrahydro-folate reductase is a necessary, but not sufficient, condition for the development of hyperhomocysteinemia. Homozygotes with a high folate intake have plasma concentrations of homocysteine as low as heterozygotes or people who are homozygous for the normal (stable) form of the enzyme (Jacques et al., 1996). Two possible mechanisms have been proposed to explain how a relatively high intake of folate can mask the effect of being homozygous for the thermolabile variant of methylene-tetrahydrofolate reductase:

1. Most dietary folate is reduced and methylated to methyl-tetrahydro-folate in the intestinal mucosa (Section 10.2.1). Intestinal mucosal cells have a rapid turnover, typically 48 hours from proliferation in the crypt to shedding at the tip of the villus. This means that an unstable variant of the enzyme, which loses activity over a shorter time than the normal enzyme, is probably irrelevant in cells that have such a rapid turnover. A high intake of folate would therefore result in a relatively high rate of supply of methyl-tetrahydrofolate to cells, arising from newly absorbed folate, so that impaired turnover of folate within cells would be less important.

2. In common with a number of enzymes, methylene-tetrahydrofolate reductase may be more resistant to thermal denaturation (and hence possibly more stable) in the presence of its substrate. Hence it is possible that high tissue levels of methylene-tetrahydrofolate (resulting from a high folate status) may protect the enzyme and enhance its stability (Guenther et al., 1999; Yamada et al., 2001a). However, it is unlikely that a high intake of folate would lead to a sufficient accumulation of methylene-tetrahydrofolate to stabilize the enzyme in this way because, as discussed in Section 10.3.2.2, there is rapid interconversion between the various one-carbon substituted folates. Any excess is converted to formyl-tetrahydrofolate, then the formyl group is oxidized to carbon dioxide to maintain a pool of free folate for the collection of one-carbon fragments in catabolic reactions.

Methylene-tetrahydrofolate reductase is a flavoprotein. There is some evidence that riboflavin also stabilizes the thermolabile variant and that riboflavin supplements may lower plasma homocysteine in people who are homozygous for the variant enzyme (McNulty et al., 2002).

There is a second polymorphism of methylene-tetrahydrofolate reductase in about 10% of the population, in which adenosine [1298] is replaced by cytosine. Like the thermolabile variant, this results in about 50% of normal activity of the enzyme in lymphocytes from homozygotes and the development of hyperhomocysteinemia which, in this case, does not seem to be responsive to high intakes of folate (Chango et al., 2000a, 2000b).

10.3.2.2 Disposal of Surplus One-Carbon Fragments   With the exception of serine hydroxymethyltransferase (Section 10.3.1.1), all of the reactions shown in Figure 10.4 as sources of one-carbon substituted folates are essentially catabolic reactions. When there is a greater entry of single carbon units into the folate pool than is required for biosynthetic reactions, the surplus can be oxidized to carbon dioxide by way of 10-formyl-tetrahydrofolate, thus ensuring the availability of tetrahydrofolate for catabolic reactions.

10-Formyl-tetrahydrofolate dehydrogenase has a high K_m relative to the normal intracellular concentration of its substrate and only has significant activity when there is a relative excess of one-carbon substituted tetrahydrofolate. The product, free tetrahydrofolate, is a poor leaving group, and much remains bound to the enzyme, resulting in significant inhibition. The activity of the dehydrogenase is thus strictly regulated by the ratio of formyl-tetrahydrofolate:free tetrahydrofolate in the tissue (Min et al., 1988).

There is an alternative pathway disposal of surplus one-carbon fragments, in which glycine N-methyltransferase catalyzes the methylation of glycine to sarcosine, using S-adenosylmethionine (Section 10.3.4) as the methyl donor. The resultant S-adenosylhomocysteine is then remethylated at the expense of methyl-tetrahydrofolate, thus regenerating free tetrahydrofolate. Factors such as vitamin A – which induce the synthesis of glycine N-methyltransferase – may act to reduce the tissue pool of one-carbon substituted folates for methylation reactions, leading to undermethylation of DNA and possible cancer development (Rowling et al., 2002).

10.3.3 Utilization of One-Carbon Substituted Folates
As shown in Figure 10.4, 10-formyl-tetrahydrofolate and methylene-tetrahydrofolate are donors of one-carbon fragments in a number of biosynthetic reactions, including the synthesis of purines, pyrimidines, porphyrins, and

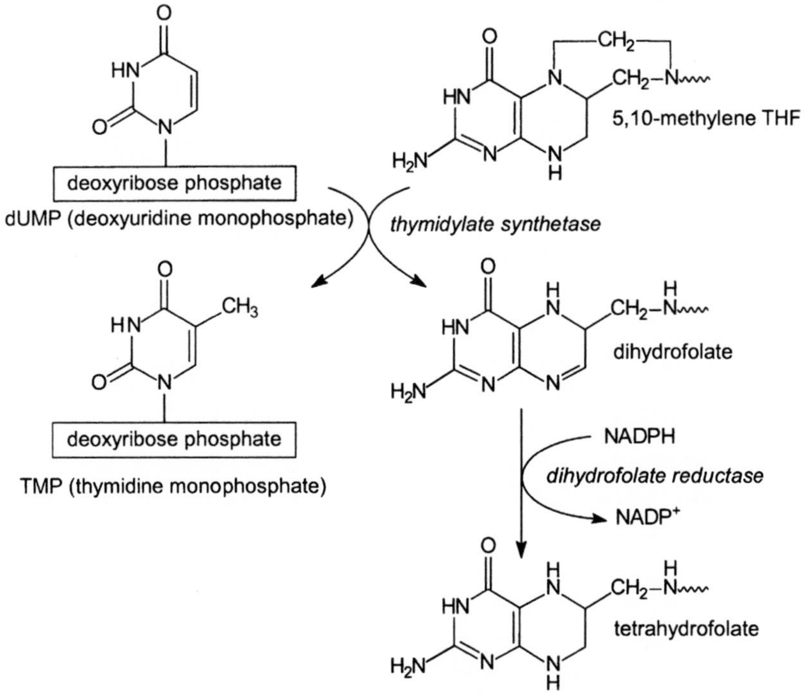

Figure 10.8. Synthesis of thymidine monophosphate. Thymidylate synthetase, EC 2.1.1.45; and dihydrofolate reductase, EC 1.5.1.3. THF, tetrahydrofolate.

the methylation of homocysteine to methionine. In most cases, the reaction is a simple transfer of the one-carbon group from substituted tetrahydrofolate onto the acceptor substrate.

Two reactions are of special interest: thymidylate synthetase (Section 10.3.3) and remethylation of homocysteine to methionine (Section 10.3.4). This latter reaction is central to the control of the metabolism of one-carbon compounds and folate.

10.3.3.1 **Thymidylate Synthetase and Dihydrofolate Reductase** Methylation of deoxyuridine monophosphate (dUMP) to thymidine monophosphate (TMP; see Figure 10.8) is essential for the synthesis of DNA, although preformed TMP can be reutilized by salvage from the catabolism of DNA.

The methyl donor is methylene-tetrahydrofolate. The reaction involves formation of a methylene bridge between N-5 of the coenzyme and C-5 of dUMP, followed by transfer of hydrogen from the pyrazine ring of tetrahydrofolate, thus forming dihydrofolate that is released from the enzyme. Dihydrofolate is a product of this reaction, whereas in methylene-tetrahydrofolate reductase (Section 10.3.2.1), the dihydrofolate is a transient enzyme-bound intermediate and is reduced back to tetrahydrofolate during the reaction.

5-Fluorouracil is widely used in cancer chemotherapy. It is a precursor of 5-fluoro-dUMP, which is a mechanism-dependent inhibitor of thymidylate synthetase. It forms a stable methylene-bridged complex with methylene-tetrahydrofolate on the enzyme catalytic site that cannot undergo reductive cleavage.

10.3.3.2 Dihydrofolate Reductase Inhibitors Under normal conditions, the dihydrofolate formed by thymidylate synthetase is rapidly reduced to tetrahydrofolate by dihydrofolate reductase. Thymidylate synthetase and dihydrofolate reductase are especially active in tissues with a high rate of cell division, and thus a high rate of DNA replication and a high requirement for thymidylate. As cells enter the S-phase of the cell cycle, there is a 7-fold increase in the rate of transcription of the dihydrofolate reductase gene (Farnham and Schimke, 1985). Because of this role in actively dividing tissue, inhibitors of dihydrofolate reductase have been exploited as anticancer drugs. The most successful of these is methotrexate, the 4-amino analog of 10-methyl-tetrahydrofolate (see Figure 10.1).

Methotrexate is a potent inhibitor of dihydrofolate reductase, with an affinity 1,000-fold greater than that of dihydrofolate. Chemotherapy consists of alternating periods of administration of methotrexate and folate (normally as 5-formyl-tetrahydrofolate, leucovorin) to replete the normal tissues and avoid induction of folate deficiency – so-called leucovorin rescue. As well as depleting tissue pools of tetrahydrofolate, methotrexate leads to the accumulation of relatively large amounts of 10-formyl-dihydrofolate, which is a potent inhibitor of both thymidylate synthetase and glycinamide ribotide transformylase, an intermediate step in purine nucleotide synthesis. It is likely that this, rather than simple depletion of tetrahydrofolate, is the basis of the cytotoxic action of methotrexate (Baram et al., 1988).

Methotrexate is also a substrate for conjugation with glutamate (Section 10.2.2.1) and a variety of methotrexate polyglutamates, which are potent inhibitors of dihydrofolate reductase are formed and retained in the cells. Susceptible tumor cells show greater conjugation, and greater accumulation, of methotrexate than bone marrow cells or the gastrointestinal mucosa, thus providing some degree of tumor specificity for the drug. Methotrexate polyglutamates inhibit the reactivation of dihydrofolate reductase by 5-formyl-tetrahydrofolate so that leucovorin rescue has less effect on (tumor) cells that accumulate and conjugate methotrexate (Goldman and Matherly, 1987).

The antibacterial agent trimethoprim also acts as an inhibitor of dihydrofolate reductase. It binds to the bacterial enzyme with much higher affinity than the mammalian enzyme. It provides a considerable degree of selectivity, permitting use of doses low enough to have little effect on the host's folate metabolism. A number of trimethoprim-resistant strains of bacteria have been isolated; at least three different plasmid-associated dihydrofolate reductases have been identified, including the following:

1. low K_i for trimethoprim and thus insensitive to inhibition;
2. sensitive to trimethoprim inhibition, but with a lower K_m for dihydrofolate, which therefore competes with the drug more effectively;
3. sensitive to trimethoprim and with a high K_m for dihydrofolate, but induced by trimethoprim, thus increasing the amount of enzyme available.

10.3.3.3 The dUMP Suppression Test Rapidly dividing cells can either use preformed TMP or can synthesize it de novo from dUMP. Isolated bone marrow cells or stimulated lymphocytes incubated with [³H]TMP will incorporate label into DNA. In the presence of adequate amounts of methylene-tetrahydrofolate, the addition of dUMP as a substrate for thymidylate synthetase reduces the incorporation of [³H]TMP as a result of dilution of the pool of labeled material by newly synthesized TMP.

The extent to which dUMP suppresses the incorporation of [³H]TMP into DNA thus reflects folate nutritional status (Section 10.10.5).

10.3.4 The Role of Folate in Methionine Metabolism

In addition to its role in the synthesis of proteins and the polyamines spermidine and spermine, the main metabolic role of methionine is as a methyl donor in a wide variety of biosynthetic reactions.

As shown in Figure 10.9, the methyl donor is *S*-adenosyl methionine, which is demethylated to *S*-adenosyl homocysteine. After removal of the adenosyl group, homocysteine may undergo one of two metabolic fates: remethylation to methionine or condensation with serine to form cystathionine, followed by cleavage to yield cysteine – the transsulfuration pathway (Section 9.5.5). Cystathionine synthetase has a relatively low K_m compared with normal intracellular concentrations of homocysteine. It functions at a relatively constant rate, and under normal conditions, most homocysteine will be remethylated to methionine.

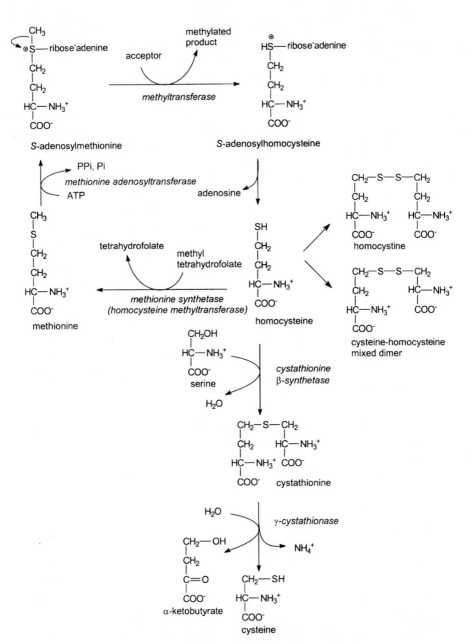

Figure 10.9. Metabolism of methionine. Methionine adenosyltransferase, EC 2.5.1.6; methionine synthetase, EC 2.1.1.13 (vitamin B$_{12}$-dependent) and EC 2.1.1.5 (betaine as a methyl donor); cystathionine β-synthetase, EC 4.2.1.22; and γ-cystathionase, EC 4.4.1.1.

There are two separate homocysteine methyltransferases in most tissues. One uses methyl-tetrahydrofolate as the methyl donor and has vitamin B_{12} (cobalamin; Section 10.8.1) as its prosthetic group. This enzyme is also known as methionine synthetase; it is the only homocysteine methyltransferase in the central nervous system. The other enzyme utilizes betaine (an intermediate in the catabolism of choline; Section 14.2.1) as the methyl donor and does not require vitamin B_{12}.

Unlike most enzymes utilizing or metabolizing tetrahydrofolate, methionine synthetase has equal activity toward methyl-tetrahydrofolate mono- and polyglutamates. As discussed in Section 10.2.2, demethylation of methyl-tetrahydrofolate is essential for the polyglutamylation and intracellular accumulation of folate.

10.3.4.1 **The Methyl Folate Trap Hypothesis**   The reduction of methylene-tetrahydrofolate to methyl-tetrahydrofolate is irreversible (Section 10.3.2.1), and the major source of folate for tissues is methyl-tetrahydrofolate. The only metabolic role of methyl-tetrahydrofolate is the methylation of homocysteine to methionine, and this is the only way in which methyl-tetrahydrofolate can be demethylated to yield free tetrahydrofolate in tissues. Methionine synthetase thus provides the link between the physiological functions of folate and vitamin B_{12}.

Impairment of methionine synthetase activity, for example, in vitamin B_{12} deficiency or after prolonged exposure to nitrous oxide (Section 10.9.7), will result in the accumulation of methyl-tetrahydrofolate. This can neither be utilized for any other one-carbon transfer reactions nor demethylated to provide free tetrahydrofolate.

Experimental animals that have been exposed to nitrous oxide to deplete vitamin B_{12} show an increase in the proportion of liver folate present as methyl-tetrahydrofolate (85% rather than the normal 45%), largely at the expense of unsubstituted tetrahydrofolate and increased urinary loss of methyl-tetrahydrofolate (Horne et al., 1989). Tissue retention of folate is impaired because methyl-tetrahydrofolate is a poor substrate for polyglutamyl-folate synthetase, compared with unsubstituted tetrahydrofolate (Section 10.2.2.1). As a result of this, vitamin B_{12} deficiency is frequently accompanied by biochemical evidence of functional folate deficiency, including impaired metabolism of histidine (excretion of formiminoglutamate; Section 10.3.1.2) and impaired thymidylate synthetase activity (as shown by abnormally low dUMP suppression; Section 10.3.3.3), although plasma concentrations of methyl-tetrahydrofolate are normal or elevated.

This functional deficiency of folate is exacerbated by the associated low concentrations of methionine and S-adenosyl methionine, although most tissues (apart from the central nervous system) also have betaine-homocysteine methyltransferase that may be adequate to maintain tissue pools of methionine. Under normal conditions S-adenosyl methionine inhibits methylene-tetrahydrofolate reductase and prevents the formation of further methyl-tetrahydrofolate. Relief of this inhibition results in increased reduction of one-carbon substituted tetrahydrofolates to methyl-tetrahydrofolate.

Both in vivo and with isolated tissue preparations from vitamin B₁₂-deficient animals, additional methionine can alleviate some of the effects on folate metabolism, elevating the liver concentration of folate and restoring a more normal proportion of unsubstituted tetrahydrofolate; increasing the metabolism of histidine, thus reducing the excretion of formiminoglutamate; reducing the excretion of formate and intermediates of purine synthesis; and restoring normal suppression of the incorporation of [³H]TMP into DNA by dUMP. The additional methionine increases the availability of S-adenosyl methionine, thus increasing reactivation of the residual methionine synthetase and permitting increased demethylation of methyl-tetrahydrofolate. The increased concentration of S-adenosyl methionine will also restore normal inhibition of methylene-tetrahydrofolate reductase.

The activity of 10-formyl-tetrahydrofolate dehydrogenase, which catalyzes the oxidation of 10-formyl tetrahydrofolate to CO_2 and tetrahydrofolate, is reduced at times of low methionine availability as a means of conserving valuable one-carbon fragments. Therefore, there is no sink for one-carbon substituted tetrahydrofolate, and increasing amounts of folate are trapped as methyl-tetrahydrofolate that cannot be used because of the lack of vitamin B₁₂ (Krebs et al., 1976).

This has been called the methyl folate trap and appears to explain many of the similarities between the symptoms and metabolic effects of folate and vitamin B₁₂ deficiency, although it does not provide a completely satisfactory explanation (Chanarin et al., 1985).

10.3.4.2 Hyperhomocysteinemia and Cardiovascular Disease

Children with homocystinuria caused by a genetic defect of cystathionine synthetase suffer multiple thromboses, and if untreated commonly die in their teens. A number of epidemiological studies during the 1990s identified moderate elevation of plasma homocysteine (significantly lower than seen in homocystinuric children) as an independent risk factor for cardiovascular disease (D'Angelo and Selhub, 1997; Selhub et al., 2000; Herrmann, 2001). As shown in Table 10.1, homocysteine has a variety of potential atherogenic, hypertensive,

Table 10.1 Adverse Effects of Hyperhomocysteinemia

Atherogenic
- Especially in the presence of transition metal ions, homocysteine can undergo redox cycling → O_2^-, thus a possibility of oxidative damage to LDL
- May react with cysteine–SH groups and modify apolipoproteins, thus impaired uptake by LDL receptors

Hypertensive

Reacts with $\cdot$NO → S-nitrosohomocysteine – this reacts with O_2 → homocysteine + $NO_3^- + O_2^-$
- Generation of O_2^- and loss of vasodilation action of $\cdot$NO

Procoagulant

Inhibits or downregulates anticoagulants
- Decreases prostacyclin synthesis
- Decreases activation of protein C
- Decreases thrombomodulin expression
- Suppresses the expression of heparan sulfate
- Decreases fibrinolysis

Activates procoagulants
- Increased Factor V activity
- Increased tissue clotting factor activity

Causes desquamation of vascular endothelium and impairs regeneration
- Endothelium has anticoagulant activity

Causes proliferation of vascular smooth muscle
- Vascular smooth muscle, when exposed, has potent procoagulant activity because of tissue clotting factor

Increases platelet coagulability and so activates platelet aggregation

Connective tissue abnormalities

Inhibition of lysyl oxidase → disturbance of collagen and elastin cross-linking (see Section 13.3.3)

In the presence of homocysteine, fibroblasts produce excessively sulfated proteoglycans
- The product is granular rather than fibrillar
- Excessively sulfated proteoglycans will attract and bind ϵ-amino groups of lysine in lipoproteins

LDL, low-density lipoprotein; $\cdot$NO, nitric oxide; SH, sulfhydryl.

and procoagulant actions (Stamler and Slivka, 1996; Welch and Loscalzo, 1998; Herrmann, 2001).

Deficiency of vitamins B_6, B_{12}, or folate are all associated with elevated plasma homocysteine, with vitamin B_6 deficiency as a result of impaired activity of cystathionine synthetase (Section 9.5.5) and folate and vitamin B_{12} as a result of impaired activity of methionine synthetase (Section 10.3.4). In subjects with apparently adequate intakes of vitamins B_6 and B_{12}, supplements of these two vitamins have little or no effect on fasting plasma homocysteine, although additional vitamin B_6 reduces the plasma concentration of homocysteine after a test dose of methionine. By contrast, supplements of folic acid do reduce plasma homocysteine in hyperhomocysteinemic subjects, despite apparently adequate folate status (Ubbink et al., 1994; Ubbink, 1997; Homocysteine Lowering Trialists' Collaboration, 1998; Refsum et al., 1998). This is presumably as a result of overcoming the effect of being homozygous for the thermolabile variant of methylene-tetrahydrofolate reductase (Section 10.3.2.1). As discussed in Section 10.12, this has led to mandatory enrichment of cereal products with folic acid in the United States and some other countries, although it remains to be demonstrated whether or not there is a causative relationship between hyperhomocysteinemia and cardiovascular disease, and therefore whether lowering homocysteine will have any beneficial effect. There is a stronger relationship between homocysteine and cardiovascular disease in cross-sectional and retrospective case-control studies than in prospective studies (Meleady and Graham, 1999).

Although, as shown in Table 10.1, there are plausible mechanisms to suggest that homocysteine is a causative factor in cardiovascular disease, it is possible that hyperhomocysteinemia is a result of the renal damage that is an early event in cardiovascular disease, thus a proxy marker of disease rather than a causative factor (Jacobsen, 1998; Langman and Cole, 1999; Kircher and Sinzinger, 2000). In chronic renal failure, hyperhomocysteinemia is associated with cardiovascular disease, probably because of both impaired excretion of homocysteine and impaired activity of betaine homocysteine methyltransferase; elevated plasma dimethylglycine predicts plasma homocysteine (McGregor et al., 2001).

10.4 TETRAHYDROBIOPTERIN

Tetrahydrobiopterin is not a vitamin, because it can be synthesized from GTP, as shown in Figure 10.2 (Thony et al., 2000). It is the coenzyme for mixed-function oxidases: phenylalanine, tyrosine, and tryptophan hydroxylases; alkyl glycerol monoxygenase, which catalyzes the cleavage of alkyl glycerol ethers; and nitric oxide synthase in the formation of nitric oxide. In addition to its coenzyme role, tetrahydrobiopterin has a direct effect on neurons, acting to stimulate dopamine release via a cAMP-dependent protein kinase and a calcium channel (Koshimura et al., 2000).

10.4.1 The Role of Tetrahydrobiopterin in Aromatic Amino Acid Hydroxylases

Most aromatic hydroxylases are either cytochrome- or flavin-dependent enzymes; the three enzymes that catalyze hydroxylation of the aromatic amino acids phenylalanine, tyrosine, and tryptophan are apparently unique in

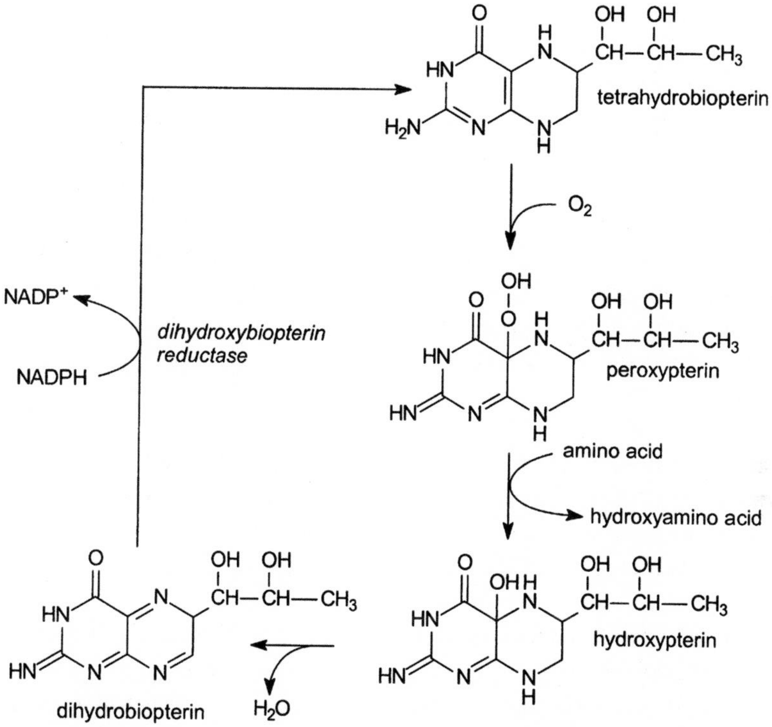

Figure 10.10. Role of tetrahydrobiopterin in aromatic amino acid hydroxylases. Phenylalanine hydroxylase, EC 1.14.16.1; tyrosine hydroxylase, EC 1.14.16.2; tryptophan hydroxylase, EC 1.14.16.4; and dihydrobiopterin reductase (dihydropteridine reductase), EC 1.6.99.7.

utilizing tetrahydrobiopterin, which acts as a reducing substrate rather than a coenzyme. They are monooxygenases; one atom of oxygen is incorporated into the substrate as a hydroxyl group, and the other is reduced to water.

As shown in Figure 10.10, tetrahydrobiopterin activates molecular oxygen by forming a peroxypterin that reacts with an iron atom in the active site, yielding Fe=O that reacts with the amino acid substrate, and hydroxypterin, which then undergoes dehydration to yield dihydrobiopterin. Dihydrobiopterin is reduced back to tetrahydrobiopterin by dihydrobiopterin reductase; dihydrofolate reductase (Section 10.3.3) does not have any significant activity toward dihydrobiopterin (Fitzpatrick, 1999).

The same pool of tetrahydrobiopterin and the same dihydrobiopterin reductase are involved in the central nervous system in the hydroxylation of all three aromatic amino acids. Classical phenylketonuria, which involves a defect of phenylalanine hydroxylase, responds well to dietary restriction of phenyl-alanine. However, so-called malignant or atypical phenylketonuria involves either a defect in dihydrobiopterin reductase, or a failure of biopterin bio-synthesis. In either case, there are disturbances of phenylalanine, tyrosine, and tryptophan metabolism, and deficits of catecholamines and 5-hydroxy-tryptamine, as well as in dopamine release and nitric oxide formation (Sec-tion 10.4.2) so that the neurological problem is more serious than in classical phenylketonuria. Dietary restriction of phenylalanine has little beneficial ef-fect, and there seems to be little that can be done for patients with dihydrofolate reductase deficiency. For patients whose problem is an inability to synthesize biopterin, the outlook is good because synthetic biopterin derivatives are well utilized.

10.4.2 The Role of Tetrahydrobiopterin in Nitric Oxide Synthase

Nitric oxide ($\cdot$NO) was discovered as the endothelium-derived relaxation fac-tor. It is produced in the central nervous system, lungs, macrophages, and vascular endothelial cells in response to a variety of stimuli. It initiates a cGMP signaling cascade.

In smooth muscle, nitric oxide activates a cytosolic guanylate cyclase by reacting with a heme-containing regulatory subunit and causing a conforma-tional change. This results in smooth muscle relaxation and hence vasodilata-tion as a result of reduced intracellular calcium in response to the increased concentration of cGMP. This role of nitric oxide in stimulating guanyl cyclase explains the clinically useful vasodilatory action of nitroglycerine and aryl ni-trites, which decompose in vivo to yield nitric oxide. Endothelium-derived nitric oxide also inhibits platelet aggregation, again acting by way of activation of guanylate cyclase.

In the central nervous system, nitric oxide formation from arginine is the im-mediate response of the glutamate-activated N-methyl-D-aspartate (NMDA) receptors. Again, it acts intercellularly, being released from neurons and acting on neighboring glial cells by stimulation of guanylate cyclase. These actions of nitric oxide are transient; activated macrophages continue to produce nitric oxide from arginine, in considerably higher concentrations, as a part of their cytotoxic action.

As shown in Figure 10.11, the reaction catalyzed by nitric oxide synthase is hydroxylation of arginine; N-hydroxy-arginine then decomposes to citrulline and nitric oxide. It is unclear whether the immediate product of the enzyme is nitric oxide itself or the nitroxyl anion NO$^-$; the addition of superoxide dis-mutase in vitro increases the formation of nitric oxide, but this could be the

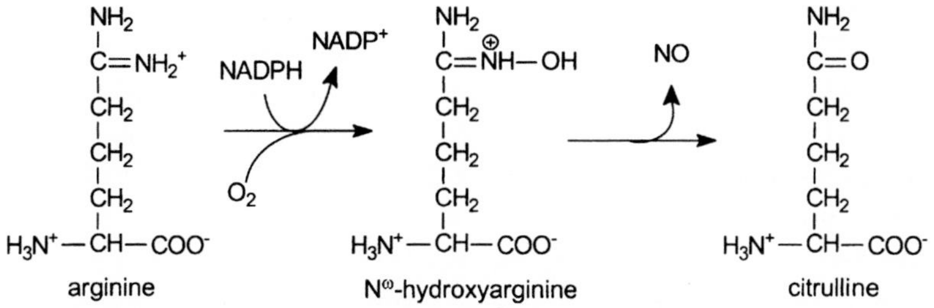

Figure 10.11. Reaction of nitric oxide synthase (EC 1.14.13.39).

result of either oxidation of nitroxyl to nitric oxide or the removal of superoxide, which reacts with nitric oxide to yield peroxynitrite (Alderton et al., 2001).

Although the enzyme requires tetrahydrobiopterin for activity, the role of pterin is not clear. Nitric oxide synthase is a heme-containing enzyme that reacts with a reduced flavin; by analogy with cytochrome P_{450}-dependent enzymes, it could catalyze the hydroxylation of arginine without requiring a pterin cofactor. There is no evidence of cycling between tetrahydrobiopterin and dihydrobiopterin, as in the case of the aromatic amino acid hydroxylases (Section 10.4.1), although there is evidence of formation of the trihydrobiopterin radical as an intermediate in the reduction of heme. In addition to a possible direct role in hydroxylation of arginine, tetrahydrobiopterin may also act in the following ways:

1. to inhibit the formation of superoxide and hydrogen peroxide;
2. to promote formation of, and stabilize, the active dimer (the monomer of the enzyme is inactive);
3. to be an allosteric modifier of either the arginine binding site or the environment of the reactive heme group; and
4. to protect the enzyme against autoinactivation.

10.5 MOLYBDOPTERIN

Three human redox enzymes, and a variety of bacterial enzymes, contain molybdenum chelated by two sulfur atoms in a modified pterin: molybdopterin (see Figure 10.1). In sulfite oxidase, the other two chelation sites of the molybdenum are occupied by oxygen; in xanthine oxidase/dehydrogenase (Section 7.3.7) and aldehyde oxidase, one site is occupied by oxygen and one by sulfur. In some bacterial enzymes, molybdopterin occurs as a guanine dinucleotide rather than free. In others, tungsten rather than molybdopterin is the chelated metal; there is no evidence that any mammalian enzymes contain tungsten.

Unusually for redox reactions involving metal ions, the molybdenum undergoes a two-electron reaction, cycling between Mo^{VI} and ^{IV}; however, the enzymes also have a reactive heme and/or an iron–sulfur cluster (which is a single-electron acceptor/donor). Thus, there must be intermediate formation of Mo^V (Kisker et al., 1997; Rajagopalan, 1997; Nishino and Okamoto, 2000).

A very small number of children have been reported who are unable to synthesize molybdopterin; they show severe neurological abnormalities shortly after birth and fail to survive more than a few days. As expected from the metabolic roles of molybdopterin, they have low blood concentrations of uric acid and sulfate, and abnormally high levels of xanthine and sulfite. The neurological damage is probably caused by sulfite, because similar abnormalities are seen in children with isolated sulfite oxidase deficiency (Reiss, 2000).

10.6 VITAMIN B_{12} VITAMERS AND NOMENCLATURE

The structure of vitamin B_{12} is shown in Figure 10.12. The corrin ring is a tetrapyrrole with fused A to D rings; the term corrinoid is used as a generic descriptor for cobalt-containing compounds of this general structure, which, depending on the substituents in the pyrrole rings, may or may not have vitamin activity. Cobalamins are corrinoids that have a dimethylbenzimidazole nucleotide attached to the D ring and chelating the central cobalt atom. The term vitamin B_{12} is used as a generic descriptor for the cobalamins – those corrinoids having the biological activity of the vitamin.

A number of noncobalamin corrinoids have activity in microbiological assays and thus appear to be vitamin B_{12}, although they have no vitamin activity and may have antimetabolite activity. The common method of measuring vitamin B_{12} and that required by law in some countries (including the United States) is a microbiological growth assay using *Lactobacillus leichmanii*, for which a number of noncobalamin corrinoids are growth factors. As a result, a number of foods that contain only noncobalamin corrinoids are nonetheless lawfully described as containing vitamin B_{12} (Herbert, 1988).

As shown in Figure 10.12, four of the six chelation sites of the cobalt atom of cobalamin are occupied by the nitrogens of the corrin ring and one by the nitrogen of the dimethylbenzimidazole side chain. The sixth site may be occupied by the following ligands in biologically active vitamers:

- CN^- (cyanocobalamin)
- OH^- (hydroxocobalamin) or H_2O (aquocobalamin), depending on pH
- $-CH_3$ (methylcobalamin)
- 5'-deoxy-5'adenosine (adenosylcobalamin)

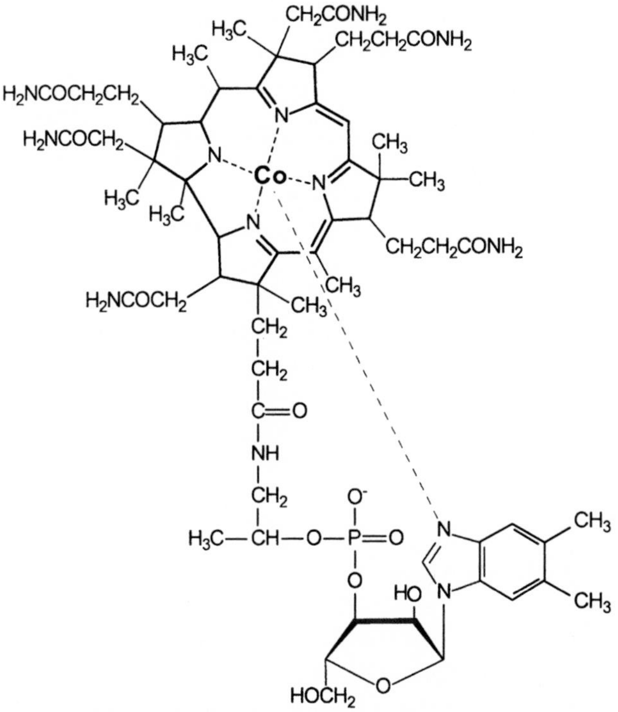

Figure 10.12. Vitamin B$_{12}$. Four coordination sites on the central cobalt atom are occupied by the nitrogen atoms of the corrin ring, and one by the nitrogen of the dimethylbenzimidazole nucleotide. The sixth coordination site may be occupied by: CN$^-$ cyanocobalamin, $M_r = 1355.4$; OH$^-$ hydroxocobalamin, $M_r = 1346.4$; H$_2$O aquocobalamin, $M_r = 1347.4$; –CH$_3$ methylcobalamin, $M_r = 1344.4$; and 5′-deoxyadenosine adenosylcobalamin, $M_r = 1579.6$.

Sulfitocobalamin, with a sulfite ligand, occurs in some foods as a result of processing, but is poorly absorbed.

The cobalt atom is in the Co^{3+} oxidation state in hydroxo-, aquo-, methyl-, and cyanocobalamins; in the Co$^+$ oxidation state in adenosylcobalamin; and, transiently, in the demethylated prosthetic group of methionine synthetase (Section 10.8.1).

Although cyanocobalamin was the first form in which vitamin B$_{12}$ was isolated, it is not an important naturally occurring vitamer, but rather an artifact caused by the presence of cyanide in the charcoal used in the extraction procedure. It is more stable to light than the other vitamers, and is commonly used in pharmaceutical preparations. Photolysis of cyanocobalamin in solution leads to the formation of aquocobalamin or hydroxocobalamin, depending on pH.

Hydroxocobalamin is also used in pharmaceutical preparations and is better retained after parenteral administration than is cyanocobalamin.

Small amounts of cyanocobalamin are found in the bloodstream (about 2% of total plasma vitamin B$_{12}$) apparently as part of the metabolism of cyanide derived from food (and tobacco smoke), but not in erythrocytes or tissues. If it is not converted to aquo- or hydroxocobalamin, cyanocobalamin may have antivitamin action and has been implicated in the neurological damage associated with chronic cyanide intoxication seen in parts of west Africa, where the dietary staple, cassava, is rich in cyanogenic glycosides.

The major plasma vitamer is methylcobalamin, accounting for 60% to 80% of plasma vitamin B$_{12}$, with up to 20% as adenosylcobalamin and the remainder mainly hydroxocobalamin. In tissues, the major vitamer is adenosylcobalamin (about 70% in liver), with about 25% as hydroxocobalamin and less than 5% as methylcobalamin.

10.7 METABOLISM OF VITAMIN B$_{12}$

Very small amounts of vitamin B$_{12}$ can be absorbed by diffusion across the intestinal mucosa. But, under normal conditions, this is insignificant, accounting for less than 1% of large oral doses. The major route of vitamin B$_{12}$ absorption is by way of attachment to a specific binding protein in the intestinal lumen.

This binding protein is intrinsic factor, so-called because in the early studies of pernicious anemia (Section 10.9.2), it was found that two factors were required – an extrinsic or dietary factor, which we now know to be vitamin B$_{12}$, and an intrinsic or endogenously produced factor. Intrinsic factor is a glycoprotein (M_r 44,000, containing 15% carbohydrate) secreted by the gastric parietal cells, which also secrete hydrochloric acid. The secretion of both intrinsic factor and gastric acid is stimulated by vagus nerve stimulation, histamine, gastrin, and insulin.

10.7.1 Digestion and Absorption of Vitamin B$_{12}$

Both gastric acid and pepsin are important in vitamin B$_{12}$ nutrition, serving to release the vitamin from protein binding and to make it available. Between 10% to 15% of people aged more than 60 years show some degree of vitamin B$_{12}$ deficiency as a result of impaired absorption due to atrophic gastritis. In the early stages, there is failure of acid secretion, resulting in failure to release the vitamin from dietary proteins. The absorption of free crystalline vitamin B$_{12}$, as opposed to vitamin that is bound to dietary proteins, is normal. As the condition progresses, there is also failure of the secretion of intrinsic factor.

In the stomach, vitamin B$_{12}$ binds to cobalophilin, a binding protein secreted in the saliva. The cobalophilins are a group of antigenically related, relatively unspecific, corrinoid binding proteins, formerly known as R-proteins because of their rapid mobility, compared with other cobalamin binding proteins, on electrophoresis.

In the duodenum, cobalophilin is hydrolyzed, releasing vitamin B$_{12}$ for binding to intrinsic factor. Pancreatic insufficiency can therefore be a factor in the development of vitamin B$_{12}$ deficiency, because failure to hydrolyze cobalophilin will result in the vitamin remaining bound to cobalophilin and being excreted, rather than being transferred to intrinsic factor (Gueant et al., 1990).

Intrinsic factor binds the various vitamin B$_{12}$ vitamers with equal affinity, but not other corrinoids. There is one vitamin B$_{12}$ binding site per mole of intrinsic factor. On binding the vitamin, the protein undergoes a conformational change, resulting in dimerization and greatly enhanced resistance to proteolysis.

Vitamin B$_{12}$ is absorbed from the distal third of the ileum by receptor-mediated endocytosis. There are intrinsic factor–vitamin B$_{12}$ binding sites on the brush border of the mucosal cells in this region. Free intrinsic factor does not interact with the receptors (Seetharam, 1999).

The absorption of vitamin B$_{12}$ is limited by the number of intrinsic factor–vitamin B$_{12}$ binding sites in the ileal mucosa, so that not more than about 0.7 to 1.1 nmol (1 to 1.5 μg) of a single oral dose of the vitamin can be absorbed. The absorption is also slow; peak blood concentrations of the vitamin are not achieved for 6 to 8 hours after an oral dose.

Within the ileal mucosal cell, the vitamin is released by lysosomal proteolysis of intrinsic factor, and is bound to transcobalamin II, a vitamin B$_{12}$ binding protein synthesized in the enterocytes. Transcobalamin II is in vesicles destined for export from the enterocytes, and it is assumed that vitamin B$_{12}$ binds to the apoprotein in these vesicles rather than in the lysosomes, because otherwise newly synthesized transcobalamin would be hydrolyzed by lysosomal proteases (Seetharam, 1999).

10.7.2 Plasma Vitamin B$_{12}$ Binding Proteins and Tissue Uptake

Vitamin B$_{12}$ enters the circulation bound to transcobalamin II; there is a relatively large amount of apotranscobalamin II in the circulation, and parenterally administered vitamin is also protein-bound. The half-life of holo-transcobalamin II in plasma is of the order of 1.5 hours, and all cells have surface receptors for holotranscobalamin II. Tissue uptake is by receptor-mediated endocytosis, followed by lysosomal proteolysis to release hydrox-ocobalamin. Functional deficiency, despite adequate intake and absorption, occurs in rare patients with a genetic lack of the transport protein for the release of the vitamin from lysosomes into the cytosol; they accumulate large amounts of the vitamin in lysosomes. Endocytosis of transcobalamin-bound vitamin B_{12} is by way of the same cubilin/megalin cell surface receptor system as for plasma lipoproteins (Seetharam and Li, 2000; Moestrup and Verroust, 2001). Hydroxocobalamin may either undergo methylation to methylcobal-amin in the cytosol, or it may enter the mitochondria and undergo a two-step reduction, catalyzed by a flavoprotein reductase, to yield Co^+-cobalamin, and reaction with ATP, catalyzed by adenosyltransferase, to form adenosylcobal-amin and tripolyphosphate.

Although transcobalamin II is the metabolically important pool of plasma vitamin B_{12}, it accounts for only 10% to 15% of the total circulating vitamin. The majority is bound to haptocorrin (also known as transcobalamin I). The function of haptocorrin is not well understood; it has a relatively long half-life (7 to 10 days), and does not seem to be involved in tissue uptake or intertissue transport of the vitamin. Although genetic lack of transcobalamin II results in severe (and fatal) vitamin B_{12} deficiency, genetic lack of haptocorrin seems to have no adverse effects.

There is a third plasma vitamin B_{12} binding protein, transcobalamin III, which is rapidly cleared by the liver, with a plasma half-life of the order of 5 minutes. This seems to provide a mechanism for returning vitamin B_{12} and its metabolites from peripheral tissues to the liver, as well as for clearance of other corrinoids without vitamin activity, which may arise either from foods or the products of intestinal bacterial action and be absorbed passively from the lower gut. These corrinoids are then secreted into the bile, bound to cobalophilins.

Considerably more intrinsic factor is secreted than is needed for the binding and absorption of dietary vitamin B_{12}, which requires only about 1% of the total intrinsic factor available. There is a considerable enterohepatic circulation of vitamin B_{12}, variously estimated as between 1 to 9 μg per day, about the same as the dietary intake. Like dietary vitamin B_{12} bound to salivary cobalophilin, the biliary cobalophilins are hydrolyzed in the duodenum, and the vitamin released binds to intrinsic factor, thus permitting reabsorption in the ileum. Whereas cobalophilins and transcorrin III have low specificity, and will bind a variety of corrinoids, intrinsic factor only binds cobalamins, and only the biologically active vitamin will be reabsorbed to any significant extent.

Estimates of requirements based on parenteral administration to subjects with pernicious anemia as a result of the lack of intrinsic factor (Section 10.9.2) are almost certainly erroneously high for subjects with normal enterohepatic circulation of the vitamin; in pernicious anemia, the biliary vitamin B$_{12}$ will not be reabsorbed to any significant extent, and requirements will therefore be considerably higher than normal.

10.7.3 Bacterial Biosynthesis of Vitamin B$_{12}$

Vitamin B$_{12}$ is synthesized only by bacteria and possibly some algae. There are no plant sources of the vitamin, and no plant enzymes are known to require vitamin B$_{12}$ as a coenzyme. A number of reports have suggested that vitamin B$_{12}$ occurs in some algae, but this may be the result of bacterial contamination of the water in which they were grown. Nori, made from the edible seaweed *Porphyra tenera*, has been reported to contain biologically active cobalamin when it is fresh; but, on drying, there is a considerable loss of the vitamin as a result of the formation of inactive corrinoids (Yamada et al., 1999).

The precursor for vitamin B$_{12}$ synthesis is uroporphyrinogen III, the common precursor for all porphyrins, including heme and chlorophyll. Uroporphyrinogen III is synthesized by condensation between succinyl coenzyme A (CoA) and glycine to yield δ-aminolevulinic acid. Two molecules of δ-aminolevulinic acid then condense to form the pyrrole phorphobilinogen, and four molecules of porphobilinogen condense to yield uroporphobilinogen III.

Uroporphobilinogen III then undergoes the following sequence of reactions (Raux et al., 2000; Roessner et al., 2001):

1. successive methylations, in which *S*-adenosyl methionine is the methyl donor;
2. excision of C-20 to give the direct fused link between the A and D pyrrole rings;
3. insertion of the central cobalt atom;
4. attachment of 5′-deoxyadenosine to the cobalt atom, yielding cobyrinic acid;
5. amidation of the acidic side chains to yield cobinamide; and
6. attachment of the dimethylbenzimidazole nucleotide, which is either formed from riboflavin, or synthesized by a similar pathway to that of riboflavin synthesis (see Figure 7.3), yielding adenosylcobalamin.

10.8 METABOLIC FUNCTIONS OF VITAMIN B$_{12}$

In microorganisms, vitamin B$_{12}$ is involved in a variety of reactions, including methyl transfer; the reduction of carbon dioxide to methane, a number of isomerase, mutase, and aminomutase reactions (1,2-migration reactions), diol dehydrogenase, ethanolamine ammonia-lyase, and in some organisms ribonucleotide reductase, which catalyzes the formation of deoxyribonucleotides from ribonucleotides. In eukaryotes, this reaction is catalyzed by an iron-containing enzyme.

In mammals, there are only three vitamin B$_{12}$-dependent enzymes: methionine synthetase, methylmalonyl CoA mutase, and leucine aminomutase. The enzymes use different coenzymes: methionine synthetase uses methylcobalamin, and cobalt undergoes oxidation during the reaction; methylmalonyl CoA mutase and leucine aminomutase use adenosylcobalamin and catalyze the formation of a 5'-deoxyadenosyl radical as the catalytic intermediate.

In addition, vitamin B$_{12}$ has a role in the metabolism of cyanide, forming cyanocobalamin. This prevents the binding of cyanide to cytochrome oxidase and permits (relatively slow) metabolism to yield thiocyanate.

10.8.1 Methionine Synthetase

As shown in Figure 10.9, the overall reaction of methionine synthetase is the transfer of the methyl group from methyl-tetrahydrofolate to homocysteine. However, the enzyme also requires S-adenosyl methionine and a flavoprotein reducing system in addition to the cobalamin prosthetic group. A common polymorphism of methionine synthetase, in which aspartate[919] is replaced by glycine, is associated with elevated plasma homocysteine in some cases, although it is less important than methylene-tetrahydrofolate reductase polymorphisms (Section 10.3.2.1; Harmon et al., 1999).

Cobalt accepts a methyl group from methyl-tetrahydrofolate, forming methyl Co^{3+}-cobalamin. Transfer of the methyl group onto homocysteine results in the formation of Co$^+$-cobalamin, which can accept a methyl group from methyl-tetrahydrofolate to reform methyl Co^{3+}-cobalamin. However, except under strictly anaerobic conditions, demethylated Co$^+$-cobalamin is susceptible to oxidation to Co^{2+}-cobalamin, which is catalytically inactive. Reactivation of the enzyme requires reductive methylation, with S-adenosyl methionine as the methyl donor, and a flavoprotein linked to NADPH. For this reductive reactivation to occur, the dimethylbenzimidazole group of the coenzyme must be displaced from the cobalt atom by a histidine residue in the enzyme (Ludwig and Matthews, 1997).

Methionine synthetase also catalyzes the reduction of nitrous oxide to nitrogen and in so doing generates a hydroxyl radical that results in irreversible inactivation of the enzyme (Frasca et al., 1986). Inactivation of methionine synthetase by nitrous oxide has been used as an acute model of vitamin B$_{12}$

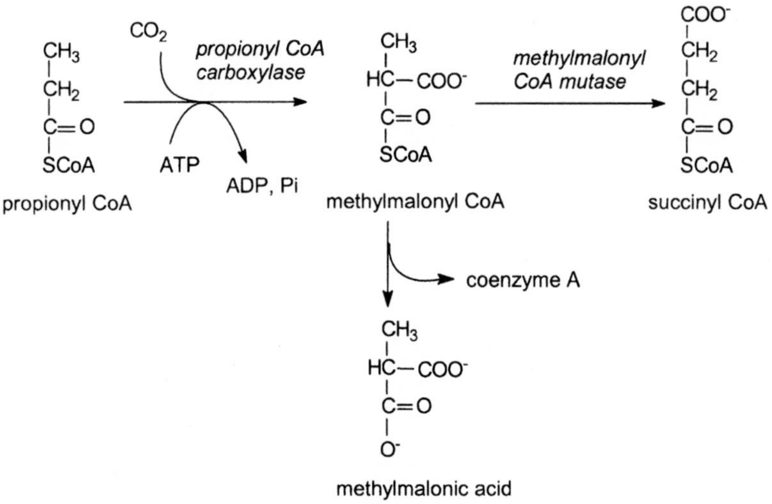

Figure 10.13. Reactions of propionyl CoA carboxylase (EC 6.4.1.3) and methylmalonyl CoA mutase (EC 5.4.99.2).

deficiency in experimental animals, and chronic exposure to nitrous oxide may cause vitamin B_{12} deficiency in human beings (Section 10.9.7).

10.8.2 Methylmalonyl CoA Mutase

Methylmalonyl CoA arises directly as an intermediate in the catabolism of valine, and is formed by the carboxylation of propionyl CoA arising in the catabolism of isoleucine, cholesterol, and fatty acids with an odd number of carbon atoms. Normally, as shown in Figure 10.13, it undergoes an adenosylcobalamin-dependent rearrangement to succinyl CoA, catalyzed by methylmalonyl CoA mutase. In vitamin B_{12} deficiency, the activity of this enzyme is greatly reduced, although there is induction of the apoenzyme to some 1.5- to 5-fold above that seen in control animals.

The sequence of the methylmalonyl CoA mutase reaction is as follows (Ludwig and Matthews, 1997; Frey, 2001):

1. Cleavage of the Co–C bond to the deoxyadenosyl group, with the probable formation of a 5′-deoxyadenosyl radical.
2. Removal of hydrogen from the substrate by the 5′-deoxyadenosyl radical, generating a substrate radical. It is not clear whether the dehydrogenation of the substrate occurs simultaneously with the cleavage of the Co–C bond or whether the 5′-deoxyadenosyl radical catalyzes this step.

3. Rearrangement of the substrate radical to give the product radical. It is not clear whether there is intermediate transfer of the carboxyl group of the substrate onto the cobalt of the coenzyme or direct carbon-to-carbon transfer in the substrate radical.
4. Removal of hydrogen from the deoxyadenosine by the product radical, forming the product and the 5′-deoxyadenosine radical.
5. Reformation of the C–Co bond of the intact coenzyme by reaction between the 5′-deoxyadenosine radical and the central cobalt atom.

As a result of the reduced activity of the mutase in vitamin B$_{12}$ deficiency, there is an accumulation of methylmalonyl CoA, some of which is hydrolyzed to yield methylmalonic acid, which is excreted in the urine. As discussed in Section 10.10.3, this can be exploited as a means of assessing vitamin B$_{12}$ nutritional status. There may also be some general metabolic acidosis, which has been attributed to depletion of CoA because of the accumulation of methylmalonyl CoA. However, vitamin B$_{12}$ deficiency seems to result in increased synthesis of CoA to maintain normal pools of metabolically useable coenzyme. Unlike coenzyme A and acetyl CoA, neither methylmalonyl CoA nor propionyl CoA (which also accumulates in vitamin B$_{12}$ deficiency) inhibits pantothenate kinase (Section 12.2.1). Thus, as CoA is sequestered in these metabolic intermediates, there is relief of feedback inhibition of its de novo synthesis. At the same time, CoA may be spared by the formation of short-chain fatty acyl carnitine derivatives (Section 14.1.1), which are excreted in increased amounts in vitamin B$_{12}$ deficiency. In vitamin B$_{12}$-deficient rats, the urinary excretion of acyl carnitine increases from 10 to 11 nmol per day to 120 nmol per day (Brass et al., 1990).

Methylmalonyl CoA inhibits the synthesis of fatty acids from acetyl CoA at concentrations of the order of those found in tissues of vitamin B$_{12}$-deficient animals. It is a substrate for fatty acid synthetase, leading to the formation of branched-chain and odd-carbon fatty acids.

Propionyl CoA inhibits *N*-acetylglutamate synthetase competitively with respect to acetyl CoA, forming *N*-propionylglutamate and reducing the synthesis of *N*-acetylglutamate. This is an obligatory activator of carbamyl phosphate synthetase, the first enzyme of urea synthesis. Vitamin B$_{12}$ deficiency may result in some degree of protein intolerance and hyperammonemia.

10.8.3 Leucine Aminomutase

Leucine aminomutase catalyzes isomerization of leucine and *β*-leucine. As with methylmalonyl CoA mutase, the reaction involves cleavage of the Co–C bond to form the 5′-deoxyadenosyl radical, which then removes hydrogen from the β-carbon of leucine (or the α-carbon of β-leucine), followed by migration of the amino group and attack by the product radical on deoxyadenosine. In addition to adenosylcobalamin, the reaction is pyridoxal phosphate-dependent (Section 9.3.1); presumably, migration of the amino group involves intermediate formation of pyridoxamine phosphate.

Poston (1984) showed that, in isolated rat tissues, about 5% of the catabolic flux of leucine was by way of aminomutase action to yield β-leucine, and then isobutyryl CoA, with the remainder provided by the more conventional α-transamination pathway leading to the formation of isovaleryl CoA. In patients suffering from vitamin B$_{12}$ deficiency, there is an elevation of plasma β-leucine, suggesting that the aminomutase may act to metabolize β-leucine arising from intestinal bacteria, rather than as a pathway for leucine catabolism.

10.9 DEFICIENCY OF FOLIC ACID AND VITAMIN B$_{12}$

Deficiency of either folic acid or vitamin B$_{12}$ results in a clinically similar megaloblastic anemia; because of the neurological damage that accompanies the megaloblastic anemia of vitamin B$_{12}$ deficiency, the condition is generally known as pernicious anemia. Suboptimal folate status is also associated with increased incidence of neural tube defects (Section 10.9.4), hyperhomocysteinemia leading to increased risk of cardiovascular disease (Section 10.3.4.2), and undermethylation of DNA leading to increased cancer risk (Section 10.9.5).

Both vitamin B$_{12}$ and folate deficiencies are associated with psychiatric illness. Folate deficiency is most commonly associated with depression, whereas cognitive impairment and dementia are seen in about 25% of patients with either deficiency (Bottiglieri, 1996; Green and Miller, 1999). Herbert (1962) noted insomnia, forgetfulness, and irritability during the development of self-imposed folate deficiency, which responded well to the administration of the vitamin.

Folate deficiency is relatively common; 8% to 10% of the population of developed countries have low or marginal folate stores. By contrast, dietary deficiency of vitamin B$_{12}$ is rare, and deficiency is most often the result of impaired absorption (Section 10.7.1).

Dietary deficiency of vitamin B$_{12}$ does occur, rarely, in strict vegetarians, because there are no plant foods that are sources of vitamin B$_{12}$. The small amounts that have been reported in some plants and algae are almost certainly from bacterial contamination.

10.9.1 Megaloblastic Anemia

Deficiency of either folic acid or vitamin B$_{12}$ results in megaloblastic anemia –
the release into circulation of immature erythrocytes because of failure of the
normal process of maturation in the bone marrow (Wickramasinghe, 1995,
1999). There may also be low white cell and platelet counts, as well as increased numbers of hypersegmented neutrophils. Iron deficiency may mask
the megaloblastic anemia.

The cause of megaloblastosis is depressed DNA synthesis, as a result of impaired methylation of dCDP to TDP, catalyzed by thymidylate synthetase, but
more or less normal synthesis of RNA. As discussed in Section 10.3.3, thymidylate synthetase uses methylene tetrahydrofolate as the methyl donor; it is obvious that folic acid deficiency will result in impaired thymidylate synthesis. It
is less easy to see how vitamin B$_{12}$ deficiency results in impaired thymidylate
synthesis without invoking the methyl folate trap hypothesis (Section 10.3.4.1).
The main circulating form of folic acid is methyl-tetrahydrofolate; before this
can be used for other reactions in tissues, it must be demethylated to yield free
folic acid. The only reaction that achieves this is the reaction of methionine
synthetase (Section 10.8.1). Thus, vitamin B$_{12}$ deficiency results in a functional
deficiency of folate.

The megaloblastic response to vitamin B$_{12}$ deficiency seems to be unique
to human beings; deficient animals develop neuropathy, but have unimpaired
hemopoeisis. It may be that human beings are more reliant on the de novo
synthesis of TMP and less able to salvage it from DNA breakdown than other
species. The normal suppression of the incorporation of [^{3}H]thymidine into
DNA by added dUMP (Section 10.3.3.3) is less than 3% in the fruit bat, 23% in
the rat, and 65% in humans.

10.9.2 Pernicious Anemia

Pernicious anemia is the megaloblastic anemia due specifically to vitamin
B$_{12}$ deficiency, in which there is also spinal cord degeneration and peripheral
neuropathy. It is a disease of later life. Only about 10% of patients are under
age 40; by the age of 60, about 1% of the population is affected, rising to 2%
to 5% of people over age 65, as a result of atrophic gastritis and thus impaired
absorption of vitamin B$_{12}$ (Section 10.7.1).

Early studies suggested that folate supplements exacerbate or hasten the
development of the neurological damage in vitamin B$_{12}$ deficiency. There is
little evidence that this is so, but high intakes of folate will prevent the development of megaloblastic anemia in vitamin B$_{12}$ deficiency. In up to one-third of

patients, the neurological signs appear without development of megaloblastic anemia (Dickinson, 1995).

Failure of intrinsic factor secretion is commonly a result of autoimmune disease; 90% of patients with pernicious anemia have complement-fixing antibodies in the cytosol of the gastric parietal cells. Similar autoantibodies are found in 30% of the relatives of pernicious anemia patients, suggesting that there is a genetic basis for the condition.

About 70% of patients also have antiintrinsic factor antibodies in plasma, saliva, and gastric juice. These can be either blocking antibodies, which prevent the binding of vitamin B$_{12}$ to intrinsic factor, or precipitating antibodies, which precipitate both free intrinsic factor and intrinsic factor–vitamin B$_{12}$ complex. Some patients have both types of antiintrinsic factor antibody. Although the oral administration of partially purified preparations of intrinsic factor will restore the absorption of vitamin B$_{12}$ in many patients with pernicious anemia, this can eventually result in the production of antiintrinsic factor antibodies, so parenteral administration of vitamin B$_{12}$ is the preferred means of treatment.

For patients who secrete antiintrinsic factor antibodies in the saliva or gastric juice, oral intrinsic factor will be of no benefit.

10.9.3 Neurological Degeneration in Vitamin B$_{12}$ Deficiency

Vitamin B$_{12}$ deficiency is accompanied by neurological degeneration in about two-thirds of cases – either peripheral neuropathy or subacute combined degeneration of the spinal cord. Folic acid deficiency is only rarely associated with similar neurological damage.

Subacute combined degeneration of the spinal cord is from demyelination of the corticospinal tracts and posterior columns of the spinal cord, leading to gait ataxia and loss of position sense and vibratory sense. Peripheral neuropathy leads to loss of cutaneous sensation and tendon reflexes (Savage and Lindenbaum, 1995).

Demyelination is because of failure of the methylation of arginine[107] of myelin basic protein. The nervous system is especially vulnerable to depletion of S-adenosylmethionine in vitamin B$_{12}$ deficiency because, unlike other tissues, it contains only methionine synthetase, which is vitamin B$_{12}$-dependent and not vitamin B$_{12}$-independent homocysteine methyltransferase that uses betaine as the methyl donor (Section 10.3.4; Weir and Scott, 1995).

Vitamin B$_{12}$ deficiency is associated with increased synthesis of tumor necrosis factor-α and decreased synthesis of epidermal growth factor in the spinal cord; in experimental animals, injection of tumor necrosis factor-α

causes spinal cord lesions. The role of vitamin B$_{12}$ is unclear, but there is some evidence that methylcobalamin has a role in regulation of the expression of cytokine genes in nerve tissue (Buccellato et al., 1999; Scalabrino et al., 2000; Peracchi et al., 2001).

Although the accumulation of methylmalonic acid can lead to increased synthesis of odd-carbon and branched-chain fatty acids (Section 10.8.2), which might become incorporated into myelin lipids, this does not seem to be responsible for neuropathy. Methylmalonic aciduria can also occur without any evidence of vitamin B$_{12}$ deficiency, as a result of a genetic defect of either methylmalonyl CoA mutase or the synthesis of adenosylcobalamin. In some cases, the condition is a vitamin dependency syndrome and responds to very high intakes of vitamin B$_{12}$. Although patients show mental retardation, failure to thrive, intermittent hypo- or hyperglycemia, and protein intolerance, they do not develop either megaloblastic anemia or neurological degeneration. Similarly, in the rare condition of combined methylmalonic aciduria with homocystinuria, caused by a failure of the synthesis of both adenosylmethionine and methylcobalamin, although there is megaloblastic anemia with failure to thrive and mental retardation, there is no evidence of neurological degeneration. This suggests that neither the accumulation of methylmalonyl CoA and methylmalonic acid nor the formation of odd-carbon and branched-chain fatty acids has any significant effect on myelin synthesis.

10.9.4 Folate Deficiency and Neural Tube Defects

The development of the brain and spinal cord begins around day 18 of gestation, with growth of the neural crest that folds and fuses to form the neural tube; closure begins about day 21 and is complete by day 24 – before the woman knows she is pregnant. The closed neural tube stimulates the development of the bony structures that will become the spinal cord and skull. Bone formation does not occur over unclosed regions of the neural tube, and this leads to the congenital defects that are collectively known as neural tube defects – anencephaly and spina bifida, which affect between 0.5 to 8 per 1,000 live births, depending on a variety of genetic and environmental factors.

Neural tube defects are seen in 3% to 6% of fetuses that abort spontaneously during the first trimester; many of these are from chromosomal abnormalities, whereas among infants born with neural tube defects chromosomal abnormalities are rare. Siblings are at a 10-fold greater risk of giving birth to infants with neural tube defects than are unrelated women from the same region; thus, there is a clear genetic factor. However, migration from areas with a high incidence to those with a low incidence reduces the number of neural tube defects in infants of women genetically at risk, so there are also environmental factors involved.

A number of observational studies suggested that low folate status was a factor in the etiology of neural tube defects, and in 1991 an intervention study (MRC Vitamin Study Research Group, 1991) showed that supplements of 4,000 μg per day of folic acid begun before conception considerably reduced the risk of recurrence of neural tube defects in women who had had a previous affected pregnancy. Later studies showed that more modest supplements of 400 μg per day of folic acid, again begun preconceptually, halve the incidence of neural tube defects among women who are not at high risk (Department of Health, 2000).

It is unlikely that an increase in folate intake equivalent to 400 μg of free folic acid per day could be achieved from unfortified foods. Women who are planning a pregnancy are advised to take supplements. As discussed in Section 10.12, enrichment of cereal products with folic acid is now mandatory in some countries, mainly to reduce the incidence of neural tube defects.

A number of mechanisms have been proposed to account for the role of folate in the development of neural tube defects (Moyers and Bailey, 2001):

1. Reduced synthesis of purines and pyrimidines as a result of folate deficiency may lead to a general impairment of DNA replication and cell division in the developing fetus.
2. Reduced synthesis of thymidine (Section 10.3.3) may lead to incorporation of uracil into DNA, thus leading to genomic instability.
3. Hyperhomocysteinemia (Section 10.3.4.2) may cause vascular impairment leading to ischemia and hypoxia.
4. Homocysteine or its metabolite homocysteic acid may have a direct neurotoxic effect in the developing neural tube; both bind to the N-methyl-D-aspartate subtype of glutamate receptor.
5. A wide variety of methylation reactions, including developmentally important methylation of DNA, may be inhibited by S-adenosylhomocysteine, which accumulates because homocysteine is a product inhibitor of S-adenosylhomocysteine hydrolase.

10.9.5 Folate Deficiency and Cancer Risk

Epidemiological studies suggest that suboptimal folate status is associated with an increased risk of colorectal and other cancers. It is difficult to determine the importance of folate per se, because the dietary sources of folate (mainly green leafy vegetables) are also sources of a variety of other compounds that have potentially protective effects against the development of cancer, including carotenoids (Section 2.5.2.3), vitamin C (Section 13.4.7) and vitamin E (Section 4.6.2.1), allyl sulfur compounds (Section 14.7.1), glucosinolates (Section 14.7.3), and flavonoids (Section 14.7.2).

There is, however, evidence that folate deficiency leads to impaired site-specific methylation of DNA, which is a factor in oncogenesis. About 4% of the total cytosines in DNA are methylated to 5-methylcytosine. Most of these are in palindromic regions of DNA that are promoters, leading to altered gene expression; methylation is important in the silencing of genes during development and tissue differentiation, and under-methylation may well result in dedifferentiation (Choi and Mason, 2000; Goodman and Watson, 2002; van den Veyver, 2002).

10.9.6 Drug-Induced Folate Deficiency

As discussed in Section 10.3.3.1, a number of folate antimetabolites are used clinically, as cancer chemotherapy (e.g., methotrexate), and as antibacterial (trimethoprim) and antimalarial (pyrimethamine) agents. Although drugs such as trimethoprim and pyrimethamine owe much of their clinical usefulness to considerably higher affinity of the dihydrofolate reductase of the target organism than the human enzyme, their prolonged use can result in (iatrogenic) folate deficiency.

A number of the older antiepileptic drugs – including diphenylhydantoin (phenytoin), phenobarbital, and primidone – can also cause folate deficiency. Although overt megaloblastic anemia affects only 0.75% of treated epileptics, there is some degree of macrocytosis in 40%. The megaloblastosis responds to folic acid supplements; but, in about 50% of patients treated with relatively large folate supplements for 1 to 3 years, there is an increase in the frequency of fits (Reynolds, 1967). The concentration of folate in the cerebrospinal fluid is normally two- to three-fold higher than in plasma; in epileptics taking diphenylhydantoin, the cerebrospinal fluid and plasma concentrations of folate are approximately equal, suggesting that the mechanism of action of the anticonvulsants may involve folate antagonism.

The mechanism of this drug-induced folic acid deficiency is not clear, although a number of effects have been reported:

1. Diphenylhydantoin and other anticonvulsants impair the intestinal absorption of folates. This may be by inhibition of intestinal conjugase; however, the evidence from various studies is conflicting.
2. Diphenylhydantoin causes an increased rate of catabolism of folate and increased excretion of folate metabolites (Kelly et al., 1979).

3. Chronic therapy of experimental animals with primidone depletes liver folate pentaglutamates, suggesting inhibition of folate polyglutamate synthetase (Carl et al., 1987). This would be expected to lead to increased excretion of folate metabolites.

4. Administration of diphenylhydantoin leads to decreased activity of methylene tetrahydrofolate reductase and an increased rate of oxidation of formyl tetrahydrofolate (increased oxidation of formate and histidine), with a fall in methylene- and methyl-tetrahydrofolate – the reverse of the effect of the methyl folate trap (Billings, 1984a, 1984b).

More modern anticonvulsants, such as valproic acid and carbamazepine, do not inhibit folate metabolism directly, and are not associated with megaloblastic anemia, although they are associated with increased risk of neural tube defects. They inhibit the glycine cleavage system (Section 10.3.1.1) and hence both reduce the pool of one-carbon substituted folates and also cause hyperglycinemia (Mortensen et al., 1980).

10.9.7 Drug-Induced Vitamin B$_{12}$ Deficiency

It was noted in Section 10.8.1 that nitrous oxide causes inhibition of methionine synthetase, as a result of irreversible oxidation of the cobalt of methylcobalamin. Patients with hitherto undiagnosed vitamin B$_{12}$ deficiency can develop neurological signs after surgery when nitrous oxide is used as the anesthetic agent. There are a number of reports of neurological damage because of vitamin B$_{12}$ depletion among people occupationally exposed to nitrous oxide (especially dental surgeons).

The histamine H$_2$ receptor antagonists and proton pump inhibitors used to treat gastric ulcers and gastroesophageal reflux act by reducing the secretion of gastric acid considerably. Prolonged use will result in impairment of protein-bound vitamin B$_{12}$ absorption. A number of studies have shown that even prolonged use of these drugs does not lead to significant depletion of vitamin B$_{12}$ reserves.

10.10 ASSESSMENT OF FOLATE AND VITAMIN B$_{12}$ NUTRITIONAL STATUS

A number of methods have been developed to permit assessment of folate and vitamin B$_{12}$ nutritional status and to differentiate between deficiency of the vitamins as a cause of megaloblastic anemia. Obviously, detection of antibodies to intrinsic factor or gastric parietal cells will confirm autoimmune pernicious anemia rather than nutritional deficiency of either vitamin.

In addition to the methods described here, measurement of urinary ac-
etamido p-aminobenzoyl glutamate will reflect folate turnover (Section 10.2.3)
and incorporation of uracil, instead of thymidine into the DNA in leukocytes
or lymphocytes may provide a sensitive index of folate status.

10.10.1 Plasma and Erythrocyte Concentrations of Folate and Vitamin B$_{12}$

Measurement of plasma concentrations of the two vitamins is probably the
method of choice, and a number of simple and reliable radioligand binding
assays have been developed, some of which permit simultaneous determination of both vitamins. Nevertheless, there are a number of problems involved in
radioligand binding assays, especially for folate. In some centers, microbiological determination of plasma or whole blood folates is the preferred technique.
The ligand binding assays that are normally used to determine erythrocyte folate are specific for 5-methyl-folate. They may not detect formyl-folates that
are present in significant amounts in erythrocytes from people who are homozygous for the thermolabile variant of methylene-tetrahydrofolate reductase (Bagley and Selhub, 1998). Radioligand binding assays for vitamin B$_{12}$
may give falsely high values if the binding protein is cobalophilin, which binds
a number of metabolically inactive corrinoids, as well as cobalamins. More
precise determination of true vitamin B$_{12}$ comes from assays in which the
binding protein is purified intrinsic factor, although this may still detect some
corrinoids without vitamin activity.

Erythrocytes accumulate a considerably higher concentration of folate than
is present in plasma (Section 10.2.2). Folate is incorporated into erythrocytes
during erythropoeisis and does not enter the cells in the circulation to any
significant extent. Therefore, erythrocyte folate is generally considered to give
an indication of folate status over 1 to 3 months (the life span of erythrocytes
in the circulation is 120 days) and not to be subject to variations in recent
intake, as is the plasma concentration of most vitamins. However, folate binds
to deoxyhemoglobin considerably more tightly than to oxyhemoglobin, and
the degree of oxygenation of hemoglobin in vitro, in the sample being used for
assay, will affect the amount of folate that is free in solution and accessible for
determination. This means that, without standardization of assay conditions,
it is difficult to compare the results of erythrocyte folate determinations from
different laboratories (Wright et al., 1998).

A serum concentration of vitamin B$_{12}$ below 110 pmol per L is associated
with megaloblastic bone marrow, incipient anemia, and myelin damage. Below

Table 10.2 Indices of Folate and Vitamin B$_{12}$ Nutritional Status

	Reference Range		Deficiency	
	nmol/L	µg/L	nmol/L	µg/L
Serum folate	9.8–16.2	4.4–7.2	<6.8	<3
Erythrocyte folate	420–620	185–270	<320	<140
Whole blood vitamin B$_{12}$	0.22–0.65	0.29–0.87	—	—
Serum vitamin B$_{12}$	0.14–0.52	0.19–0.69	<0.075	<0.10
Erythrocyte vitamin B$_{12}$	0.06–0.21	0.08–0.28	—	—
Transcorrin II bound vitamin B$_{12}$	—	—	<0.15	<0.22
Mean cell volume	—		>100 fL	
Serum methylmalonic acid	—		>1 µmol/L	
Serum homocysteine	—		>20 µmol/L	
Urine FIGLU more than 8 h after histidine load	—		>50 µg /mL	
Excretion of radiolabeled vitamin B$_{12}$ (Schilling test)	16–45%		<5%	

Sources: From data reported by Herbert, 1987a,1987b; van den Berg, 1993; Bailey and Gregory, 1999.

150 pmol per L, there are early bone marrow changes, abnormalities of the dUMP suppression test (Section 10.10.5), and methylmalonic aciduria after a valine load (Section 10.10.3). As shown in Table 10.2, this is considered to be the lower limit of adequacy.

Serum folate below 7 nmol per L or erythrocyte folate below 320 nmol per L indicates negative folate balance and early depletion of body reserves. At this stage, the first bone marrow changes are detectable.

About 30% of vitamin B$_{12}$-deficient subjects have elevated serum folate. This is mainly methyl-tetrahydrofolate, the result of the methyl folate trap (Section 10.3.4.1). About one-third of folate-deficient subjects have low serum vitamin B$_{12}$; the reason for this is not clear, but it responds to the administration of folate supplements.

10.10.2 The Schilling Test for Vitamin B$_{12}$ Absorption

The absorption of vitamin B$_{12}$ can be determined by the Schilling test. An oral dose of [^{57}Co] or [^{58}Co]vitamin B$_{12}$ is given with a parenteral flushing dose of 1 mg of nonradioactive vitamin, and the urinary excretion of radioactivity is followed as an index of absorption of the oral material. Normal subjects excrete 16% to 45% of the radioactivity over 24 hours, whereas patients lacking intrinsic factor or with antiintrinsic factor antibodies excrete less than 5%.

The test can be repeated, giving intrinsic factor orally together with the radioactive vitamin B_{12} – if the impaired absorption was because of a simple lack of intrinsic factor, and not to antiintrinsic factor antibodies in saliva or gastric juice, then a normal amount of the radioactive material should be absorbed and excreted.

A modified technique permits determination of the absorption in the presence and absence of exogenous intrinsic factor at the same time, by giving intrinsic factor–$[^{57}Co]$vitamin B_{12} complex and free $[^{58}Co]$vitamin B_{12} together, and measuring the relative amounts of each isotope excreted in the urine.

Atrophic gastritis will cause decreased secretion of gastric acid before there is any impairment of intrinsic factor secretion. This means that the absorption of crystalline vitamin B_{12}, as used in the Schilling test, is normal but the absorption of protein-bound vitamin B_{12} from foods will be impaired (Section 10.7.1), and the Schilling test will give a false-negative result.

10.10.3 Methylmalonic Aciduria and Methylmalonic Acidemia

As discussed in Section 10.8.2, moderate vitamin B_{12} deficiency results in increased accumulation of methylmalonyl CoA, and methylmalonic aciduria and methylmalonic acidemia. This can be exploited as both a means of detecting subclinical deficiency and monitoring vitamin B_{12} status in patients with pernicious anemia who have been treated with parenteral vitamin. As they become depleted, the excretion of methylmalonic acid, especially after a loading dose of valine, will provide a sensitive index of depletion of vitamin B_{12} reserves.

Methylmalonyl CoA mutase is especially sensitive to vitamin B_{12} depletion, so methylmalonic aciduria is the most sensitive index of vitamin B_{12} status. Folate deficiency does not cause methylmalonic aciduria. However, up to 25% of patients with confirmed pernicious anemia excrete normal amounts of methylmalonic acid, even after a loading dose of valine (Chanarin et al., 1973).

10.10.4 Histidine Metabolism – the FIGLU Test

The ability to metabolize a test dose of histidine provides a sensitive functional test of folate nutritional status; as shown in Figure 10.6, formiminoglutamate (FIGLU) is an intermediate in histidine catabolism and is metabolized by the tetrahydrofolate-dependent enzyme FIGLU formiminotransferase. In folate deficiency, the activity of this enzyme is impaired, and FIGLU accumulates and is excreted in the urine, especially after a test dose of histidine – the FIGLU test.

Although the FIGLU test depends on folate nutritional status, the metabolism of histidine will also be impaired and a positive result obtained, in vitamin B_{12} deficiency, because of the secondary deficiency of folate (Section 10.3.4.1). About 60% of vitamin B_{12}-deficient subjects show increased FIGLU excretion after a histidine load.

In experimental animals and with isolated tissue preparations and organ cultures, the test can be refined by measuring the production of $^{14}CO_2$ from [^{14}C]histidine in the presence and absence of added methionine. If the impairment of histidine metabolism is the result of primary folate deficiency, the addition of methionine will have no effect. By contrast, if the problem is trapping of folate as methyl-tetrahydrofolate, the addition of methionine will restore normal histidine oxidation as a result of restoring the inhibition of methylene-tetrahydrofolate reductase by S-adenosylmethionine and restoring the activity of 10-formyl-tetrahydrofolate dehydrogenase, thus permitting more normal folate metabolism (Section 10.3.4.1).

10.10.5 The dUMP Suppression Test

The ability of deoxyuridine to suppress the incorporation of [^3H]thymidine into DNA in rapidly dividing cells (Section 10.3.3.3) can also be used to give an index of functional folate nutritional status. Bone marrow biopsy samples provide the best source, and this has been generally a research tool rather than a screening test; however, transformed lymphocytes can also be used. The dUMP suppression test is probably the most sensitive index of folate depletion; abnormalities are apparent within 5 weeks of initiating folate deprivation, whereas detectably high urinary FIGLU occurs only after 13 weeks depletion, and bone marrow is overtly megaloblastic at 19 weeks (Herbert, 1962, 1987a).

Cells that have been preincubated with deoxyuridine, then exposed to [^3H]thymidine, incorporate little or none of the labeled material into DNA. This is because of both dilution of the labeled material in the larger intracellular pool of newly synthesized TMP and also inhibition of thymidylate kinase by thymidine triphosphate.

In normal cells, the incorporation of [^3H]thymidine into DNA after preincubation with dUMP is 1.4% to 1.8% of that without preincubation. By contrast, cells that are deficient in folate form little or no thymidine from dUMP, and incorporate nearly as much of the [^3H]thymidine after incubation with dUMP as they do without preincubation.

Again, either a primary deficiency of folic acid or functional deficiency secondary to vitamin B_{12} deficiency will have the same effect. In folate deficiency, addition of any biologically active form of folate, but not vitamin B_{12}, will normalize the dUMP suppression of [^{3}H]thymidine incorporation. In vitamin B_{12} deficiency, the addition of vitamin B_{12} or methylene-tetrahydrofolate, but not methyl-tetrahydrofolate, will normalize dUMP suppression (Killman, 1964; Pelliniemi and Beck, 1980).

10.11 FOLATE AND VITAMIN B_{12} REQUIREMENTS AND REFERENCE INTAKES

10.11.1 Folate Requirements

At the time that the U.K. and European Union reference intakes of folate shown in Table 10.3 were being discussed, the results of intervention trials for the prevention of neural tube defects (Section 10.9.4) were only just becoming available. At that time, there was no information concerning the effects of folate status on hyperhomocysteinemia (Section 10.3.4.2). The U.S./Canadian report (Institute of Medicine, 1998) notes specifically that protective effects with respect to neural tube defects were not considered relevant to the determination of the Dietary Reference Intake of folate, and there was insufficient evidence to associate higher intakes of folate (and lower plasma concentrations of homocysteine) with reduced risk of cardiovascular disease.

The total body pool of folate in adults is 17 μmol (7.5 mg), with a biological half-life of 101 days. This suggests a minimum requirement for replacement of 85 nmol (37 μg) per day (Herbert, 1987a). Studies of the urinary excretion of acetamido-p-aminobenzoyl glutamate in subjects maintained on folate-free diets suggest that there is catabolism of 170 nmol (80 μg) of folate per day.

Depletion/repletion studies to determine folate requirements using methyl-tetrahydrofolate suggest a requirement of the order of 170 to 220 nmol (80 to 100 μg) per day. However, because of the problems of determining the biological availability of the various folates found in foods (Section 10.2.1), reference intakes allow a wide margin of safety and are generally based on an allowance of 3 to 6 μg (7 to 14 nmol) per kg of body weight (200 to 400 μg per day for adults). In pregnancy and lactation, an additional 200 μg per day is generally recommended; this is probably more than can be obtained from foods, and therefore may require supplementation. As discussed in Section 10.9.4, supplements of 400 μg of folic acid per day in addition to normal intake are recommended for reduction of the risk of neural tube defects.

10.11.2 Vitamin B_{12} Requirements

The total body pool of vitamin B_{12} is of the order of 1.8 μmol (2.5 mg), with a minimum desirable body pool of about 0.3 μmol (1 mg).

Table 10.3 Reference Intakes of Folate (μg/day)

Age	U.K. 1991	EU 1993	U.S./Canada 1998	FAO 2001
0–6 m	50	—	65	80
7–12 m	50	50	80	80
1–3 y	70	100	150	160
4–6 y	100	130	200	200
7–8 y	150	150	200	300
Males				
9–10 y	150	150	300	300
11–13 y	200	180	300	400
>14 y	200	200	400	400
Females				
9–10 y	150	150	300	300
11–13 y	200	180	300	400
>14 y	200	200	400	400
Pregnant	300	400	600	600
Lactating	260	350	500	500

EU, European Union; FAO, Food and Agriculture Organization; WHO, World Health Organization.
Sources: Department of Health, 1991; Scientific Committee for Food, 1993; Institute of Medicine 1998; FAO/WHO, 2001.

The daily loss is about 0.1% of the body pool in subjects with normal intrinsic factor secretion and enterohepatic circulation of the vitamin (Section 10.7.2). On this basis, the requirement is 0.3 to 1.8 nmol (1 to 2.5 μg) per day (Herbert, 1987b). This is probably a considerable overestimate of requirements, because parenteral administration of less than 0.3 nmol per day is adequate to maintain normal hematology in patients with pernicious anemia, in whom the enterohepatic recycling of the vitamin is grossly impaired.

Requirements are probably between 0.1 to 1 μg per day; as shown in Table 10.4, reference intakes range between 1 to 2.4 μg per day, which is considerably lower than the average intake of 5 μg per day by nonvegetarians in most countries.

10.11.3 Upper Levels of Folate Intake
There are two potential problems associated with widespread enrichment of foods with folic acid or the indiscriminate use of folic acid supplements:

1. Intakes of folic acid in excess of about 5,000 μg per day antagonize the anticonvulsants used in treatment of epilepsy, leading to an increase in fit frequency (Section 10.9.6).

Table 10.4 Reference Intakes of Vitamin B$_{12}$ (μg/day)

Age	U.K. 1991	EU 1993	U.S./Canada 1998	FAO 2001
0–3 m	0.3	—	0.4	0.4
4–6 m	0.3	—	0.4	0.4
7–9 m	0.4	0.5	0.5	0.5
10–12 m	0.4	0.5	0.5	0.5
1–3 y	0.5	0.7	0.9	0.9
4–6 y	0.8	0.9	1.2	1.2
7–8 y	1.0	1.0	1.2	1.8
Males				
9–10 y	1.0	1.0	1.8	1.8
11–13 y	1.2	1.3	1.8	2.4
14–15 y	1.5	1.3	2.4	2.4
>16 y	1.5	1.4	2.4	2.4
Females				
9–10 y	1.0	1.0	1.8	1.8
11–13 y	1.2	1.3	1.8	2.4
14–15 y	1.2	1.3	2.4	2.4
>16 y	1.5	1.4	2.4	2.4
Pregnant	1.5	1.6	2.6	2.6
Lactating	2.0	1.9	2.8	2.8

EU, European Union; FAO, Food and Agriculture Organization; WHO, World Health Organization.

Sources: Department of Health, 1991; Scientific Committee for Food, 1993; Institute of Medicine 1998; FAO/WHO, 2001.

2. High intakes of folic acid mask the development of megaloblastic anemia from vitamin B$_{12}$ deficiency and result in the development of subacute combined degeneration of the spinal cord as the first sign of deficiency (Section 10.9.3). This means that elderly people are especially vulnerable, because of atrophic gastritis. It has been suggested that vitamin B$_{12}$ should be added to cereal products, as well as folate; as long as intrinsic factor secretion is unimpaired, crystalline vitamin B$_{12}$ will be absorbed normally, despite atrophic gastritis.

The U.S./Canadian upper level of folic acid intake (Institute of Medicine, 1998) is set at 1,000 μg per day, which is considered to be unlikely to mask the development of megaloblastic anemia in elderly people. The United Kingdom (Department of Health, 2000) considered the number of people over age 50 who would be exposed to intakes greater than 1,000 μg per day and the number of neural tube defects that would be prevented at various levels of folic acid enrichment of flour. It was concluded that fortification at 240 μg per 100 g of flour would have a significant beneficial effect without resulting in unacceptably high intakes by any population group. After public consultation in the United Kingdom, it was decided in May 2002 not to require fortification of flour with folic acid, pending surveillance of the effects of mandatory fortification in other countries.

10.12 PHARMACOLOGICAL USES OF FOLATE AND VITAMIN B$_{12}$

The only pharmacological use of vitamin B$_{12}$, other than for the treatment of deficiency or for rare children with vitamin dependency diseases affecting the binding of the coenzyme to methylmalonyl CoA mutase (Section 10.8.2), is as an antidote for cyanide poisoning. Supplements of vitamin B$_{12}$ are available for strict vegetarians who might be at risk of deficiency. There is no evidence of any adverse effects of high intakes of vitamin B$_{12}$.

Supplements of 400 μg per day of folic acid, begun before conception, halve the risk of neural tube defect (Section 10.9.4), and similar supplements reduce the plasma concentration of homocysteine in people homozygous for the thermolabile variant of methylene-tetrahydrofolate reductase (Section 10.3.4.2), although it is not known whether or not this will reduce their risk of cardiovascular disease. A number of manufacturers voluntarily enrich foods with folic acid. In the United States and other countries, there is mandatory enrichment of cereal products with folic acid.

It remains to be seen whether mandatory enrichment of cereal products with folic acid will reduce death from cardiovascular disease. But this, and the widespread voluntary enrichment of foods in other countries, means that intervention studies with folic acid supplements for cancer prevention are unlikely to yield useful results, because the control group will also be receiving a high intake of folic acid.

FURTHER READING

Baik HW and Russell RM (1999) Vitamin B$_{12}$ deficiency in the elderly. *Annual Reviews of Nutrition* **19,** 357–77.

Bailey LB and Gregory JF, 3rd (1999) Folate metabolism and requirements. *Journal of Nutrition* **129,** 779–82.

Blakley RL (1995) Eukaryotic dihydrofolate reductase. *Advances in Enzymology and Related Areas of Molecular Biology* **70,** 23–102.

Boers GHJ (1997) Hyperhomocysteinemia as a risk factor for arterial and venous disease. A review of evidence and relevance. *Thrombosis and Haemostasis* **78,** 520–2.

Carmel R (2000) Current concepts in cobalamin deficiency. *Annual Reviews of Medicine* **51,** 357–75.

Chanarin I and Metz J (1997) Diagnosis of cobalamin deficiency: the old and the new. *British Journal of Haematology* **97,** 695–700.

Chanarin I, Deacon R, Lumb M, Muir M, and Perry J (1985) Cobalamin-folate interrelations: a critical review. *Blood* **66,** 179–89.

Choi SW and Mason JB (2000) Folate and carcinogenesis: an integrated scheme. *Journal of Nutrition* **130,** 129–32.

Cortese C and Motti C (2001) MTHFR gene polymorphism, homocysteine and cardiovascular disease. *Public Health Nutrition* **4,** 493–7.

D'Angelo A and Selhub J (1997) Homocysteine and thrombotic disease. *Blood* **90,** 1–11.

Department of Health (2000) *Folic Acid and the Prevention of Disease.* London: The Stationery Office.

Duthie SJ (1999) Folic acid deficiency and cancer: mechanisms of DNA instability. *British Medical Bulletin* **55,** 578–92.

Fenech M (2001) The role of folic acid and vitamin B$_{12}$ in genomic stability of human cells. *Mutation Research* **475,** 57–67.

Fitzpatrick PF (1998) The aromatic amino acid hydroxylases. *Advances in Enzymology and Related Areas of Molecular Biology* **74B,** 235–94.

Fitzpatrick PF (1999) Tetrahydropterin-dependent amino acid hydroxylases. *Annual Reviews of Biochemistry* **68,** 355–81.

Fleming A (2001) The role of folate in the prevention of neural tube defects: human and animal studies. *Nutrition Reviews* **59,** S13–S20; discussion S13–S20.

Glusker JP (1995) Vitamin B$_{12}$ and the B$_{12}$ coenzymes. *Vitamins and Hormones* **50,** 1–76.

Gregory IJ and Quinlivan EP (2002) In vivo kinetics of folate metabolism. *Annual Reviews of Nutrition* **22,** 199–220.

Kapadia CR (1995) Vitamin B$_{12}$ in health and disease: part I – inherited disorders of function, absorption, and transport. *Gastroenterologist* **3,** 329–44.

Kruger WD (2000) Vitamins and homocysteine metabolism. *Vitamins and Hormones* **60,** 333–52.

Marsh EN (1999) Coenzyme B$_{12}$ (cobalamin)-dependent enzymes. *Essays in Biochemistry* **34,** 139–54.

McGuire JJ and Bertino JR (1981) Enzymatic synthesis and function of folylpolyglutamates. *Molecular and Cellular Biochemistry* **38,** 19–48.

Moyers S and Bailey LB (2001) Fetal malformations and folate metabolism: review of recent evidence. *Nutrition Reviews* **59,** 215–24.

Rajagopalan KV (1991) Novel aspects of the biochemistry of the molybdenum cofactor. *Advances in Enzymology and Related Areas of Molecular Biology* **64,** 215–90.

Rajagopalan KV (1997) Biosynthesis and processing of the molybdenum cofactors. *Biochemical Society Transactions* **25,** 757–61.

Refsum H, Ueland PM, Nygard O, and Vollset SE (1998) Homocysteine and cardiovascular disease. *Annual Reviews of Medicine* **49,** 31–62.

Rozen R (1997) Genetic predisposition to hyperhomocysteinemia: deficiency of methylenetetrahydrofolate reductase (MTHFR). *Thrombosis and Haemostasis* **78,** 523–6.

Scott J and Weir D (1994) Folate/vitamin B$_{12}$ inter-relationships. *Essays in Biochemistry* **28,** 63–72.

Scott JM (1999) Folate and vitamin B$_{12}$. *Proceedings of the Nutrition Society* **58,** 441–8.

Seetharam B (1999) Receptor-mediated endocytosis of cobalamin (vitamin B_{12}). *Annual Reviews of Nutrition* **19**, 173–95.

Seetharam B and Alpers DH (1994) Cobalamin binding proteins and their receptors. In *Vitamin Receptors: Vitamins as Ligands in Cell Communication*, K Dakshinamurti (ed.), pp. 78–105. Cambridge, UK: Cambridge University Press.

Seetharam B and Li N (2000) Transcobalamin II and its cell surface receptor. *Vitamins and Hormones* **59**, 337–66.

Selhub J (1999) Homocysteine metabolism. *Annual Reviews of Nutrition* **19**, 217–46.

Selhub J and D'Angelo A (1997) Hyperhomocysteinemia and thrombosis: acquired conditions. *Thrombosis and Haemostasis* **78**, 527–31.

Stamler JS and Slivka A (1996) Biological chemistry of thiols in the vasculature and in vascular-related disease. *Nutrition Reviews* **54**, 1–30.

Verhoef P and Stamler MJ (1995) Prospective studies of homocysteine and cardiovascular disease. *Nutrition Reviews* **53**, 283–8.

Weitman S, Anderson RGW, and Kamen BA (1994) Folate binding proteins. In *Vitamin Receptors: Vitamins as Ligands in Cell Communication*, K Dakshinamurti (ed.), pp. 106–36. Cambridge, UK: Cambridge University Press.

Wu G and Meininger CJ (2002) Regulation of nitric oxide synthesis by dietary factors. *Annual Reviews of Nutrition* **22**, 61–86.

References cited in the text are listed in the Bibliography.

ELEVEN

Biotin (Vitamin H)

Biotin was originally discovered as part of the complex called bios, which promoted the growth of yeast, and separately, as vitamin H, the protective or curative factor in egg white injury – the disease caused by diets containing large amounts of uncooked egg white. The glycoprotein avidin in egg white binds biotin with high affinity. This has been exploited to provide a variety of extremely sensitive assay systems.

Dietary deficiency of biotin sufficient to cause clinical signs is extremely rare in human beings, although it may be a problem in intensively reared poultry. However, there is increasing evidence that suboptimal biotin status may be relatively common, despite the fact that the vitamin is widely distributed in many foods, is synthesized by intestinal flora, and there is an efficient mechanism for conserving biotin after the catabolism of biotin-containing enzymes.

Metabolically, biotin is of central importance in lipogenesis, gluconeogenesis, and the catabolism of branched-chain (and other) amino acids. There are two well-characterized biotin-responsive inborn errors of metabolism, which are fatal if untreated: holocarboxylase synthetase deficiency and biotinidase deficiency. In addition, biotin induces a number of enzymes, including glucokinase and other key enzymes of glycolysis. Biotinylation of histones may be important in regulation of the cell cycle.

11.1 METABOLISM OF BIOTIN

As shown in Figure 11.1, biotin is a bicyclic compound with fused ureido (imidazolidone) and thiophene rings, and an aliphatic carboxylate side chain. It is bound covalently to enzymes by the formation of a peptide bond between the carboxyl group of the side chain and the ε-amino group of a lysine residue forming biocytin (biotinyl-lysine).

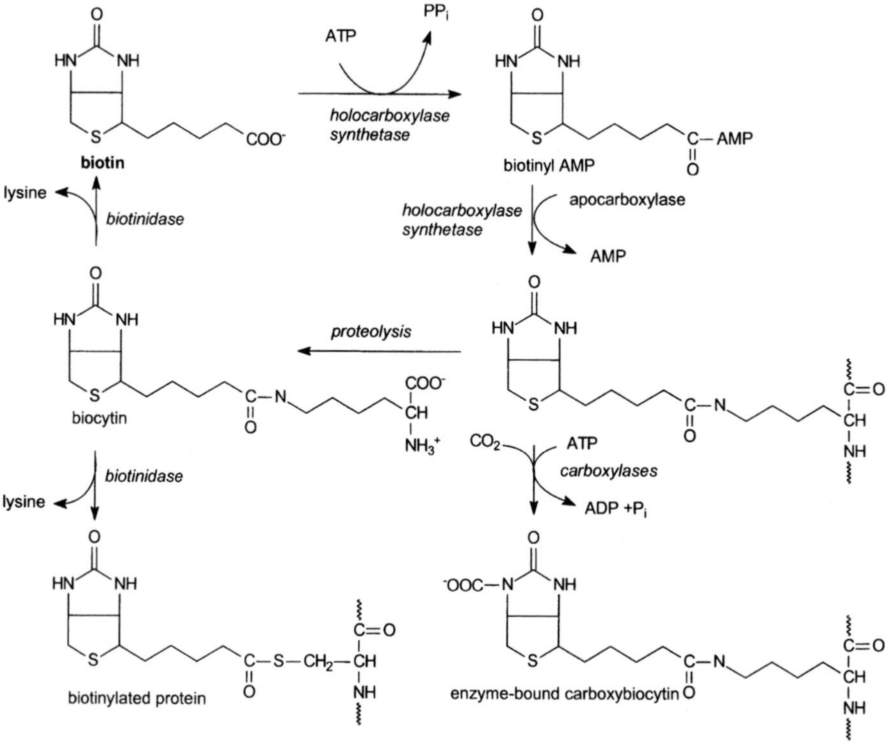

Figure 11.1. Metabolism of biotin. Holocarboxylase synthetase (biotin protein ligase), EC 6.3.4.10; and biotinidase (biotinamide amidohydrolase), EC 3.5.1.12. Relative molecular mass (M_r): biotin, 244.3; and biocytin, 372.5.

Most biotin in foods is present as biocytin, incorporated into enzymes, which is released on proteolysis, then hydrolyzed by biotinidase in the pancreatic juice and intestinal mucosal secretions to yield free biotin. Biocytin is not absorbed to any significant extent.

Biotin uptake into enterocytes is by a sodium-dependent carrier, which also transports pantothenic acid (Section 12.2) and lipoic acid, but is inhibited by biocytin and dethiobiotin. The carrier is found in both the small intestine and the colon, so both biotin and pantothenic acid synthesized by intestinal bacteria can be absorbed (Chatterjee et al., 1999; Ramaswamy, 1999; Said, 1999; Prasad and Ganapathy, 2000). Even at relatively high intakes (up to 80 μmol), biotin is more-or-less completely absorbed (Zempleni and Mock, 1999b).

Most biotin circulates in the bloodstream bound to a serum glycoprotein, biotinidase (Section 11.2.3), which not only acts as a transport protein, but also catalyzes the hydrolysis of biocytin, and the transfer of biotin from biocytin

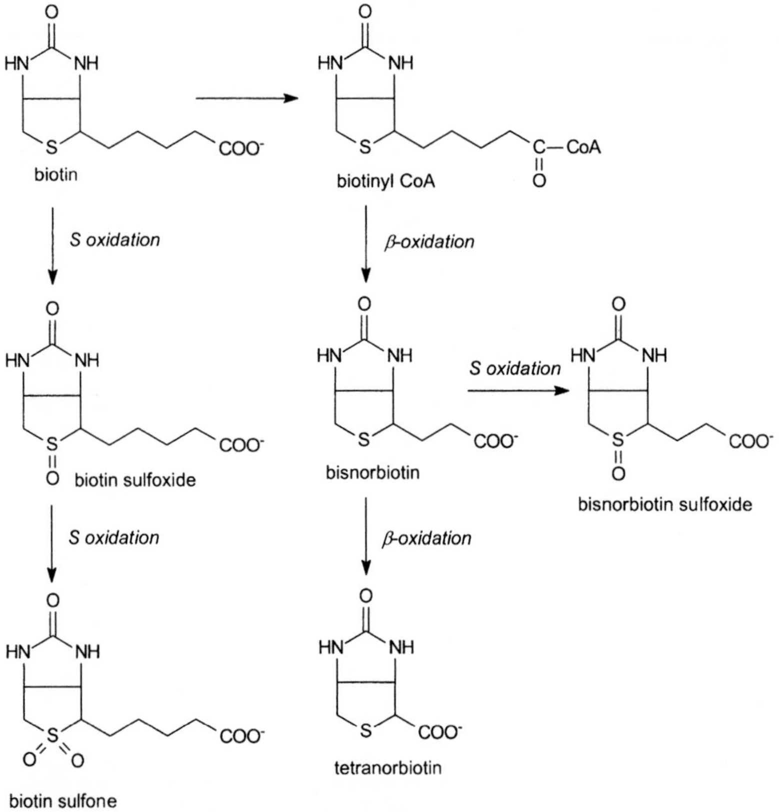

Figure 11.2. Biotin metabolites. Relative molecular masses (M_r): biotin, 244.3; biotin sulfoxide, 260.3; biotin sulfone, 276.3; bisnorbiotin, 212.3; tetranorbiotin, 180.3; and bisnorbiotin sulfoxide, 228.3.

onto sulfhydryl groups of histones and other proteins (Section 11.2.5). Some biotin is also nonspecifically bound to albumin and α- and β-globulins.

Dietary biotin bound to avidin (Section 11.6) is unavailable, but intravenously administered avidin–biotin is biologically active. Cells in culture are not inhibited by the addition of avidin to the culture medium, and can take up the avidin–biotin complex by pinocytosis followed by lysosomal hydrolysis, releasing free biotin. Unlike other B vitamins, for which concentrative uptake into tissues is achieved by facilitated diffusion, followed by metabolic trapping, the incorporation of biotin into enzymes is slow and cannot be considered part of the uptake process.

As discussed in Section 11.2.2, biotin is incorporated covalently into biotin-dependent enzymes as the ε-amino-lysine peptide, biocytin. On catabolism of the enzymes, biocytin is hydrolyzed by biotinidase, permitting reutilization of the biotin.

As shown in Figure 11.2, the side chain of biotin can undergo mitochondrial or peroxisomal β-oxidation to yield bisnorbiotin and tetranorbiotin. In the microsomes, both biotin and bisnorbiotin undergo S-oxidation to the sulfoxides, and biotin sulfoxide can undergo further oxidation to the sulfone. At physiological levels of intake, about 30% of biotin is excreted unchanged, and 50% to 60% as bisnorbiotin and bisnorbiotin methyl ketone; sulfoxides and biotin sulfone make up the remainder (Zempleni and Mock, 1999a).

The brush border of the kidney cortex has a sodium–biotin cotransport system similar to that in the intestinal mucosa, thus providing for reabsorption of free biotin filtered into the urine. It is only when this mechanism is saturated (it has a relatively low K_m) that there will be a significant excretion of biotin.

As a result of this resorption and the protein binding of plasma biotin, which reduces filtration at the glomerulus, renal clearance of biotin is only 40% of that of creatinine. This efficient conservation of biotin, together with the recycling of biocytin released from the catabolism of biotin-containing enzymes, may be as important as intestinal bacterial synthesis of the vitamin in explaining the rarity of deficiency.

11.1.1 Bacterial Synthesis of Biotin

Biotin is synthesized in microorganisms from pimelate by the pathway shown in Figure 11.3 (Marquet et al., 2001; Schneider and Lindqvist, 2001). The first committed step is the condensation of pimeloyl CoA with alanine, with release of the carboxyl group of alanine, catalyzed by keto-aminopelargonic acid synthase. It is a pyridoxal phosphate-dependent reaction, similar to the condensation of succinyl CoA and glycine to form δ-aminolevulinic acid in porphyrin synthesis. There is a considerable degree of structural homology between keto-aminopelargonic acid and δ-aminolevulinic acid synthases.

The second nitrogen of biotin is incorporated by transamination of keto-aminopelargonic acid, with S-adenosylmethionine – an apparently unique metabolic role for this amino acid derivative that is normally a methyl donor. The immediate product of the deamination of S-adenosylmethionine, S-adenosyl-2-oxo-4-methylthiobutyric acid, is unstable and decomposes non-enzymically to 2-oxo-3-butenoic acid and 5'-methyl thioadenosine.

Completion of the ureido ring of biotin, yielding dethiobiotin, is by a carboxylation reaction using CO_2 and ATP. The reaction proceeds by the formation of a monocarbamate by reaction between diaminopelargonic acid and CO_2, followed by formation of a substituted carbamyl phosphate, which then

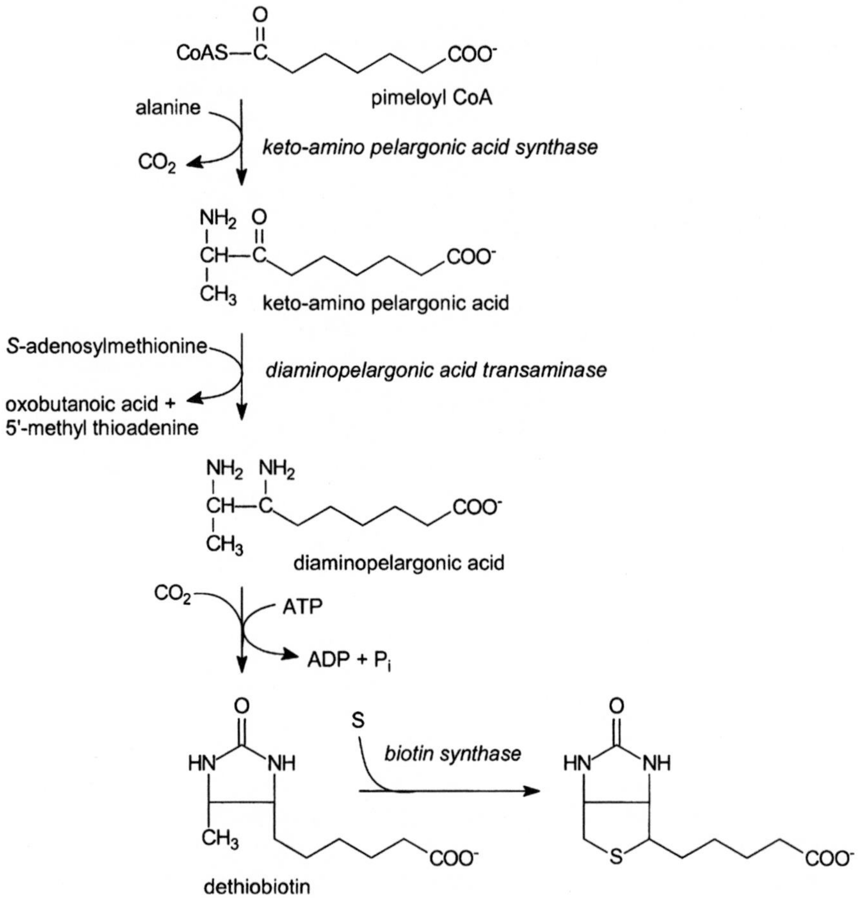

Figure 11.3. Biosynthesis of biotin. Keto-aminopelargonic acid synthase, EC 2.3.1.47; diaminopelargonic acid synthase (aminotransferase), EC 2.6.1.62; dethiobiotin synthase, EC 6.3.3.3; and biotin synthase, EC 2.8.1.6.

cyclizes by elimination of the phosphate group. The two amide bonds are thus formed using a single mole of ATP $\rightarrow$ ADP + P_i.

The final reaction, catalyzed by biotin synthase, involves the insertion of sulfur between the unreactive methyl and methylene carbons of dethiobiotin. The enzyme has an iron–sulfur box, and requires NADPH and a ferredoxin or flavodoxin reducing system. S-Adenosylmethionine is also required, and is cleaved to yield methionine and a 5'-deoxyadenosyl radical during the reaction. Biotin synthase is a member of the radical SAM family of enzymes, in which the catalytic 5'-deoxyadenosyl radical is formed from S-adenosylmethionine, rather than from adenosylcobalamin as in methylmalonyl CoA mutase and similar vitamin B_{12}-dependent enzymes (Section 10.8.2). Two moles of S-adenosylmethionine are required: one 5'-deoxyadenosyl radical abstracts hydrogen from the methyl group of dethiobiotin and the other from the methylene group.

It is possibly incorrect to consider biotin synthase an enzyme in the true sense of the word; it has a turnover number of 1. It only catalyzes the synthesis of a single molecule of biotin from dethiobiotin before being inactivated. This is because the iron–sulfur cluster of the protein is the source of the sulfur that is incorporated into biotin. There is some evidence that the enzyme can be reactivated by incorporation of sulfur from cysteine, but in vitro addition of the enzymes believed to catalyze this reaction has no effect on the turnover number of the enzyme (Frey, 2001; Marquet et al., 2001).

11.1.1.1.1 The Importance of Intestinal Bacterial Synthesis of Biotin   It was noted in Section 11.1 that biotin is absorbed throughout the intestinal tract, including the colon, and synthesis by intestinal bacteria may make a significant contribution to biotin nutrition. In balance studies, the total output of biotin in urine plus feces is three to six times greater than the intake; most of this excess is in the feces, reflecting bacterial synthesis. In experimental animals maintained on biotin- and cellulose-free diets, the addition of cellulose or sorbitol as a substrate for bacterial fermentation can alleviate the vitamin deficiency.

11.2 THE METABOLIC FUNCTIONS OF BIOTIN
Biotin is the coenzyme in a small number of carboxylation reactions in mammalian metabolism and some decarboxylation and transcarboxylation reactions in bacteria. Although the biotin-dependent enzymes are cytosolic and mitochondrial, about 25% of tissue biotin is found in the nucleus, much of it bound as thioesters to histones. Biotin has two noncoenzyme functions: induction of enzyme synthesis and regulation of the cell cycle.

The biotin-dependent decarboxylases of anerobic microorganisms are transmembrane proteins. In addition to their roles in the metabolism of oxaloacetate, methylmalonyl CoA, and glutaconyl CoA, they serve as energy transducers. They transport 2 mol of sodium out of the cell for each mole of substrate decarboxylated. The resultant sodium gradient is then used for active transport of substrates by sodium cotransport systems, or may be used to drive ATP synthesis in a similar manner to the proton gradient in mammalian mitochondria (Buckel, 2001).

11.2.1 The Role of Biotin in Carboxylation Reactions

The reactive intermediate is 1-N-carboxy-biotin (see Figure 11.1) bound to a lysine residue of the enzyme as biocytin, which is formed from enzyme-bound biocytin by reaction with bicarbonate.

The biotin-dependent carboxylases catalyze a two-step reaction:

1. enzyme-biotin + ATP + HCO_3^- → enzyme-biotin-COOH + ADP + P_i
2. enzyme-biotin- COOH + acceptor → enzyme-biotin + acceptor-COOH.

In the bacterial biotin-dependent decarboxylases, reaction 2 proceeds from right to left, followed by decomposition of the carboxy-biotin to biotin and CO_2.

The role of ATP in the carboxylation of biotin is unclear. It is possible that biotin is O-phosphorylated during the carboxylation reaction. However, evidence suggests that the immediate reactive species that carboxylates biotin is carboxyphosphate, as in the (biotin-independent) reaction of carbamyl phosphate synthetase in urea and pyrimidine synthesis.

Steady-state kinetic analysis shows that biotin-dependent reactions proceed by way of a two-site ping-pong mechanism; the two-part reactions are catalyzed at distinct sites in the enzyme. These sites may be on the same or different polypeptide chains in different biotin-dependent enzymes. The ε-amino linkage of lysine to the side chain of biotin in biocytin allows considerable movement of the coenzyme – the distance from C-2 of lysine to C-5 of biotin is 14Å, thus allowing movement of biotin between the carboxylation and carboxyltransfer sites.

In mammals and birds, there are four biotin-dependent carboxylases: acetyl CoA carboxylase, pyruvate carboxylase, propionyl CoA carboxylase, and methylcrotonyl CoA carboxylase. Congenital deficiency of three of the four human biotin-dependent carboxylases has been reported.

11.2.1.1 Acetyl CoA Carboxylase

Acetyl CoA carboxylase catalyzes the first and rate-limiting step of fatty acid synthesis: carboxylation of acetyl CoA to malonyl CoA. The mammalian enzyme is activated allosterically by citrate and isocitrate, and inhibited by long-chain fatty acyl CoA derivatives. It is also activated in response to insulin and inactivated in response to glucagon.

Tissues that oxidize fatty acids, but do not synthesize them, such as muscle, also have acetyl CoA carboxylase and form malonyl CoA to regulate the activity of carnitine palmitoyltransferase, and thus control the uptake of fatty acids into the mitochondria for β-oxidation.

There are no unequivocal reports of acetyl CoA carboxylase deficiency; presumably impairment of this key enzyme in lipogenesis would not be compatible with intrauterine development.

11.2.1.2 Pyruvate Carboxylase

Pyruvate carboxylase catalyzes the carboxylation of pyruvate to oxaloacetate – both the first committed step of gluconeogenesis from pyruvate and also an important anaplerotic reaction, permitting repletion of tricarboxylic acid cycle intermediates and hence fatty acid synthesis. The mammalian enzyme is activated allosterically by acetyl CoA, which accumulates when there is a need for increased activity of pyruvate carboxylase to synthesize oxaloacetate to permit increased citric acid cycle activity or for gluconeogenesis (Attwood, 1995; Jitrapakdee and Wallace, 1999).

Pyruvate carboxylase is also important in lipogenesis. Citrate is transported out of mitochondria and cleaved in the cytosol to provide acetyl CoA for fatty acid synthesis; the resultant oxaloacetate is reduced to malate, which undergoes oxidative decarboxylation to pyruvate, a reaction that provides at least half of the NADPH required for fatty acid synthesis. Pyruvate reenters the mitochondria and is carboxylated to oxaloacetate to maintain the process.

Mammalian pyruvate carboxylase has four identical subunits, and the isolated monomer will catalyze the complete reaction. By contrast, three distinct subunits can be isolated from acetyl CoA carboxylase of *Escherichia coli* and spinach chloroplasts: a biotinyl carrier protein, biotin carboxylase, and carboxyl transferase.

Genetic deficiency of pyruvate carboxylase does not cause the expected hypoglycemia. Rather, it seems that depletion of tissue pools of oxaloacetate results in impaired activity of citrate synthase, and a slowing of citric acid cycle activity, leading to accumulation of lactate, pyruvate, and alanine, and also increased accumulation of acetyl CoA, resulting in ketosis. Affected infants have serious neurological problems and rarely survive. A less severe variant of the disease is associated with low residual activity of pyruvate carboxylase.

11.2.1.3 Propionyl CoA Carboxylase

Propionyl CoA carboxylase catalyzes the carboxylation of propionyl CoA to methylmalonyl CoA, which undergoes a vitamin B_{12}-dependent isomerization to succinyl CoA (see Figure 10.13). This reaction provides a pathway for the oxidation, through the tricarboxylic acid cycle, of propionyl CoA arising from the catabolism of isoleucine, valine, odd-carbon fatty acids, and the side chain of cholesterol.

Propionic acidemia caused by propionyl CoA carboxylase deficiency causes severe ketosis and acidosis, resulting in failure to thrive and mental retardation, and is generally fatal in infancy. Some reports of ketotic hyperglycinemia may also, with hindsight, be attributed to propionyl CoA carboxylase deficiency.

11.2.1.4 Methylcrotonyl CoA Carboxylase Methylcrotonyl CoA carboxylase catalyzes the conversion of methylcrotonyl CoA, arising from the catabolism of leucine, to methylglutaconyl CoA. This in turn undergoes hydroxylation catalyzed by crotonase, yielding hydroxymethyl-glutaryl CoA, which is cleaved to acetyl CoA and acetoacetate.

Methylcrotonyl CoA carboxylase deficiency is the least severe of the carboxylase deficiencies. Maintenance on a low-protein diet, to minimize the burden of leucine that must be catabolized, prevents the development of metabolic acidosis. At higher intakes of protein, the affected infants become hypoglycemic and comatose.

11.2.2 Holocarboxylase Synthetase

Biotin is bound covalently to enzymes by a peptide link to the ε-amino group of a lysine residue, forming biotinyl-ε-amino-lysine or biocytin (see Figure 11.1). This postsynthetic modification is catalyzed by holocarboxylase synthetase with the intermediate formation of biotinyl-5′-AMP. In bacteria, this intermediate also acts as a potent repressor of all four enzymes of biotin synthesis.

A single holocarboxylase synthetase (biotin protein ligase, EC 6.3.4.10) acts on the apoenzymes of acetyl CoA, pyruvate, propionyl CoA, and methylcrotonyl CoA carboxylases. Acetyl CoA carboxylase is a cytosolic enzyme, whereas the other three enzymes are mitochondrial. Although holocarboxylase synthetase is found in both the cytosol and mitochondria, it is not clear whether biotin is incorporated into the mitochondrial enzymes before or after they are translocated into the mitochondria.

Holocarboxylase synthetase from a wide variety of species will act on all four apocarboxylases from other species and on a variety of bacterial biotin-dependent apoenzymes. In all the biotin-dependent enzymes investigated to date, the reactive lysine residue is flanked by methionine residues on both sides, and there is a high degree of conservation of the amino sequence around this Met-Lys-Met sequence (Chapman-Smith and Cronan, 1999a, 1999b).

11.2.2.1 Holocarboxylase Synthetase Deficiency Genetic deficiency of holocarboxylase synthetase leads to the neonatal form of multiple carboxylase

---

Table 11.1 Abnormal Urinary Organic Acids in Biotin Deficiency and Multiple Carboxylase Deficiency from Lack of Holocarboxylase Synthetase or Biotinidase

---

Arising from impaired activity of acetyl CoA carboxylase
   2-Ethyl-3-hydroxyhexanoic acid
   2-Ethylhexanedioic acid
Arising from impaired activity of pyruvate carboxylase
   Lactate
   Pyruvate
   Alanine
Arising from impaired activity of propionyl CoA carboxylase
   Propionic acid
   3-Hydroxypropionic acid
   Propionylglycine
   Methylcitric acid
   Tiglic acid (from leucine catabolism)
   Tiglylglycine
   2-Methyl-3-hydroxybutyric acid
   Lactate
Arising from impaired activity of methylcrotonyl CoA carboxylase
   3-Methylcrotonic acid
   3-Hydroxy-isovaleric acid (hydration products of 3-methylcrotonic acid)
   3-Methylcrotonyl glycine

CoA, coenzyme A.

---

deficiency; most infants present within the first 6 weeks of life, although some may not present until 15 months of age. They have a scaly dermatitis and alopecia, as seen in biotin deficiency (Section 11.3), and develop potentially life-threatening keto-acidosis and sometimes also hyperammonemia, with mental retardation, delayed development, and acute neurological problems. They also have high blood and urine concentrations of a number of substrates of biotin-dependent enzymes, and the metabolic precursors and alternative metabolites of these substrates, as shown in Table 11.1 (Baumgartner and Suormala, 1997, 1999).

The affected infants have a normal plasma concentration of biotin and excrete normal amounts of biotin in the urine. Skin fibroblasts have extremely low activities of all four biotin-dependent carboxylases when they are cultured in media containing approximately physiological concentrations of biotin. But, culture with considerably higher concentrations of biotin results in normal activity of all four carboxylases. The defect is in the affinity of holocarboxylase synthetase for biotin (its K_m is 20- to 70-fold higher than normal).

The condition is a biotin-responsive genetic disease, and patients can be maintained in good health with supplements of high doses of biotin. Doses of biotin of the order of 1 mg per day prevent clinical signs in most patients, but 10 mg per day or more is required to correct the organic aciduria.

Holocarboxylase synthetase deficiency can be diagnosed prenatally by assessing the response of carboxylase activity in cultured amniocytes (obtained by amniocentesis) to the addition of biotin, or by the detection of methylcitric and hydroxyisovaleric acids in the amniotic fluid. Prenatal therapy, by giving the mother 10 mg of biotin per day, results in sufficiently elevated fetal blood concentrations of biotin to prevent the development of organic acidemia at birth.

11.2.3 Biotinidase

Proteolysis of biotin-containing enzymes releases biocytin, either as free biotinyl-lysine or as a variety of small biocytin-containing peptides; the ε-amino lysine link of biocytin is not a substrate for peptidases.

Biocytin is hydrolyzed by biotinidase, which acts on free or peptide-incorporated biocytin to release biotin, but has no general peptidase or esterase activity. Biotinidase is most active toward free biocytin, but it will also release biotin from biocytin-containing peptides. The activity decreases as the size of the peptide increases, so it is likely that in vivo the catabolism of biotin-containing enzymes is by proteolysis, followed by biotinidase action, rather than the release of biotin, leaving the apoenzyme as a substrate for proteolysis. Biotinidase is found in all tissues, including the pancreatic juice and intestinal mucosa.

Biotinidase functions both to release free biotin from biocytin in foods, and to recycle and conserve biotin after turnover of biotin-containing enzymes. As discussed in Section 11.2.3.1, rare congenital deficiency of biotinidase results in severe functional biotin deficiency.

Biotinidase is also the major plasma binding protein for biotin. The pH optimum of the enzyme is 4.5 to 5.5, and its K_m is in the micromolar range, compared with the nanomolar concentrations of biocytin, so it will have little enzymic activity in plasma. Rather, it functions as a transport protein for biotin, preventing its urinary excretion; children with biotinidase deficiency (Section 11.2.3.1) excrete large amounts of both biocytin and free biotin. Biotin is covalently bound to biotinidase in plasma, as a thioester to a cysteine residue in the active site of the enzyme (see Figure 11.1). This thioester is formed only from biocytin, not free biotin, and is presumably the (normally transient) intermediate in the hydrolysis of biocytin. Free biotin can be released from biotinidase at low pH, and the enzyme can also catalyze a biotinoyl transferase reaction, biotinoylating histones, and other nucleophilic acceptors (Hymes and Wolf, 1996, 1999).

11.2.3.1 **Biotinidase Deficiency**   Genetic lack of biotinidase results in the late-onset variant of multiple carboxylase deficiency. Patients generally present later in life than those with holocarboxylase synthetase deficiency (Section 11.2.2.1) and have a lower than normal blood concentration of biotin. Culture of fibroblasts in media containing low concentrations of biotin results in normal activities of carboxylases, and holocarboxylase synthetase activity is normal.

The problem is a functional deficiency of biotin, due both to inability to release free biotin from dietary biocytin and also to failure of the normal recovery of free biotin by biotinidase action on the biocytin released by proteolysis of biotin-containing enzymes. Normal intakes of biotin are inadequate to meet the requirements of these patients; the provision of pharmacological doses of free biotin provides an adequate amount to meet requirements without the need for reutilization. The delayed development of clinical and biochemical abnormalities is a result of the accumulation of biotin by the fetus, so that at birth the infant has adequate stores of the vitamin.

Biotinidase-deficient patients have higher than normal amounts of biocytin in plasma and urine, and excrete larger than normal amounts of biotin, reflecting the importance of protein binding of biotin to prevent urinary loss. Therapy with 10 mg of biotin per day prevents the development of most symptoms, although some patients develop neurosensory hearing loss and optic atrophy despite therapy with biotin. This reflects the role of biotinidase in tissue uptake of biotin (Section 11.1; Wolf and Feldman, 1982; Wolf and Heard, 1991; Baumgartner and Suormala, 1997, 1999).

11.2.4 Enzyme Induction by Biotin

Biotin acts to induce glucokinase, phosphofructokinase, and pyruvate kinase (key enzymes of glycolysis), phosphoenolpyruvate carboxykinase (a key enzyme of gluconeogenesis), and holocarboxylase synthetase, acting via a cell-surface receptor linked to formation of cGMP and increased activity of RNA polymerase. The activity of holocarboxylase synthetase (Section 11.2.2) falls in experimental biotin deficiency and increases with a parallel increase in mRNA during repletion (Chauhan and Dakshinamurti, 1991; Borboni et al., 1996; Rodriguez-Melendez et al., 2001).

Glucokinase is the high-K_m isoenzyme of hexokinase found in liver and pancreatic β-islet cells. In the liver, its function is to permit rapid uptake and metabolism of glucose when the concentration of glucose in the portal blood is high after a meal. In the pancreas, the increased uptake and metabolism of glucose caused by glucokinase acts as the signal for insulin release. Children with a genetic lack of glucokinase suffer from what has been termed maturity-onset diabetes of the young (MODY); although they can synthesize and secrete normal basal amounts of insulin, they are unable to secrete additional insulin in response to glucose (Froguel et al., 1993). Presumably as a result of increased activity of glucokinase, high doses of biotin have a hypoglycemic effect in insulin-dependent diabetic patients. In non-insulin-dependent spontaneously diabetic mice, the administration of 2 mg of biotin per kg of body weight (considerably in excess of vitamin requirements) lowers blood glucose and improves both oral glucose tolerance and the blood glucose response to insulin (Reddi et al., 1988).

Hyperammonemia occurs in biotin deficiency and the functional deficiency associated with lack of holocarboxylase synthetase (Section 11.2.2.1) and biotinidase (Section 11.2.3.1). In deficient rats, the activity of ornithine carbamyltransferase is two-thirds of that in control animals, as a result of decreased gene expression, although the activities of other urea cycle enzymes are unaffected (Maeda et al., 1996).

In addition to induction of specific proteins, the administration of biotin to deficient rats results in an overall two-fold stimulation of the incorporation of amino acids into proteins. The synthesis of serum albumin in liver is increased two-fold, but at least 10 other proteins show increases in amino incorporation of about five-fold, and some show an eight-fold increase, whereas others show no change (Dakshinamurti and Litvak, 1970; Boeckx and Dakshinamurti, 1974).

11.2.5 Biotin in Regulation of the Cell Cycle

Biotin is essential for cell proliferation. Peripheral blood mononuclear cells appear to take up biotin by a system that is distinct from the sodium-dependent multivitamin transporter that is responsible for intestinal and renal uptake of biotin (Section 11.1). In response to mitogenic stimuli the uptake of biotin increases several-fold, with no change in the activity of the sodium-dependent transporter. At the same time, there is an increase in the rate of expression of methylcrotonyl CoA, propionyl CoA carboxylases, and holocarboxylase synthetase, suggesting that much of the increased biotin is used for increased carboxylation reactions. However, there is also an increase in the biotinylation of histones compared with quiescent cells, suggesting that biotin has a role in regulation of the cell cycle (Zempleni and Mock, 2000b, 2001; Stanley et al., 2001; Zempleni et al., 2001).

11.3 BIOTIN DEFICIENCY

The few early reports of human biotin deficiency all concerned people who consumed large amounts of uncooked eggs and therefore had a high intake of avidin, which binds biotin and renders it unavailable (Section 11.6). Provision of biotin supplements of between 200 to 1000 μg per day cured the skin lesions despite continuing the abnormal diet providing large amounts of avidin. Unfortunately, there seems to have been no studies of provision of modest doses of biotin to such patients, and none in which their high intake of uncooked eggs was either not replaced by an equivalent intake of cooked eggs (in which avidin has been denatured by heat and the yolks of which are a good source of biotin) or continued unchanged. Thus, there is no information from these case reports on the amounts of biotin that are required for normal health.

More recently, similar signs of biotin deficiency have been observed in patients receiving total parenteral nutrition for prolonged periods, after major resection of the gut. The signs resolve after the provision of biotin, but again there have been no studies of the amounts of biotin required; intakes have ranged between 60 to 200 μg per day (Mock et al., 1985).

Biotin deficiency, and the functional deficiency associated with lack of holo-carboxylase synthetase (Section 11.2.2.1), or biotinidase (Section 11.2.3.1), causes alopecia (hair loss) and a scaly erythematous dermatitis, especially around the body orifices. The dermatitis is similar to that seen in zinc and essential fatty acid deficiency, and is commonly associated with *Candida albicans* infection. Histology of the skin shows an absence of sebaceous glands and atrophy of the hair follicles. The dermatitis is because of impaired metabolism of polyunsaturated fatty acids as a result of low activity of acetyl CoA carboxylase (Section 11.2.1.1). In biotin-deficient experimental animals, provision of supplements of long-chain ω6 polyunsaturated fatty acids prevents the development of skin lesions (Mock et al., 1988a, 1988b; Mock, 1991).

In biotin-deficient rats, the total fatty acid content of the skin is about one-third of normal, and contains a lower than normal proportion of C16 and C18 saturated and unsaturated fatty acids, and a higher than normal proportion of very long-chain fatty acids (especially C24:1 and C26:1). There are also increased amounts of odd-chain fatty acids (C15:0 to C29:0), reflecting impaired activity of propionyl CoA carboxylase (Section 11.2.1.3) and incorporation of propionyl CoA into fatty acids in competition with acetyl CoA (Proud et al., 1990).

11.3.1 Metabolic Consequences of Biotin Deficiency

The activities of biotin-dependent carboxylases fall in deficiency, resulting in impaired gluconeogenesis, with accumulation of lactate, pyruvate, and alanine, and impaired lipogenesis, with accumulation of acetyl CoA, resulting in ketosis. There are also changes in the fatty acid composition of membrane lipids. A variety of abnormal organic acids are excreted by both biotin-deficient patients and experimental animals (as shown in Table 11.1).

There is accumulation of the apoenzymes of biotin-dependent carboxylases in deficiency. Response to repletion is rapid, as a result of activation of the apoenzymes; activation of biotin-dependent apoenzymes in vitro may provide an index of status (Section 11.4).

11.3.1.1 Glucose Homeostasis in Biotin Deficiency
The impairment of pyruvate carboxylase in biotin deficiency results in impaired gluconeogenesis. Additionally, biotin deficiency results in a lowering of the NADH:NAD ratio and further reduction of gluconeogenesis by impairment of glyceraldehyde-3-phosphate dehydrogenase activity. This impairment of gluconeogenesis may result in fatal hypoglycemia in marginally biotin-deficient chicks subjected to a relatively minor metabolic stress (Section 11.3.2).

Rather than the expected hypoglycemia, biotin deficiency may sometimes be associated with hyperglycemia, because of reduced activity of glucokinase. As discussed in Section 11.2.4, this results in both decreased clearance of glucose by the liver and also decreased secretion of insulin in response to hyperglycemia. In streptozotocin diabetic animals, the administration of biotin improves glucose tolerance as a result of the induction of glucokinase (Zhang et al., 1997).

11.3.1.2 Fatty Liver and Kidney Syndrome in Biotin-Deficient Chicks
Birds are especially sensitive to biotin deficiency, at least partly because their intestinal flora make little or no contribution to biotin intake. This is of considerable commercial importance with intensively reared poultry. In adult birds, biotin deficiency does not affect egg production, but does reduce the amount of biotin in the eggs, thus impairing embryonic development. In severe deficiency, the hatchability of the eggs can fall to near zero.

In young chicks, biotin deficiency is associated with the fatal fatty liver and kidney syndrome. Apparently healthy chicks 3 to 5 weeks old become lethargic, then sink onto the sternum and become motionless, dying within 6 to 10 hours of the onset of the condition. Postmortem examination shows enlarged liver and kidneys, with extensive fatty infiltration, but none of the classical skin and feather signs of biotin deficiency. The syndrome can be induced with only a moderate degree of biotin deficiency if the birds are maintained on a high-carbohydrate, low-fat and low-protein diet; a mild stress, such as short-term fasting, will then induce the syndrome in up to 20% of the birds. Supplementing the diet with biotin prevents the problem.

Gluconeogenesis is severely impaired in birds suffering from the fatty liver and kidney syndrome; the administration of biotin rapidly restores gluconeogenesis to normal, by activating apopyruvate carboxylase. The affected animals also have impaired glucose 6-phosphatase and phosphoenolpyruvate carboxykinase activity, but increased hepatic activity of acetyl CoA carboxylase and malate dehydrogenase, with increased desaturation of long-chain fatty acids.

The problem is thus obviously not simply one of biotin deficiency, although supplementary biotin will alleviate the condition.

Birds fed the high-carbohydrate, low-fat, low-protein diet show more marked hypoglycemia on fasting than do controls, and modest hyperglycemia on refeeding. The cause of death in response to modest stress is believed to be acute hypoglycemia because of the impairment of hepatic gluconeogenesis; birds fed the same diet that do not succumb are believed to have a compensatory increase in renal gluconeogenesis, and hence are more resistant to the effects of food deprivation (Bannister, 1976a, 1976b; Whitehead et al., 1976).

11.3.1.3 **Cot Death** Cot death, or Sudden Infant Death Syndrome, when an apparently healthy child dies suddenly, and from no apparent cause, has some similarities with the fatty liver and kidney syndrome in birds. It has been suggested that it may result from marginal biotin deficiency, together with a precipitating metabolic stress.

There is circumstantial evidence to support this suggestion, because the liver content of biotin is lower in infants who have died from cot death than in infants who have died from known causes. By parallel with the fatty liver and kidney syndrome, it has been suggested that a modest metabolic stress, such as a mild fever, causes a higher requirement for gluconeogenesis than can be met, resulting in acute hypoglycemia. There are rapid postmortem changes in blood and tissue glucose, so it is unlikely that there can be any direct evidence to support this suggestion (Johnson et al., 1980; Heard et al., 1983).

11.3.2 Biotin Deficiency In Pregnancy

Biotin deficiency in experimental animals is teratogenic, and a number of the resultant birth defects resemble human birth defects. Up to half of pregnant women have elevated excretion of 3-hydroxy-isovaleric acid (Section 11.4), which responds to supplements of biotin, in the first trimester, suggesting that marginal status may be common in early pregnancy and may be a factor in the etiology of some birth defects. This may be the result of increased catabolism of biotin as a result of steroid induction of biotin catabolic enzymes; there is increased excretion of bisnorbiotin and biotin sulfoxide (Zempleni and Mock, 2000a; Mock et al., 2002).

11.4 ASSESSMENT OF BIOTIN NUTRITIONAL STATUS

The plasma concentration of the biotin does not provide a sensitive index of status, at least partly because there is increased renal reabsorption of the vitamin as intake falls. Urinary excretion of biotin and its metabolites is more sensitive, but may be confounded by changes in biotin excretion caused by glucocorticoid hormones (McMahon, 2002). There are three sensitive markers of status (Mock, 1999):

1. The activity of propionyl CoA carboxylase in lymphocytes falls, and the activation of the apoenzyme on incubation with biotin rises, in patients receiving total parenteral nutrition before there is any change in the plasma concentration of biotin (Velazquez et al., 1990). In experimental animals, the activity of lymphocyte propionyl CoA carboxylase falls early during biotin depletion, at the same time as the activity of the hepatic enzyme. There is not the expected increase in urinary excretion of hydroxypropionic acid, presumably because propionyl CoA carboxylase is not rate-limiting for propionate metabolism (Mock and Mock, 2002).

2. Reduced activity of methylcrotonyl CoA carboxylase (Section 11.2.1.4) results in the formation and excretion of 3-hydroxy-isovaleric acid; in experimental biotin depletion, significant amounts of 3-hydroxy-isovaleric acid are excreted at the same time as the excretion of biotin and bisnorbiotin falls, before there is any change in the plasma concentration of biotin (Mock et al., 1997).

3. As a result of impaired activity of acetyl CoA and propionyl CoA carboxylases, there are changes in the fatty acid composition of lipids in the lymphocytes of biotin-deficient rats. There is an increase in the proportion of long-chain fatty acids (C22:0 to C30:0) and odd-carbon fatty acids (C15:0 to C29:0), with a decrease in the proportion of unsaturated fatty acids and the ratio of *cis*-vaccenic acid (C18:1ω9): palmitoleic acid (C16:1ω6), which is indicative of impaired elongation and desaturation of fatty acids (Liu et al., 1994).

11.5 BIOTIN REQUIREMENTS

It is apparent from the discussion in Section 11.3 that there is little information concerning human biotin requirements and no evidence on which to base recommendations. Average intakes of biotin range between 15 to 70 μg per day. Such intakes are obviously adequate to prevent deficiency, and the safe and adequate range of biotin intakes is set at 10 to 200 μg per day (Department of Health, 1991; Scientific Committee for Food, 1993). The U.S./Canadian adequate intake for adults is 30 μg per day (Institute of Medicine, 1998).

On the basis of studies in patients who developed deficiency during total parenteral nutrition, and who are therefore presumably wholly reliant on an exogenous source of the vitamin – with no significant contribution from intestinal bacterial synthesis – the provision of 60 μg of biotin per day for adults receiving total parenteral nutrition is generally recommended (Bitsch et al., 1985).

11.6 AVIDIN

The original interest in avidin was because of the egg white injury that was subsequently shown to be avidin-induced biotin deficiency. Thereafter, avidin was used because of its high affinity for biotin (a dissociation constant of 10^{-15} mol per L), not only to induce experimental biotin deficiency, but also to bind to biotin in isolated enzymes and thus, by irreversible inhibition, demonstrate the coenzyme role of biotin. Because of the stability of the avidin–biotin complex, it has not been possible to use immobilized avidin as a means of purifying biotin enzymes – there seems to be no way in which the enzyme can be released from avidin binding. Because of its high affinity for biotin, avidin is used to provide an extremely sensitive system for linking reporter molecules in a variety of analytical systems.

Avidin has been found in the eggs and oviducts of many species of birds and in the egg jelly of frogs, but not in other tissues and not in the mammalian oviduct. It accounts for 0.05% of the total proteins of egg white. Avidin is synthesized in the goblet cells of the epithelium of the oviduct, whereas the other egg white proteins are synthesized in the underlying tubular gland cells. Its synthesis is induced by progesterone.

Avidin is a strongly basic glycoprotein; 10% of the molecular weight is carbohydrate – mannose and N-acetyl glucosamine linked to asparagine, with a high degree of heterogeneity in the sequence of the carbohydrate residues. These carbohydrate residues are not essential for biotin binding. Commercially available avidin consists of a mixture of glycosylated and unglycosylated forms that can be separated electrophoretically or on concanavalin A columns, but that cannot be distinguished on the basis of their biotin binding.

A closely similar protein, streptavidin, has been isolated from culture filtrates of several species of *Streptomyces*. Unlike avidin, streptavidin is not glycosylated and has an acidic isoelectric point. It binds biotin with a similarly high affinity.

Avidin is a tetrameric protein and binds 4 mol of biotin per tetramer; it also binds N-carboxybiotin with a somewhat lower affinity. The unit of avidin activity is that amount which will bind 1 μg (4.09 nmol) of biotin; commercially available avidin has an activity of 10 to 15 units per mg of protein.

The carboxyl group of the side chain of biotin is not essential for binding, and enzyme-bound biocytin will also bind to avidin. Binding is by hydrogen bonding to a hydrophobic pocket formed by two tryptophan residues at positions 70 and 110 in the peptide sequence. Adjacent to each of these tryptophan residues is a lysine that is also essential for biotin binding; there are similar conserved tryptophan-lysine sequences in streptavidin (Gitlin et al., 1988a, 1988b).

The physiological role of avidin in egg white is unknown. It is unlikely to act as a storage form of biotin, because most of the biotin of eggs is in the yolk, not the white, and most avidin occurs as the free glycoprotein, without biotin. Furthermore, biotin bound to avidin in egg white is not available to the developing chick embryo. Egg white contains 3 to 10 times more avidin than would be required to complex all the biotin in the yolk; feeding experimental animals on diets based on whole dried egg results in the development of biotin deficiency signs, despite the high biotin content of the yolk (White et al., 1992).

In *Streptomyces*, it is assumed that streptavidin has an antibiotic role; it is secreted together with a low molecular weight inhibitor of biotin synthesis, stravidin. It has been suggested that avidin in eggs has a similar role, to protect the developing embryo from (biotin-requiring) bacteria that penetrate the shell. Alternatively, because cells in culture can take up and utilize avidin–biotin, it has been suggested that the physiological role of avidin may be to facilitate the uptake of biotin by the developing embryo (Board and Fuller, 1974; Dakshinamurti et al., 1985; Bush and White, 1989).

Both avidin and the avidin–biotin complex are very stable to heat. To release biotin from avidin binding, autoclaving above 130°C is required, and free avidin is stable up to about 85°C. Avidin is also resistant to proteolysis and, as is obvious from the use of raw egg white diets to induce biotin deficiency, biotin cannot be released from avidin binding in the gastrointestinal tract. Lysosomal hydrolases do release biotin from avidin binding, and intravenously administered avidin–biotin can be a source of biotin.

Because the side chain carboxyl group of biotin is not required for avidin binding, avidin will recognize and bind biotin esterified to proteins and other molecules. This is the basis of a variety of highly sensitive analytical systems (Airenne et al., 1999). Biotin can be attached to antibodies and other ligand binding proteins, group-specific reagents to permit detection of amino acids in proteins, carbohydrates, or functional groups in DNA and RNA. This creates a biotinylated probe, with each biotin residue binding to avidin. The avidin can be labeled with a colored, fluorescent, chemiluminescent, or electron-dense group, thus permitting ready detection, or the avidin may be linked to an enzyme as a reporter molecule, thus permitting further amplification. An alternative approach is to react the avidin–biotinylated probe complex with a biotinylated reported molecule that binds to the free sites of the avidin tetramer. Such assay systems have sensitivity equal to, or better than, conventional radioligand binding assays. The carbohydrate groups of avidin result in some nonspecific binding, thus giving an undesirably high background in some systems. The high isoelectric point of avidin also causes problems with some systems. Both of the problems are overcome if bacterial streptavidin is used rather than egg white avidin, and a number of genetically modified variants of streptavidin that have been designed for specific functions are available (Sano et al., 1998; Stayton et al., 1999).

FURTHER READING

Baumgartner ER and Suormala T (1997) Multiple carboxylase deficiency: inherited and acquired disorders of biotin metabolism. *International Journal of Vitamin and Nutrition Research* **67**, 377–84.

Baumgartner ER and Suormala T (1999) Inherited defects of biotin metabolism. *Biofactors* **10**, 287–90.

Dakshinamurti K and Chauhan J (1988) Regulation of biotin enzymes. *Annual Reviews of Nutrition* **8**, 211–33.

Dakshinamurti K and Chauhan J (1989) Biotin. *Vitamins and Hormones* **45**, 337–84.

Dakshinamurti K and Chauhan J (1994) Biotin-binding proteins. In *Vitamin Receptors: Vitamins as Ligands in Cell Communication*, K Dakshinamurti (ed.), pp. 200–49. Cambridge, UK: Cambridge University Press.

Hommes FA (1986) Biotin. *World Review of Nutrition and Dietetics* **48**, 34–84.

Hymes J and Wolf B (1996) Biotinidase and its roles in biotin metabolism. *Clinica Chimica Acta* **255**, 1–11.

Knowles JR (1989) The mechanism of biotin-dependent enzymes. *Annual Reviews of Biochemistry* **58**, 195–221.

McMahon RJ (2002) Biotin in metabolism and molecular biology. *Annual Reviews of Nutrition* **22**, 221–39.

Roth KS (1981) Biotin in clinical medicine – a review. *American Journal of Clinical Nutrition* **34**, 1967–74.

Various authors (1999) Symposium proceedings: nutrition, biochemistry and molecular biology of biotin. *Journal of Nutrition* **129**, 476s–503s.

Wolf B and Heard GS (1991) Biotinidase deficiency. *Advances in Pediatrics* **38**, 1–21.

Zempleni J and Mock D (2001) Biotin homeostasis during the cell cycle. *Nutrition Research Reviews* **14**, 45–63.

References cited in the text are listed in the Bibliography.

TWELVE

Pantothenic Acid

Pantothenic acid has a central role in energy-yielding metabolism as the functional moiety of coenzyme A (CoA), in the biosynthesis of fatty acids as the prosthetic group of acyl carrier protein, and through its role in CoA in the mitochondrial elongation of fatty acids; the biosynthesis of steroids, porphyrins, and acetylcholine; and other acyl transfer reactions, including postsynthetic acylation of proteins. Perhaps 4% of all known enzymes utilize CoA derivatives. CoA is also bound by disulfide links to protein cysteine residues in sporulating bacteria, where it may be involved with heat resistance of the spores, and in mitochondrial proteins, where it seems to be involved in the assembly of active cytochrome c oxidase and ATP synthetase complexes.

Pantothenic acid is widely distributed in all foodstuffs. The name is derived from the Greek for *from everywhere*, as opposed to other vitamins that were originally isolated from individual rich sources.

Deficiency is well documented in chickens, which develop a pantothenic acid-responsive dermatitis. Other experimental animals show a variety of abnormalities from pantothenic acid deficiency. In human beings dietary deficiency has not been reliably documented, although it has been implicated in the burning foot syndrome (nutritional melalgia). Subjects maintained on pantothenic acid-deficient diets or given the antagonist ω-methyl pantothenate develop relatively unspecific symptoms that respond to repletion with the vitamin.

12.1 PANTOTHENIC ACID VITAMERS

The only naturally occurring vitamer of pantothenic acid is the D-isomer (as shown in Figure 12.1). It is the peptide of pantoic acid and β-alanine.

Free pantothenic acid and its sodium salt are chemically unstable, and therefore the usual pharmacological preparation is the calcium salt (calcium

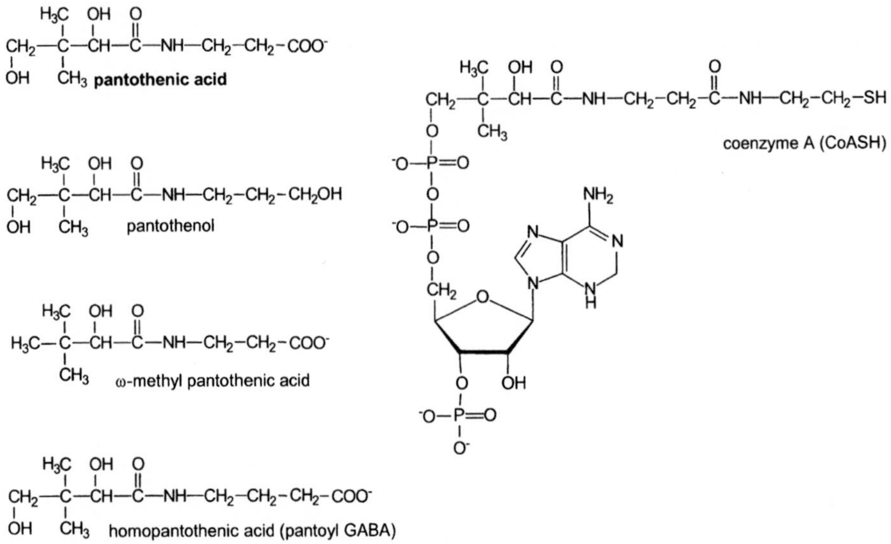

Figure 12.1. Pantothenic acid and related compounds and coenzyme A. Relative molecular masses (M_r): pantothenic acid, 219.2 (calcium dipantothenate, 476.5); pantothenol, 214.2; ω-methyl pantothenic acid, 213.6; homopantothenic acid, 233.2; and coenzyme A, 767.6. CoASH, free coenzyme A; GABA, γ-aminobutyric acid.

dipantothenate). The alcohol, pantothenol, is a synthetic compound that has biological activity because it is oxidized to pantothenic acid in vivo.

ω-Methyl pantothenic acid is a potent antagonist of the vitamin that has been used in studies of pantothenic acid deficiency, and the γ-aminobutyric acid (GABA) peptide of pantoic acid, pantoyl GABA or homopantothenic acid, has pharmacological actions in cholinergic neurotransmission and has been used in the treatment of Alzheimer's disease.

12.2 METABOLISM OF PANTOTHENIC ACID

About 85% of dietary pantothenic acid is as CoA or phosphopantetheine. In the intestinal lumen, these undergo hydrolysis to phosphopantetheine, then pantetheine (see Figure 12.2). Intestinal mucosal cells have a high pantetheinase activity and rapidly hydrolyze pantetheine to yield free pantothenic acid.

The intestinal absorption of pantothenic acid is by use of the same sodium-dependent carrier as biotin and lipoic acid (Section 11.1). The carrier is found throughout the intestinal tract, and therefore pantothenic acid synthesized by intestinal bacteria (Section 12.2.4) will, like biotin, be available for absorption (Said et al., 1998; Chatterjee et al., 1999; Ramaswamy, 1999; Said, 1999; Prasad

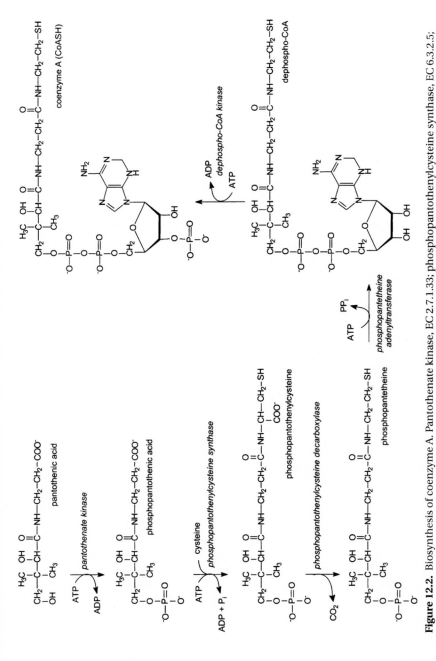

Figure 12.2. Biosynthesis of coenzyme A. Pantothenate kinase, EC 2.7.1.33; phosphopantothenylcysteine synthase, EC 6.3.2.5; phosphopantothenylcysteine decarboxylase, EC 4.1.1.36; phosphopantetheine adenyltransferase, EC 2.7.7.3; and dephospho-CoA kinase, EC 2.7.1.24. CoASH, free coenzyme A.

347

and Ganapathy, 2000). Other tissues take up pantothenic acid from the circulation by the same mechanism. The transport mechanism is not normally saturated, so pantothenate uptake into tissues will increase with plasma concentration.

The first step in pantothenic acid utilization is phosphorylation (see Figure 12.2). Pantothenate kinase is rate-limiting, so that, unlike many vitamins that are accumulated by metabolic trapping, there can be significant accumulation of free pantothenic acid in tissues. Intracellular concentrations may be as high as 200 to 500 μmol per L.

Red blood cells contain pantothenic acid, 4'-phosphopantothenic acid, and pantetheine. These seem to enter by diffusion, and their function is not known; unsurprisingly, because they contain no mitochondria, erythrocytes do not contain CoA (Annous and Song, 1995). The permeability of erythrocytes to pantothenate is normally relatively low, but in red cells infected with malaria parasites, the permeability is increased considerably; the vitamin is taken up and utilized by the parasites, which require CoA (Saliba et al., 1998).

Pantothenic acid is well conserved; over a week after the administration of tracer doses of [^{14}C]pantothenic acid to rats, less than 40% of the dose is recovered in the urine, all as the free vitamin. Pantothenic acid filtered by the kidneys is largely resorbed by a sodium-dependent system in the renal tubule brush border membrane (Barbarat and Podevin, 1986).

Pantothenic acid is largely excreted unchanged by mammals. Some phosphopantetheine may also be excreted in the urine; after administration of [^{14}C]pantothenic acid, some of the label may be recovered in exhaled CO_2. This is probably the result of intestinal bacterial metabolism, because many bacteria have pantothenase, a specific amidase that cleaves pantothenic acid to β-alanine and pantoic acid. *Pseudomonas* species are capable of using pantothenic acid as their sole carbon source.

12.2.1 The Formation of CoA from Pantothenic Acid

CoA functions as the carrier of fatty acids, as thioesters, in mitochondrial β-oxidation. The resultant two-carbon fragments, as acetyl CoA, then undergo oxidation in the citric acid cycle. CoA also functions as a carrier in the transfer of fatty acyl groups in a variety of biosynthetic and catabolic reactions, including steroidogenesis; long-chain fatty acid synthesis from palmitate in mitochondria and endoplasmic reticulum; monounsaturation of palmitoyl CoA to palmitoleyl CoA (C16:1 ω9) and stearyl CoA to oleyl CoA (C18:1 ω9); elongation of polyunsaturated fatty acids; acylation of serine, threonine, and cysteine residues on proteolipids and acetylation to form N-acetyl neuraminic acid.

All tissues are capable of forming CoA from pantothenic acid, by the pathway shown in Figure 12.2 (Tahiliani and Beinlich, 1991; Begley et al., 2001). The first three enzymes catalyzing the formation of phosphopantetheine from pantothenic acid are found only in the cytosol. Although phosphopantetheine crosses the mitochondrial inner membrane, CoA does not, but must be synthesized in situ.

The first step is phosphorylation to 4′-phosphopantothenic acid; the activity of pantothenate kinase is rate-limiting for CoA synthesis. There are two human genes for pantothenate kinase; genetic lack of the more recently discovered gene leads to an autosomal recessive neurodegenerative disease – the Hallevorden–Spatz syndrome. It is not clear how lack of pantothenate kinase leads to the accumulation of iron in the basal ganglia that is the underlying cause of the pathology (Zhou et al., 2001).

Pantothenol, the alcohol of pantothenic acid (see Figure 12.1), is frequently used in pharmaceutical preparations. Although it is a substrate for pantothenate kinase in vitro, it is more likely that it first undergoes oxidation to pantothenic acid, catalyzed by liver alcohol dehydrogenase, rather than phosphorylation to phosphopantothenol followed by oxidation.

Phosphopantothenic acid reacts with cysteine, forming 4′-phosphopantothenyl cysteine, which is decarboxylated to 4′-phosphopantetheine in a flavin-dependent reaction. In most bacteria, phosphopantetheinyl cysteine synthase and decarboxylase occur as a single bifunctional enzyme, but the human enzymes occur as two separate proteins (Daugherty et al., 2002).

Phosphopantetheine undergoes adenylyl transfer from ATP to yield dephospho-CoA, which is then phosphorylated at the 3′ position of the ribose moiety to yield CoA. Phosphopantetheine adenylyltransferase and dephospho-CoA kinase activities occur in a single bifunctional enzyme, which is found in both cytosol and mitochondria. However, in addition to the bifunctional protein, human tissues also contain a separate dephospho-CoA kinase (Begley et al., 2001; Zhyvoloup et al., 2002).

12.2.1.1 Metabolic Control of CoA Synthesis

Pantothenate kinase is rate-limiting for the synthesis of CoA, and both regulation of the activity of the existing enzyme protein and changes in its synthesis are important in the control of intracellular concentrations of CoA (Rock et al., 2000). The enzyme has a low K_m compared with the normal intracellular concentration of pantothenic acid and is thus insensitive to the availability of substrate, even in deficiency.

Short-chain fatty acyl CoA derivatives are inhibitors of pantothenate kinase; in perfused rat hearts, the addition of any of the major energy-yielding substrates (glucose, pyruvate, free fatty acids, or 3-hydroxybutyrate) to the perfusion medium results in inhibition of pantothenate kinase and thus a reduced rate of CoA synthesis (Robishaw et al., 1982).

Expression of the pantothenate kinase gene is induced by glucagon (which is secreted under conditions when there is an increased need for CoA for fatty acid oxidation) and repressed by insulin (Kirschbaum et al., 1990; Yun et al., 2000).

12.2.2 Catabolism of CoA

CoA undergoes dephosphorylation, catalyzed by lysosomal acid phosphatase, to dephospho-CoA, followed by pyrophosphatase action to release 4'-phosphopantetheine and 5'-AMP – the reverse of the final stages of CoA synthesis shown in Figure 12.2. CoA is also a substrate for direct pyrophosphatase action, at about 10% of the rate of action on dephospho-CoA. The pyrophosphatase seems to be a general nucleotide pyrophosphatase of plasma membrane rather than an enzyme specific for the degradation of CoA.

Phosphopantetheine, arising from either the catabolism of CoA or the inactivation of holo-acyl carrier protein (ACP), can be reutilized for CoA synthesis. Phosphopantetheine is a potent inhibitor of pantothenic acid kinase, the first step of de novo CoA synthesis.

Alternatively, phosphopantetheine is dephosphorylated, again by a relatively unspecific phosphatase. The resultant pantetheine is cleaved by pantetheinase, a specific amidase, to pantothenic acid and cysteamine. The resultant cysteamine may be an important precursor of taurine (Section 14.5.1). Pantetheinase is found in both the liver and kidneys. The kidney isoenzyme acts on both pantetheine and (at a lower rate) on phosphopantetheine, whereas the liver enzyme acts only on pantetheine (Dupre et al., 1973; Wittwer et al., 1983).

12.2.3 The Formation and Turnover of ACP

Although fatty acid β-oxidation is catalyzed by a series of intramitochondrial enzymes, and the fatty acyl chain is carried by CoA, fatty acid synthesis is catalyzed by a cytosolic–multienzyme complex in which the growing fatty acyl chain is bound by thioester linkage to an enzyme-bound 4'-phosphopantetheine residue. This component of the fatty acid synthetase complex is ACP.

Apo-ACP is activated by a transferase, holo-ACP synthetase, which transfers 4'-phosphopantetheine from CoA to the hydroxyl group of a serine residue in the apoprotein, releasing ADP. ACP is inactivated by a hydrolase that releases 4'-phosphopantetheine, which can be reutilized for CoA synthesis.

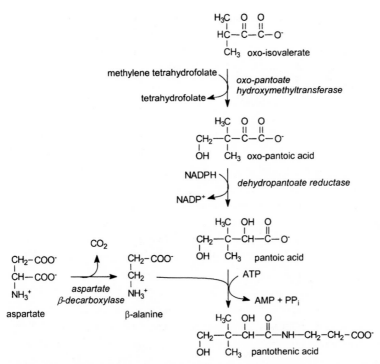

Figure 12.3. Biosynthesis of pantothenic acid. Oxo-pantoate hydroxymethyltransferase, EC 2.1.2.11; dehydropantoate reductase, EC 1.1.1.169; and aspartate β-decarboxylase, EC 4.1.1.12.

There is a rapid turnover of phosphopantetheine between ACP and CoA, in response to the metabolic state, and the need for fatty acid synthesis (and thus ACP in the fed state) or fatty acid β-oxidation (and thus CoA in the fasting state). Apo-ACP has a half-life of 6 to 7 days, whereas the prosthetic group turns over with a half-life of a few hours (Tweto and Larrabee, 1972; Volpe and Vagelos, 1973).

12.2.4 Biosynthesis of Pantothenic Acid

Plants and microorganisms are capable of the de novo synthesis of pantothenic acid from oxo-isovalerate and aspartate, by the pathway shown in Figure 12.3; animals are reliant on a preformed source of pantothenic acid.

Oxo-isovalerate may be formed by the transamination of valine; it is also the immediate precursor of valine biosynthesis and an intermediate in the synthesis of leucine (both are essential amino acids in mammals). Oxo-isovalerate undergoes a hydroxymethyl transfer reaction, in which the donor is methylene-tetrahydrofolate, yielding oxo-pantoic acid. The hydroxymethyl-transferase is subject to feedback inhibition by pantoic acid, pantothenic acid, and CoA. Oxo-pantoate is then reduced in an NADPH-dependent reaction to pantoic acid. The reductase is reversible, but the equilibrium lies greatly in favor of pantoic acid formation.

Aspartate undergoes β-decarboxylation to β-alanine; unlike most amino acid decarboxylases, aspartate decarboxylase is not pyridoxal phosphate-dependent, but has a catalytic pyruvate residue, derived by postsynthetic mod-ification of a serine residue (Section 9.8.1). Pantothenic acid results from the formation of a peptide bond between β-alanine and pantoic acid.

12.3 METABOLIC FUNCTIONS OF PANTOTHENIC ACID

The major functions of pantothenic acid are in CoA (Section 12.2.1) and as the prosthetic group for ACP in fatty acid synthesis (Section 12.2.3). In addition to its role in fatty acid oxidation, CoA is the major carrier of acyl groups for a wide variety of acyl transfer reactions. It is noteworthy that a wide variety of metabolic diseases in which there is defective metabolism of an acyl CoA derivative (e.g., the biotin-dependent carboxylase deficiencies; Sections 11.2.2.1 and 11.2.3.1), CoA is spared by formation and excretion of acyl carnitine derivatives, possibly to such an extent that the capacity to synthe-size carnitine is exceeded, resulting in functional carnitine deficiency (Section 14.1.2).

A variety of proteins are acylated by formation of thioesters to cysteine and esters to serine and threonine. Acylation may serve either to anchor the proteins in membranes (e.g., rhodopsin; Section 2.3.1) and the mannosidase of the Golgi, or to increase lipophilicity and thus enhance the solubilization of lipids being transported (e.g., the plasma apolipoproteins and milk globule proteins). Proteolipids with fatty acids esterified to threonine residues occur in the myelin sheath in nerves.

Several of the proteins of the Golgi transport system are N-acetylated at either the amino terminal or the ε-amino group of a lysine residue. Acylation may be either cotranslational or posttranslational. Amino terminal acylation protects the proteins from degradation, and various acylations are required for the assembly of multisubunit membrane proteins and transport of glyco-proteins through the Golgi.

Acetyl CoA is the donor for the 7- and 9-O-acetylation of sialic acids in the Golgi membrane. Neither free acetate nor acetyl CoA crosses the Golgi membrane, and the reaction appears to be a transmembrane process, with intermediate acetylation of a membrane component that then accumulates intravesicularly. The intermediate can transfer acetyl groups onto N-acetylneuraminic acid, but not other potential acetyl acceptors.

Acetyl CoA acetyltransferase, a key enzyme of ketogenesis, and 3-oxo-acyl CoA thiolase, involved in β-oxidation, bind CoA by formation of a disulfide bond to cysteine, a reaction that can be reversed by glutathione and other sulfhydryl reagents. The physiological significance of this reaction with CoA, which inactivates the enzymes, is not clear (Quandt and Huth, 1984, 1985; Schwerdt and Huth, 1993).

12.4 PANTOTHENIC ACID DEFICIENCY

Pantothenic acid is widely distributed in foods, and because it is absorbed throughout the small intestine, it is likely that intestinal bacterial synthesis also makes a contribution to pantothenic acid nutrition. As a result, deficiency has not been unequivocally reported in human beings except in specific depletion studies, which have also frequently used the antagonist ω-methyl pantothenic acid.

12.4.1 Pantothenic Acid Deficiency in Experimental Animals

Pantothenic acid deficiency in black and brown rats leads to a loss of fur color – at one time, pantothenic acid was known as the antigray hair factor. There is no evidence that the normal graying of hair with age is related to pantothenic acid nutrition, nor that pantothenic acid supplements have any effect on hair color.

In pantothenic acid-deficient rats, tissue CoA is depleted, affecting mainly the peroxisomal oxidation of fatty acids, which is mainly concerned with detoxication; mitochondrial β-oxidation, which is an essential energy-yielding pathway, is spared to a great extent (Youssef et al., 1997). However, relatively moderate deficiency in animals results in increased plasma triacylglycerol and nonesterified fatty acids, suggesting some impairment of lipid metabolism (Wittwer et al., 1990).

Rats on a pantothenic acid-free diet show rapid depletion of adrenal corticosteroids, and reduced production of the steroids in isolated adrenal glands in response to stimulation with adrenocorticotrophic hormone (ACTH). This presumably reflects the role of acetyl CoA in the synthesis of steroids; deficiency also results in atrophy of the seminiferous tubules of male rats and delayed sexual maturation in females. As deficiency progresses, there is enlargement, then congestion, and finally hemorrhage, of the adrenal cortex. In young animals, but not in adults, pantothenic acid deprivation eventually leads to necrosis of the adrenal cortex.

Deficient animals have an impaired ability to respond to metabolic and physical stress as a result of this decreased adrenocortical hormone synthesis, although this may be accompanied by enhanced sensitivity of target tissues to hormone action. Some strains of rat are susceptible to the development of duodenal ulcers in pantothenic acid deficiency. Ulceration can be prevented by adrenalectomy and is exacerbated by administration of glucocorticoid hormones.

Dogs develop severe and potentially fatal hypoglycemia in pantothenic acid deficiency – this responds to the administration of glucocorticoid hormones, suggesting that it is secondary to impairment of adrenal cortical function.

Deficient animals are also more susceptible to infection than are adequately nourished control animals, with impaired antibody responses. This seems to be due to a defect in the transport of proteins destined for export from the cell as a result of the impairment of acylation of Golgi proteins.

12.4.2 Human Pantothenic Acid Deficiency – The Burning Foot Syndrome

In the 1940s, prisoners of war in the Far East who were severely malnourished showed, among other signs and symptoms of vitamin deficiency diseases, a new condition of paresthesia and severe pain in the feet and toes, which was called the burning foot syndrome or nutritional melalgia. Although it was tentatively attributed to pantothenic acid deficiency, no specific trials of pantothenic acid were conducted; rather the subjects were given yeast extract and other rich sources of all vitamins as part of an urgent program of nutritional rehabilitation. There seem to be no reports of neurological damage in deficient animals which may explain the burning foot syndrome.

Experimental pantothenic acid depletion, sometimes together with the administration of ω-methyl pantothenic acid, results in the following signs and symptoms after 2 to 3 weeks:

1. Neuromotor disorders, including paresthesia of the hands and feet, hyperactive deep tendon reflexes and muscle weakness. These can be explained by the role of acetyl CoA in the synthesis of the neurotransmitter acetylcholine and the impaired formation of threonine acyl esters in myelin. Dysmyelination may explain the persistence and recurrence of neurological problems many years after nutritional rehabilitation in people who had suffered from burning foot syndrome.

2. Mental depression, which again may be related to either acetylcholine deficit or impaired myelin synthesis.
3. Gastrointestinal complaints, including severe vomiting and pain, with depressed gastric acid secretion in response to insulin and gastrin. As with the development of ulcers in deficient animals, this may reflect hypersensitivity to glucocorticoid stimulation.
4. Increased insulin sensitivity and a flattened glucose tolerance curve, which may reflect decreased antagonism by glucocorticoids.
5. Decreased serum cholesterol and decreased urinary excretion of 17-ketosteroids, reflecting the impairment of steroidogenesis.
6. Decreased acetylation of p-aminobenzoic acid, sulfonamides, and other drugs, thus reflecting reduced availability of acetyl CoA for these reactions.
7. Increased susceptibility to upper respiratory tract infections, which presumably reflects the impairment of immune responses.

12.5 ASSESSMENT OF PANTOTHENIC ACID NUTRITIONAL STATUS

Urinary excretion of pantothenic acid mirrors intake, albeit with wide range of individual variation, and may provide a means of assessing status. Urinary excretion of less than 1 mg (4.5 μmol) of pantothenic acid per 24 hours is considered to be abnormally low (Sauberlich et al., 1974).

Sauberlich (1974) suggested that a whole blood total pantothenic acid below 4.5 μmol per L was indicative of inadequate intake. However, few studies have reported mean blood concentrations of pantothenic acid as high as 4.5 μmol per L in normal subjects. Eissenstat and coworkers (1986) showed that serum or plasma free pantothenic acid was not a good index of nutritional status.

There are no functional tests of pantothenic acid nutritional status that are generally applicable. Deficiency of pantothenic acid impairs the ability to acetylate a variety of drugs, such as p-aminobenzoic acid, but this has not been developed as an index of vitamin status. The capacity to acetylate drugs is genetically determined; neither experimental pantothenate deficiency nor the administration of supplements affects the determination of fast or slow acetylator status (Pietrzik et al., 1975; Vas et al., 1990).

12.6 PANTOTHENIC ACID REQUIREMENTS

From the limited studies that have been performed it is not possible to establish requirements for pantothenic acid. Average intakes are between 2 to 7 mg per day. This is obviously adequate, because, as discussed previously, deficiency is unknown under normal conditions. The U.S./Canadian adequate intake for adults is 5 mg per day (Institute of Medicine, 1998).

12.7 PHARMACOLOGICAL USES OF PANTOTHENIC ACID

Fibroblasts in culture undergo faster proliferation and migration when the concentration of pantothenic acid is high, and this has led to the topical use of pantothenol in skin disorders and wound healing. There is no evidence that oral supplements have any effect on wound healing (Vaxman et al., 1995, 1996; Egger et al., 1999; Weimann and Hermann, 1999; Ebner et al., 2002).

Some of the side effects of valproate administration to young children to control seizures (ketosis and liver damage) are associated with sequestration of CoA as valproyl CoA, which is poorly metabolized, and the administration of pantothenate supplements (generally together with carnitine; Section 14.1) prevents depletion of CoA and reduces the risk of liver damage (Thurston and Hauhart, 1992). In the same way, pantothenate supplements protect mice against neural tube defects caused by valproate (Sato et al., 1995).

Homopantothenic acid (pantoyl-GABA or hopanthate; see Figure 12.1) has been reported to enhance cholinergic function in the central nervous system. It seems to act by binding to GABA receptors and stimulating the release of acetylcholine in the cerebral cortex and hippocampus, rather than by any direct effect on acetylcholine synthesis or cholinergic receptors. It appears to have some beneficial effect in Alzheimer's disease, reducing loss of memory and cognitive impairment in some patients (Nakahiro et al., 1985).

Pantothenic acid seems to have very low toxicity. Intakes of up to 10 g of calcium pantothenate per day (compared with a normal dietary intake of 2 to 7 mg per day) have been given for up to 6 weeks, with no apparent ill effects.

FURTHER READING

Begley TP, Kinsland C, and Strauss E (2001) The biosynthesis of coenzyme A in bacteria. *Vitamins and Hormones* **61**, 157–71.

Plesofsky-Vig N and Brambl R (1988) Pantothenic acid and coenzyme A in cellular modification of proteins. *Annual Reviews of Nutrition* **8**, 461–82.

Tahiliani AG and Beinlich CJ (1991) Pantothenic acid in health and disease. *Vitamins and Hormones* **46**, 165–228.

References cited in the text are listed in the Bibliography.

THIRTEEN

Vitamin C (Ascorbic Acid)

Vitamin C is a vitamin for only a limited number of vertebrate species: humans and the other primates, the guinea pig, bats, the passeriform birds, and most fishes. Most insects and invertebrates are also incapable of ascorbate synthesis. Ascorbate is synthesized as an intermediate in the gulonolactone pathway of glucose metabolism; in those vertebrate species for which ascorbate is a vitamin, one enzyme of the pathway, gulonolactone oxidase, is absent.

The vitamin C deficiency disease, scurvy, has been known for many centuries, and was described in the Ebers papyrus of 1500 B.C. and by Hippocrates. The Crusaders are said to have lost more men through scurvy than were killed in battle; in some of the long voyages of exploration of the fourteenth and fifteenth centuries, up to 90% of the crew died from scurvy. Cartier's expedition to Quebec in 1535 was struck by scurvy; the local native Americans taught him to use infusion of swamp spruce leaves to prevent or cure the condition. Recognition that scurvy was the result of a dietary deficiency came relatively early. James Lind demonstrated in 1757 that orange and lemon juice were protective, and Cook maintained his crew in good health during his circumnavigation of the globe (1772 to 1775) by stopping frequently to take on fresh fruit and vegetables. In 1804, the British Navy decreed a daily ration of lemon or lime juice for all ratings, a requirement that was extended to the merchant navy in 1865.

Ascorbic acid was isolated from cabbage, lemon juice, and adrenal glands by Szent-Györgi in 1928, and identified as the antiscorbutic factor by Waugh and King in 1932. Its structure was established by Haworth and coworkers in 1933, and the same year Haworth, in Birmingham, and Reichstein, in Switzerland, succeeded in synthesizing the vitamin.

Ascorbate is a reducing sugar; in addition to its specific role as cofactor for a variety of redox reactions, it also functions as a relatively nonspecific reducing agent. Some of these nonspecific reactions are physiologically important;

others have led to confusion in the literature because in vitro it will enhance the activity of a number of enzymes for which it is not a cofactor.

Ascorbate is synthesized in large amounts in plants. It can reach 20 to 300 mmol per L in chloroplasts, where its function is mainly to remove hydrogen peroxide formed during photosynthesis. Ascorbate-deficient mutant plants are especially sensitive to ozone- and ultraviolet-induced stress (Smirnoff, 2000).

13.1 VITAMIN C VITAMERS AND NOMENCLATURE

The physiologically important compound is L-ascorbic acid (Figure 13.1). It can undergo oxidation to the monodehydroascorbate free radical and onward to dehydroascorbate, both of which have vitamin activity because they can be reduced to ascorbate. Further oxidation in the presence of oxygen, and especially under alkaline conditions, or in the presence of transition metal ions that undergo reduction, results in the formation of diketogulonic acid (dioxogulonic acid; see Figure 13.2), which has no biological activity.

D-Iso-ascorbic acid (erythorbic acid; see Figure 13.1) also has vitamin activity. In vivo and in cell culture, it has only about 5% of the biological activity of ascorbate, but this seems to be from poor intestinal absorption and tissue uptake. In vitro with purified enzymes, it has the same cofactor activity as ascorbate. Although it is not a naturally occurring compound, erythorbic acid is widely used interchangeably with ascorbic acid, in cured meats and as an antioxidant in a variety of foods.

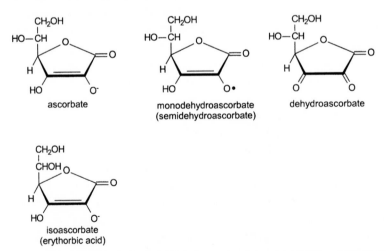

Figure 13.1. Vitamin C vitamers. Relative molecular masses (M_r): ascorbate and iso-ascorbate, 176.1; monodehydroascorbate, 175.1; and dehydroascorbate, 174.1.

Ascorbic acid 2-phosphate and triphosphate are more stable to atmospheric oxidation than ascorbate and are used in food processing. They have the same biological activity as ascorbic acid on a molar basis, because they are substrates for intestinal phosphatases. Ascorbic acid 2-sulfate is a metabolite of the vitamin in some species and has little or no biological activity.

Vitamin C is used in food processing as the free acid (E-300), the sodium (E-301) and calcium (E-302) salts, and as ascorbyl palmitate or stearate (E-304), a lipid-soluble antioxidant. The palmitate and stearate have low biological activity. Although most of the ascorbate used as a flour improver in bread making is destroyed in baking, a considerable number of other processed foods provide significant amounts of the vitamin because of its use as an antioxidant and in meat curing.

13.1.1 Assay of Vitamin C
Because it is a potent reducing agent, vitamin C is commonly determined by titrimetric or potentiometric redox methods. Such methods underestimate the amount of the vitamin present because dehydroascorbate – which has vitamin activity – is formed by atmospheric oxidation of ascorbate in the sample, especially under neutral conditions, and is not detected by redox assay methods.

Vitamin C can also be determined colorimetrically, after oxidation to dehydroascorbate, by reaction with dinitrophenylhydrazine. Under appropriate conditions, neither ascorbic acid itself nor potentially interfering sugars react with dinitrophenylhydrazine. However, diketogulonate, which has no vitamin activity, also reacts with dinitrophenylhydrazine under the same conditions. Unless diketogulonate is determined separately after reduction of dehydroascorbate to ascorbate, this method overestimates the vitamin.

These problems can be overcome by using more specific assay methods: either high-performance liquid chromatography or a fluorescence assay (Brubacher et al., 1985).

13.2 METABOLISM OF VITAMIN C
As shown in Figure 13.2, ascorbate is an intermediate in the gulonolactone pathway of glucuronic acid metabolism. In those species for which ascorbate is not a vitamin, this is a major pathway of glucuronic acid catabolism, and ascorbate is a metabolic intermediate whose rate of synthesis and turnover bear no relation to physiological requirements for ascorbate per se. In these species, rates of ascorbate synthesis and turnover range between 5 mg per kg of body weight per day (cats and dogs) and 30 to 40 mg per kg per day (goats, rats, and mice). Metabolic stress and the administration of xenobiotics

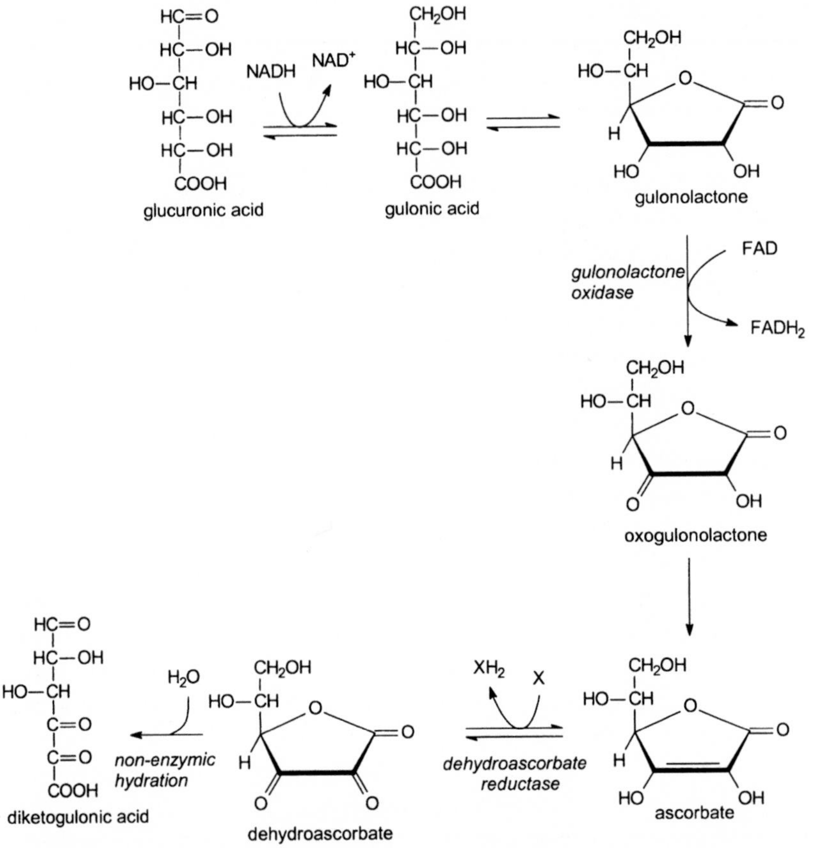

Figure 13.2. Biosynthesis of ascorbate. Glucuronate reductase, EC 1.1.1.19; glucono-lactone 3-lactonase, EC 3.1.1.17; gulonolactone oxidase, EC 1.1.3.8; NADPH-dependent dehydroascorbate reductase, EC 1.6.5.4; and glutathione-dependent dehydroascorbate reductase, EC 1.8.5.1.

(Section 13.3.8) can increase the rate of ascorbate turnover several-fold in species for which it is not a vitamin.

Species for which ascorbate is a vitamin lack gulonolactone oxidase, and metabolize gulonic acid by reduction and decarboxylation directly to xylulose. The loss of gulonolactone oxidase seems to be the result of nonexpression of the gene rather than a gene deletion (Sato and Udenfriend, 1978).

An autosomal recessive mutant strain of rat, which lacks gulonolactone oxidase and hence is unable to synthesize ascorbic acid, has been described (Mizushima et al., 1984). The animals have an osteogenic disorder akin to scurvy in human infants, and homozygotes are sterile. The addition of ascorbate to their diet restores normal growth and fertility, but because, like all species for which it is not normally a vitamin, they lack the intestinal active transport carrier for the ascorbate (Section 13.2.1), and they require relatively large amounts of the vitamin.

13.2.1 Intestinal Absorption and Secretion of Vitamin C

In rats and hamsters (for which ascorbate is not a vitamin), intestinal absorption is passive, whereas in guinea pigs and human beings there is sodium-dependent active transport of the vitamin at the brush border membrane, with sodium independent transport at the basolateral membrane, throughout the intestinal tract. Isolated guinea pig intestine can concentrate ascorbate up to five-fold, compared with the incubation medium, representing a considerable electrochemical gradient. Intestinal absorption of dehydroascorbate is carrier-mediated, linked to intracellular reduction to ascorbate before transport across the basolateral membrane (Malo and Wilson, 2000). Ascorbate is secreted in gastric juice, and the ratio of plasma:gastric juice ascorbate is 4 to 5:1 (Mowat and McColl, 2001).

Some 80% to 95% of dietary ascorbate is absorbed at intakes up to about 100 mg per day; the absorption of larger amounts of the vitamin is lower, falling from 50% of a 1.5-g dose to 25% of a 6-g dose, and 16% of a 12 g-dose (Rivers, 1987). Unabsorbed ascorbate from high doses is a substrate for intestinal bacterial metabolism.

13.2.2 Tissue Uptake of Vitamin C

Ascorbate and dehydroascorbate are taken up into tissues by separate mechanisms, and there is little or no competition between them (Welch et al., 1995):

1. Ascorbate enters cells by way of sodium-dependent transporters.
2. Dehydroascorbate enters cells by way of the (insulin-dependent) glucose transporters (GLUT), and is reduced to ascorbate intracellularly.

The relative importance of uptake of dehydroascorbate and dehydroascorbate by tissues is unclear. It has been suggested that normal physiological concentrations of glucose will inhibit uptake of dehydroascorbate (Liang et al., 2001). Functional signs of deficiency may develop in poorly controlled diabetes mellitus, despite an adequate intake and adequate plasma concentrations, suggesting that hyperglycemia and insulin insensitivity – and thus uptake of dehydroascorbate in competition with glucose – are important. Some of the adverse effects of poor glycemic control in diabetes mellitus (especially the development of cataract) may be related to this impairment of vitamin C uptake, and supplements of vitamin C may be beneficial (Cunningham, 1998a, 1998b).

With cells in culture, high concentrations of flavonoids (Section 14.7.2) inhibit the uptake of both ascorbate and dehydroascorbate, although it is not clear whether inhibitory concentrations of flavonoids occur in vivo (Park and Levine, 2000).

About 70% of blood ascorbate is in plasma and erythrocytes (which do not concentrate the vitamin from plasma). The remainder is in white cells, which have a marked ability to concentrate ascorbate: mononuclear leukocytes achieve 80-fold, platelets 40-fold, and granulocytes 25-fold concentration, compared with plasma concentration. In adequately nourished subjects, and those receiving supplements, the ascorbate concentration in erythrocytes, platelets, and granulocytes, but not in mononuclear leukocytes, is correlated with plasma concentration. Mononuclear leukocytes concentrate ascorbate independently of plasma concentration (Evans et al., 1982). In deficiency, as plasma concentrations of ascorbate fall, mononuclear leukocyte, granulocyte, and platelet concentrations of ascorbate are protected to a considerable extent. As discussed in Section 13.5.2, the leukocyte content of ascorbate is used as an index of vitamin C nutritional status, but in view of the differing capacity of different cell types to accumulate the vitamin, differential white cell counts are essential to interpret the results.

There is no specific storage organ for ascorbate; apart from leukocytes (which account for only 10% of total blood ascorbate), the only tissues showing a significant concentration of the vitamin are the adrenal and pituitary glands. Although the concentration of ascorbate in muscle is relatively low, skeletal muscle contains much of the body pool of 5 to 8.5 mmol (900 to 1,500 mg) of ascorbate.

13.2.3 Oxidation and Reduction of Ascorbate

As shown in Figure 13.3, oxidation of ascorbic acid, for example, by the reduction of superoxide to hydrogen peroxide or Fe^{3+} to Fe^{2+}, and similar reduction of other transition metal ions, proceeds by a one-electron process, forming the monodehydroascorbate radical. The radical rapidly disproportionates into ascorbate and dehydroascorbate. Most tissues also have both nicotinamide adenine dinucleotide phosphate (NADPH) and glutathione-dependent monodehydroascorbate reductases, which reduce the radical back to ascorbate. Ascorbate is thus an effective quencher of singlet oxygen and other radicals.

Dehydroascorbate is unstable in solution, undergoing hydrolytic ring opening to yield diketogulonic acid. However, in vivo, it is normally reduced to

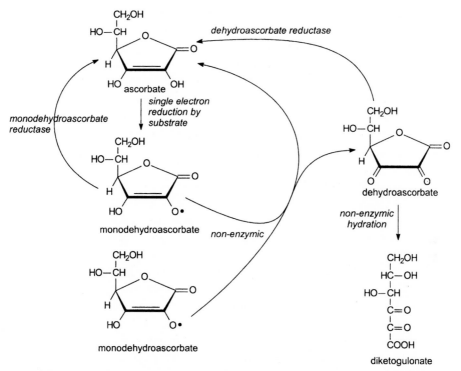

Figure 13.3. Redox reactions of ascorbate. Monodehydroascorbate reductase, EC 1.8.5.3; NADPH-dependent dehydroascorbate reductase, EC 1.6.5.4; and glutathione-dependent dehydroascorbate reductase, EC 1.8.5.1.

ascorbate by either NADPH or glutathione-dependent reductases. Dehydroascorbate may also be reduced by reaction with homocysteine, forming homocysteic acid; this may be an important source of homocysteic acid for the synthesis of phosphoadenosine phosphosulfate (PAPS) for sulfation reactions (McCully, 1971).

In plants, ascorbate oxidase reduces oxygen to water, in a series of four single-electron steps, forming monodehydroascorbate. This enzyme, and onward nonenzymic oxidation to diketogulonic acid, is responsible for the oxidative loss of much of the vitamin C in vegetables after harvesting. In animals, where the role of ascorbate seems to be mainly as a reducing agent, there is no specific ascorbate oxidase.

13.2.4 Metabolism and Excretion of Ascorbate

The major fate of ascorbic acid in human metabolism is excretion in the urine, either unchanged or as dehydroascorbate and diketogulonate. Both ascorbate and dehydroascorbate are filtered at the glomerulus, then reabsorbed, by a sodium-independent process. Reabsorbed dehydroascorbate is reduced to ascorbate in the kidneys. At plasma concentrations above about 85 μmol per L, the renal transport system is saturated, and ascorbate is excreted quantitatively with increasing intake.

Ascorbate catabolism is increased in subjects with iron overload, probably as a result of nonenzymic reactions with iron that is not protein-bound. The transferrin polymorphisms that are associated with susceptibility to iron overload result in higher vitamin C requirements for those subjects with high iron status (Kasvosve et al., 2002).

As shown in Figure 13.3, dehydroascorbate can undergo hydration to diketogulonate, followed by decarboxylation to xylose, thus providing a route for entry into central carbohydrate metabolic pathways via the pentose phosphate pathway. This is the major metabolic fate of ascorbate in those species for which it is not vitamin and also in the guinea pig. However, oxidation to carbon dioxide is only a minor fate of ascorbate in humans. At intakes up to about 100 mg per day, less than 1% of the radioactivity from [^{14}C]ascorbate is recovered as carbon dioxide. Although more $^{14}CO_2$ is recovered from subjects receiving high intakes of the vitamin, this may be the result of bacterial metabolism of unabsorbed vitamin in the intestinal lumen (Kallner et al., 1985).

Although a number of studies have suggested that high intakes of ascorbate lead to synthesis and excretion of oxalate (Section 13.6.5.1), this seems to be the result of nonenzymic formation of oxalate in urine samples after collection. There is no known pathway for oxalate synthesis from ascorbate.

Some species (but not primates) excrete ascorbate 2-sulfate, and in vitro ascorbic acid is a substrate for catechol O-methyltransferase, forming 2-methyl ascorbate.

13.3 METABOLIC FUNCTIONS OF VITAMIN C

Ascorbic acid has specific and well-defined roles in two classes of enzymes: the copper-containing hydroxylases (such as dopamine β-hydroxylase and peptidyl glycine hydroxylase) and the 2-oxoglutarate–linked iron-containing hydroxylases, of which the best studied are the proline and lysine hydroxylases involved in maturation of connective tissue (and other) proteins.

In addition to its coenzyme role in postsynthetic modification of collagen and other connective tissue proteins, there is evidence that vitamin C is involved in the regulation of connective tissue protein gene expression (Mahmoodian and Peterkofsky, 1999). The expression of a number of other genes has also been reported to be modulated by vitamin C, including

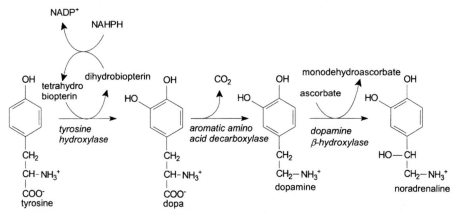

Figure 13.4. Synthesis of the catecholamines. Tyrosine hydroxylase, EC 1.14.16.2 (see also Figure 10.10); aromatic amino acid decarboxylase, EC 4.1.1.26; and dopamine β-hydroxylase, EC 1.14.17.1.

cytochromes P_{450}, and ubiquitin (Mizutani et al., 1997; Mori et al., 1997; Catani et al., 2001). The mechanism by which ascorbate affects gene expression is not clear, but may involve changes in the intracellular redox state (Lopez-Lluch et al., 2001).

Ascorbate also increases the activity of a number of other enzymes in vitro, although this is a nonspecific reducing action rather than reflecting any metabolic function of the vitamin. In addition, it has a number of relatively unspecific actions as a reducing agent and oxygen radical quencher. It is a potentially important antioxidant nutrient acting to recycle oxidized vitamin E. It also enhances absorption of inorganic iron and inhibits the formation of nitrosamines in the stomach.

13.3.1 Dopamine β-Hydroxylase

Dopamine β-hydroxylase is a copper-containing enzyme involved in the synthesis of the catecholamines noradrenaline and adrenaline from tyrosine in the adrenal medulla and central nervous system (see Figure 13.4). The active enzyme contains Cu^+, which is oxidized to Cu^{2+} during the hydroxylation of the substrate. Reduction back to Cu^+ specifically requires ascorbate, which is oxidized to monodehydroascorbate.

In intact adrenal medullary chromaffin cells in culture, the amount of noradrenaline formed is considerably greater than the amount of ascorbate present, and there is little or no detectable loss of ascorbate with enzyme action. However, in isolated chromaffin granules, there is stoichiometric oxidation of

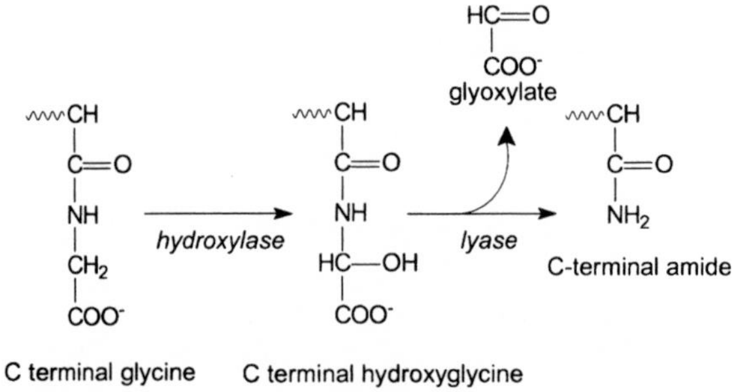

Figure 13.5. Reactions of peptidyl glycine hydroxylase (EC 1.14.17.3) and peptidyl hydroxyglycine α-amidating lyase (EC 4.3.2.5).

ascorbate and hydroxylation of dopamine. The small pool of ascorbate in the granules is maintained by a transmembrane electron transport system (probably involving cytochrome b_{561}), at the expense of the considerably larger cytosolic pool of ascorbate. In turn, cytosolic monodehydroascorbate is reduced by monodehydroascorbate reductase in the outer mitochondrial membrane. This is an NADH-dependent enzyme; therefore, overall, the tissue shows stoichiometric oxidation of NADH with dopamine hydroxylation, although it is ascorbate that is the immediate electron donor (Diliberto et al., 1982; Menniti et al., 1986).

13.3.2 Peptidyl Glycine Hydroxylase (Peptide α-Amidase)

More than half of the peptide hormones undergo postsynthetic modification to form a carboxy terminal amide, which is essential for biological activity. One function of this amidation is to render the peptides more hydrophobic and enhance receptor binding. The amide group is derived from a glycine residue that is to the carboxyl side of the amino acid which will become the amidated terminal of the mature peptide.

The initial reaction is proteolysis of a precursor peptide to leave a carboxy terminal glycine. This is hydroxylated on the α-carbon by peptidyl glycine hydroxylase, a copper-containing enzyme that has considerable sequence homology with dopamine β-hydroxylase, and also uses ascorbate as the reductant. A second enzyme, peptidyl hydroxyglycine α-amidating lyase, catalyzes cleavage of the hydroxyglycine to glyoxylate and the amide of the carboxy terminal amino acid, as shown in Figure 13.5. In animals, the two activities occur in a single bifunctional enzyme, although in invertebrates there are separate hydroxylase and lyase proteins (Prigge et al., 2000).

Table 13.1 Vitamin C-Dependent
2-Oxoglutarate–Linked Hydroxylases

Aspartate β-hydroxylase	EC 1.14.11.16
γ-Butyrobetaine hydroxylase	EC 1.14.11.1
p-Hydroxyphenylpyruvate hydroxylase	EC 1.14.11.27
Procollagen lysine hydroxylase	EC 1.14.11.4
Procollagen proline 3-hydroxylase	EC 1.14.11.7
Procollagen proline 4-hydroxylase	EC 1.14.11.2
Pyrimidine deoxynucleotide dioxygenase	EC 1.14.11.3
Thymidine dioxygenase	EC 1.14.11.10
Thymine dioxygenase	EC 1.14.11.6
Trimethyllysine hydroxylase	EC 1.14.11.8

13.3.3 2-Oxoglutarate–Linked Iron-Containing Hydroxylases

As shown in Table 13.1, a number of iron-containing hydroxylases share an unusual reaction mechanism in which hydroxylation of the substrate is linked to decarboxylation of 2-oxoglutarate. Proline and lysine hydroxylases are required for the postsynthetic modification of collagen, and proline hydroxylase also for the postsynthetic modification of osteocalcin (Section 5.3.3) and other proteins. Aspartate β-hydroxylase is required for the postsynthetic modification of protein C, the vitamin K-dependent protease that hydrolyzes activated Factor V in the blood clotting cascade (Section 5.3.2). Trimethyllysine and γ-butyrobetaine hydroxylases are required for the synthesis of carnitine (Section 14.1.1).

Procollagen proline 4-hydroxylase is the best studied of this class of enzymes; it is assumed that the others have essentially the same mechanism, although proline and lysine hydroxylases show very little sequence homology (Kivirikko and Pihlajaniemi, 1998). Although 3-hydroxyproline is found only in collagen, 4-hydroxyproline and hydroxylysine are found in a variety of other proteins, including the C1q component of complement, osteocalcin, macrophage receptor proteins, and a variety of transmembrane and intercellular proteins and proteins of the cytoskeleton, as well as some enzymes. 4-Hydroxyproline, but not hydroxylysine, also occurs in elastin.

In collagen, hydroxyproline stabilizes the triple helix structure by forming hydrogen bonds via water between adjacent chains or regions of the same chain. Hydroxylysine provides sites for glycosylation of proteins, and is essential for stabilization of intermolecular cross-links formed by reaction between lysine or hydroxylysine aldehyde and the ε-amino group of lysine or hydroxylysine.

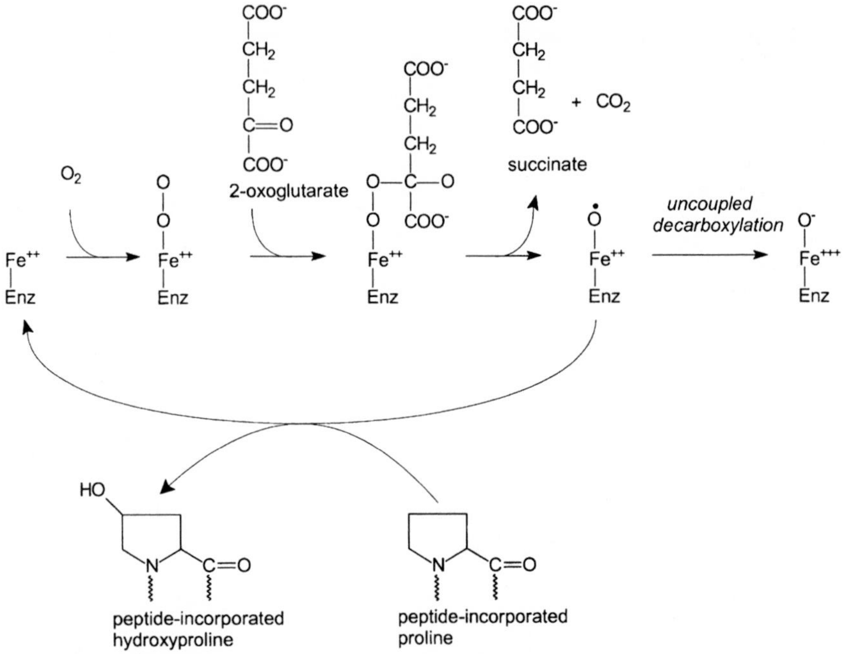

Figure 13.6. Reaction sequence of prolyl hydroxylase (EC 1.14.11.2). Enz, enzyme.

Proline and lysine hydroxylases are found in the lumen of rough endoplasmic reticulum. Hydroxylation of the peptide substrate occurs both cotranslationally and later as a postsynthetic modification. The enzymes act only on peptides and not on free amino acids.

As shown in Figure 13.6, the first step is binding of oxygen to the enzyme-bound iron, followed by attack on the 2-oxoglutarate substrate, resulting in decarboxylation to succinate, leaving a ferryl radical at the active site of the enzyme. This catalyzes the hydroxylation of proline, restoring the free iron to undergo further reaction with oxygen.

It has long been known that ascorbate is oxidized during the reaction, but not stoichiometrically with hydroxylation of proline and decarboxylation of 2-oxoglutarate. The purified enzyme is active in the absence of ascorbate; but, after 5 to 10 seconds (about 15 to 30 cycles of enzyme action), the rate of reaction begins to fall. The loss of activity is from a side reaction of the highly reactive ferryl radical in which the iron is oxidized to Fe^{3+}, which is catalytically inactive – so-called uncoupled decarboxylation of 2-oxoglutarate. Activity is only restored by ascorbate, which reduces the iron back to Fe^{2+} (Kivirikko and Pihlajaniemi, 1998).

13.3.4 Stimulation of Enzyme Activity by Ascorbate In Vitro

Over the years, a number of enzymes have been assumed to be ascorbate-dependent, because their activity is stimulated in vitro by the addition of ascorbate to the incubation medium. In general, these reactions are not ascorbate-dependent; ascorbate is one of a variety of reducing reagents that enhance the reaction.

Ascorbate is also frequently added to the incubation medium to remove hydrogen peroxide formed during a variety of reactions. Again, a number of reducing agents have the same action, and in vivo this role would presumably be performed by catalase.

In vitro, ascorbate and Fe^{2+} ions are frequently used as a source of superoxide for such enzymes as indoleamine dioxygenase. Although ascorbate does have prooxidant and superoxide generating activity (Section 13.3.7), there is no evidence that it is the physiological source of this radical for superoxide-utilizing enzymes.

There is a long-standing myth that ascorbate is required for the hydroxylation of tyrosine to dihydroxyphenylalanine (see Figure 13.4) and the similar reactions of phenylalanine and tryptophan hydroxylases. This belief arose as a result of early studies of a nonenzymic reaction to synthesize the hydroxylated amino acids for further study. It became established that ascorbate was required for these hydroxylations, and it is still common to include it in the incubation buffer. So far from requiring ascorbate, the addition of relatively low concentrations of ascorbate to preparations of tyrosine hydroxylase that has been activated by cAMP-dependent protein kinase results in irreversible loss of activity, although the unactivated form of the enzyme is unaffected by ascorbate (Wilgus and Roskoski, 1988). As discussed in Section 10.4.1, these enzymes are biopterin-dependent, and require dihydrobiopterin reductase and NADPH for activity. There is, however, evidence that, in some nerve cell lines in culture, tyrosine hydroxylase may be induced by ascorbate (Seitz et al., 1998).

13.3.5 The Role of Ascorbate in Iron Absorption and Metabolism

Nonheme iron is absorbed as Fe^{2+}, and not as Fe^{3+}; ascorbic acid in the intestinal lumen will both maintain iron in the reduced state and also chelate it, thus increasing absorption. A dose of 25 mg of vitamin C taken together with a semisynthetic meal increases the absorption of iron 65%, whereas a 1-g dose gives a nine-fold increase (Hallberg, 1982).

This is an effect of ascorbic acid present together with the test meal; neither intravenous administration of vitamin C nor supplements several hours before the test meal has any significant effect on iron absorption, although the ascorbate secreted in gastric juice should be effective (Mowat and McColl, 2001). A variety of other reducing agents, including alcohol and fructose, also enhance the absorption of inorganic iron.

Ascorbate is also active in the reduction of Fe^{3+} in the plasma transport protein, transferrin, to Fe^{2+} for storage in ferritin in the liver or for heme synthesis. It is not clear to what extent this represents specific actions of ascorbate, because other reducing reagents, including glutathione, also enhance heme synthesis, and the NADH-dependent flavoprotein ferriductase is the major factor controlling the transfer of iron between transferrin and ferritin.

13.3.6 Inhibition of Nitrosamine Formation by Ascorbate

In addition to dietary sources, a significant amount of nitrate is formed endogenously by the metabolism of nitric oxide – 1 mg per kg of body weight per day (about the same as the average dietary intake), increasing 20-fold in response to inflammation and immune stimulation. There is considerable secretion of nitrate in saliva, and up to 20% of this may be reduced to nitrite by oral bacteria. Under the acidic conditions of the stomach, nitrite can react with amines in foods to form carcinogenic N-nitrosamines, although it is not known to what extent this occurs in vivo.

Ascorbate reacts with nitrite forming NO, NO_2, and N_2, thus preventing the formation of nitrosamines. In addition to ascorbate in foods, there is considerable secretion of ascorbate in the gastric juice, and inhibition of gastric secretion for treatment of gastric ulcers, as well as reducing vitamin B_{12} absorption (Section 10.9.7), also inhibits this presumably protective gastric secretion of ascorbate (Mowat and McColl, 2001). As a result of secretion in gastric juice, the ratio of ascorbate in gastric juice to that in plasma is normally 4 to 5:1, but infection with *Helicobacter pylori* reduces this to 1:1. Loss of the protective effect against nitrosamine formation may be part of the mechanism by which *H. pylori* causes gastric cancer (Banerjee et al., 1994).

Although ascorbate can deplete nitrosating compounds under anaerobic conditions, the situation may be reversed in the presence of oxygen. Nitric oxide reacts with oxygen to form N_2O_3 and N_2O_4, both of which are nitrosating reagents, and can also react with ascorbate to form NO and monodehydroascorbate. It is thus possible for ascorbate to be depleted, with no significant effect on the total concentration of nitrosating species (Tannenbaum and Wishnok, 1987). There is, however, some evidence of a protective effect of vitamin C supplements against the development of gastric cancer (Feiz and Mobarhan, 2002).

13.3.7 Pro- and Antioxidant Roles of Ascorbate

Ascorbate can act as a radical-trapping antioxidant, reacting with superoxide and a proton to yield hydrogen peroxide, or with the hydroxy radical to yield water. In each case, the product is monodehydroascorbate.

The antioxidant activity of ascorbate is variable. From consideration of the chemistry involved, it would be expected that, overall, 2 moles of peroxyl radical would be trapped per mole of ascorbate, because of the reaction of 2 moles of monodehydroascorbate to regenerate ascorbate and yield dehydroascorbate (see Figure 13.3). However, as the concentration of ascorbate increases, so the molar ratio decreases, and it is only at very low concentrations of ascorbate that it tends toward the theoretical 2:1.

As well as its antioxidant role, ascorbate can be a source of hydroxyl and superoxide radicals. At high concentrations, it can reduce molecular oxygen to superoxide, being oxidized to monodehydroascorbate. At lower concentrations of ascorbate, both Fe^{3+} and Cu^{2+} ions are reduced by ascorbate, yielding monodehydroascorbate. Fe^{2+} and Cu^+ are readily reoxidized by reaction with hydrogen peroxide to yield hydroxide ions and hydroxyl radicals. Cu^+ also reacts with molecular oxygen to yield superoxide.

It seems likely that the prooxidant actions of ascorbate are of relatively little importance in vivo. Except in cases of iron overload, there are almost no transition metal ions in free solution. They are all bound to proteins, and because the renal transport system is readily saturated, plasma and tissue concentrations of ascorbate are unlikely to rise to a sufficient extent to lead to radical formation (Halliwell, 1996; Carr and Frei, 1999a).

13.3.7.1 Reduction of the Vitamin E Radical by Ascorbate
As discussed in Section 4.3.1, one of the major roles of vitamin E is as a radical-trapping antioxidant in membranes and lipoproteins. α-Tocopherol reacts with lipid peroxides forming the α-tocopheroxyl radical, which reacts with ascorbate in the aqueous phase, regenerating α-tocopherol, and forming monodehydroascorbate. Vitamin C may have a vitamin E-sparing antioxidant action, coupling lipophilic and hydrophilic reactions.

13.3.8 Ascorbic Acid in Xenobiotic and Cholesterol Metabolism

A number of xenobiotics – such as polychlorinated biphenyls, DDT, and aminopyrine – increase the urinary excretion and tissue concentrations of ascorbate in rats, and increase the incorporation of label from [^{14}C]glucose into ascorbate. The rate of ascorbate turnover can increase 5- to 10-fold under these conditions. Although this might be interpreted as suggesting a role for ascorbate in the metabolism of these compounds, it is more likely that it is a response to increased requirement for uridine diphosphate (UDP)-glucuronic acid for conjugation. The effect of the xenobiotics is to increase the activity of hepatic UDP-glucose dehydrogenase activity, with no change in gulonolactone oxidase. The same compounds also increase UDP-glucose dehydrogenase activity in the guinea pig. The increased formation of ascorbate is thus a result of increased availability of glucuronic acid; in rats, this excess glucuronic acid can then be catabolized by way of ascorbate, as shown in Figure 13.2 (Horio and Yoshida, 1982).

There is impairment of drug metabolism in ascorbate-deficient guinea pigs, which is normalized on repletion (Zannoni et al., 1972), possibly reflecting the effects of ascorbate on expression of cytochrome P_{450} (Mori et al., 1997). This may also account for the hypercholesterolemia and impaired synthesis of bile acids that is seen in vitamin C-deficient guinea pigs. Cholesterol 7-hydroxylase, the first enzyme of bile acid synthesis, is cytochrome P_{450}-dependent, and its activity is reduced in deficiency.

13.4 VITAMIN C DEFICIENCY – SCURVY

Although there is no specific site of vitamin C storage in the body, signs of deficiency do not develop until previously adequately nourished subjects have been deprived of the vitamin for 4 to 6 months, by which time plasma and tissue concentrations have fallen considerably.

The term scurvy is derived from the Italian *scorbutico*, meaning an irritable, neurotic, discontented, whining, and cranky person. The deficiency disease is certainly associated with listlessness and general malaise, and sometimes changes in personality and psychomotor performance and a lowering of the general level of arousal. The behavioral effects can presumably be attributed to impaired synthesis of catecholamines as a result of reduced activity of dopamine β-hydroxylase (Section 13.3.1).

Most of the other clinical signs of scurvy can be accounted for by effects of deficiency on collagen synthesis as a result of impaired proline and lysine hydroxylase activity (Section 13.3.3).

In general, the effects on collagen synthesis are more marked and more important than those of decreased formation of carnitine (as a result of impaired activity of trimethyllysine and γ-butyrobetaine hydroxylases; Section 14.1.1), impaired xenobiotic metabolism, or hypercholesterolemia (Section 13.3.8). However, depletion of muscle carnitine may account for the lassitude and fatigue that precede clinical signs of scurvy.

The earliest signs of scurvy in volunteers maintained on a vitamin C-free diet are skin changes, beginning with plugging of hair follicles by horny material, followed by enlargement of the hyperkeratotic follicles and petechial hemorrhage, with significant extravasation of red cells – presumably the result of increased fragility of blood capillaries from impaired collagen synthesis (Chatterjee, 1978).

Vascular fragility may also result from reduced sulfation of proteoglycans in connective tissue, as may also occur in hyperhomocysteinemia (Section 10.3.4.2). Dehydroascorbate catalyzes the oxidation of homocysteine to homocysteic acid, which is the precursor of PAPS, the sulfate donor for sulfation reactions (McCully, 1971).

At a later stage in deficiency, there is also hemorrhage of the gums, beginning in the interdental papillae, and progressing to generalized sponginess and bleeding of the gums. This is frequently accompanied by secondary bacterial infection and considerable withdrawal of the gum from the necks of the teeth. As the condition progresses, there is loss of dental cement, and the teeth become loose in the alveolar bone, and may be lost.

Wounds show only superficial healing in scurvy, with little or no formation of (collagen-rich) scar tissue, so that healing is delayed and wounds can readily be reopened. The scorbutic scar tissue has only about half the tensile strength of that normally formed.

Advanced scurvy is accompanied by intense pain in the bones, which can be attributed to changes in bone mineralization and demineralization as a result of abnormal collagen synthesis. Bone formation ceases and the existing bone becomes rarefied, so that the bones fracture with minimal trauma.

Some scorbutic patients develop chest pains, and acute cardiac emergency in response to exercise has been reported in some studies. Postmortem examination of patients and experimental animals shows thickening of the pericardium and accumulation of fluid in the pericardial cavity. Thrombosis may also occur, presumably because of hyperhomocysteinemia (Section 10.3.4.2), and hypercholesterolemia (Section 13.4.8) may also be a factor.

13.4.1 Anemia in Scurvy

Anemia is frequently associated with scurvy, and may be either macrocytic, indicative of folate deficiency, or hypochromic, indicative of iron deficiency.

Folate deficiency may be epiphenomenal, because the major dietary sources of folate are the same as those of ascorbate. However, some patients with clear megaloblastic anemia respond to the administration of vitamin C alone, suggesting that there may be a role of ascorbate in the maintenance of normal pools of reduced folates. There is no evidence that any of the reactions of folate is ascorbate-dependent.

Iron deficiency in scurvy may well be secondary to reduced absorption of inorganic iron and impaired mobilization of tissue iron reserves (Section 3.4.5). At the same time, the hemorrhages of advanced scurvy can cause a considerable loss of blood.

There is also evidence that erythrocytes have a shorter half-life than normal in scurvy, possibly as a result of peroxidative damage to membrane lipids from impairment of the reduction of tocopheroxyl radical by ascorbate (Section 13.3.7.1).

13.5 ASSESSMENT OF VITAMIN C STATUS

Vitamin C status is generally assessed by estimating the saturation of body reserves or measuring plasma and leukocyte concentrations of the vitamin. Urinary excretion of hydroxyproline-containing peptides is reduced in people with inadequate vitamin C status, but a number of other factors that affect bone and connective tissue turnover confound interpretation of the results (Bates, 1977). The ratio of deoxypyridinoline:pyridinoline compounds derived from collagen cross-links provides a more useful index, but is potentially affected by copper status (Tsuchiya and Bates, 1997).

13.5.1 Urinary Excretion of Vitamin C and Saturation Testing

Urinary excretion of ascorbate falls to undetectably low levels in deficiency; therefore, very low excretion will indicate deficiency. However, no guidelines for the interpretation of urinary ascorbate have been established, and basal urinary excretion of ascorbate is rarely used in the assessment of status. During depletion/repletion studies, urinary excretion increases before tissue saturation has been achieved (Sauberlich, 1975).

It is relatively easy to assess the state of body reserves of vitamin C by measuring the excretion after a test dose. A subject whose reserves are saturated will excrete the whole of a test dose of 500 mg of ascorbate over 6 hours. A more precise method involves repeating the loading test daily until complete excretion is achieved, thus giving an indication of how depleted the body stores were.

13.5.2 Plasma and Leukocyte Concentrations of Ascorbate

The plasma concentration of vitamin C falls relatively rapidly during depletion studies, to undetectably low levels within 4 weeks of initiating a vitamin C-free diet, although clinical signs of scurvy may not develop for a further 3 to

Table 13.2 Plasma and Leukocyte Ascorbate Concentrations as Criteria of Vitamin C Nutritional Status

		Deficient	Marginal	Adequate
Whole blood	mmol/L	<17	17–28	>28
	mg/L	<3.0	3.0–5.0	>5.0
Plasma	mmol/L	<11	11–17	>17
	mg/L	<2.0	2.0–3.0	>3.0
Leukocytes	pmol/10^6 cells	<1.1	1.1–2.8	>2.8
	μg/10^6 cells	<0.2	0.2–0.5	>0.5

4 months and tissue concentrations of the vitamin may be as high as 50% of saturation. In field studies and surveys, subjects with plasma ascorbate below 11 μmol per L are considered to be at risk of developing scurvy (see Table 13.2), whereas anyone with a plasma concentration below 6 μmol per L would be expected to show clinical signs. At intakes above about 100 mg per day, the plasma concentration of ascorbate reaches a plateau around 70 to 80 μmol per L, because of quantitative excretion of the vitamin as the renal threshold is exceeded (Section 13.2.4).

The concentration of ascorbate in leukocytes is well correlated with the concentrations in other tissues and falls more slowly than plasma concentration in depletion studies. The reference range of leukocyte ascorbate is 1.1 to 2.8 pmol per 10^6 cells; a significant loss of leukocyte ascorbate coincides with the development of clear clinical signs of scurvy. Predictably, at high levels of ascorbate intake, although the plasma concentration continues to increase with intake, the leukocyte content does not, because the cells, like other tissues, are saturated.

There is a problem in the interpretation of leukocyte concentrations of ascorbate as an index of vitamin C nutritional status. As discussed in Section 13.2.2, the different types of leukocytes have different capacities to accumulate ascorbate. This means that a change in the proportion of granulocytes, platelets, and mononuclear leukocytes will result in a change in the total concentration of ascorbate per 10^6 cells, although there may well be no change in vitamin nutritional status. Stress, myocardial infarction, infection, burns, and surgical trauma all result in changes in leukocyte distribution, with an increase in the proportion of granulocytes that achieve saturation at a lower concentration of ascorbate than other leukocytes, and thus an apparent change in leukocyte ascorbate. This has been misinterpreted to indicate an increased requirement for vitamin C in these conditions (Schorah et al., 1986). Without a differential white cell count, leukocyte ascorbate concentration cannot be considered to give a meaningful reflection of vitamin C status.

13.5.3 Markers of DNA Oxidative Damage

There is increased formation of 8-hydroxyguanine (a marker of oxidative radical damage) in DNA during (short-term) vitamin C depletion (Fraga et al., 1991). In addition, the rate of removal of 8-hydroxyguanine from DNA by excision repair, and thus the urinary excretion of 8-hydroxyguanine, is affected by vitamin C status (Cooke et al., 1998). These results suggest that measurement of 8-hydroxyguanine in DNA, or its urinary excretion, may provide a way of estimating requirements to meet a biomarker of optimum status.

13.6 VITAMIN C REQUIREMENTS AND REFERENCE INTAKES

There have been two major studies of ascorbate requirements in depletion/repletion studies, one in Sheffield during the 1940s (Medical Research Council, 1948) and the other in Iowa during the 1960s (Baker et al., 1969, 1971; Hodges et al., 1969, 1971). In addition, Kallner and coworkers (1979, 1981) have determined the body pool of ascorbate and the fractional rate of turnover under various conditions. Levine and coworkers (1995, 1996, 1999) have measured plasma and leukocyte ascorbate in studies of subjects maintained on more than minimally adequate amounts of vitamin C for relatively prolonged periods of time to determine optimum, rather than minimum, requirements.

Although the minimum requirement for vitamin C is firmly established, there are considerable discrepancies between the reference intakes published by different national and international authorities (see Table 13.3), with figures ranging between 30 to 90 mg per day. This is the result of the use of different criteria of adequacy and reflects differences of opinion as to what represents an adequate intake of vitamin C. It is possible to produce arguments to support reference intakes of between 30 to 100 mg per day.

Carr and Frei (1999b) reviewed studies of vitamin C intake associated with reduced risks of cancer and cardiovascular disease and suggested that, by this criterion, the average requirement was 90 to 100 mg per day, giving a reference intake of 120 mg per day.

13.6.1 The Minimum Requirement for Vitamin C

The minimum requirement for vitamin C was established in the Sheffield study (Medical Research Council, 1948), which showed that an intake of marginally less than 10 mg per day was adequate to prevent the development of scurvy or to cure the clinical signs. Results from the Iowa study (Baker et al., 1969,

Table 13.3 Reference Intakes of Vitamin C (mg/day)

Age	U.K. 1991	EU 1993	U.S./Canada 2000	FAO 2001
0–3 m	25	—	40	25
4–6 m	25	—	40	25
7–9 m	25	20	50	30
10–12 m	25	20	50	30
1–3 y	30	25	15	30
4–6 y	30	25	25	30
7–8 y	30	30	25	35
Males				
9–10 y	30	30	45	40
11–13 y	35	35	45	40
14–15 y	40	35	75	40
16–18 y	40	40	75	40
19–30 y	40	45	90	45
31–50 y	40	45	90	45
51–70 y	40	45	90	45
>70 y	40	45	90	45
Females				
9–10 y	30	30	45	40
11–13 y	35	35	45	40
14–15 y	40	35	65	40
16–18 y	40	40	75	40
19–30 y	40	45	75	45
31–50 y	40	45	75	45
51–70 y	40	45	75	45
>70 y	40	45	75	45
Pregnant	50	55	85	55
Lactating	70	70	120	70

EU, European Union; FAO, Food and Agriculture Organization; WHO, World Health Organization.
Sources: Department of Health, 1991; Scientific Committee for Food, 1993; Institute of Medicine, 2000; FAO/WHO, 2001.

1971; Hodges et al., 1969, 1971) suggested that as little as 6.5 mg per day were adequate, and studies in India show that intakes as low as 10 mg per day are compatible with health.

However, at this level of intake, the subjects in the Sheffield study had impaired wound healing, as assessed by the tensile strength of scar tissue. Optimum wound healing required a mean intake of 20 mg per day. Allowing for individual variation, this gives a reference intake of 30 mg per day, which was the British reference intake until 1991 and the World Health Organization reference intake until 2001.

13.6.2 Requirements Estimated from the Plasma and Leukocyte Concentrations of Ascorbate

The plasma concentration of ascorbate shows a sigmoidal relationship with intake. Below about 30 mg per day, the plasma concentration is extremely low and does not reflect increasing intake to any significant extent. As the intake rises above 30 mg per day, so the plasma concentration begins to increase sharply, reaching a plateau of 70 to 85 μmol per L, at intakes between 70 to 100 mg per day, when the renal threshold is reached and the vitamin is excreted quantitatively with increasing intake.

The midpoint of the steep region of the curve, where the plasma concentration increases linearly with increasing intake, represents a state in which tissue reserves are adequate and plasma ascorbate is available for transfer between tissues. This corresponds to an intake of 40 mg per day, and is the basis of the U.K., European Union, and Food and Agriculture Organization/World Health Organization figures shown in Table 13.3. At this level of intake, the total body pool is about 900 mg (5.1 mmol).

Levine and coworkers (1995, 1999) have argued that setting requirements and reference intakes on the basis of the steep part of a sigmoidal curve is undesirable. They suggested that a more appropriate point would be where the plasma concentration reaches a plateau, at an intake of around 100 to 200 mg per day.

The U.S./Canadian reference intakes of 75 mg for women and 90 mg for men are based on studies of leukocyte saturation (Levine et al., 1996; Institute of Medicine, 2000).

13.6.3 Requirements Estimated from Maintenance of the Body Pool of Ascorbate

A priori, the best means of determining vitamin C requirement would seem to be determination of the total body pool and its fractional rate of loss or catabolism. An appropriate intake would then be that to replace losses and maintain the body pool. Clinical signs of scurvy are seen when the total body pool of ascorbate is below 1.7 mmol (300 mg). The pool increases with intake, reaching a maximum of about 8.5 mmol (1,500 mg) in adults – 114 μmol (20 mg) per kg of body weight. The fractional turnover rate of ascorbate is 3% to 4% daily, suggesting a need for 45 to 60 mg per day for replacement. The basis for the 1989 U.S. Recommended Daily Allowance (RDA) of 60 mg (National Research Council, 1989) was the observed mean fractional turnover rate of 3.2% of a body pool of 20 mg per kg of body weight per day, with allowances for incomplete absorption of dietary ascorbate and individual variation.

Olson and Hodges (1987) suggested that a total body pool of 900 mg (5.1 mmol) is adequate; it is three-fold higher than the minimum pool required to prevent scurvy. There was no evidence of health benefits from a body pool greater than 600 mg. They noted that the figure of 1,500 mg – which was the basis of the U.S. and other reference intakes – was found in subjects consuming a self-selected diet, with a relatively high intake of vitamin C, and therefore could not be considered to represent any index of requirement. On the basis of a mean catabolic rate of 2.7 per day, and allowing for efficiency of absorption and individual variation, they proposed an RDA of 40 mg.

Because the mean fractional turnover rate of 3.2% per day was observed during a depletion study, and the rate of ascorbate catabolism varies with intake, it has been suggested that this implies a rate of 3.6% per day before depletion. On this basis, and allowing for incomplete absorption and individual variation, various national authorities arrive at a reference intake of 80 mg per day.

The rate of ascorbate catabolism is not constant. If it were, more or less complete depletion of the body pool would be expected within 25 to 33 days; yet, in the Sheffield study, in which the subjects were initially maintained on 70 mg of ascorbate per day, they received a diet essentially free from the vitamin; no changes were apparent for 17 weeks (Medical Research Council, 1948). In the Iowa study, the subjects were not initially saturated with vitamin C; the first skin lesions did not develop for 5 to 6 weeks after the depletion period (Baker et al., 1969, 1971; Hodges et al., 1969, 1971). Kallner and coworkers (1979) showed that the turnover time of body ascorbate varied between 56 days at low intake (about 15 mg per day) and 14 days (at intakes of 80 mg per day). It is thus apparent that the rate of ascorbate catabolism is affected markedly by the intake, and the requirement to maintain the body pool cannot be estimated as an absolute value. A habitual low intake, with a consequent low rate of catabolism, will maintain the same body pool as a habitual higher intake with a higher rate of catabolism.

13.6.4 Higher Recommendations

There is a school of thought that human requirements for vitamin C are considerably higher than the reference intakes discussed previously. Pauling (1970) measured the vitamin C intake of gorillas in captivity, assumed that this was the same as their intake in the wild (where they eat considerably less fruit than under zoo conditions), and then assumed that because they had this intake, it was their requirement – an unjustified assumption. Scaling this to humans, he suggested a requirement of 1 to 2 g per day. He also quoted the rate of endogenous synthesis of ascorbate in the rat (26 mg per day) and assumed that this represented a physiological requirement. This would lead to an apparent requirement for a 70 kg human being of 2 g per day. Again, it is unjustified to extrapolate from the rate of synthesis of ascorbate in the rat, where it is an intermediate in carbohydrate metabolism (Section 13.2) to metabolic requirements for animals that are not capable of de novo synthesis.

At intakes in excess of about 100 mg per day, there is quantitative urinary excretion of unmetabolized vitamin C with increasing intake, indicating that tissue reserves are saturated and the renal threshold has been exceeded. It is difficult to justify a requirement in excess of tissue storage capacity.

13.6.4.1 **The Effect of Smoking on Vitamin C Requirements** There is evidence that smokers have a higher requirement for vitamin C than nonsmokers. A number of studies have shown lower plasma and leukocyte concentrations of vitamin C in smokers, but many also report lower intake of the vitamin by smokers. The rate of catabolism of ascorbate is up to 40% greater in smokers than nonsmokers (Kallner et al., 1981), and therefore their vitamin C requirement may be almost twice that of nonsmokers.

13.6.5 **Safety and Upper Levels of Intake of Vitamin C**
Regardless of whether or not high intakes of ascorbate have any beneficial effects, large numbers of people habitually take between 1 to 5 g per day of vitamin C supplements, and some take considerably more. The U.S./Canadian tolerable upper level of intake is 2 g per day (Institute of Medicine, 2000).

There is little evidence of any significant toxicity from these high intakes, although there are a number of potential problems (Rivers, 1987; Johnston, 1999). Because ascorbate is largely absorbed by active transport, absorption is saturable, and a decreasing proportion of high doses is absorbed. Similarly, once the plasma concentration has reached the renal threshold, the vitamin is excreted quantitatively with increasing intake (Section 13.2.1).

Unabsorbed ascorbate in the intestinal lumen is a substrate for bacterial fermentation, which may explain the diarrhea and intestinal discomfort reported in some studies with high doses of the vitamin.

13.6.5.1 **Renal Stones** Up to 5% of the population are at risk from the development of renal oxalate stones, as a result of both ingested oxalate and that formed endogenously. The process of stone formation is not well understood, and the presence of compounds in urine that retard crystallization, rather than concentration of oxalate, is the main factor. People who form renal oxalate stones may well have a lower urine concentration of oxalate than those who do not.

A number of reports have suggested that high intakes of vitamin C are associated with increased excretion of oxalate; however, much of the oxalate may be the result of nonenzymic formation from ascorbate under alkaline conditions, occurring either in the bladder or after collection, and thus not a risk factor for renal stone formation (Chalmers et al., 1986). Gerster (1997) suggested that people who are recurrent oxalate stone formers should, as a matter of prudence, restrict their intake of vitamin C to 100 mg per day, but noted that the risk of stone formation in the population at large is inversely related to vitamin C intake.

Ascorbate may also increase the urinary excretion of uric acid, and be possibly protective against the development of gout, but may raise the urinary concentration above the low solubility threshold, thus increasing the likelihood of developing urate renal stones.

13.6.5.2 **False Results in Urine Glucose Testing**   High concentrations of ascorbate in the urine inhibit the development of color using glucose oxidase test strips, because ascorbate will reduce the hydrogen peroxide formed by glucose oxidase before it reacts with the dye and will also reduce the dye back to its colorless form. This presents a potential problem for the management of diabetes, with false-negative results in people taking high-dose supplements of ascorbate (Mayson et al., 1973).

By contrast, ascorbate gives a positive result when urine is tested with alkaline copper reagents for reducing compounds, and a result that can falsely be interpreted as indicating glucosuria.

13.6.5.3 **Rebound Scurvy**   It was noted in Section 13.6.3 that the rate of ascorbate catabolism increases with increasing intake. This has led to the suggestion that abrupt cessation of high intakes of ascorbate may result in rebound scurvy, because of metabolic conditioning and a greatly increased rate of catabolism, so that lower intakes are now inadequate. Although there have been a number of anecdotal reports, there is no evidence that this occurs to any significant extent; the effect of increased ascorbate catabolism in accelerating the development of deficiency is minimal, at least partly because as intake falls, so does catabolism (Johnston, 1999).

13.6.5.4 Ascorbate and Iron Overload A common polymorphism of the transferrin gene puts 10% of the population (and significantly more in some ethnic groups) at risk of iron overload and the development of hemochromatosis. Patients with iron overload suffer from scurvy (Section 13.4), as a result of greatly increased ascorbate oxidation caused by reaction with the Fe^{3+} metalloproteins that accumulate in this condition. Although there is a school of thought that high vitamin C intakes should be encouraged to improve iron nutrition, this would have an adverse effect on those genetically at risk of iron overload, both increasing their absorption of iron (Section 13.3.1) and also increasing nonenzymic formation of oxygen radicals by nonenzymic reaction between ascorbate and iron (Gerster, 1999; Kasvosve et al., 2002).

13.7 PHARMACOLOGICAL USES OF VITAMIN C

As discussed in Section 13.3.5, ascorbate enhances the intestinal absorption of inorganic iron, and therefore it is frequently prescribed together with iron supplements. It is also used when it is desired to acidify the urine (e.g., in conjunction with some antibiotics). Supplements of vitamin C (often of the order of grams per day) are widely consumed to protect against cancer, cardiovascular disease, and viral infections, although (as discussed below) the evidence of efficacy is poor.

13.7.1 Vitamin C in Cancer Prevention and Therapy

Epidemiological evidence shows that diets that are rich in vitamin C are associated with lower incidence of cancer and cardiovascular disease. However, such diets are rich in fruits and vegetables, and thus a wide variety of other potentially protective factors; studies of 8-hydroxyguanine excretion as a marker of oxidative damage to DNA do not provide evidence of a protective effect of vitamin C per se, except in people whose intake is low (Halliwell, 2001).

As discussed in Section 13.3.6, ascorbate may act to inhibit the formation of carcinogenic nitrosamines from nitrite and dietary amines, and thus reduce the risk of cancer. There is some evidence that vitamin C supplements are protective against the development of gastric cancer.

A number of studies have reported low ascorbate status in patients with advanced cancer – perhaps an unsurprising finding in seriously ill patients. However, there is also some evidence that a number of tumors can accumulate ascorbate at the expense of the host. Cameron and Pauling (1974a, 1974b) published a lengthy review of the supposed roles of ascorbate in enhancing host resistance to cancer, and suggested, on the basis of an uncontrolled open trial in terminally ill patients, that 10-g daily doses of vitamin C resulted in increased survival. In a controlled study, patients matched for age, sex, site, and stage of primary tumors and metastases (and for previous chemotherapy; Creagan et al. (1979) were unable to demonstrate any beneficial effects of high-dose ascorbic acid in the treatment of advanced cancer.

13.7.2 Vitamin C in Cardiovascular Disease
Vitamin C deficiency is associated with an increased risk of atherosclerosis, but there is little evidence of protective effects at intakes greater than needed to meet requirements (Jacob, 1998). A systematic review (Ness et al., 1996) found limited evidence of benefits of high intakes of vitamin C in reducing the incidence of stroke, but inconsistent evidence with respect to coronary heart disease.

Scorbutic guinea pigs develop hypercholesterolemia, which may lead to the development of cholesterol-rich gallstones. This is largely the result of impaired activity of cholesterol 7-hydroxylase, which is an ascorbate-dependent enzyme (Section 13.3.8), resulting in reduced oxidation of cholesterol to bile acids. There is no evidence that increased intakes of vitamin C above requirements result in increased cholesterol catabolism.

13.7.3 Vitamin C and the Common Cold
High doses of vitamin C are popularly recommended for the prevention and treatment of the common cold. Evidence from controlled trials is unconvincing. Chalmers (1975) reviewed 15 reports and considered that only 8 reports met the basic criteria of well-conducted scientific research. Assessment of these 8 reports gave no evidence of any beneficial effects. Similarly, Dykes and Meier (1975), reviewing only those reports that had been published in peer-reviewed journals, concluded that there was no evidence of any significant benefit.

Hemila (1992) reviewed a number of studies and again concluded that there was no evidence of a protective effect against the incidence of colds. He did, however, note that there is consistent evidence of a beneficial effect in reducing the severity and duration of symptoms (a notoriously difficult subject to research). He suggested that this might be because of the antioxidant actions of ascorbate against the oxidizing agents produced by, and released from, activated phagocytes – thus a decreased inflammatory response. A systematic review (Douglas et al., 2000) similarly concluded that there was no beneficial effect in terms of preventing infection, but a modest benefit in terms of reducing the duration of symptoms.

FURTHER READING

Benzie IF (1999) Vitamin C: prospective functional markers for defining optimal nutritional status. *Proceedings of the Nutrition Society* **58,** 469–76.

Carr AC, Zhu BZ, and Frei B (2000) Potential antiatherogenic mechanisms of ascorbate (vitamin C) and alpha-tocopherol (vitamin E). *Circulation Research* **87,** 349–54.

Englard S and Seifter S (1986) The biochemical functions of ascorbic acid. *Annual Reviews of Nutrition* **6,** 365–406.

Sato P and Udenfriend S (1978) Studies on ascorbic acid related to the genetic basis of scurvy. *Vitamins and Hormones* **36,** 33–52.

Sauberlich HE (1994) Pharmacology of vitamin C. *Annual Reviews of Nutrition* **14,** 371–91.

Smirnoff N (2000) Ascorbic acid: metabolism and functions of a multi-facetted molecule. *Current Opinion Plant Biology* **3,** 229–35.

Smirnoff N (2001) L-Ascorbic acid biosynthesis. *Vitamins and Hormones* **61,** 241–66.

Szent-György A (1963) Lost in the twentieth century. *Annual Reviews of Biochemistry* **32,** 1–15.

References cited in the text are listed in the Bibliography.

FOURTEEN

Marginal Compounds and Phytonutrients

In addition to the established vitamins, a number of organic compounds have clear metabolic functions; they can be synthesized in the body, but it is possible that under some circumstances (as in premature infants and patients maintained on long-term total parenteral nutrition) endogenous synthesis may not be adequate to meet requirements. These compounds include biopterin (Section 10.4), carnitine (Section 14.1), choline (Section 14.2), creatine (Section 14.3), inositol (Section 14.4), molybdopterin (Section 10.5), taurine (Section 14.5), and ubiquinone (Section 14.6).

A number of compounds found in foods of plant origin have potentially protective effects, although they cannot be considered to be dietary essentials; they are variously known as phytonutrients, phytoceuticals, or nutraceuticals. Such compounds include allyl sulfur compounds, flavonoids, glucosinolates, and phytoestrogens.

14.1 CARNITINE

Carnitine (3-hydroxy,4-N-trimethylaminobutyric acid) has a central role in the transport of fatty acids across the mitochondrial membrane for β-oxidation. At the outer face of the outer mitochondrial membrane, carnitine acyltransferase I catalyzes the reaction shown in Figure 14.1, the transfer of fatty acids from coenzyme A (CoA) to form acyl carnitine esters that cross into the mitochondrial matrix. At the inner face of the inner mitochondrial membrane, carnitine acyltransferase II catalyzes the reverse reaction.

Acyl carnitine can cross only the inner mitochondrial membrane on a countertransport system that takes in acyl carnitine in exchange for free carnitine being returned to the intermembrane space. Once inside the mitochondrial inner membrane, acyl carnitine transfers the acyl group onto CoA ready to undergo β-oxidation. This countertransport system provides regulation of the

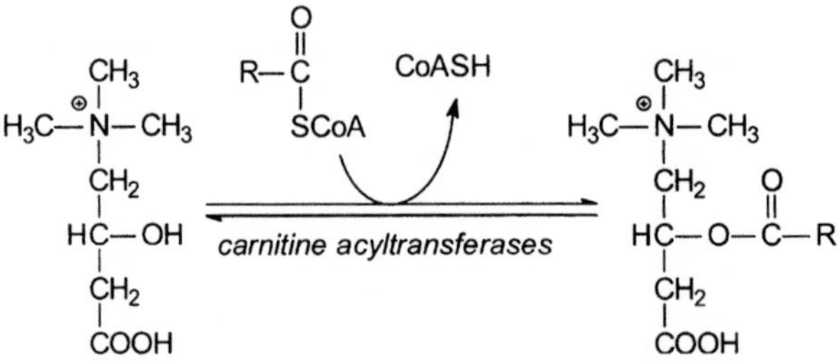

Figure 14.1. Reaction of carnitine acyltransferase (carnitine palmitoyltransferase, EC 2.3.1.21).

uptake of fatty acids into the mitochondrion for oxidation. As long as there is free CoA available in the mitochondrial matrix, fatty acids can be taken up and the carnitine returned to the outer membrane for uptake of more fatty acids. However, if most of the CoA in the mitochondrion is acylated, then there is no need for further fatty uptake immediately, and it is not possible.

Carnitine acyltransferase I is strongly inhibited by malonyl CoA, and muscle has both acetyl CoA carboxylase, which forms malonyl CoA, and malonyl CoA decarboxylase, which acts to remove malonyl CoA and relieve the inhibition of carnitine acyl transferase. The two enzymes are regulated in opposite directions in response to insulin, which stimulates fatty acid synthesis and reduces β-oxidation, and glucagon that reduces fatty acid synthesis and increases β-oxidation (Kerner and Hoppel, 2000; Louet et al., 2001; Eaton, 2002).

Fatty acids are the major fuel for red muscle fibers, which are the main type involved in moderate exercise. Children who lack one or the other of the enzymes required for carnitine synthesis, and are therefore reliant on a dietary intake, have poor exercise tolerance, because they have an impaired ability to transport fatty acids into the mitochondria for β-oxidation. Provision of supplements of carnitine to the affected children overcomes the problem. Extrapolation from this rare clinical condition has led to the use of carnitine as a so-called ergogenic aid to improve athletic performance.

14.1.1 Biosynthesis and Metabolism of Carnitine

Carnitine is synthesized from lysine and methionine by the pathway shown in Figure 14.2 (Vaz and Wanders, 2002). The synthesis of carnitine involves the stepwise methylation of a protein-incorporated lysine residue at the expense of methionine to yield a trimethyllysine residue. Free trimethyllysine is then released by proteolysis. It is not clear whether there is a specific precursor protein for carnitine synthesis, because trimethyllysine occurs in a number of proteins, including actin, calmodulin, cytochrome c, histones, and myosin.

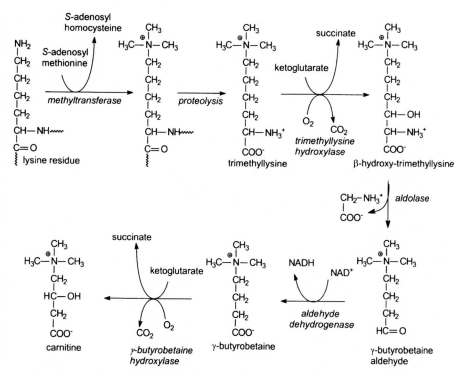

Figure 14.2. Biosynthesis of carnitine. Trimethyllysine hydroxylase, EC 1.14.11.8; aldolase, EC 4.1.2."x"; aldehyde dehydrogenase, EC 1.2.1.47; γ-butyrobetaine hydroxylase, EC 1.14.11.1. Relative molecular mass (M_r): carnitine, 161.2.

Both hydroxylation reactions in the synthesis of carnitine from trimethyllysine are ascorbic acid-dependent, 2-oxoglutarate–linked, reactions (Section 13.3.3), and impaired synthesis of carnitine probably accounts for the muscle fatigue associated with vitamin C deficiency.

The total body content of carnitine is about 100 mmol, and about 5% of this turns over daily. Plasma total carnitine is between 36 to 83 μmol per L in men and 28 to 75 μmol per L in women, mainly as free carnitine. Although both free carnitine and acyl carnitine esters are excreted in the urine, much is oxidized to trimethylamine and trimethylamine oxide. It is not known whether the formation of trimethylamine and trimethylamine oxide is caused by endogenous enzymes or intestinal bacterial metabolism of carnitine.

Total urinary excretion of carnitine is between 300 to 530 μmol (men) or 200 to 320 μmol (women); 30% to 50% of this is free carnitine; the remainder is a variety of acyl carnitine esters. Acyl carnitine esters are readily cleared in the kidney, whereas free carnitine and acetyl carnitine are reabsorbed until the plasma concentration exceeds the renal threshold.

Urinary excretion of acyl carnitine esters increases considerably in a variety of conditions involving organic aciduria; carnitine acts to spare CoA and pantothenic acid (Section 12.2), by releasing the coenzyme from otherwise nonmetabolizable esters that would trap the coenzyme and cause functional pantothenic acid deficiency.

14.1.2 The Possible Essentiality of Carnitine

Meat and fish contain relatively large amounts of carnitine, and average dietary intakes by omnivores are 100 to 300 mg (2 to 12 μmol) per day, compared with endogenous synthesis of about 1.2 μmol per day. There are few plant sources, and strict vegetarians have lower plasma concentrations of carnitine and are reliant on endogenous synthesis. Even in strict vegetarians, tissue carnitine depletion is only seen together with general protein-energy malnutrition (and hence deficiency of methionine and lysine).

Although endogenous synthesis of carnitine can meet normal metabolic demands, administration of the anticonvulsant valproic acid, which is excreted as the carnitine ester, or metabolic organic acidemias that result in considerable excretion of acyl carnitine esters, can lead to carnitine depletion. This results in impaired β-oxidation of fatty acids and ketogenesis in the liver, and thus nonketotic hypoglycemia, with elevated plasma nonesterified fatty acids and triacylglycerol. Because hepatocytes rely on fatty acid oxidation for their own energy-yielding metabolism in fasting, there may also be signs of liver dysfunction, with hyperammonemia and encephalopathy. The administration of carnitine supplements in these conditions has a beneficial effect (Arrigoni-Martelli and Caso, 2001).

Although carnitine is not generally nutritionally important, it may be required for premature infants, because they have a limited capacity to synthesize it. However, there is inadequate evidence to support routine carnitine supplementation of parenterally fed premature infants (Cairns and Stalker, 2000). Carnitine depletion occurs in patients undergoing hemodialysis, but there is little evidence that supplements are effective in treating the associated dyslipidemia (Raskind and El-Chaar, 2000; Hurot et al., 2002).

14.1.3 Carnitine as an Ergogenic Aid

By extrapolation from the muscle weakness and fatigue seen in children with genetic defects of carnitine biosynthesis or metabolism, it has been assumed that supplementary carnitine may have a performance-enhancing or ergogenic action, and supplements are commonly taken by athletes and body-builders. There is little evidence that supplements have any effect on muscle work output. Although supplements increase plasma carnitine, they do not significantly increase the muscle content (Brass, 2000).

14.2 CHOLINE

Choline is an essential component of phospholipids – phosphatidylcholine (lecithin) is the major phospholipid in cell membranes and sphingomyelin is important in the nervous system. Acetylcholine is a transmitter in the central and parasympathetic nervous systems and at neuromuscular junctions, and has a role in the regulation of differentiation and development of the nervous system (Biagioni et al., 2000). Acetylcholine is also synthesized in mononuclear lymphocytes, where it has an autocrine or paracrine role in regulating immune function (Fujii and Kawashima, 2001).

In addition to being used as a trivial name for phosphatidylcholine, the name lecithin is used for phospholipid fractions relatively rich in phosphatidylcholine, which are widely used as emulsifying agents in food manufacture (E-322). Such preparations typically contain some 40% to 80% phosphatidylcholine, together with a variety of other phospholipids.

14.2.1 Biosynthesis and Metabolism of Choline

Phosphatidylcholine can be synthesized by the pathway shown in Figure 14.3. Decarboxylation of phosphatidylserine to phosphatidylethanolamine (cephalin) is followed by methylation in which S-adenosylmethionine is the methyl donor to yield successively the relatively rare mono- and dimethyl derivatives, then phosphatidylcholine.

Sequential removal of the fatty acids by phospholipase action results in the formation of lysolecithin (glycerophosphorylcholine), then hydrolysis to release choline. Acetylcholine is synthesized in neurons using acetyl CoA.

About 30% of dietary phosphatidylcholine is absorbed intact into the lymphatic system; the remainder is hydrolyzed to lysolecithin in the intestinal mucosa and to free choline in the liver. Free choline in the diet is largely metabolized by intestinal bacteria, forming trimethylamine, which is absorbed and excreted in the urine. Only about 30% of free choline is absorbed intact.

Choline can be used for synthesis of phosphatidylcholine by reaction between CDP-choline and diacylglycerol. Under normal conditions, the major pathway of phosphatidylcholine synthesis is by the incorporation of

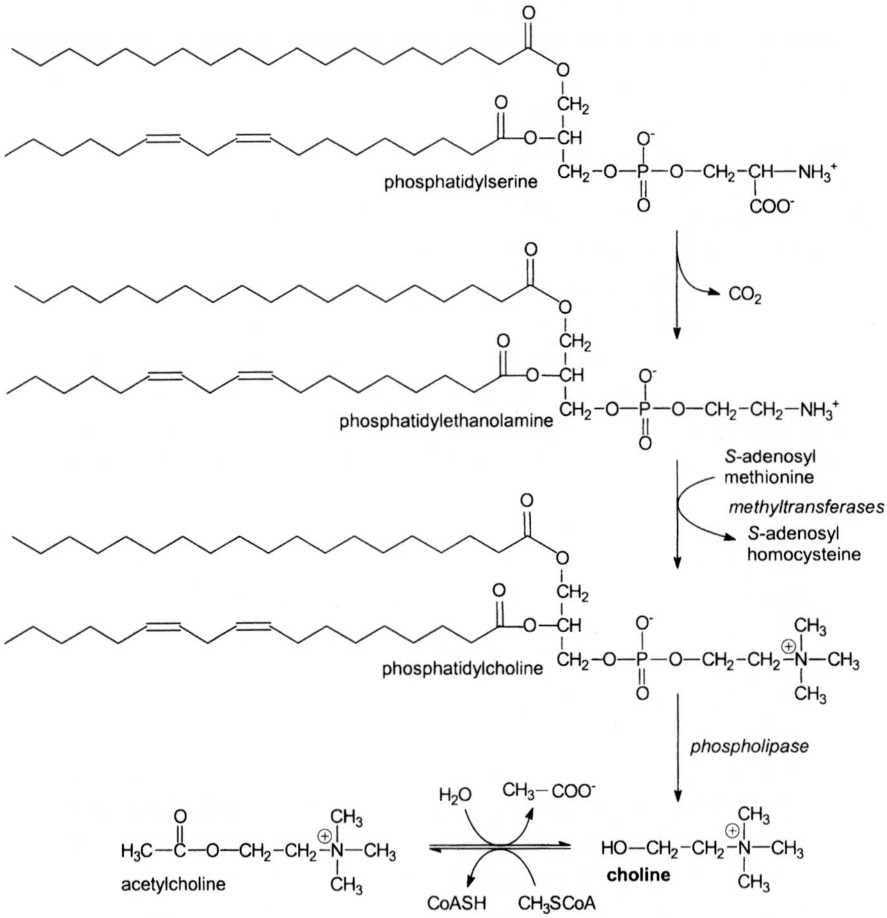

Figure 14.3. Biosynthesis of choline, and acetylcholine. Relative molecular masses (M_r): choline, 104.2 (chloride, 139.6); and acetylcholine, 146.3 (chloride, 181.7). CoASH, free coenzyme A.

preformed choline rather than methylation of phosphatidylethanolamine. The activities of the two pathways are coordinately regulated, so that increased choline availability reduces the methylation of phosphatidylethanolamine, whereas decreased availability of preformed choline results in increased de novo synthesis (Lykidis and Jackowski, 2001).

As shown in Figure 14.4, choline catabolism involves two oxidation reactions to form betaine (trimethylglycine), followed by three successive demethylations. As discussed in Section 10.9.3, the remethylation of homocysteine to methionine catalyzed by the betaine-dependent methyltransferase

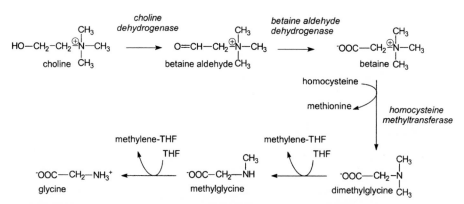

Figure 14.4. Catabolism of choline. Choline dehydrogenase, EC 1.1.99.1; betaine aldehyde dehydrogenase, EC 1.2.1.8; and homocysteine methyltransferase, EC 2.1.1.5. Relative molecular masses (M_r): choline, 104.2; betaine, 117.2; dimethylglycine, 102.2; methylglycine, 88.2; and glycine, 74.2. THF, tetrahydrofolate.

can maintain adequate concentrations of methionine in tissues other than the central nervous system when the activity of methionine synthetase is impaired because of vitamin B_{12} deficiency.

14.2.2 The Possible Essentiality of Choline

In many animals, dietary deprivation of choline leads to liver dysfunction and growth retardation, and some patients maintained on choline-free total parenteral nutrition develop liver damage that resolves when choline is provided, suggesting that endogenous synthesis may be inadequate to meet requirements (Zeisel, 2000). There is inadequate information to permit the setting of reference intakes, but the Acceptable Intake for adults is 550 mg (for men) or 425 mg (for women) per day (Institute of Medicine, 1998). In experimental animals choline deficiency is exacerbated by deficiency of methionine, folic acid, or vitamin B_{12}, which impairs the capacity for de novo synthesis.

There is some evidence that the availability of choline may be limiting for the synthesis of acetylcholine in the central nervous system under some conditions, and supplements of phosphatidylcholine increase the rate of acetylcholine turnover. One systematic review concludes that phosphatidylcholine supplements result in some improvement in cognitive function in patients with dementia, especially when this is secondary to cerebrovascular disorder (Fioravanti and Yanagi, 2000), but another concludes that there is no evidence to support its use in the treatment of dementia (Higgins and Flicker, 2000). Although phosphatidylcholine has been used to treat tardive dyskinesia associated with neuroleptic medication, there is little evidence to support its use (McGrath and Soares, 2000).

14.3 CREATINE

Creatine functions as a phosphagen in muscle. Neither the small amount of ATP in muscle nor the speed with which metabolic activity can be increased, and hence ADP be rephosphorylated, matches the demand for ATP for rapid or sustained muscle contraction. Muscle contains a relatively large amount

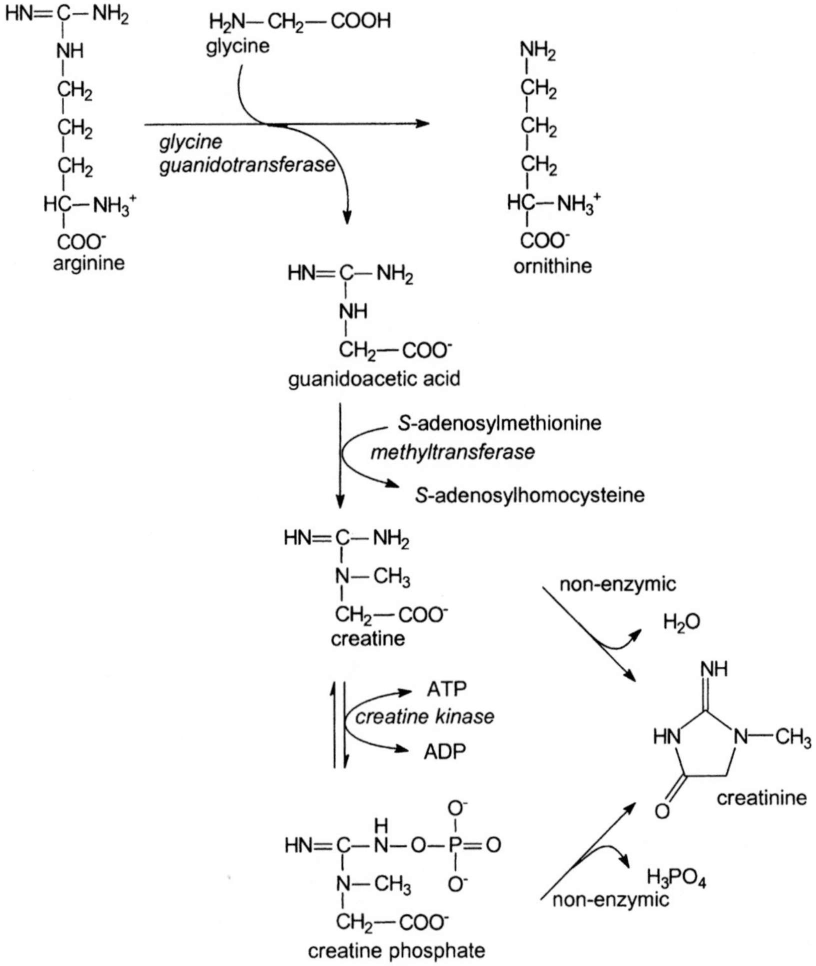

Figure 14.5. Synthesis of creatine. Glycine guanidotransferase (amidinotransferase), EC 2.1.4.1; guanidinoacetate methyltransferase, EC 2.1.1.2; and creatine kinase, EC 2.7.3.2.

of creatine phosphate (about four-fold higher than ATP). This acts as a reservoir or buffer to maintain a supply of ATP for muscle contraction until metabolic activity increases.

Creatine is not a dietary essential; as shown in Figure 14.5, it is synthesized from the amino acids glycine, arginine, and methionine. However, a single serving of meat will provide about 1 g of preformed creatine, whereas the average daily rate of de novo synthesis is 1 to 2 g, and endogenous synthesis is inhibited by a dietary intake.

Both creatine and creatine phosphate undergo a nonenzymic reaction to yield creatinine, which is metabolically useless and is excreted in the urine. Because the formation of creatinine is a nonenzymic reaction, the rate at which it is formed, and the amount excreted each day, depends mainly on muscle mass, and is therefore relatively constant from day to day in any one individual. This is commonly exploited in clinical chemistry; urinary metabolites are commonly expressed per mole of creatinine, and the excretion of creatinine is measured to assess the completeness of a 24-hour urine collection. There is normally little or no excretion of creatine in urine; significant amounts are only excreted when there is breakdown of muscle tissue.

Creatine supplements are often used as a so-called ergogenic aid to enhance athletic performance. Supplements of 3 to 20 g of creatine per day increase muscle creatine and creatine phosphate by 10% to 15% in people whose muscle creatine is initially relatively low, and have some effect on muscle work output and athletic performance, with little evidence of adverse effects (Casey and Greenhaff, 2000; Poortmans and Francaux, 2000; Hespel et al., 2001).

14.4 INOSITOL

Inositol is a hexahydric sugar alcohol. Its main function is in phospholipids; phosphatidylinositol constitutes some 5% to 10% of the total membrane phospholipids. In addition to its structural role in membranes, phosphatidylinositol has a major function in the intracellular responses to peptide hormones and neurotransmitters, yielding two intracellular second messengers: inositol trisphosphate and diacylglycerol.

Plant foods contain relatively large amounts of inositol phosphates, including the hexaphosphate, phytic acid. Phytate chelates minerals, such as calcium, zinc, and magnesium, forming insoluble complexes that are not absorbed. However, both intestinal phosphatases and endogenous phosphatases (phytase) in many foods dephosphorylate a significant proportion of dietary phytate. The inositol released can be absorbed and utilized for phosphatidylinositol synthesis.

Inositol can also be synthesized endogenously; inositol 1-phosphate is formed by isomerization of glucose 6-phosphate, catalyzed by an NAD-dependent enzyme, although overall there is no change in redox state. Phosphatidylinositol is formed by a reaction between CDP-inositol and diacylglycerol. Most inositol is catabolized by oxidation to glucuronic acid.

14.4.1 Phosphatidylinositol in Transmembrane Signaling

As shown in Figure 14.6, a proportion of the phosphatidylinositol in membranes undergoes two successive phosphorylations to yield phosphatidylinositol bisphosphate. This is a substrate for hormone-sensitive phospholipase C, which is activated by the G-protein–GTP complex released into the cell membrane by a hormone receptor following binding of the hormone at the outer surface of the membrane. Phospholipase C hydrolyzes phosphatidylinositol bisphosphate to release diacylglycerol and inositol trisphosphate.

Inositol trisphosphate opens a calcium transport channel in the membrane of the endoplasmic reticulum. This leads to an influx of calcium from storage in the endoplasmic reticulum and a 10-fold increase in the cytosolic concentration of calcium ions. Calmodulin is a small calcium binding protein found in all cells. Its affinity for calcium is such that, at the resting concentration of calcium in the cytosol (of the order of 0.1 μmol per L), little or none is bound to calmodulin. When the cytosolic concentration of calcium rises to about 1 μmol per L, as occurs in response to opening of the endoplasmic reticulum calcium transport channel, calmodulin binds 4 mol of calcium per mol of protein. When this occurs, calmodulin undergoes a conformational change, and calcium-calmodulin binds to, and activates, cytosolic protein kinases, which in turn phosphorylate target enzymes.

The diacylglycerol released by phospholipase C action remains in the membrane, where it activates a membrane-bound protein kinase. It may also diffuse into cytosol, where it enhances the binding of calcium-calmodulin to cytosolic protein kinase.

Inositol trisphosphate is inactivated by further phosphorylation to inositol tetrakisphosphate, and diacylglycerol is inactivated by hydrolysis to glycerol and fatty acids.

14.4.2 The Possible Essentiality of Inositol

There is no evidence that inositol is a dietary essential, because it is synthesized by all eukaryotic cells. Infants may have a higher requirement than can be met by endogenous synthesis, and dietary inositol is a growth factor for the newborn mouse. In female gerbils, inositol is a dietary essential, and deficiency

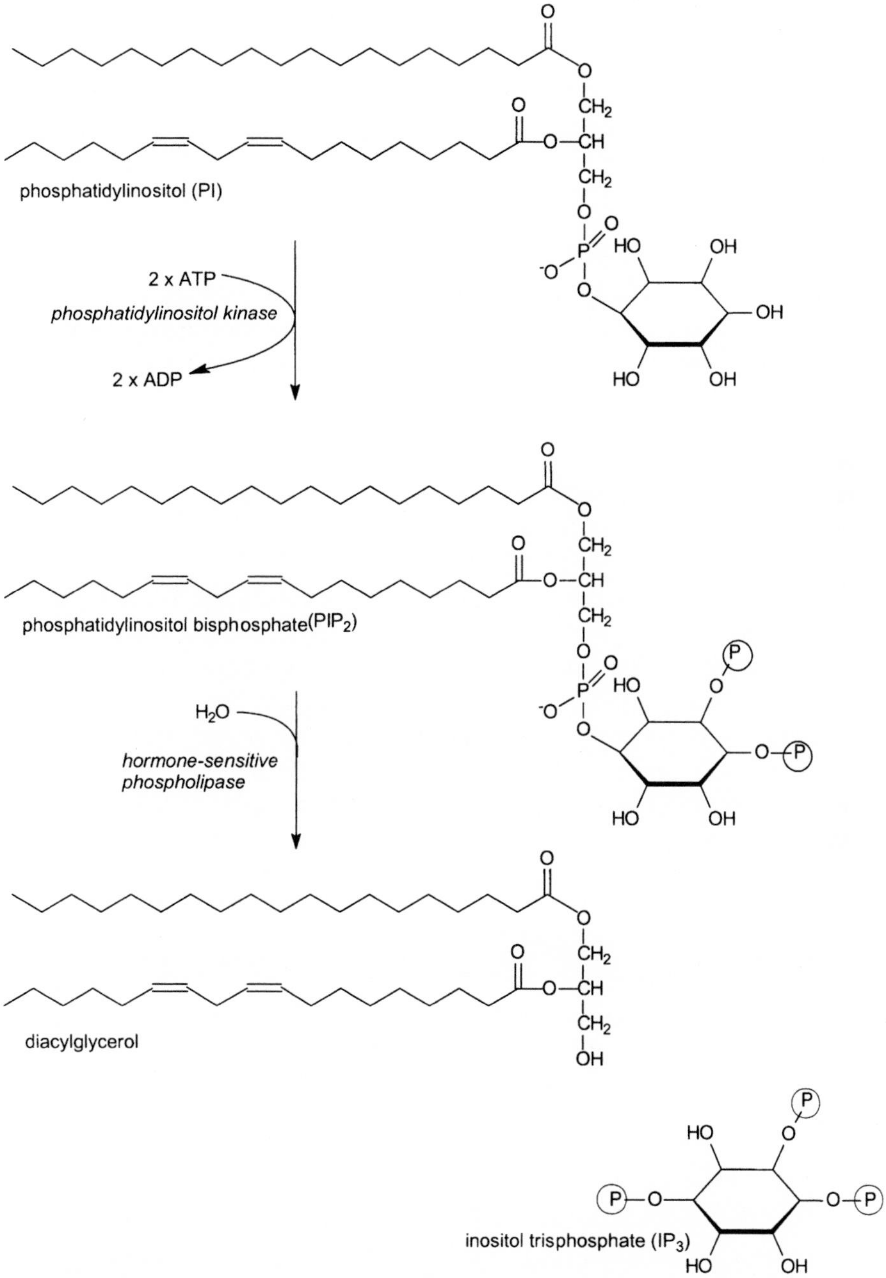

Figure 14.6. Formation of inositol trisphosphate and diacylglycerol. Phosphatidylinositol kinase, EC 2.7.1.67; and hormone-sensitive phospholipase, EC 3.1.4.3.

leads to lipodystrophy, mainly as a result of impaired synthesis and secretion of plasma lipoproteins.

People with untreated diabetes have high plasma concentrations of free inositol, and high urinary excretion of inositol, associated with relatively low intracellular concentrations of inositol, suggesting that elevated plasma glucose may inhibit the uptake of inositol. There is some evidence that impaired nerve conduction velocity in diabetic neuropathy is associated with low intracellular concentrations of inositol and that inositol supplements improve nerve conduction velocity. However, high intracellular concentrations of inositol also impair nerve conduction velocity, and supplements may have a deleterious effect.

14.5 TAURINE

Taurine was discovered in 1827 in ox bile, where it is conjugated with the bile acids. It was later shown to be a major excretory product of the sulfur amino acids methionine and cysteine. Until about 1976, it was assumed that it was a metabolic end-product whose only function was the conjugation of bile acids. In the rat, taurine synthesis accounts for 70% to 85% of total cysteine catabolism.

Kittens fed on diets with little preformed taurine develop retinal degeneration and blindness, which is prevented by taurine supplements. In the early stages, electroretinography shows changes similar to those seen in human retinitis pigmentosa. However, there is no evidence that retinitis pigmentosa is associated with taurine deficiency, and patients have normal plasma concentrations of the amino acid. Electron microscopy of the retinae of deficient kittens shows early disorientation of the cone photoreceptor outer segments, followed by extensive degeneration of both rod and cone outer segments. Children maintained on long-term total parenteral nutrition without added taurine show changes in the electrical activity of the retina, suggesting that endogenous synthesis may be inadequate to meet requirements. The function of taurine in the retina (and possibly other tissues) seems to be mainly as an osmolyte, maintaining normal intracellular osmolarity.

14.5.1 Biosynthesis of Taurine

As shown in Figure 14.7, taurine is a β-amino sulfonic acid (2-aminoethane sulfonic acid) and can be synthesized from cysteine by three pathways:

1. Oxidation to cysteine sulfinic acid, followed by decarboxylation to hypotaurine and oxidation to taurine. In most tissues, it is the decarboxylation

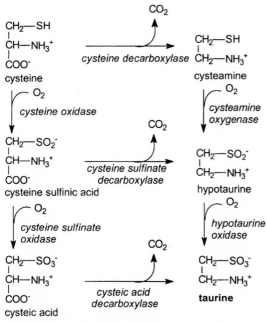

Figure 14.7. Pathways for the synthesis of taurine from cysteine. Cysteine sulfinate decarboxylase, EC 4.1.1.29; cysteic acid decarboxylase, EC 4.1.1.29 (glutamate decarboxylase, EC 4.1.1.15); cysteine oxidase, EC 1.13.11.20; cysteamine oxygenase, EC 1.13.11.19; and hypotaurine oxidase, EC 1.8.1.3. Relative molecular masses (M_r): cysteine, 121.2; cysteamine, 77.2; cysteine sulfinic acid, 153.2; cysteic acid, 169.2; hypotaurine, 109.1; and taurine, 125.1.

of cysteine sulfinic acid that is rate-limiting for taurine synthesis, not the oxidation of cysteine.

2. Oxidation to cysteic acid, followed by decarboxylation to taurine. Cysteic acid and cysteine sulfinic acid decarboxylase activities occur in constant ratio in various tissues, and it is likely that both substrates are decarboxylated by the same enzyme. In general, cysteine sulfinic acid is the preferred substrate, and there is little formation of taurine by way of cysteic acid.

3. S-Oxidation of cysteamine released by the catabolism of pantothenic acid (Section 12.2.2) or formed by the decarboxylation of cysteine.

In the liver and brain, the main pathway is by way of cysteine sulfinic acid, whereas in tissues with low cysteine sulfinic acid decarboxylase activity the main precursor of taurine is cysteamine.

The central nervous system has at least three enzymes capable of decarboxylating cysteine sulfonic acid, one of which is glutamate decarboxylase. Glutamate and cysteine sulfinic acid are mutually competitive. In some brain regions, more than half the total cysteine sulfinic acid decarboxylase activity may be from glutamate decarboxylase.

In addition to taurocholic acid in the bile, free taurine is excreted in the urine. At times of low intake or when synthesis is impaired, for example by vitamin B_6 deficiency, the renal tubular resorption of taurine is increased, thus reducing urinary losses.

Feeding experimental animals on high taurine diets results in increased urinary excretion, but has little or no effect on endogenous synthesis.

14.5.2 Metabolic Functions of Taurine

14.5.2.1 Taurine Conjugation of Bile Acids   The bile acids are conjugated with either taurine or glycine to increase their polarity. Increased availability of taurine results in decreased glycine conjugation and an increase in biliary taurocholic acid. Conversely, glycine overload results in an increase in the plasma concentration of taurine, apparently as a result of increased glycocholic acid formation and reduced utilization of taurine for bile acid conjugation.

Although supplements of taurine alter the ratio of taurocholic:glycocholic acids, they have no effect on the total output of bile salts or on fat absorption in normal subjects. There is some evidence that patients with cystic fibrosis have improved fat absorption if given taurine supplements. This may be because taurine-conjugated bile acids are generally reabsorbed lower down the small intestine than glycine conjugates; in patients whose intestinal absorption is compromised, this may give a beneficial increase in the total length of intestinal tract available for fat absorption.

In addition to bile acid conjugation, a variety of other compounds may also be excreted as taurine conjugates, including retinoic acid (Section 2.2.1.3) and a number of xenobiotics.

14.5.2.2 Taurine in the Central Nervous System   There is a relatively high concentration of taurine in the central nervous system – higher than would be expected for a neurotransmitter and without a specific anatomical localization. As in the retina, the main function of taurine in the central nervous system seems to be as an osmolyte (Hussy et al., 2000; Saransaari and Oja, 2000).

The concentration of taurine in the developing brain is three- to four-fold higher than in the adult brain, and falls rapidly between birth and weaning. Unlike other amino acids, taurine is transported within axons to a greater extent in young animals than in adults. The highest concentrations of taurine in the brain and the greatest rates of axonal transport occur before and during the process of synaptic development, suggesting that it may have a role in the development of the central nervous system and the postnatal development of synaptic connections (Lima et al., 2001).

14.5.2.3 **Taurine and Heart Muscle** Cardiomyopathy is a major problem in taurine-deficient cats, and after prolonged deficiency, there is a failure of contractility, leading to heart failure. Heart muscle concentrates taurine from the bloodstream, and the heart can synthesize taurine by oxidation of cysteamine, although not by the cysteine sulfinic acid decarboxylase pathway. Pharmacologically, taurine affects drug-induced cardiac arrhythmias by depressing the hyperirritability caused by loss of potassium – a digitalis-like action that suggests an effect on membrane permeability and ion flux, and perhaps especially on the maintenance of stable intracellular concentrations of calcium (Nittynen et al., 1999; Militante et al., 2000).

14.5.3 **The Possible Essentiality of Taurine**

Taurine is a dietary essential in the cat, which is an obligate carnivore with a limited capacity for taurine synthesis from cysteine. On a taurine-free diet, neither supplementary methionine nor cysteine will maintain normal plasma concentrations of taurine, because cats have an alternative pathway of cysteine metabolism: reaction with mevalonic acid to yield felinine (3-hydroxy-1,1-dimethylpropyl-cysteine), which is excreted in the urine. The activity of cysteine sulfinic acid decarboxylase in cat liver is very low.

It is not known to what extent taurine may be a dietary essential for human beings. There is little cysteine sulfinic acid decarboxylase activity in the human liver and, like the cat, loading doses of methionine and cysteine do not result in any significant increase in plasma taurine. This may be because cysteine sulfinic acid can also undergo transamination to β-sulfhydryl pyruvate, which then loses sulfur dioxide nonenzymically to form pyruvate, thus regulating the amount of taurine that is formed from cysteine. There is no evidence of the development of any taurine deficiency disease under normal conditions.

There are very few plant sources of taurine, and strict vegetarians have a very low intake of preformed taurine. Nevertheless, the plasma concentration of taurine in strict vegetarians is generally between 40 to 50 μmol per L, compared with concentrations between 55 to 70 μmol per L in omnivores.

In children undergoing long-term total parenteral nutrition without taurine supplements, there are changes in the electroretinogram similar to those seen in the taurine-deficient cat, suggesting that there is a requirement for some preformed taurine and that endogenous synthesis may be inadequate.

It has been suggested that preterm infants may require a dietary source of preformed taurine; breast milk initially contains a high concentration (about 300 μmol per L), and breast-fed infants maintain a higher plasma concentration of taurine than those fed on formula without added taurine (Chesney et al., 1998). Although milk from vegan mothers has a low concentration of taurine, and their infants have lower plasma concentrations and urinary excretion of taurine than the infants of omnivore mothers, there is no evidence that (full-term) infants of vegan mothers show any signs of taurine deficiency.

14.6 UBIQUINONE (COENZYME Q)

Ubiquinone functions as a carrier in the mitochondrial electron transport chain; it is responsible for the proton pumping associated with complex I (Brandt, 1999) and is directly reduced by the citric acid cycle enzyme succinate dehydrogenase (Lancaster, 2002). As shown in Figure 14.8, it undergoes two single-electron reduction reactions to form the relatively stable semiquinone radical, then the fully reduced quinol. In addition to its role in the electron transport chain, it has been implicated as a coantioxidant in membranes and plasma lipoproteins, acting together with vitamin E (Section 4.3.1; Thomas et al., 1995, 1999).

There is no evidence that ubiquinone is a dietary essential, because it is synthesized in the body from mevalonate; indeed, dietary ubiquinone is relatively poorly absorbed (Dallner and Sindelar, 2000). However, because of its potential antioxidant action, supplements have been used, with little evidence

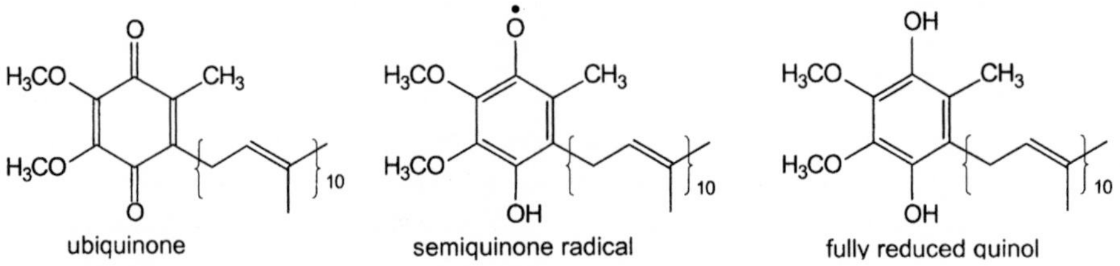

Figure 14.8. Ubiquinone. Relative molecular mass (M_r): 863.3.

of efficacy, in the hope of preventing cancer and cardiovascular and neurode-generative diseases (Beal, 1999; Hodges et al., 1999; Langsjoen and Langsjoen, 1999; Overvad et al., 1999; Thomas et al., 1999).

14.7 PHYTONUTRIENTS: POTENTIALLY PROTECTIVE COMPOUNDS IN PLANT FOODS

There is overwhelming epidemiological evidence that diets rich in fruit and vegetables are associated with a lower incidence of cancer, cardiovascular, and other degenerative diseases. To some extent, this may be because such diets provide less fat, and especially saturated fat, than diets that are richer in meat. The relatively high content of vitamins C and E and carotenoids in plant foods may also be important. In addition, fruits and vegetables contain a wide variety of compounds that have (potential) protective actions. These compounds are not strictly nutrients, in that they are not dietary essentials and have no physiological function.

Many fruits contain salicylates, which inhibit the synthesis of thrombox-ane A_2, and have an anticoagulant action, in amounts that provide the same intake as the low dose of aspirin used as prophylaxis against thrombosis.

Plant sterols inhibit the intestinal absorption of cholesterol and so have a useful hypocholesterolemic action. They also inhibit endogenous synthesis of cholesterol, by inhibiting and repressing the regulatory enzyme of cholesterol synthesis, hydroxymethylglutaryl (HMG)-CoA reductase. Other compounds synthesized from mevalonate also inhibit and repress HMG-CoA reductase and have a hypocholesterolemic action, including squalene (found in relati-vely large amounts in olive oil), ubiquinone (Section 14.6), and the tocotrienols (Section 4.1).

A number of the terpenes in aromatic oils of citrus peel, herbs, and spices inhibit the isoprenylation of the P21-*ras* oncogene product. Isoprenylation is essential for the biological action of the *ras* protein, which is associated with pancreatic cancer.

14.7.1 Allyl Sulfur Compounds

Members of the allium family (onions, garlic, and leeks) contain cysteine sulfoxide derivatives (allyl sulfur compounds), such as allicin and alliin (see Figure 14.9). When the plant cells are damaged, the enzyme alliinase is re-leased from vacuoles and catalyzes the formation of thiosulfinates and thiols, including the lachrymator thiopropanal S-oxide. Their function in the plant is presumably to provide protection against attack by pests.

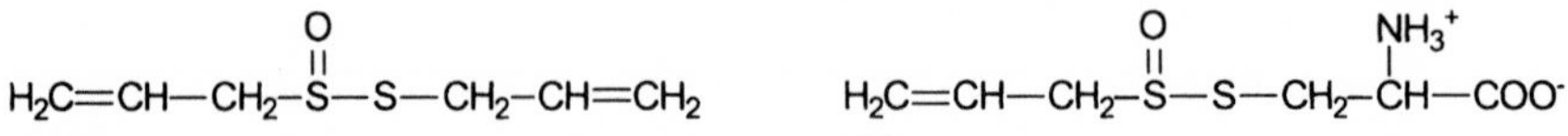

$$\text{H}_2\text{C}{=}\text{CH}{-}\text{CH}_2{-}\overset{\overset{\displaystyle O}{\|}}{\text{S}}{-}\text{S}{-}\text{CH}_2{-}\text{CH}{=}\text{CH}_2$$
allicin

$$\text{H}_2\text{C}{=}\text{CH}{-}\text{CH}_2{-}\overset{\overset{\displaystyle O}{\|}}{\text{S}}{-}\text{S}{-}\text{CH}_2{-}\overset{\overset{\displaystyle \text{NH}_3^+}{|}}{\text{CH}}{-}\text{COO}^-$$
alliin

Figure 14.9. Allyl sulfur compounds allicin and alliin. Relative molecular masses (M_r): allicin, 162.3; and alliin, 177.2.

The allyl sulfur compounds have two actions that may protect against the development of cancer:

1. They reduce the activation of many procarcinogens to the active carcinogen by lowering the activity of microsomal cytochrome P_{450} enzymes. They achieve this by acting as partial substrates of the enzyme, leading to mechanism-dependent inhibition; antagonizing the induction of cytochrome P_{450} by ethanol and various other compounds; and by decreasing the translation of mRNA, with no effect on transcription.
2. They increase the metabolic clearance of potential carcinogens and their metabolites, by induction of glutathione S-transferases.

The allyl sulfur compounds of garlic also have an anticoagulant action (by inhibiting platelet coagulability) and inhibit cholesterol synthesis by inactivating HMG-CoA reductase.

14.7.2 Flavonoids and Polyphenols

A wide variety of compounds collectively known as flavonoids (or sometimes as bioflavonoids) occur in plants as glycosides; some function to defend the plants against attack, others are pigments in flowers and fruits. Many of the glycosides are hydrolyzed by intestinal bacterial glycosidases, and the aglycones are absorbed. Figure 14.10 shows the six main types of flavonoid aglycone.

In the 1940s, the flavonoids were known as vitamin P (for permeability), because they were shown to have effects on the permeability of blood capillaries. They were not shown to be dietary essentials. By the 1970s, they were regarded as hazardous mutagens and potential carcinogens, because they can undergo redox cycling reactions and generate oxygen radicals. By the 1990s, they were regarded as potentially protective compounds with three types of action:

1. Although they can undergo redox cycling, they can also act as radical-trapping antioxidants.
2. They reduce the activation of many procarcinogens to the active proximate carcinogen by lowering the activity of microsomal cytochrome P_{450} enzymes.

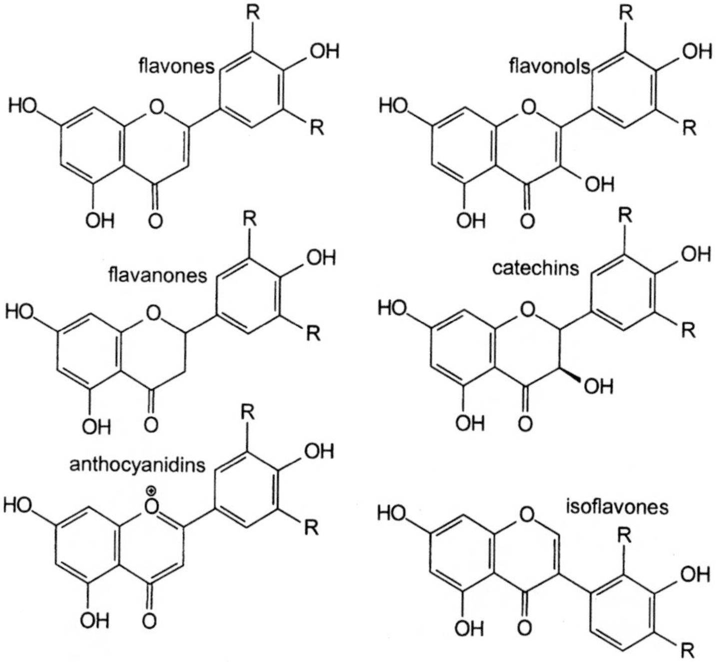

Figure 14.10. Major classes of flavonoids.

3. They increase the metabolic clearance of potential carcinogens and their metabolites by inducing the enzymes involved in conjugation for excretion.

Some of the flavonoids also have estrogenic and antiestrogenic actions (Section 14.7.4).

14.7.3 Glucosinolates

Glucosinolates are glucosides of sulfur-containing amino acid derivatives found mainly in brassicas; the enzyme myrosinase in the plant and similar intestinal bacterial enzymes catalyze cleavage of the glycoside to form a variety of isothiocyanates, thiocyanates, and the aglycone. Figure 14.11 shows the general structure of glucosinolates.

The glucosinolates have two actions that may protect against the development of cancer:

1. They reduce the activation of many procarcinogens to the active proximate carcinogen by both inhibiting and reducing the synthesis of microsomal cytochrome P_{450} enzymes.

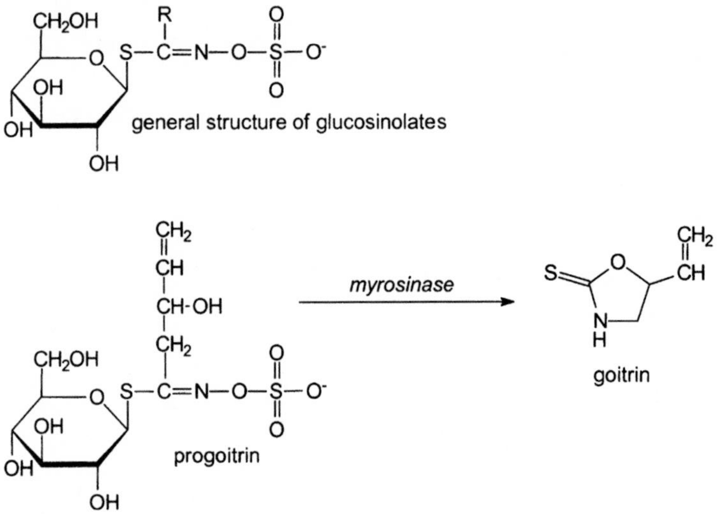

Figure 14.11. Glucosinolates.

2. They increase the metabolic clearance of potential carcinogens and their metabolites, by the induction of glutathione S-transferases and quinone reductases.

Myrosinase action on the glucosinolate progoitrin yields goitrin (see Figure 14.11), which inhibits the synthesis of thyroid hormones by inhibition of the iodination of monoiodotyrosine to diiodotyrosine and thus has a goitrogenic action. In addition, the thiocyanate released by myrosinase competes with iodide for uptake into the thyroid and may be goitrogenic when iodine status is marginal. There is no evidence that normal consumption of vegetables has any effect on thyroid hormone status, although goiter is a problem when cattle are fed on large amounts of brassicas.

14.7.4 Phytoestrogens

A number of compounds shown in Figure 14.12 that occur in plant foods as glycosides and other conjugates have weak estrogenic/antiestrogenic actions, and are collectively known as phytoestrogens. They all have two hydroxyl groups that are the same distance apart as the hydroxyl groups of estradiol and can bind to estrogen receptors. The amounts of phytoestrogens produced increase in response to microbial and insect attack, suggesting that they have antibacterial or antifungal actions in the plant. They produce typical estrogen responses in animals, with a biological activity 1/500 to 1/1,000 of that of estradiol.

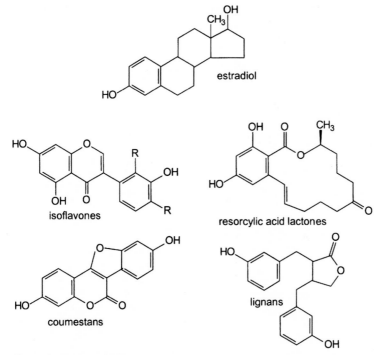

Figure 14.12. Estradiol and the major phytoestrogens.

   High consumption of legumes, especially soya beans, which are rich sources of phytoestrogens, is associated with lower incidence of breast and uterine cancer, as well as a lower incidence of osteoporosis. The estrogenic action is probably responsible for the effects on the development of osteoporosis, whereas three factors may be involved in the effect on hormone-dependent cancer:

1. The isoflavones are mainly antiestrogenic, because they compete with estradiol for receptor binding, but the phytoestrogen–receptor complex does not undergo normal activation, thus it has only a weak effect on hormone response elements on DNA. Even those phytoestrogens that have a mainly estrogenic action will reduce responsiveness to estradiol because they compete for receptor binding but have lower biological activity.
2. The phytoestrogens increase the synthesis of sex hormone binding globulin in the liver by stabilizing mRNA, leading to a lower circulating concentration of free estradiol.

3. Some of the phytoestrogens inhibit aromatase and therefore reduce the endogenous synthesis of estradiol, especially the unregulated synthesis that occurs in adipose tissue.

FURTHER READING

Bingham SA, Atkinson C, Liggins J, Bluck L, and Coward A (1998) Phyto-oestrogens: where are we now? *British Journal of Nutrition* **79**, 393–406.

Crane FL (2001) Biochemical functions of coenzyme Q10. *Journal of the American College of Nutrition* **20**, 591–8.

Dallner G and Sindelar PJ (2000) Regulation of ubiquinone metabolism. *Free Radicals in Biology and Medicine* **29**, 285–94.

Davies SR, Dalais FS, Simpson ER, and Murkies AL (1999) Phytoestrogens in health and disease. *Recent Progress in Hormone Research* **54**, 185–211.

Johnson I, Williamson G, and Musk SRR (1994) Anticarcinogenic factors in plant foods: a new class of nutrients? *Nutrition Research Reviews* **7**, 175–204.

Kurzer MS and Xu X (1997) Dietary phytoestrogens. *Annual Reviews of Nutrition* **17**, 353–81.

Liao S, Kao YH, and Hiipakka RA (2001) Green tea: biochemical and biological basis for health benefits. *Vitamins and Hormones* **62**, 1–94.

Liu L and Yeh YY (2002) *S*-alk(en)yl cysteines of garlic inhibit cholesterol synthesis by deactivating HMG-CoA reductase in cultured rat hepatocytes. *Journal of Nutrition* **132**, 1129–34.

Nugon-Baudon L and Rabot S (1994) Glucosinolates and glucosinolate derivatives: implications for protection against chemical carcinogenesis. *Nutrition Research Reviews* **7**, 205–32.

Overvad K, Diamant B, Holm L, Holmer G, Mortensen SA, and Stender S (1999) Coenzyme Q10 in health and disease. *European Journal of Nutrition* **53**, 764–70.

Persky AM and Brazeau GA (2001) Clinical pharmacology of the dietary supplement creatine monohydrate. *Pharmacology Reviews Reviews* **53**, 161–76.

Rebouche CJ and Seim H (1998) Carnitine metabolism and its regulation in microorganisms and mammals. *Annual Reviews of Nutrition* **18**, 39–61.

Redmond HP, Stapleton PP, Neary P, and Bouchier-Hayes D (1998) Immunonutrition: the role of taurine. *Nutrition* **14**, 599–604.

Ross JA and Kasum CM (2002) Dietary flavonoids: bioavailability, metabolic effects and safety. *Annual Reviews of Nutrition* **22**, 19–34.

Schaffer S, Takahashi K, and Azuma J (2000) Role of osmoregulation in the actions of taurine. *Amino Acids* **19**, 527–46.

Schwartz B and Klinman JP (2001) Mechanisms of biosynthesis of protein-derived redox cofactors. *Vitamins and Hormones* **61**, 219–39.

Various authors (1996) Colloquium proceedings: bioactive components of food. *Biochemical Society Transactions* **24**, 771–835.

Various authors (1996) Symposium proceedings: physiologically active substances in plant foods. *Proceedings of the Nutrition Society* **55**, 371–446.

Vaz FM and Wanders RJ (2002) Carnitine biosynthesis in mammals. *Biochemical Journal* **361**, 417–29.

Wiseman H (1999) The bioavailability of non-nutrient plant factors: dietary flavonoids and phyto-oestrogens. *Proceedings of the Nutrition Society* **58,** 139–46.

Wyss M and Kaddurah-Daouk R (2000) Creatine and creatinine metabolism. *Physiological Reviews* **80,** 1107–1213.

Yang CS, Landau JM, Huang MT, and Newmark HL (2001) Inhibition of carcinogenesis by dietary polyphenolic compounds. *Annual Reviews of Nutrition* **21,** 381–406.

Zeisel SH (2000) Choline: an essential nutrient for humans. *Nutrition* **16,** 669–71.

References cited in the text are listed in the Bibliography.

Bibliography

Achkar CC, Derguini F, Blumberg B, Langston A, Levin AA, Speck J, Evans RM, Bolado J Jr., Nakanishi K, Buck J, et al. (1996) 4-Oxoretinol, a new natural ligand and transactivator of the retinoic acid receptors. *Proceedings of the National Academy of Sciences of the USA* **93,** 4879–84.

Adams PW, Wynn V, Folkard J, and Seed M (1976) Influence of oral contraceptives, pyridoxine (vitamin B_6) and tryptophan on carbohydrate metabolism. *Lancet* **(i),** 759–64.

Adiga PR (1994) Riboflavin carrier protein in reproduction. In *Vitamin Receptors: Vitamins as Ligands in Cell Communication*, K Dakshinamurti (ed.), pp. 137–76. Cambridge, UK: Cambridge University Press.

Adiga PR, Subramanian S, Rao J, and Kumar M (1997) Prospects of riboflavin carrier protein (RCP) as an antifertility vaccine in male and female mammals. *Human Reproduction Update* **3,** 325–34.

Ahmed F, Jones DB, and Jackson AA (1990) The interaction of vitamin A deficiency and rotavirus infection in the mouse. *British Journal of Nutrition* **63,** 363–73.

Airenne KJ, Marjomaki VS, and Kulomaa MS (1999) Recombinant avidin and avidin-fusion proteins. *Biomolecular Engineering* **16,** 87–92.

Akompong T, Eksi S, Williamson K, and Haldar K (2000a) Gametocytocidal activity and synergistic interactions of riboflavin with standard antimalarial drugs against growth of *Plasmodium falciparum* in vitro. *Antimicrobial Agents and Chemotherapy* **44,** 3107–11.

Akompong T, Ghori N, and Haldar K (2000b) In vitro activity of riboflavin against the human malaria parasite *Plasmodium falciparum. Antimicrobial Agents and Chemotherapy* **44,** 88–96.

Alderton WK, Cooper CE, and Knowles RG (2001) Nitric oxide synthases: structure, function and inhibition. *Biochemical Journal* **357,** 593–615.

Allgood VE and Cidlowski JA (1992) Vitamin B_6 modulates transcriptional activation by multiple members of the steroid hormone receptor superfamily. *Journal of Biological Chemistry* **267,** 3819–24.

Alpha-Tocopherol Beta-Carotene Cancer Prevention Study Group (1994) The effect of vitamin E and beta carotene on the incidence of lung and other cancers in male smokers. *New England Journal of Medicine* **330,** 1029–35.

Alston TA and Abeles RH (1987) Enzymatic conversion of the antibiotic metronidazole to an analog of thiamine. *Archives of Biochemistry and Biophysics* **257**, 357–62.

Altman K and Greengard O (1966) Correlation of kynurenine excretion with liver tryptophan pyrrolase levels in disease and after hydrocortisone induction. *Journal of Clinical Investigation* **45**, 1525–34.

American Academy of Pediatrics Committee on Infectious Diseases (1993) American Academy of Pediatrics Committee on Infectious Diseases: vitamin A treatment of measles. *Pediatrics* **91**, 1014–15.

American College of Rheumatology Ad Hoc Committee on Glucocorticoid-Induced Osteoporosis (2001) Recommendations for the prevention and treatment of glucocorticoid-induced osteoporosis: 2001 update. *Arthritis and Rheumatism* **44**, 1496–1503.

Amin S, LaValley MP, Simms RW, and Felson DT (1999) The role of vitamin D in corticosteroid-induced osteoporosis: a meta-analytic approach. *Arthritis and Rheumatism* **42**, 1740–51.

Annous KF and Song WO (1995) Pantothenic acid uptake and metabolism by red blood cells of rats. *Journal of Nutrition* **125**, 2586–93.

Appling DR and Chytil F (1981) Evidence of a role for retinoic acid (vitamin A-acid) in the maintenance of testosterone production in male rats. *Endocrinology* **108**, 2120–24.

Arnhold T, Nau H, Meyer S, Rothkoetter HJ, and Lampen AD (2002) Porcine intestinal metabolism of excess vitamin A differs following vitamin A supplementation and liver consumption. *Journal of Nutrition* **132**, 197–203.

Arrigoni-Martelli E and Caso V (2001) Carnitine protects mitochondria and removes toxic acyls from xenobiotics. *Drugs Experimental Clinical Research* **27**, 27–49.

Attwood PV (1995) The structure and the mechanism of action of pyruvate carboxylase. *International Journal of Biochemistry and Cell Biology* **27**, 231–49.

Audi L, Garcia-Ramirez M, and Carrascosa A (1999) Genetic determinants of bone mass. *Hormone Research* **51**, 105–23.

Aw TY, Jones DP, and McCormick DB (1983) Uptake of riboflavin by isolated rat liver cells. *Journal of Nutrition* **113**, 1249–54.

Awumey EM, Mitra DA, Hollis BW, Kumar R, and Bell NH (1998) Vitamin D metabolism is altered in Asian Indians in the southern United States: a clinical research center study. *Journal of Clinical Endocrinology and Metabolism* **83**, 169–73.

Azzi A, Breyer I, Feher M, Pastori M, Ricciarelli R, Spycher S, Staffieri M, Stocker A, Zimmer S, and Zingg JM (2000) Specific cellular responses to alpha-tocopherol. *Journal of Nutrition* **130**, 1649–52.

Azzi A, Breyer I, Feher M, Ricciarelli R, Stocker A, Zimmer S, and Zingg J (2001) Nonantioxidant functions of alpha-tocopherol in smooth muscle cells. *Journal of Nutrition* **131**, 378S–81S.

Azzi A, Ricciarelli R, and Zingg JM (2002) Non-antioxidant molecular functions of alpha-tocopherol (vitamin E). *FEBS Letters* **519**, 8–10.

Babior BM (1992) The respiratory burst oxidase. *Advances in Enzymology and Related Areas of Molecular Biology* **65**, 49–95.

Bacher A, Eberhardt S, Fischer M, Kis K, and Richter G (2000) Biosynthesis of vitamin B_2 (riboflavin). *Annual Reviews of Nutrition* **20**, 153–67.

Bacher A, Eberhardt S, Eisenreich W, Fischer M, Herz S, Illarionov B, Kis K, and Richter G (2001) Biosynthesis of riboflavin. *Vitamins and Hormones* **61**, 1–49.

Badawy AA (1977) The functions and regulation of tryptophan pyrrolase. *Life Sciences* **21**, 755–68.

Badawy AA and Evans M (1975) Regulation of rat liver tryptophan pyrrolase by its cofactor haem: experiments with haematin and 5-aminolaevulinate and comparison with the substrate and hormonal mechanisms. *Biochemical Journal* **150**, 511–20.

Badawy AA-B (2002) Tryptophan metabolism in alcoholism. *Nutrition Research Reviews* **15**, 123–52.

Baggenstoss AH, Christensen NA, Berge KG, Baldus WP, Spieckerman RE, and Ellefson RD (1967) Fine structural changes in the liver in hypercholesterolemic patients receiving long-term nicotinic acid therapy. *Mayo Clinic Proceedings* **42**, 385–99.

Baggott JE, Tamura T, and Baker H (2001) Re-evaluation of the metabolism of oral doses of racemic carbon-6 isomers of formyltetrahydrofolate in human subjects. *British Journal of Nutrition* **85**, 653–7.

Bagley PJ and Selhub J (1998) A common mutation in the methylenetetrahydrofolate reductase gene is associated with an accumulation of formylated tetrahydrofolates in red blood cells. *Proceedings of the National Academy of Sciences of the USA* **95**, 13217–20.

Bailey LB and Gregory JF, 3rd (1999) Folate metabolism and requirements. *Journal of Nutrition* **129**, 779–82.

Baker EM, Hodges RE, Hood J, Sauberlich HE, and March SC (1969) Metabolism of ascorbic-1-^{14}C acid in experimental human scurvy. *American Journal of Clinical Nutrition* **22**, 549–58.

Baker EM, Hodges RE, Hood J, Sauberlich HE, March SC, and Canham JE (1971) Metabolism of ^{14}C- and ^{3}H-labeled L-ascorbic acid in human scurvy. *American Journal of Clinical Nutrition* **24**, 444–54.

Ball GFM (1998) *Bioavailability and Analysis of Vitamins in Foods.* London: Chapman and Hall.

Banerjee S, Hawksby C, Miller S, Dahill S, Beattie AD, and McColl KE (1994) Effect of *Helicobacter pylori* and its eradication on gastric juice ascorbic acid. *Gut* **35**, 317–22.

Bannister DW (1976a) The biochemistry of fatty liver and kidney syndrome. Biotin-mediated restoration of hepatic gluconeogenesis in vitro and its relationship to pyruvate carboxylase activity. *Biochemical Journal* **156**, 167–73.

Bannister DW (1976b) Hepatic gluconeogenesis in chicks: effect of biotin on gluconeogenesis in biotin-deficiency and fatty liver and kidney syndrome. *Comparative Biochemistry and Physiology B* **53**, 575–9.

Baram J, Chabner BA, Drake JC, Fitzhugh AL, Sholar PW, and Allegra CJ (1988) Identification and biochemical properties of 10-formyldihydrofolate, a novel folate found in methotrexate-treated cells. *Journal of Biological Chemistry* **263**, 7105–11.

Barbarat B and Podevin RA (1986) Pantothenate-sodium cotransport in renal brush-border membranes. *Journal of Biological Chemistry* **261**, 14455–60.

Barclay LL, Gibson GE, and Blass JP (1981) Impairment of behavior and acetylcholine metabolism in thiamine deficiency. *Journal of Pharmacology and Experimental Therapeutics* **217**, 537–43.

Barnard HC, de Kock JJ, Vermaak WJ, and Potgieter GM (1987) A new perspective in the assessment of vitamin B-6 nutritional status during pregnancy in humans. *Journal of Nutrition* **117**, 1303–6.

Bar-Shavit Z, Teitelbaum SL, Reitsma P, Hall A, Pegg LE, Trial J, and Kahn AJ (1983) Induction of monocytic differentiation and bone resorption by 1,25-dihydroxyvitamin D$_3$. *Proceedings of the National Academy of Sciences of the USA* **80**, 5907–11.

Barua AB (1997) Retinoyl beta-glucuronide: a biologically active form of vitamin A. *Nutrition Reviews* **55**, 259–67.

Barua AB and Olson JA (2000) Beta-carotene is converted primarily to retinoids in rats in vivo. *Journal of Nutrition* **130**, 1996–2001.

Bates CJ (1977) Proline and hydroxyproline excretion and vitamin C status in elderly human subjects. *Clinical Science and Molecular Medicine* **52**, 535–43.

Bates CJ (1987a) Human requirements for riboflavin. *American Journal of Clinical Nutrition* **46**, 122–3.

Bates CJ (1987b) Human riboflavin requirements, and metabolic consequences of deficiency in man and animals. *World Review of Nutrition and Dietetics* **50**, 215–65.

Bates CJ (1989) Metabolism of [^{14}C]adipic acid in riboflavin-deficient rats: a test in vivo for fatty acid oxidation. *Journal of Nutrition* **119**, 887–91.

Bates CJ (1990) Liberation of ^{14}CO$_2$ from [^{14}C]adipic acid and [^{14}C]octanoic acid by adult rats during riboflavin deficiency and its reversal. *British Journal of Nutrition* **63**, 553–62.

Bates C (1993) Flair concerted action no 10 status papers: riboflavin. *International Journal of Vitamin and Nutrition Research* **3**, 274–7.

Bates C and Heseker H (1994) Human bioavailability of vitamins. *Nutrition Research Reviews* **7**, 93–127.

Bates CJ, Pentieva KD, and Prentice A (1999a) An appraisal of vitamin B$_6$ status indices and associated confounders, in young people aged 4–18 years and in people aged 65 years and over, in two national British surveys. *Public Health Nutrition* **2**, 529–35.

Bates CJ, Pentieva KD, Prentice A, Mansoor MA, and Finch S (1999b) Plasma pyridoxal phosphate and pyridoxic acid and their relationship to plasma homocysteine in a representative sample of British men and women aged 65 years and over. *British Journal of Nutrition* **81**, 191–201.

Batres RO and Olson JA (1987) A marginal vitamin A status alters the distribution of vitamin A among parenchymal and stellate cells in rat liver. *Journal of Nutrition* **117**, 874–9.

Baumgartner ER and Suormala T (1997) Multiple carboxylase deficiency: inherited and acquired disorders of biotin metabolism. *International Journal of Vitamin and Nutrition Research* **67**, 377–84.

Baumgartner ER and Suormala T (1999) Inherited defects of biotin metabolism. *Biofactors* **10**, 287–90.

Baylor D (1996) How photons start vision. *Proceedings of the National Academy of Sciences of the USA* **93**, 560–5.

Beal MF (1999) Coenzyme Q10 administration and its potential for treatment of neurodegenerative diseases. *Biofactors* **9**, 261–6.

Beatty S, Koh H, Phil M, Henson D, and Boulton M (2000) The role of oxidative stress in the pathogenesis of age-related macular degeneration. *Survey of Ophthalmology* **45**, 115–34.

Beers KW, Chini EN, and Dousa TP (1995) All-*trans*-retinoic acid stimulates synthesis of cyclic ADP-ribose in renal LLC-PK1 cells. *Journal of Clinical Investigation* **95,** 2385–90.

Begley TP, Downs DM, Ealick SE, McLafferty FW, Van Loon AP, Taylor S, Campobasso N, Chiu HJ, Kinsland C, Reddick JJ, and Xi J (1999) Thiamin biosynthesis in prokaryotes. *Archives of Microbiology* **171,** 293–300.

Begley TP, Kinsland C, and Strauss E (2001) The biosynthesis of coenzyme A in bacteria. *Vitamins and Hormones* **61,** 157–71.

Bellovino D, Morimoto T, Tosetti F, and Gaetani S (1996) Retinol binding protein and transthyretin are secreted as a complex formed in the endoplasmic reticulum in HepG2 human hepatocarcinoma cells. *Experimental Cell Research* **222,** 77–83.

Bender DA (1980) Effects of benserazide, carbidopa and isoniazid administration on tryptophan-nicotinamide nucleotide metabolism in the rat. *Biochemical Pharmacology* **29,** 2099–2104.

Bender DA (1983) Effects of a dietary excess of leucine on the metabolism of tryptophan in the rat: a mechanism for the pellagragenic action of leucine. *British Journal of Nutrition* **50,** 25–32.

Bender DA (1987) Oestrogens and vitamin B_6 – actions and interactions. *World Review of Nutrition and Dietetics* **51,** 140–88.

Bender DA (1989a) Effects of a dietary excess of leucine and of the addition of leucine and 2-oxo-isocaproate on the metabolism of tryptophan and niacin in isolated rat liver cells. *British Journal of Nutrition* **61,** 629–40.

Bender DA (1989b) Vitamin B_6 requirements and recommendations. *European Journal of Nutrition* **43,** 289–309.

Bender DA and McCreanor GM (1985) Kynurenine hydroxylase: a potential rate-limiting enzyme in tryptophan metabolism. *Biochemical Society Transactions* **13,** 441–3.

Bender DA and Olufunwa R (1988) Utilization of tryptophan, nicotinamide and nicotinic acid as precursors for nicotinamide nucleotide synthesis in isolated rat liver cells. *British Journal of Nutrition* **59,** 279–87.

Bender DA and Totoe L (1984a) High doses of vitamin B_6 in the rat are associated with inhibition of hepatic tryptophan metabolism and increased uptake of tryptophan into the brain. *Journal of Neurochemistry* **43,** 733–6.

Bender DA and Totoe L (1984b) Inhibition of tryptophan metabolism by oestrogens in the rat: a factor in the aetiology of pellagra. *British Journal of Nutrition* **51,** 219–24.

Bender DA, Magboul BI, and Wynick D (1982) Probable mechanisms of regulation of the utilization of dietary tryptophan, nicotinamide and nicotinic acid as precursors of nicotinamide nucleotides in the rat. *British Journal of Nutrition* **48,** 119–27.

Berdowska A and Zwirska-Korczala K (2001) Neopterin measurement in clinical diagnosis. *Journal of Clinical Pharmacology and Therapeutics* **26,** 319–29.

Berg JP and Haug E (1999) Vitamin D: a hormonal regulator of the cAMP signalling pathway. *Critical Reviews in Biochemistry and Molecular Biology* **34,** 315–23.

Berthon HA, Kuchel PW, and Nixon PF (1992) High control coefficient of transketolase in the nonoxidative pentose phosphate pathway of human erythrocytes: NMR, antibody, and computer simulation studies. *Biochemistry* **31,** 12792–8.

Bertoldi M and Voltattorni CB (2000) Reaction of dopa decarboxylase with L-aromatic amino acids under aerobic and anaerobic conditions. *Biochemical Journal* **352**(Pt 2), 533–8.

Bertoldi M and Voltattorni CB (2001) Dopa decarboxylase exhibits low pH half-transaminase and high pH oxidative deaminase activities toward serotonin (5-hydroxytryptamine). *Protein Science* **10**, 1178–86.

Bertram JS (1999) Carotenoids and gene regulation. *Nutrition Reviews* **57**, 182–91.

Bessey OA, Adam DJD, and Hansen AE (1957) Intake of vitamin B_6 and infantile convulsions: a first approximation to the requirements of pyridoxine in infants. *Pediatrics* **20**, 33–44.

Bettendorff L (1996) A non-cofactor role of thiamine derivatives in excitable cells? *Archives of Physiology and Biochemistry* **104**, 745–51.

Bettendorff L, Hennuy B, De Clerck A, and Wins P (1994) Chloride permeability of rat brain membrane vesicles correlates with thiamine triphosphate content. *Brain Research* **652**, 157–60.

Beynon RJ, Bartram C, Hopkins P, Toescu V, Gibson H, Phoenix J, and Edwards RH (1995) McArdle's disease: molecular genetics and metabolic consequences of the phenotype. *Muscle Nerve* **3**, S18–S22.

Biagioni S, Tata AM, De Jaco A, and Augusti-Tocco G (2000) Acetylcholine synthesis and neuron differentiation. *International Journal of Developmental Biology* **44**, 689–97.

Bikle DD (1995) 1,25$(OH)_2D_3$-regulated human keratinocyte proliferation and differentiation: basic studies and their clinical application. *Journal of Nutrition* **125**, 1709S–14S.

Bikle DD, Ng D, Tu CL, Oda Y, and Xie Z (2001) Calcium- and vitamin D-regulated keratinocyte differentiation. *Molecular and Cellular Endocrinology* **177**, 161–71.

Billings RE (1984a) Decreased hepatic 5,10-methylenetetrahydrofolate reductase activity in mice after chronic phenytoin treatment. *Molecular Pharmacology* **25**, 459–66.

Billings RE (1984b) Interactions between folate metabolism, phenytoin metabolism, and liver microsomal cytochrome P450. *Drug Nutrient Interactions* **3**, 21–32.

Binkley N and Krueger D (2000) Hypervitaminosis A and bone. *Nutrition Reviews* **58**, 138–44.

Binkley NC, Krueger DC, Engelke JA, Foley AL, and Suttie JW (2000) Vitamin K supplementation reduces serum concentrations of under-gamma-carboxylated osteocalcin in healthy young and elderly adults. *American Journal of Clinical Nutrition* **72**, 1523–8.

Bitsch R (1993) Vitamin B_6. *International Journal of Vitamin and Nutrition Research* **63**, 278–82.

Bitsch R, Toth-Dersi A, and Hoetzel D (1985) Biotin deficiency and biotin supply. *Annals of the New York Academy of Sciences* **447**, 133–9.

Black AL, Guirard BM, and Snell EE (1978) The behavior of muscle phosphorylase as a reservoir for vitamin B_6 in the rat. *Journal of Nutrition* **108**, 670–77.

Blair PV, Kobayashi R, Edwards HM 3rd, Shay NF, Baker DH, and Harris RA (1999) Dietary thiamin level influences levels of its diphosphate form and thiamin-dependent enzymic activities of rat liver. *Journal of Nutrition* **129**, 641–8.

Blaner WS, Obunike JC, Kurlandsky SB, al-Haideri M, Piantedosi R, Deckelbaum RJ, and Goldberg IJ (1994) Lipoprotein lipase hydrolysis of retinyl ester. Possible implications for retinoid uptake by cells. *Journal of Biological Chemistry* **269**, 16559–65.

Blansjaar BA, Zwang R, and Blijenberg BG (1991) No transketolase abnormalities in Wernicke-Korsakoff patients. *Journal of Neurological Science* **106**, 88–90.

Blass JP and Gibson GE (1977) Abnormality of a thiamine-requiring enzyme in patients with Wernicke-Korsakoff syndrome. *New England Journal of Medicine* **297**, 1367–70.

Blomhoff R, Green MH, Green JB, Berg T, and Norum KR (1991) Vitamin A metabolism: new perspectives on absorption, transport, and storage. *Physiological Reviews* **71,** 951–90.

Blot WJ, Li JY, Taylor PR, Guo W, Dawsey S, Wang GQ, Yang CS, Zheng SF, Gail M, Li GY, et al. (1993) Nutrition intervention trials in Linxian, China: supplementation with specific vitamin/mineral combinations, cancer incidence, and disease-specific mortality in the general population. *Journal of the National Cancer Institute* **85,** 1483–92.

Blumberg B, Bolado J Jr., Derguini F, Craig AG, Moreno TA, Chakravarti D, Heyman RA, Buck J, and Evans RM (1996) Novel retinoic acid receptor ligands in Xenopus embryos. *Proceedings of the National Academy of Sciences of the USA* **93,** 4873–8.

Blutt SE and Weigel NL (1999) Vitamin D and prostate cancer. *Proceedings of the Society for Experimental Biology and Medicine* **221,** 89–98.

Board RG and Fuller R (1974) Non-specific antimicrobial defences of the avian egg, embryo and neonate. *Biological Reviews of the Cambridge Philosophical Society* **49,** 15–49.

Bock A, Forchhammer K, Heider J, Leinfelder W, Sawers G, Veprek B, and Zinoni F (1991) Selenocysteine: the 21st amino acid. *Molecular Microbiology* **5,** 515–20.

Boeckx RL and Dakshinamurti K (1974) Biotin-mediated protein biosynthesis. *Biochemical Journal* **140,** 549–56.

Boerman MH and Napoli JL (1996) Cellular retinol-binding protein-supported retinoic acid synthesis. Relative roles of microsomes and cytosol. *Journal of Biological Chemistry* **271,** 5610–16.

Bollinger Bollag W and Bollag RJ (2001) 1,25-Dihydroxyvitamin D(3), phospholipase D and protein kinase C in keratinocyte differentiation. *Molecular and Cellular Endocrinology* **177,** 173–82.

Bolton-Smith CPR, Fenton ST, Harrington DJ, and Shearer MJ (2000) Compilation of a provisional UK database for the phylloquinone (vitamin K_1) content of foods. *British Journal of Nutrition* **83,** 389–99.

Booth SL and Suttie JW (1998) Dietary intake and adequacy of vitamin K. *Journal of Nutrition* **128,** 785–8.

Borboni P, Magnaterra R, Rabini RA, Staffolani R, Porzio O, Sesti G, Fusco A, Mazzanti L, Lauro R, and Marlier LN (1996) Effect of biotin on glucokinase activity, mRNA expression and insulin release in cultured beta-cells. *Acta Diabetologica* **33,** 154–8.

Borschel MW, Kirksey A, and Hannemann RE (1986) Effects of vitamin B_6 intake on nutriture and growth of young infants. *American Journal of Clinical Nutrition* **43,** 7–15.

Bottiglieri T (1996) Folate, vitamin B_{12}, and neuropsychiatric disorders. *Nutrition Reviews* **54,** 382–90.

Boucher BJ (1998) Inadequate vitamin D status: does it contribute to the disorders comprising syndrome 'X'? *British Journal of Nutrition* **79,** 315–27.

Bowry VW, Ingold KU, and Stocker R (1992) Vitamin E in human low-density lipoprotein. When and how this antioxidant becomes a pro-oxidant. *Biochemical Journal* **288,** 341–4.

Boyan BD, Schwartz Z, Bonewald LF, and Swain LD (1989) Localization of 1,25-$(OH)_2D_3$-responsive alkaline phosphatase in osteoblast-like cells (ROS 17/2.8, MG 63, and MC

3T3) and growth cartilage cells in culture. *Journal of Biological Chemistry* **264,** 11879–86.

Boyan BD, Sylvia VL, Dean DD, Pedrozo H, Del Toro F, Nemere I, Posner GH, and Schwartz Z (1999) 1,25-(OH)$_2$D$_3$ modulates growth plate chondrocytes via membrane receptor-mediated protein kinase C by a mechanism that involves changes in phospholipid metabolism and the action of arachidonic acid and PGE$_2$. *Steroids* **64,** 129–36.

Boyan BD, Sylvia VL, Dean DD, and Schwartz Z (2001) 24,25-(OH)(2)D(3) regulates cartilage and bone via autocrine and endocrine mechanisms. *Steroids* **66,** 363–74.

Brandt U (1999) Proton translocation in the respiratory chain involving ubiquinone – a hypothetical semiquinone switch mechanism for complex I. *Biofactors* **9,** 95–101.

Brass EP (2000) Supplemental carnitine and exercise. *American Journal of Clinical Nutrition* **72,** 618S–23S.

Brass EP, Allen RH, Ruff LJ, and Stabler SP (1990) Effect of hydroxycobalamin[c-lactam] on propionate and carnitine metabolism in the rat. *Biochemical Journal* **266,** 809–15.

Brin M (1964) Erythrocyte as a biopsy tool for functional evaluation of thiamine adequacy. *Journal of the American Medical Association* **187,** 762–6.

Brouwer A and van den Berg KJ (1986) Binding of a metabolite of 3,4,3′,4′-tetrachlorobiphenyl to transthyretin reduces serum vitamin A transport by inhibiting the formation of the protein complex carrying both retinol and thyroxin. *Toxicology and Applied Pharmacology* **85,** 301–12.

Brown AJ (1998) Vitamin D analogues. *American Journal of Kidney Disease* **32,** S25–S39.

Brown AJ (2001) Therapeutic uses of vitamin D analogues. *American Journal of Kidney Disease* **38,** S3–S19.

Brown WV (1995) Niacin for lipid disorders. Indications, effectiveness, and safety. *Postgraduate Medicine* **98,** 185–89, and 192–3.

Broze GJ Jr. (2001) Protein Z-dependent regulation of coagulation. *Thrombosis and Haemostasis* **86,** 8–13.

Brubacher G, Muller-Mulot W, and Southgate DAT (1985) *Methods for the Determination of Vitamins in Food: Recommended by COST 91.* London: Elsevier Applied Science Publishers.

Buccellato FR, Miloso M, Braga M, Nicolini G, Morabito A, Pravettoni G, Tredici G, and Scalabrino G (1999) Myelinolytic lesions in spinal cord of cobalamin-deficient rats are TNF-alpha-mediated. *FASEB Journal* **13,** 297–304.

Buckel W (2001) Sodium ion-translocating decarboxylases. *Biochimica et Biophysica Acta* **1505,** 15–27.

Burton G and Ingold K (1984) β-Carotene, an unusual type of lipid antioxidant. *Science* **224,** 569–73.

Bush L and White HB 3rd (1989) Avidin traps biotin diffusing out of chicken egg yolk. *Comparative Biochemistry and Physiology B* **93,** 543–7.

Butterworth RF and Heroux M (1989) Effect of pyrithiamine treatment and subsequent thiamine rehabilitation on regional cerebral amino acids and thiamine-dependent enzymes. *Journal of Neurochemistry* **52,** 1079–84.

Cabezas A, Pinto RM, Fraiz F, Canales J, Gonzalez-Santiago S, and Cameselle JC (2001) Purification, characterization, and substrate and inhibitor structure-activity studies of rat liver FAD-AMP lyase (cyclizing): preference for FAD and specificity for splitting ribonucleoside diphosphate-X into ribonucleotide and a five-atom cyclic phosphodiester of X, either a monocyclic compound or a cis-bicyclic phosphodiester-pyranose fusion. *Biochemistry* **40**, 13710–22.

Cabrera-Valladares G, German MS, Matschinsky FM, Wang J, and Fernandez-Mejia C (1999) Effect of retinoic acid on glucokinase activity and gene expression and on insulin secretion in primary cultures of pancreatic islets. *Endocrinology* **140**, 3091–6.

Cairns PA and Stalker DJ (2000) Carnitine supplementation of parenterally fed neonates. *Cochrane Database Systematic Reviews*, CD000950.

Calingasan NY and Gibson GE (2000) Vascular endothelium is a site of free radical production and inflammation in areas of neuronal loss in thiamine-deficient brain. *Annals of the New York Academy of Sciences* **903**, 353–6.

Calingasan NY, Gandy SE, Baker H, Sheu KF, Smith JD, Lamb BT, Gearhart JD, Buxbaum JD, Harper C, Selkoe DJ, Price DL, Sisodia SS, and Gibson GE (1996) Novel neuritic clusters with accumulations of amyloid precursor protein and amyloid precursor-like protein 2 immunoreactivity in brain regions damaged by thiamine deficiency. *American Journal of Pathology* **149**, 1063–71.

Cameron E and Pauling L (1974a) The orthomolecular treatment of cancer. I. The role of ascorbic acid in host resistance. *Chemical Biological Interactions* **9**, 273–83.

Cameron E and Pauling L (1974b) The orthomolecular treatment of cancer. II. Clinical trial of high dose ascorbic acid supplements in advanced human cancer. *Chemical Biological Interactions* **9**, 285–315.

Campbell CH (1984) The severe lacticacidosis of thiamine deficiency: acute pernicious or fulminating beriberi. *Lancet* **2**, 446–9.

Cancela JM (2001) Specific Ca^{2+} signaling evoked by cholecystokinin and acetylcholine: the roles of NAADP, cADPR, and IP_3. *Annual Reviews of Physiology* **63**, 99–117.

Canham JE, Baker EM, Harding RS, Sauberlich HE, and Plough IC (1969) Dietary protein – its relationship to vitamin B_6 requirements and function. *Annals of the New York Academy of Sciences* **166**, 16–29.

Canton MC and Cremin FM (1990) The effect of dietary zinc depletion and repletion on rats: Zn concentration in various tissues and activity of pancreatic gamma-glutamyl hydrolase (EC 3.4.22.12) as indices of Zn status. *British Journal of Nutrition* **64**, 201–9.

Capuzzi DM, Morgan JM, Brusco OA Jr., and Intenzo CM (2000) Niacin dosing: relationship to benefits and adverse effects. *Current Atherosclerosis Report* **2**, 64–71.

Carl GF, Eto I, and Krumdieck CL (1987) Chronic treatment of rats with primidone causes deletion of pteroylpentaglutamates in liver. *Journal of Nutrition* **117**, 970–5.

Carlberg C (1996) The concept of multiple vitamin D signaling pathways. *Journal of Investigative Dermatology Symposium Proceedings* **1**, 10–14.

Carlberg C, Quack M, Herdick M, Bury Y, Polly P, and Toell A (2001) Central role of VDR conformations for understanding selective actions of vitamin D(3) analogues. *Steroids* **66**, 213–21.

Carpenter KJ and Lewin WJ (1985) A reexamination of the composition of diets associated with pellagra. *Journal of Nutrition* **115**, 543–52.

Carr A and Frei B (1999a) Does vitamin C act as a pro-oxidant under physiological conditions? *FASEB Journal* **13**, 1007–24.

Carr AC and Frei B (1999b) Toward a new recommended dietary allowance for vitamin C based on antioxidant and health effects in humans. *American Journal of Clinical Nutrition* **69**, 1086–1107.

Carr AC, Zhu BZ, and Frei B (2000) Potential antiatherogenic mechanisms of ascorbate (vitamin C) and alpha-tocopherol (vitamin E). *Circulation Research* **87,** 349–54.

Carter EG and Carpenter KJ (1982) The bioavailability for humans of bound niacin from wheat bran. *American Journal of Clinical Nutrition* **36,** 855–61.

Casey A and Greenhaff PL (2000) Does dietary creatine supplementation play a role in skeletal muscle metabolism and performance? *American Journal of Clinical Nutrition* **72,** 607S–17S.

Cassady IA, Budge MM, Healy MJ, and Nixon PF (1980) An inverse relationship of rat liver folate polyglutamate chain length to nutritional folate sufficiency. *Biochimica et Biophysica Acta* **633,** 258–68.

Casteels K, Bouillon R, Waer M, and Mathieu C (1995) Immunomodulatory effects of 1,25-dihydroxyvitamin D_3. *Current Opinion in Nephrology and Hypertension* **4,** 313–18.

Castenmiller JJ and West CE (1998) Bioavailability and bioconversion of carotenoids. *Annual Reviews of Nutrition* **18,** 19–38.

Catani MV, Rossi A, Costanzo A, Sabatini S, Levrero M, Melino G, and Avigliano L (2001) Induction of gene expression via activator protein-1 in the ascorbate protection against UV-induced damage. *Biochemical Journal* **356,** 77–85.

Chabré M and Deterre P (1989) Molecular mechanisms of visual transduction. *European Journal of Biochemistry* **179,** 255–66.

Chalmers AH, Cowley DM, and Brown JM (1986) A possible etiological role for ascorbate in calculi formation. *Clinical Chemistry* **32,** 333–6.

Chalmers TC (1975) Effects of ascorbic acid on the common cold. An evaluation of the evidence. *American Journal of Medicine* **58,** 532–6.

Chambers I, Frampton J, Goldfarb P, Affara N, McBain W, and Harrison PR (1986) The structure of the mouse glutathione peroxidase gene: the selenocysteine in the active site is encoded by the 'termination' codon, TGA. *EMBO Journal* **5,** 1221–7.

Chanarin I, Deacon R, Lumb M, Muir M, and Perry J (1985) Cobalamin-folate interrelations: a critical review. *Blood* **66,** 179–89.

Chanarin I, England JM, Mollin C, and Perry J (1973) Methylmalonic acid excretion studies. *British Journal of Haematology* **25,** 45–53.

Chango A, Boisson F, Barbe F, Quilliot D, Droesch S, Pfister M, Fillon-Emery N, Lambert D, Fremont S, Rosenblatt DS, and Nicolas JP (2000a) The effect of 677C → T and 1298A → C mutations on plasma homocysteine and 5,10-methylenetetrahydrofolate reductase activity in healthy subjects. *British Journal of Nutrition* **83,** 593–6.

Chango A, Potier De Courcy G, Boisson F, Guilland JC, Barbe F, Perrin MO, Christides JP, Rabhi K, Pfister M, Galan P, Hercberg S, and Nicolas JP (2000b) 5,10-Methylenetetrahydrofolate reductase common mutations, folate status and plasma homocysteine in healthy French adults of the Supplementation en Vitamines et Mineraux Antioxydants (SU.VI.MAX) cohort. *British Journal of Nutrition* **84,** 891–6.

Chanock SJ, el Benna J, Smith RM, and Babior BM (1994) The respiratory burst oxidase. *Journal of Biological Chemistry* **269,** 24519–22.

Chapman-Smith A and Cronan JE Jr. (1999a) The enzymatic biotinylation of proteins: a post-translational modification of exceptional specificity. *Trends in Biochemical Science* **24,** 359–63.

Chapman-Smith A and Cronan JE Jr. (1999b) Molecular biology of biotin attachment to proteins. *Journal of Nutrition* **129,** 477S–84S.

Chatterjee IB (1978) Ascorbic acid metabolism. *World Review of Nutrition and Dietetics* **30,** 69–87.

Chatterjee NS, Kumar CK, Ortiz A, Rubin SA, and Said HM (1999) Molecular mechanism of the intestinal biotin transport process. *American Journal of Physiology* **277,** C605–13.

Chauhan J and Dakshinamurti K (1991) Transcriptional regulation of the glucokinase gene by biotin in starved rats. *Journal of Biological Chemistry* **266,** 10035–8.

Chen H, Howald WN, and Juchau MR (2000) Biosynthesis of all-*trans*-retinoic acid from all-*trans*-retinol: catalysis of all-*trans*-retinol oxidation by human P-450 cytochromes. *Drug Metabolism and Disposal* **28,** 315–22.

Chen P, Hao W, Rife L, Wang XP, Shen D, Chen J, Ogden T, Van Boemel GB, Wu L, Yang M, and Fong HK (2001) A photic visual cycle of rhodopsin regeneration is dependent on Rgr. *Nature Genetics* **28,** 256–60.

Chesney RW (1990) Requirements and upper limits of vitamin D intake in the term neonate, infant, and older child. *Journal of Pediatrics* **116,** 159–66.

Chesney RW, Helms RA, Christensen M, Budreau AM, Han X, and Sturman JA (1998) The role of taurine in infant nutrition. *Advances in Experimental Medicine and Biology* **442,** 463–76.

Chia CP, Addison R, and McCormick DB (1978) Absorption, metabolism, and excretion of 8-alpha-(amino acid) riboflavins in the rat. *Journal of Nutrition* **108,** 373–81.

Chini EN, de Toledo FG, Thompson MA, and Dousa TP (1997) Effect of estrogen upon cyclic ADP ribose metabolism: beta-estradiol stimulates ADP ribosyl cyclase in rat uterus. *Proceedings of the National Academy of Sciences of the USA* **94,** 5872–6.

Choi SW and Mason JB (2000) Folate and carcinogenesis: an integrated scheme. *Journal of Nutrition* **130,** 129–32.

Choi SY, Churchich JE, Zaiden E, and Kwok F (1987) Brain pyridoxine-5-phosphate oxidase. Modulation of its catalytic activity by reaction with pyridoxal 5-phosphate and analogs. *Journal of Biological Chemistry* **262,** 12013–17.

Christakos S, Raval-Pandya M, Wernyj RP, and Yang W (1996) Genomic mechanisms involved in the pleiotropic actions of 1,25-dihydroxyvitamin D_3. *Biochemical Journal* **316,** 361–71.

Christensen E, Kolvraa S, and Gregersen N (1984) Glutaric aciduria type II: evidence for a defect related to the electron transfer flavoprotein or its dehydrogenase. *Pediatric Research* **18,** 663–7.

Chytil F (1984) Retinoic acid: biochemistry, pharmacology, toxicology, and therapeutic use. *Pharmacology Reviews* **36,** 93S–100S.

Cichowicz DJ and Shane B (1987) Mammalian folylpoly-gamma-glutamate synthetase. 2. Substrate specificity and kinetic properties. *Biochemistry* **26,** 513–21.

Cidlowski JA and Thanassi JW (1981) Pyridoxal phosphate: a possible cofactor in steroid hormone action. *Journal of Steroid Biochemistry* **15,** 11–16.

Cimino JA, Jhangiani S, Schwartz E, and Cooperman JM (1987) Riboflavin metabolism in the hypothyroid human adult. *Proceedings of the Society for Experimental Biology and Medicine* **184,** 151–3.

Clements JE and Anderson BB (1980) Glutathione reductase activity and pyridoxine (pyridoxamine) phosphate oxidase activity in the red cell. *Biochimica et Biophysica Acta* **632,** 159–63.

Coburn SP (1990) Location and turnover of vitamin B_6 pools and vitamin B_6 requirements of humans. *Annals of the New York Academy of Sciences* **585,** 76–85.

Coburn SP (1994) A critical review of minimal vitamin B_6 requirements for growth in various species with a proposed method of calculation. *Vitamins and Hormones* **48,** 259–300.

Coburn SP (1996) Modeling vitamin B_6 metabolism. *Advances in Food and Nutrition Research* **40,** 107–32.

Coburn SP, Reynolds RD, Mahuren JD, Schaltenbrand WE, Wang Y, Ericson KL, Whyte MP, Zubovic YM, Ziegler PJ, Costill DL, Fink WJ, Pearson DR, Pauly TA, Thampy KG, and Wortsman J (2002) Elevated plasma 4-pyridoxic acid in renal insufficiency. *American Journal of Clinical Nutrition* **75,** 57–64.

Coburn SP, Thampy KG, Lane HW, Conn PS, Ziegler PJ, Costill DL, Mahuren JD, Fink WJ, Pearson DR, Schaltenbrand WE, et al. (1995) Pyridoxic acid excretion during low vitamin B-6 intake, total fasting, and bed rest. *American Journal of Clinical Nutrition* **62,** 979–83.

Collins M and Mao G (1999) Teratology of retinoids. *Annual Reviews of Pharmacology and Toxicology* **39,** 399–430.

Cooke MS, Evans MD, Podmore ID, Herbert KE, Mistry N, Mistry P, Hickenbotham PT, Hussieni A, Griffiths HR, and Lunec J (1998) Novel repair action of vitamin C upon in vivo oxidative DNA damage. *FEBS Letters* **439,** 363–7.

Cooke NE and Haddad JG (1989) Vitamin D binding protein (Gc-globulin). *Endocrine Reviews* **10,** 294–307.

Coon WW and Nagler E (1969) The tryptophan load as a test for pyridoxine deficiency in hospitalized patients. *Annals of the New York Academy of Sciences* **166,** 30–43.

Crane FL and Navas P (1997) The diversity of coenzyme Q function. *Molecular Aspects of Medicine* **18**(Suppl), S1–S6.

Creagan ET, Moertel CG, O'Fallon JR, Schutt AJ, O'Connell MJ, Rubin J, and Frytak S (1979) Failure of high-dose vitamin C (ascorbic acid) therapy to benefit patients with advanced cancer. A controlled trial. *New England Journal of Medicine* **301,** 687–90.

Crespi F and Jouvet M (1982) Sleep and indolamine alterations induced by thiamine deficiency. *Brain Research* **248,** 275–83.

Crowe DL (1993) Retinoic acid mediates post-transcriptional regulation of keratin 19 mRNA levels. *Journal of Cell Science* **106,** 183–8.

Cunningham JJ (1998a) The glucose/insulin system and vitamin C: implications in insulin-dependent diabetes mellitus. *Journal of the American College of Nutrition* **17,** 105–8.

Cunningham JJ (1998b) Micronutrients as nutriceutical interventions in diabetes mellitus. *Journal of the American College of Nutrition* **17,** 7–10.

Czerniecki J and Czygier M (2001) Cooperation of divalent ions and thiamin diphosphate in regulation of the function of pig heart pyruvate dehydrogenase complex. *Journal of Nutritional Scence and Vitaminology* (Tokyo) **47,** 385–6.

Dakshinamurti K and Dakshinamurti S (2001) blood pressure regulation and micronutrients. *Nutrition Research Reviews* **14,** 3–43.

Dakshinamurti K and Lal KJ (1992) Vitamins and hypertension. *World Review of Nutrition and Dietetics* **69,** 40–73.

Dakshinamurti K and Litvak S (1970) Biotin and protein synthesis in rat liver. *Journal of Biological Chemistry* **245,** 5600–5.

Dakshinamurti K, Chalifour L, and Bhullar RP (1985) Requirement for biotin and the function of biotin in cells in culture. *Annals of the New York Academy of Sciences* **447,** 38–55.

Dallner G and Sindelar PJ (2000) Regulation of ubiquinone metabolism. *Free Radicals in Biology and Medicine* **29,** 285–94.

D'Amours D, Desnoyers S, D'Silva I, and Poirier GG (1999) Poly(ADP-ribosyl)ation reactions in the regulation of nuclear functions. *Biochemical Journal* **342,** 249–68.

D'Angelo A and Selhub J (1997) Homocysteine and thrombotic disease. *Blood* **90,** 1–11.

Daugherty M, Polanuyer B, Farrell M, Scholle M, Lykidis A, De Crecy-Lagard V. and Osterman A (2002) Complete reconstitution of the human coenzyme A biosynthetic pathway via comparative genomics. *Journal of Biological Chemistry* **27,** 21431–9.

Davis BA and Cowing BE (2000) Pyridoxal supplementation reduces cell proliferation and DNA synthesis in estrogen-dependent and -independent mammary carcinoma cell lines. *Nutrition and Cancer* **38,** 281–6.

de Lange PJ and Joubert CP (1964) Assessment of nicotinic acid status of population groups. *American Journal of Clinical Nutrition* **15,** 169–74.

DeLuca H (1977) The direct involvement of vitamin A in glycosyl transfer reactions of mammalian membranes. *Vitamins and Hormones* **35,** 1–57.

DeLuca HF and Zierold C (1998) Mechanisms and functions of vitamin D. *Nutrition Reviews* **56,** S4–S10; discussion S54–S75.

Delva L, Bastie JN, Rochette-Egly C, Kraiba R, Balitrand N, Despouy G, Chambon P, and Chomienne C (1999) Physical and functional interactions between cellular retinoic acid binding protein II and the retinoic acid-dependent nuclear complex. *Molecular and Cell Biology* **19,** 7158–67.

De Miranda J, Santoro A, Engelender S, and Wolosker H (2000) Human serine racemase: molecular cloning, genomic organization and functional analysis. *Gene* **256,** 183–8.

Department of Health (1991) *Dietary Reference Values for Food Energy and Nutrients for the United Kingdom.* London: Her Majesty's Stationery Office.

Department of Health (2000) *Folic Acid and the Prevention of Disease.* London: The Stationery Office.

Department of Health and Human Services (2001) *Adverse Event Reporting for Dietary Supplements: An Inadequate Safety Valve.* Boston MA: Office of the Inspector General.

Devaraj S, Harris A, and Jialal I (2002) Modulation of monocyte-macrophage function with alpha-tocopherol: implications for atherosclerosis. *Nutrition Reviews* **60,** 8–14.

Devery J and Milborrow BV (1994) Beta-carotene-15,15′-dioxygenase (EC 1.13.11.21) isolation, reaction mechanism and an improved assay procedure. *British Journal of Nutrition* **72,** 397–414.

Dialameh GH, Yekundi KG, and Olson RE (1970) Enzymatic alkylation of menaquinone-0 to menaquinones microsomes from chick liver. *Biochimica et Biophysica Acta* **223,** 332–8.

Dickinson CJ (1995) Does folic acid harm people with vitamin B_{12} deficiency? *Quarterly Journal of Medicine* **88,** 357–64.

Dickson IR, Walls J, and Webb S (1989) Vitamin A and bone formation. Different responses to retinol and retinoic acid of chick bone cells in organ culture. *Biochimica et Biophysica Acta* **1013**, 254–8.

Dierkes J, Kroesen M, and Pietrzik K (1998) Folic acid and vitamin B_6 supplementation and plasma homocysteine concentrations in healthy young women. *International Journal of Vitamin and Nutrition Research* **68**, 98–103.

Diliberto EJ Jr, Dean G, Carter C, and Allen PL (1982) Tissue, subcellular, and submitochondrial distributions of semidehydroascorbate reductase: possible role of semidehydroascorbate reductase in cofactor regeneration. *Journal of Neurochemistry* **39**, 563–8.

Dillon JC, Malfait P, Demaux G, and Foldi-Hope C (1992) The urinary metabolites of niacin during the course of pellagra. *Annals of Nutrition and Metabolism* **36**, 181–5.

Doucette MM and Stevens VL (2001) Folate receptor function is regulated in response to different cellular growth rates in cultured mammalian cells. *Journal of Nutrition* **131**, 2819–25.

Douglas CE, Chan AC, and Choy PC (1986) Vitamin E inhibits platelet phospholipase A2. *Biochimica et Biophysica Acta* **876**, 639–45.

Douglas RM, Chalker EB, and Treacy B (2000) Vitamin C for preventing and treating the common cold. *Cochrane Database Systematic Reviews*, CD000980.

Dousa TP, Chini EN, and Beers KW (1996) Adenine nucleotide diphosphates: emerging second messengers acting via intracellular Ca^{2+} release. *American Journal of Physiology* **271**, C1007–24.

Drewke C and Leistner E (2001) Biosynthesis of vitamin B_6 and structurally related derivatives. *Vitamins and Hormones* **61**, 121–55.

Drezner MK (2000) PHEX gene and hypophosphatemia. *Kidney International* **57**, 9–18.

Dubick MA, Gretz D, and Majumdar AP (1995) Overt vitamin B-6 deficiency affects rat pancreatic digestive enzyme and glutathione reductase activities. *Journal of Nutrition* **125**, 20–5.

Ducy P, Desbois C, Boyce B, Pinero G, Story B, Dunstan C, Smith E, Bonadio J, Goldstein S, Gundberg C, Bradley A, and Karsenty G (1996) Increased bone formation in osteocalcin-deficient mice. *Nature* **382**, 448–52.

Dudeja PK, Tyagi S, Kavilaveettil RJ, Gill R, and Said HM (2001) Mechanism of thiamine uptake by human jejunal brush-border membrane vesicles. *American Journal of Physiology Cell Physiology* **281**, C786–92.

Duerden JM and Bates CJ (1985) Effect of riboflavin deficiency on lipid metabolism of liver and brown adipose tissue of sucking rat pups. *British Journal of Nutrition* **53**, 107–15.

Duester G (2000) Families of retinoid dehydrogenases regulating vitamin A function: production of visual pigment and retinoic acid. *European Journal of Biochemistry* **267**, 4315–24.

Duester G (2001) Genetic dissection of retinoid dehydrogenases. *Chemical Biological Interactions* **130–132**, 469–80.

Dunnigan MG and Henderson JB (1997) An epidemiological model of privational rickets and osteomalacia. *Proceedings of the Nutrition Society* **56**, 939–56.

Dupre S, Rosei MA, Bellussi L, Del Grosso E, and Cavallini D (1973) The substrate specificity of pantetheinase. *European Journal of Biochemistry* **40**, 103–7.

During A, Nagao A, Hoshino C, and Terao J (1996) Assay of beta-carotene 15,15'-dioxygenase activity by reverse-phase high-pressure liquid chromatography. *Analytical Biochemistry* **241**, 199–205.

Dutta P (1991) Enhanced uptake and metabolism of riboflavin in erythrocytes infected with *Plasmodium falciparum. Journal of Protozoology* **38**, 479–83.

Dutta P, Pinto J, and Rivlin R (1985) Antimalarial effects of riboflavin deficiency. *Lancet* **2**, 1040–3.

Dykes MH and Meier P (1975) Ascorbic acid and the common cold. Evaluation of its efficacy and toxicity. *JAMA* **231**, 1073–9.

Eaton S (2002) Control of mitochondrial beta-oxidation flux. *Progress in Lipid Research* **41**, 197–239.

Ebner F, Heller A, Rippke F, and Tausch I (2002) Topical use of dexpanthenol in skin disorders. *American Journal of Clinical Dermatology* **3**, 427–33.

Edwin EE and Jackman R (1970) Thiaminase I in the development of cerebrocortical necrosis in sheep and cattle. *Nature* **228**, 772–4.

Egger SF, Huber-Spitzy V, Alzner E, Scholda C, and Vecsei VP (1999) Corneal wound healing after superficial foreign body injury: vitamin A and dexpanthenol versus a calf blood extract. A randomized double-blind study. *Ophthalmologica* **213**, 246–9.

Eisman JA (1999) Genetics of osteoporosis. *Endocrine Reviews* **20**, 788–804.

Eissenstat BR, Wyse BW, and Hansen RG (1986) Pantothenic acid status of adolescents. *American Journal of Clinical Nutrition* **44**, 931–7.

Erin AN, Gorbunov NV, Brusovanik VI, Tyurin VA, and Prilipko LL (1986) Stabilization of synaptic membranes by alpha-tocopherol against the damaging action of phospholipases. Possible mechanism of biological action of vitamin E. *Brain Research* **398**, 85–90.

Evans RM, Currie L, and Campbell A (1982) The distribution of ascorbic acid between various cellular components of blood, in normal individuals, and its relation to the plasma concentration. *British Journal of Nutrition* **47**, 473–82.

FAO/WHO (2001) *Human Vitamin and Mineral Requirements: Report of a Joint FAO/WHO Expert Consultation, Bankok, Thailand.* Rome: Food and Nutrition Division of the United Nations Food and Agriculture Organization.

Farach-Carson MC and Ridall AL (1998) Dual 1,25-dihydroxyvitamin D_3 signal response pathways in osteoblasts: cross-talk between genomic and membrane-initiated pathways. *American Journal of Kidney Disease* **31**, 729–42.

Farnham PJ and Schimke RT (1985) Transcriptional regulation of mouse dihydrofolate reductase in the cell cycle. *Journal of Biological Chemistry* **260**, 7675–80.

Feiz HR and Mobarhan S (2002) Does vitamin C intake slow the progression of gastric cancer in *Helicobacter pylori*-infected populations? *Nutrition Reviews* **60**, 34–6.

Felsted RL and Chaykin S (1967) N^1-methylnicotinamide oxidation in a number of mammals. *Journal of Biological Chemistry* **242**, 1274–9.

Fenech M (2001) Recommended dietary allowances (RDAs) for genomic stability. *Mutation Research* **480–481**, 51–4.

Feskanich D, Weber P, Willett WC, Rockett H, Booth SL, and Colditz GA (1999) Vitamin K intake and hip fractures in women: a prospective study. *American Journal of Clinical Nutrition* **69**, 74–9.

Finglass P (1993) Flair concerted action no 10 status papers: thiamin. *International Journal of Vitamin and Nutrition Research* **63,** 270–4.

Fioravanti M and Yanagi M (2000) Cytidinediphosphocholine (CDP choline) for cognitive and behavioural disturbances associated with chronic cerebral disorders in the elderly. *Cochrane Database Systematic Reviews,* CD000269.

Fitzpatrick PF (1999) Tetrahydropterin-dependent amino acid hydroxylases. *Annual Reviews of Biochemistry* **68,** 355–81.

Fraga CG, Motchnik PA, Shigenaga MK, Helbock HJ, Jacob RA, and Ames BN (1991) Ascorbic acid protects against endogenous oxidative DNA damage in human sperm. *Proceedings of the National Academy of Sciences of the USA* **88,** 11003–6.

Fraiz FJ, Pinto RM, Costas MJ, Aavalos M, Canales J, Cabezas A, and Cameselle JC (1998) Enzymic formation of riboflavin 4′,5′-cyclic phosphate from FAD: evidence for a specific low-Km FMN cyclase in rat liver. *Biochemical Journal* **330,** 881–8.

Franceschi RT, Romano PR, and Park KY (1988) Regulation of type I collagen synthesis by 1,25-dihydroxyvitamin D_3 in human osteosarcoma cells. *Journal of Biological Chemistry* **263,** 18938–45.

Frasca V, Riazzi BS, and Matthews RG (1986) In vitro inactivation of methionine synthase by nitrous oxide. *Journal of Biological Chemistry* **261,** 15823–6.

Freedman JE and Keaney JF Jr. (2001) Vitamin E inhibition of platelet aggregation is independent of antioxidant activity. *Journal of Nutrition* **131,** 374S–7S.

Fregly MJ and Cade JR (1995) Effect of pyridoxine and tryptophan, alone and combined, on the development of deoxycorticosterone acetate-induced hypertension in rats. *Pharmacology* **50,** 298–306.

Frey PA (2001) Radical mechanisms of enzymatic catalysis. *Annual Reviews of Biochemistry* **70,** 121–48.

Friedman PA (2000) Mechanisms of renal calcium transport. *Experimental Nephrology* **8,** 343–50.

Frimpter GW, Andelman RJ, and George WF (1969) Vitamin B_6-dependency syndromes. New horizons in nutrition. *American Journal of Clinical Nutrition* **22,** 794–805.

Froguel P, Zouali H, Vionnet N, Velho G, Vaxillaire M, Sun F, Lesage S, Stoffel M, Takeda J, Passa P, et al. (1993) Familial hyperglycemia due to mutations in glucokinase. Definition of a subtype of diabetes mellitus. *New England Journal of Medicine* **328,** 697–702.

Frot-Coutaz J, Letoublon R, Degiuli A, Fayet Y, Audiger-Petit C, and Gor R (1985) Spatial aspects of mannosyl phosphoryl retinol formation. *Biochimica et Biophysica Acta* **841,** 299–305.

Fu CS, Swendseid ME, Jacob RA, and McKee RW (1989) Biochemical markers for assessment of niacin nutritional status in young men: levels of erythrocyte niacin coenzymes and plasma tryptophan. *Journal of Nutrition* **119,** 1945–9.

Fujii T and Kawashima K (2001) An independent non-neuronal cholinergic system in lymphocytes. *Japanese Journal of Pharmacology* **85,** 11–15.

Fujita A and Mitsuhashi T (1999) Differential regulation of ligand-dependent and ligand-independent functions of the mouse retinoid X receptor beta by alternative splicing. *Biochemical and Biophysical Research Communications* **255,** 625–30.

Furie B and Furie BC (1988) The molecular basis of blood coagulation. *Cell* **53,** 505–18.

Furie BC and Furie B (1997) Structure and mechanism of action of the vitamin K-dependent gamma-glutamyl carboxylase: recent advances from mutagenesis studies. *Thrombosis and Haemostasis* **78,** 595–8.

Gale EA (1996a) Molecular mechanisms of beta-cell destruction in IDDM: the role of nicotinamide. *Hormone Research* **45**(Suppl 1), 39–43.

Gale EA (1996b) Theory and practice of nicotinamide trials in pre-type 1 diabetes. *Journal of Pediatrics Endocrinology and Metabolism* **9,** 375–9.

Gamble MV, Mata NL, Tsin AT, Mertz JR, and Blaner WS (2000) Substrate specificities and 13-*cis*-retinoic acid inhibition of human, mouse and bovine *cis*-retinol dehydrogenases. *Biochimica et Biophysica Acta* **1476,** 3–8.

Gann PH, Ma J, Giovannucci E, Willett W, Sacks FM, Hennekens CH, and Stampfer MJ (1999) Lower prostate cancer risk in men with elevated plasma lycopene levels: results of a prospective analysis. *Cancer Research* **59,** 1225–30.

Gardill SL and Suttie JW (1990) Vitamin K epoxide and quinone reductase activities. Evidence for reduction by a common enzyme. *Biochemical Pharmacology* **40,** 1055–61.

Garg R, Malinow M, Pettinger M, Upson B, and Hunninghake D (1999) Niacin treatment increases plasma homocyst(e)ine levels. *American Heart Journal* **138,** 1082–7.

Gastaldi G, Ferrari G, Verri A, Casirola D, Orsenigo MN, and Laforenza U (2000) Riboflavin phosphorylation is the crucial event in riboflavin transport by isolated rat enterocytes. *Journal of Nutrition* **130,** 2556–61.

Gerber GB and Deroo J (1970) Metabolism of labeled nicotinamide coenzyme in different organs of mice and rats. *Proceedings of the Society for Experimental Biology and Medicine* **134,** 689–93.

Gerster H (1997) No contribution of ascorbic acid to renal calcium oxalate stones. *Annals of Nutrition and Metabolism* **41,** 269–82.

Gerster H (1999) High-dose vitamin C: a risk for persons with high iron stores? *International Journal of Vitamin and Nutrition Research* **69,** 67–82.

Gey KF (1995) Cardiovascular disease and vitamins. Concurrent correction of 'suboptimal' plasma antioxidant levels may, as important part of 'optimal' nutrition, help to prevent early stages of cardiovascular disease and cancer, respectively. *Biblio Nutritio et Dieta* **52,** 75–91.

Giasuddin AS and Diplock AT (1981) The influence of vitamin E on membrane lipids of mouse fibroblasts in culture. *Archives of Biochemistry and Biophysics* **210,** 348–62.

Gitlin G, Bayer EA, and Wilchek M (1988a) Studies on the biotin-binding site of avidin. Tryptophan residues involved in the active site. *Biochemical Journal* **250,** 291–4.

Gitlin G, Bayer EA, and Wilchek M (1988b) Studies on the biotin-binding site of streptavidin. Tryptophan residues involved in the active site. *Biochemical Journal* **256,** 279–82.

Glass C, Rosenfeld M, Rose D, Kurokawa R, Kamei Y, Xu L, Torchia J, Ogliastro M, and Westin S (1997) Mechanisms of transcriptional activation by retinoic acid receptors. *Biochemical Society Transactions* **25,** 602–5.

Glass CK (1996) Some new twists in the regulation of gene expression by thyroid hormone and retinoic acid receptors. *Journal of Endocrinology* **150,** 349–57.

Goldman ID and Matherly LH (1987) Biochemical factors in the selectivity of leucovorin rescue: selective inhibition of leucovorin reactivation of dihydrofolate reductase and leucovorin utilization in purine and pyrimidine biosynthesis by methotrexate and dihydrofolate polyglutamates. *National Cancer Institute Monograph* **5**, 17–26.

Goodman JI and Watson RE (2002) Altered DNA methylation: a secondary mechanism involved in carcinogenesis. *Annual Reviews of Pharmacology and Toxicology* **42**, 501–25.

Goodman SI (1981) Organic aciduria in the riboflavin-deficient rat. *American Journal of Clinical Nutrition* **34**, 2434–7.

Goodrich RP (2000) The use of riboflavin for the inactivation of pathogens in blood products. *Vox Sang* **78**(Suppl 2), 211–15.

Gopalan C and Rao KSJ (1975) Pellagra and amino acid imbalance. *Vitamins and Hormones* **33**, 505–28.

Gontzea L, Rujinski A, and Sutzesco P (1976) Rapide evaluation de l'etat de nutrition niacinique. *Bibliotheca Nutritio et Dieta* **23**, 95–104.

Green MH, Green JB, Berg T, Norum KR, and Blomhoff R (1993) Vitamin A metabolism in rat liver: a kinetic model. *American Journal of Physiology* **264**, G509–21.

Green R and Miller JW (1999) Folate deficiency beyond megaloblastic anemia: hyperhomocysteinemia and other manifestations of dysfunctional folate status. *Seminars in Hematology* **36**, 47–64.

Gregersen N, Christensen MF, Christensen E, and Kolvraa S (1986) Riboflavin responsive multiple acyl-CoA dehydrogenation deficiency. Assessment of 3 years of riboflavin treatment. *Acta Paediatrica Scandinavica* **75**, 676–81.

Gregory J, Foster K, Tyler H, and Wiseman M (1990) *The Dietary and Nutritional Survey of British Adults*. London: Her Majesty's Stationery Office.

Gregory JF 3rd (1980a) Effects of epsilon-pyridoxyllysine and related compounds on liver and brain pyridoxal kinase and liver pyridoxamine (pyridoxine) 5′-phosphate oxidase. *Journal of Biological Chemistry* **255**, 2355–9.

Gregory JF 3rd (1980b) Effects of epsilon-pyridoxyllysine bound to dietary protein on the vitamin B-6 status of rats. *Journal of Nutrition* **110**, 995–1005.

Gregory JF 3rd (1998) Nutritional properties and significance of vitamin glycosides. *Annual Reviews of Nutrition* **18**, 277–96.

Gregory JF 3rd (2001) Case study: folate bioavailability. *Journal of Nutrition* **131**, 1376S–82S.

Greig F, Casas J, and Castells S (1989) Changes in plasma osteocalcin concentrations during treatment of rickets. *Journal of Pediatrics* **114**, 820–3.

Griffin JE and Zerwekh JE (1983) Impaired stimulation of 25-hydroxyvitamin D-24-hydroxylase in fibroblasts from a patient with vitamin D-dependent rickets, type II. A form of receptor-positive resistance to 1,25-dihydroxyvitamin D_3. *Journal of Clinical Investigation* **72**, 1190–9.

Grolier P, Duszka C, Borel P, Alexandre-Gouabau MC, and Azais-Braesco V (1997) In vitro and in vivo inhibition of beta-carotene dioxygenase activity by canthaxanthin in rat intestine. *Archives of Biochemistry and Biophysics* **348**, 233–8.

Gromisch DS, Lopez R, Cole HS, and Cooperman JM (1977) Light (phototherapy)-induced riboflavin deficiency in the neonate. *Journal of Pediatrics* **90**, 118–22.

Grundman M (2000) Vitamin E and Alzheimer disease: the basis for additional clinical trials. *American Journal of Clinical Nutrition* **71**, 630S–6S.

Gueant JL, Champigneulle B, Gaucher P, and Nicolas JP (1990) Malabsorption of vitamin B_{12} in pancreatic insufficiency of the adult and of the child. *Pancreas* **5,** 559–67.

Guenther BD, Sheppard CA, Tran P, Rozen R, Matthews RG, and Ludwig ML (1999) The structure and properties of methylenetetrahydrofolate reductase from *Escherichia coli* suggest how folate ameliorates human hyperhomocysteinemia. *Nature Structural Biology* **6,** 359–65.

Guilarte TR and Wagner HN Jr (1987) Increased concentrations of 3-hydroxykynurenine in vitamin B_6 deficient neonatal rat brain. *Journal of Neurochemistry* **49,** 1918–26.

Gupta RN, Hemscheidt T, Sayer BG, and Spenser ID (2001) Biosynthesis of vitamin B(6) in yeast: incorporation pattern of glucose. *Journal of the American Chemical Society* **123,** 11353–9.

Guyton KZ, Kensler TW, and Posner GH (2001) Cancer chemoprevention using natural vitamin D and synthetic analogs. *Annual Reviews of Pharmacology and Toxicology* **41,** 421–42.

Haddad JG (1995) Plasma vitamin D-binding protein (Gc-globulin): multiple tasks. *Journal of Steroid Biochemistry and Molecular Biology* **53,** 579–82.

Haddad JG, Jennings AS, and Aw TC (1988) Vitamin D uptake and metabolism by perfused rat liver: influences of carrier proteins. *Endocrinology* **123,** 498–504.

Hageman GJ and Stierum RH (2001) Niacin, poly(ADP-ribose) polymerase-1 and genomic stability. *Mutation Research* **475,** 45–56.

Hagen I, Nesheim BI, and Tuntland T (1985) No effect of vitamin B-6 against premenstrual tension. A controlled clinical study. *Acta Obstetrica et Gynecologica Scandinavica* **64,** 667–70.

Hallberg L (1982) Iron absorption and iron deficiency. *Human Nutrition: Clinical Nutrition* **36,** 259–78.

Halliwell B (1996) Vitamin C: antioxidant or pro-oxidant in vivo? *Free Radical Research* **25,** 439–54.

Halliwell B (2001) Vitamin C and genomic stability. *Mutation Research* **475,** 29–35.

Halloran BP (1989) Is 1,25-dihydroxyvitamin D required for reproduction? *Proceedings of the Society for Experimental Biology and Medicine* **191,** 227–32.

Handelman GJ (2001) The evolving role of carotenoids in human biochemistry. *Nutrition* **17,** 818–22.

Hannah SS and Norman AW (1994) 1 alpha, $25(OH)_2$ vitamin D3-regulated expression of the eukaryotic genome. *Nutrition Reviews* **52,** 376–82.

Hansen CM, Leklem JE, and Miller LT (1996a) Vitamin B-6 status indicators decrease in women consuming a diet high in pyridoxine glucoside. *Journal of Nutrition* **126,** 2512–18.

Hansen CM, Leklem JE, and Miller LT (1996b) Vitamin B-6 status of women with a constant intake of vitamin B-6 changes with three levels of dietary protein. *Journal of Nutrition* **126,** 1891–1901.

Hansen CM, Shultz TD, Kwak HK, Memon HS, and Leklem JE (2001) Assessment of vitamin B-6 status in young women consuming a controlled diet containing four levels of vitamin B-6 provides an estimated average requirement and recommended dietary allowance. *Journal of Nutrition* **131,** 1777–86.

Hao W and Fong HK (1999) The endogenous chromophore of retinal G protein-coupled receptor opsin from the pigment epithelium. *Journal of Biological Chemistry* **274**, 6085–90.

Harmon DL, Shields DC, Woodside JV, McMaster D, Yarnell JW, Young IS, Peng K, Shane B, Evans AE, and Whitehead AS (1999) Methionine synthase D919G polymorphism is a significant but modest determinant of circulating homocysteine concentrations. *Genetics and Epidemiology* **17**, 298–309.

Harper C (1979) Wernicke's encephalopathy: a more common disease than realised. A neuropathological study of 51 cases. *Journal of Neurology Neurosurgery and Psychiatry* **42**, 226–31.

Harris RA, Kobayashi R, Murakami T, and Shimomura Y (2001) Regulation of branched-chain alpha-keto acid dehydrogenase kinase expression in rat liver. *Journal of Nutrition* **131**, 841S–5S.

Harris S (2002) Can vitamin D supplementation in infancy prevent type 1 diabetes? *Nutrition Reviews* **60**, 118–21.

Harrison EH and Hussain MM (2001) Mechanisms involved in the intestinal digestion and absorption of dietary vitamin A. *Journal of Nutrition* **131**, 1405–8.

Haussler MR, Jurutka PW, Hsieh JC, Thompson PD, Selznick SH, Haussler CA, and Whitfield GK (1995) New understanding of the molecular mechanism of receptor-mediated genomic actions of the vitamin D hormone. *Bone* **17**, 33S–8S.

Hayaishi O and Ueda K (1977) Poly(ADP-ribose) and ADP-ribosylation of proteins. *Annual Reviews of Biochemistry* **46**, 95–116.

Hayakawa M and Shibata M (1991) The in vitro and in vivo inhibition of protein glycosylation and diabetic vascular basement membrane thickening by pyridoxal-5'-phosphate. *Journal of Nutritional Science and Vitaminology* (Tokyo) **37**, 149–59.

Hayes CE (2000) Vitamin D: a natural inhibitor of multiple sclerosis. *Proceedings of the Nutrition Society* **59**, 531–5.

Hayes CE, Cantorna MT, and DeLuca HF (1997) Vitamin D and multiple sclerosis. *Proceedings of the Society for Experimental Biology and Medicine* **216**, 21–7.

Hazell AS, Todd KG, and Butterworth RF (1998) Mechanisms of neuronal cell death in Wernicke's encephalopathy. *Metabolic Brain Diseases* **13**, 97–122.

Heard GS, Hood RL, and Johnson AR (1983) Hepatic biotin and the sudden infant death syndrome. *Medical Journal of Australia* **2**, 305–6.

Heinonen OP, Albanes D, Virtamo J, Taylor PR, Huttunen JK, Hartman AM, Haapakoski J, Malila N, Rautalahti M, Ripatti S, Maenpaa H, Teerenhovi L, Koss L, Virolainen M, and Edwards BK (1998) Prostate cancer and supplementation with alpha-tocopherol and beta-carotene: incidence and mortality in a controlled trial. *Journal of the National Cancer Institute* **90**, 440–6.

Hemila H (1992) Vitamin C and the common cold. *British Journal of Nutrition* **67**, 3–16.

Hennekens CH, Buring JE, Manson JE, Stampfer M, Rosner B, Cook NR, Belanger C, LaMotte F, Gaziano JM, Ridker PM, Willett W, and Peto R (1996) Lack of effect of long-term supplementation with beta carotene on the incidence of malignant neoplasms and cardiovascular disease. *New England Journal of Medicine* **334**, 1145–9.

Henry HL (2001) The 25(OH)D(3)/1alpha,25(OH)(2)D(3)-24R-hydroxylase: a catabolic or biosynthetic enzyme? *Steroids* **66**, 391–8.

Henry HL and Norman AW (1978) Vitamin D: two dihydroxylated metabolites are required for normal chicken egg hatchability. *Science* **201**, 835–7.

Henry HL, Taylor AN, and Norman AW (1977) Response of chick parathyroid glands to the vitamin D metabolites, 1,25-dihydroxycholecalciferol and 24,25-dihydroxycholecalciferol. *Journal of Nutrition* **107**, 1918–26.

Herbert V (1962) Experimental nutritional folate deficiency in man. *Transactions of the Association of American Physicians* **75**, 307–20.

Herbert V (1987a) Recommended dietary intakes (RDI) of folate in humans. *American Journal of Clinical Nutrition* **45**, 661–70.

Herbert V (1987b) Recommended dietary intakes (RDI) of vitamin B-12 in humans. *American Journal of Clinical Nutrition* **45**, 671–78.

Herbert V (1988) Vitamin B-12: plant sources, requirements, and assay. *American Journal of Clinical Nutrition* **48**, 852–8.

Heroux M and Butterworth RF (1995) Regional alterations of thiamine phosphate esters and of thiamine diphosphate-dependent enzymes in relation to function in experimental Wernicke's encephalopathy. *Neurochemistry Research* **20**, 87–93.

Herrmann W (2001) The importance of hyperhomocysteinemia as a risk factor for diseases: an overview. *Clinical Chemistry Laboratory Medicine* **39**, 666–74.

Hespel P, Eijnde BO, Derave W, and Richter EA (2001) Creatine supplementation: exploring the role of the creatine kinase/phosphocreatine system in human muscle. *Canadian Journal of Applied Physiology* **26**(suppl), S79–102.

Hewison M, Zehnder D, Bland R, and Stewart PM (2000) 1alpha-Hydroxylase and the action of vitamin D. *Journal of Molecular Endocrinology* **25**, 141–8.

Hickenbottom SJ, Follett JR, Lin Y, Dueker SR, Burri BJ, Neidlinger TR, and Clifford AJ (2002) Variability in conversion of beta-carotene to vitamin A in men as measured by using a double-tracer study design. *American Journal of Clinical Nutrition* **75**, 900–7.

Hicks R and Turton J (1986) Retinoids and cancer. *Biochemical Society Transactions* **14**, 939–42.

Higgins JP and Flicker L (2000) Lecithin for dementia and cognitive impairment. *Cochrane Database Systematic Reviews*, CD001015.

Hildebrandt EF, Preusch PC, Patterson JL, and Suttie JW (1984) Solubilization and characterization of vitamin K epoxide reductase from normal and warfarin-resistant rat liver microsomes. *Archives of Biochemistry and Biophysics* **228**, 480–92.

Hodges R, Sauberlich H, Canham J, Wallace D, Rucker R, Mejia L, and Mohanram M (1978) Hematopoietic studies in vitamin A deficiency. *American Journal of Clinical Nutrition* **31**, 876–85.

Hodges RE, Baker EM, Hood J, Sauberlich HE, and March SC (1969) Experimental scurvy in man. *American Journal of Clinical Nutrition* **22**, 535–48.

Hodges RE, Hood J, Canham JE, Sauberlich HE, and Baker EM (1971) Clinical manifestations of ascorbic acid deficiency in man. *American Journal of Clinical Nutrition* **24**, 432–43.

Hodges S, Hertz N, Lockwood K, and Lister R (1999) CoQ10: could it have a role in cancer management? *Biofactors* **9**, 365–70.

Hoffer A, Osmond H, Callbeck MJ, and Kahan I (1957) Treatment of schizophrenia with nicotinic acid and nicotinamide. *Journal of Clinical and Experimental Psychopathology* **18**, 131–58.

Hohmann S and Meacock PA (1998) Thiamin metabolism and thiamin diphosphate-dependent enzymes in the yeast *Saccharomyces cerevisiae*: genetic regulation. *Biochimica et Biophysica Acta* **1385**, 201–19.

Holick MF (1990) The use and interpretation of assays for vitamin D and its metabolites. *Journal of Nutrition* **120**(Suppl 11), 1464–9.

Holick MF (1995) Environmental factors that influence the cutaneous production of vitamin D. *American Journal of Clinical Nutrition* **61**, 638S–45S.

Holland B, Welch AA, Unwin ID, Buss DH, Paul AA, and Southgate DAT (1991) *McCance and Widdowson's The Composition of Foods*, 5th Edition. London: Royal Society of Chemistry and Ministry of Agriculture, Fisheries and Food.

Homewood J and Bond NW (1999) Thiamin deficiency and Korsakoff's syndrome: failure to find memory impairments following nonalcoholic Wernicke's encephalopathy. *Alcohol* **19**, 75–84.

Homocysteine Lowering Trialists' Collaboration (1998) Lowering blood homocysteine with folic acid based supplements: meta-analysis of randomized trials. *British Medical Journal* **316**, 894–8.

Hoppe PP and Krennrich G (2000) Bioavailability and potency of natural-source and all-racemic alpha-tocopherol in the human: a dispute. *European Journal of Nutrition* **39**, 183–93.

Horio F and Yoshida A (1982) Effects of some xenobiotics on ascorbic acid metabolism in rats. *Journal of Nutrition* **112**, 416–25.

Horne DW, Patterson D, and Cook RJ (1989) Effect of nitrous oxide inactivation of vitamin B_{12}-dependent methionine synthetase on the subcellular distribution of folate coenzymes in rat liver. *Archives of Biochemistry and Biophysics* **270**, 729–33.

Horst RL, Napoli JL, and Littledike ET (1982) Discrimination in the metabolism of orally dosed ergocalciferol and cholecalciferol by the pig, rat and chick. *Biochemical Journal* **204**, 185–9.

Horwitt M (1960) Vitamin E and lipid metabolism in man. *American Journal of Clinical Nutrition* **8**, 451–61.

Horwitt M (2001) Critique of the requirement for vitamin E. *American Journal of Clinical Nutrition* **73**, 1003–5.

Horwitt MK, Harvey CC, Rothwell WS, Cutler JL, and Haffron D (1956) Tryptophan-niacin relationships in man. *Journal of Nutrition* **60**(Suppl 1), 1–43.

Horwitt MK, Harvey CC, Dahm CH Jr, and Searcy MT (1972) Relationship between tocopherol and serum lipid levels for determination of nutritional adequacy. *Annals of the New York Academy of Sciences* **203**, 223–36.

Hoshi K, Nomura K, Sano Y, and Koshihara Y (1999) Nuclear vitamin K_2 binding protein in human osteoblasts: homologue to glyceraldehyde-3-phosphate dehydrogenase. *Biochemical Pharmacology* **58**, 1631–8.

Hosomi A, Arita M, Sato Y, Kiyose C, Ueda T, Igarashi O, Arai H, and Inoue K (1997) Affinity for alpha-tocopherol transfer protein as a determinant of the biological activities of vitamin E analogs. *FEBS Letters* **409**, 105–8.

Houle B, Rochette-Egly C, and Bradley WE (1993) Tumor-suppressive effect of the retinoic acid receptor beta in human epidermoid lung cancer cells. *Proceedings of the National Academy of Sciences of the USA* **90**, 985–9.

Hoyumpa AM Jr, Nichols SG, Wilson FA, and Schenker S (1977) Effect of ethanol on intestinal (Na, K)ATPase and intestinal thiamine transport in rats. *Journal of Laboratory and Clinical Medicine* **90,** 1086–95.

Huang B, Wu P, Bowker-Kinley MM, and Harris RA (2002) Regulation of pyruvate dehydrogenase kinase expression by peroxisome proliferator-activated receptor-alpha ligands, glucocorticoids, and insulin. *Diabetes* **51,** 276–83.

Hume E and Krebs H (1949) *Vitamin A Requirements of Human Adults. Report of the Vitamin A Sub-committee of the Accessory Foods Factors Committee.* London: Medical Research Council, His Majesty's Stationery Office.

Huque T (1982) A survey of human liver reserves of retinol in London. *British Journal of Nutrition* **47,** 165–72.

Hurley JB, Spencer M, and Niemi GA (1998) Rhodopsin phosphorylation and its role in photoreceptor function. *Vision Research* **38,** 1341–52.

Hurot JM, Cucherat M, Haugh M, and Fouque D (2002) Effects of L-carnitine supplementation in maintenance hemodialysis patients: a systematic review. *Journal of the American Society of Nephrology* **13,** 708–14.

Hussy N, Deleuze C, Desarmenien MG, and Moos FC (2000) Osmotic regulation of neuronal activity: a new role for taurine and glial cells in a hypothalamic neuroendocrine structure. *Progress in Neurobiology* **62,** 113–34.

Hymes J and Wolf B (1996) Biotinidase and its roles in biotin metabolism. *Clinica Chimica Acta* **255,** 1–11.

Hymes J and Wolf B (1999) Human biotinidase isn't just for recycling biotin. *Journal of Nutrition* **129,** 485S–9S.

Infante JP (1999) A function for the vitamin E metabolite alpha-tocopherol quinone as an essential enzyme cofactor for the mitochondrial fatty acid desaturases. *FEBS Letters* **446,** 1–5.

Ingold KU, Burton GW, Foster DO, Hughes L, Lindsay DA, and Webb A (1987) Biokinetics of and discrimination between dietary *RRR*- and *SRR*-alpha-tocopherols in the male rat. *Lipids* **22,** 163–72.

Institute of Medicine (1997) *Dietary Reference Intakes for Calcium, Phosphorus, Magnesium, Vitamin D and Fluoride.* Washington, DC: National Academy Press.

Institute of Medicine (1998) *Dietary Reference Values for Thiamin, Riboflavin, Niacin, Vitamin B_6, Folate, Vitamin B_{12}, Pantothenic Acid, Biotin and Choline.* Washington, DC: National Academy Press.

Institute of Medicine (2000) *Dietary Reference Values for Vitamin C, Vitamin E, Selenium and Carotenoids.* Washington, DC: National Academy Press.

Institute of Medicine (2001) *Dietary Reference Intakes for Vitamin A, Vitamin K, Arsenic, Boron, Chromium, Copper, Iodine, Iron, Manganese, Molybdenum, Nickel, Silicon, Vanadium and Zinc.* Washington, DC: National Academy Press.

Israels LG, Israels ED, and Saxena SP (1997) The riddle of vitamin K_1 deficit in the newborn. *Seminars in Perinatology* **21,** 90–6.

Ivanov VI and Karpeisky MY (1969) Dynamic three-dimensional model for enzymic transamination. *Advances in Enzymology and Related Areas of Molecular Biology* **32,** 21–53.

Jacob RA (1998) Vitamin C nutriture and risk of atherosclerotic heart disease. *Nutrition Reviews* **56,** 334–7.

Jacobsen DW (1998) Homocysteine and vitamins in cardiovascular disease. *Clinical Chemistry* **44,** 1833–43.

Jacobson MK, Ame JC, Lin W, Coyle DL, and Jacobson EL (1995) Cyclic ADP-ribose. A new component of calcium signaling. *Receptor* **5,** 43–9.

Jacques PF, Bostom AG, Williams RR, Ellison RC, Eckfeldt JH, Rosenberg IH, Selhub J, and Rozen R (1996) Relation between folate status, a common mutation in methylenetetrahydrofolate reductase, and plasma homocysteine concentrations. *Circulation* **93,** 7–9.

Jain SK and Lim G (2001) Pyridoxine and pyridoxamine inhibits superoxide radicals and prevents lipid peroxidation, protein glycosylation, and $(Na^+ + K^+)$-ATPase activity reduction in high glucose-treated human erythrocytes. *Free Radicals in Biology and Medicine* **30,** 232–7.

James SY, Williams MA, Newland AC, and Colston KW (1999) Leukemia cell differentiation: cellular and molecular interactions of retinoids and vitamin D. *General Pharmacology* **32,** 143–54.

Jang JT, Green JB, Beard JL, and Green MH (2000) Kinetic analysis shows that iron deficiency decreases liver vitamin A mobilization in rats. *Journal of Nutrition* **130,** 1291–6.

Jewell C and O'Brien NM (1999) Effect of dietary supplementation with carotenoids on xenobiotic metabolizing enzymes in the liver, lung, kidney and small intestine of the rat. *British Journal of Nutrition* **81,** 235–42.

Jhee KH, McPhie P, and Miles EW (2000) Yeast cystathionine beta-synthase is a pyridoxal phosphate enzyme but, unlike the human enzyme, is not a heme protein. *Journal of Biological Chemistry* **275,** 11541–4.

Jialal I, Devaraj S, and Kaul N (2001) The effect of alpha-tocopherol on monocyte proatherogenic activity. *Journal of Nutrition* **131,** 389S–94S.

Jiang Q, Christen S, Shigenaga MK, and Ames BN (2001) Gamma-tocopherol, the major form of vitamin E in the US diet, deserves more attention. *American Journal of Clinical Nutrition* **74,** 714–22.

Jitrapakdee S and Wallace JC (1999) Structure, function and regulation of pyruvate carboxylase. *Biochemical Journal* **340,** 1–16.

Johnson A and Chandraratna RA (1999) Novel retinoids with receptor selectivity and functional selectivity. *British Journal of Dermatology* **140**(Suppl 54), 12–17.

Johnson AR, Hood RL, and Emery JL (1980) Biotin and the sudden infant death syndrome. *Nature* **285,** 159–60.

Johnston CS (1999) Biomarkers for establishing a tolerable upper intake level for vitamin C. *Nutrition Reviews* **57,** 71–7.

Jolly DW, Craig C, and Nelson TE Jr (1977) Estrogen and prothrombin synthesis: effect of estrogen on absorption of vitamin K_1. *American Journal of Physiology* **232,** H12–H17.

Joyce EM (1994) Aetiology of alcoholic brain damage: alcoholic neurotoxicity or thiamine malnutrition? *British Medical Bulletin* **50,** 99–114.

Juan D and DeLuca HF (1977) The regulation of 24,25-dihydroxyvitamin D_3 production in cultures of monkey kidney cells. *Endocrinology* **101,** 1184–93.

Kabil O, Toaka S, LoBrutto R, Shoemaker R, and Banerjee R (2001) Pyridoxal phosphate binding sites are similar in human heme-dependent and yeast heme-independent cystathionine beta-synthases. Evidence from ^{31}P NMR and pulsed EPR spectroscopy that heme and PLP cofactors are not proximal in the human enzyme. *Journal of Biological Chemistry* **276,** 19350–5.

Kallner A, Hartmann D, and Hornig D (1979) Steady-state turnover and body pool of ascorbic acid in man. *American Journal of Clinical Nutrition* **32,** 530–9.

Kallner A, Hornig D, and Pellikka R (1985) Formation of carbon dioxide from ascorbate in man. *American Journal of Clinical Nutrition* **41,** 609–13.

Kallner AB, Hartmann D, and Hornig DH (1981) On the requirements of ascorbic acid in man: steady-state turnover and body pool in smokers. *American Journal of Clinical Nutrition* **34,** 1347–55.

Kameda T, Miyazawa K, Mori Y, Yuasa T, Shiokawa M, Nakamaru Y, Mano H, Hakeda Y, Kameda A,and Kumegawa M (1996) Vitamin K_2 inhibits osteoclastic bone resorption by inducing osteoclast apoptosis. *Biochemical and Biophysical Research Communications* **220,** 515–19.

Kang SS, Wong PW, Susmano A, Sora J, Norusis M, and Ruggie N (1991) Thermolabile methylenetetrahydrofolate reductase: an inherited risk factor for coronary artery disease. *Amerian Journal of Human Genetics* **48,** 536–45.

Karlson B, Leijd B, and Hellstrom K (1986) On the influence of vitamin K-rich vegetables and wine on the effectiveness of warfarin treatment. *Acta Medica Scandinivica* **220,** 347–50.

Kasvosve I, Delanghe JR, Gomo ZA, Gangaidzo IT, Khumalo H, Langlois MR, Moyo VM, Saungweme T, Mvundura E, Boelaert JR, and Gordeuk VR (2002) Effect of transferrin polymorphism on the metabolism of vitamin C in Zimbabwean adults. *American Journal of Clinical Nutrition* **75,** 321–5.

Katunuma N, Kominami E, and Kominami S (1971) A new enzyme that specifically inactivates apo-protein of pyridoxal enzymes. *Biochemical and Biophysical Research Communications* **45,** 70–5.

Kaul N, Devaraj S, and Jialal I (2001) Alpha-tocopherol and atherosclerosis. *Experimental Biology and Medicine* (Maywood) **226,** 5–12.

Kawada T, Kamei Y, and Sugimoto E (1996) The possibility of active form of vitamins A and D as suppressors on adipocyte development via ligand-dependent transcriptional regulators. *International Journal of Obesity and Related Metabolic Disorders* **20**(Suppl 3), S52–7.

Keiver KM, Draper HH, and Ronald K (1988) Vitamin D metabolism in the hooded seal (*Cystophora cristata*). *Journal of Nutrition* **118,** 332–41.

Kelly D, Weir D, Reed B, and Scott J (1979) Effect of anticonvulsant drugs on the rate of folate catabolism in mice. *Journal of Clinical Investigation* **64,** 1089–96.

Kelsay JL (1969) A compendium of nutritional status studies and dietary evaluation studies in the USA, 1957–1967. *Journal of Nutrition* **99 suppl 1,** 119–66.

Kelsay J, Baysal A, and Linkswiler H (1968a) Effect of vitamin B_6 depletion on the pyridoxal, pyridoxamine and pyridoxine content of the blood and urine of men. *Journal of Nutrition* **94,** 490–4.

Kelsay J, Miller LT, and Linkswiler H (1968b) Effect of protein intake on the excretion of quinolinic acid and niacin metabolites by men during vitamin B_6 depletion. *Journal of Nutrition* **94,** 27–31.

Kennedy MJ, Lee KA, Niemi GA, Craven KB, Garwin GG, Saari JC, and Hurley JB (2001) Multiple phosphorylation of rhodopsin and the in vivo chemistry underlying rod photoreceptor dark adaptation. *Neuron* **31,** 87–101.

Kerner J and Hoppel C (2000) Fatty acid import into mitochondria. *Biochimica et Biophysica Acta* **1486,** 1–17.

Khalifah RG, Baynes JW, and Hudson BG (1999) Amadorins: novel post-Amadori inhibitors of advanced glycation reactions. *Biochemical and Biophysical Research Communications* **257**, 251–8.

Kiefer C, Hessel S, Lampert JM, Vogt K, Lederer MO, Breithaupt DE, and von Lintig J (2001) Identification and characterization of a mammalian enzyme catalyzing the asymmetric oxidative cleavage of provitamin A. *Journal of Biological Chemistry* **276**, 14110–16.

Killman S-A (1964) Effect of deoxyuridine on incorporation of tritiated thymidine: difference between normoblasts and megaloblasts. *Acta Medica Scandinavica* **175**, 483–8.

Kim JH and Miller LL (1969) The functional significance of changes in activity of the enzymes, tryptophan pyrrolase and tyrosine transaminase, after induction in intact rats and in the isolated, perfused rat liver. *Journal of Biological Chemistry* **244**, 1410–16.

Kim KA, Kim S, Chang I, Kim GS, Min YK, Lee MK, Kim KW, and Lee MS (2002) IFN gamma/TNF alpha synergism in MHC class II induction: effect of nicotinamide on MHC class II expression but not on islet-cell apoptosis. *Diabetologia* **45**, 385–93.

Kircher T and Sinzinger H (2000) Homocysteine – relevant for atherogenesis? *Wiener Klinische Wochenschrift* **112**, 523–32.

Kirschbaum N, Clemons R, Marino KA, Sheedy G, Nguyen ML, and Smith CM (1990) Pantothenate kinase activity in livers of genetically diabetic mice (db/db) and hormonally treated cultured rat hepatocytes. *Journal of Nutrition* **120**, 1376–86.

Kisker C, Schindelin H, and Rees DC (1997) Molybdenum-cofactor-containing enzymes: structure and mechanism. *Annual Reviews of Biochemistry* **66**, 233–67.

Kitamura K, Yamaguchi T, Tanaka H, Hashimoto S, Yang M, and Takahashi T (1996) TPN-induced fulminant beriberi: a report on our experience and a review of the literature. *Surgery Today* **26**, 769–76.

Kivirikko KI and Pihlajaniemi T (1998) Collagen hydroxylases and the protein disulfide isomerase subunit of prolyl 4-hydroxylases. *Advances in Enzymology and Related Areas of Molecular Biology* **72**, 325–98.

Kleijnen J, Ter Riet G, and Knipschild P (1990) Vitamin B_6 in the treatment of the premenstrual syndrome–a review. *British Journal of Obstetrics and Gynaecology* **97**, 847–52.

Knip M, Douek IF, Moore WP, Gillmor HA, McLean AE, Bingley PJ, and Gale EA (2000) Safety of high-dose nicotinamide: a review. *Diabetologia* **43**, 1337–45.

Kodentsova VM, Vrzhesinskaya OA, and Spirichev VB (1995) Fluorometric riboflavin titration in plasma by riboflavin-binding apoprotein as a method for vitamin B_2 status assessment. *Annals of Nutrition and Metabolism* **39**, 355–60.

Koga H, Fujita I, and Miyazaki S (1997) Effects of all-*trans*-retinoic acid on superoxide generation in intact neutrophils and a cell-free system. *British Journal of Haematology* **97**, 300–5.

Kohlmeier M, Salomon A, Saupe J, and Shearer MJ (1996) Transport of vitamin K to bone in humans. *Journal of Nutrition* **126**, 1192S–6S.

Kolb H and Burkart V (1999) Nicotinamide in type 1 diabetes. Mechanism of action revisited. *Diabetes Care* **22**(Suppl 2), B16–B20.

Kolobova E, Tuganova A, Boulatnikov I, and Popov KM (2001) Regulation of pyruvate dehydrogenase activity through phosphorylation at multiple sites. *Biochemical Journal* **358**, 69–77.

Korotchkina LG and Patel MS (2001a) Probing the mechanism of inactivation of human pyruvate dehydrogenase by phosphorylation of three sites. *Journal of Biological Chemistry* **276,** 5731–8.

Korotchkina LG and Patel MS (2001b) Site specificity of four pyruvate dehydrogenase kinase isoenzymes toward the three phosphorylation sites of human pyruvate dehydrogenase. *Journal of Biological Chemistry* **276,** 37223–9.

Koshimura K, Murakami Y, Tanaka J, and Kato Y (2000) The role of 6R-tetrahydrobiopterin in the nervous system. *Progress in Neurobiology* **61,** 415–38.

Kotake Y, Ueda T, Mori T, Murakami E, and Hattori M (1975) The physiological significance of the xanthurenic acid-insulin complex. *Journal of Biochemistry* (Tokyo) **77,** 685–7.

Krall EA, Sahyoun N, Tannenbaum S, Dallal GE, and Dawson-Hughes B (1989) Effect of vitamin D intake on seasonal variations in parathyroid hormone secretion in postmenopausal women. *New England Journal of Medicine* **321,** 1777–83.

Krebs HA, Hems R, and Tyler B (1976) The regulation of folate and methionine metabolism. *Biochemical Journal* **158,** 341–53.

Kretsch MJ, Sauberlich HE, Skala JH, and Johnson HL (1995) Vitamin B-6 requirement and status assessment: young women fed a depletion diet followed by a plant- or animal-protein diet with graded amounts of vitamin B-6. *American Journal of Clinical Nutrition* **61,** 1091–1101.

Krill D, O'Leary L, Koehler AN, Kramer MK, Warty V, Wagner MA, and Dorman JS (1997) Association of retinol binding protein in multiple-case families with insulin-dependent diabetes. *Human Biology* **69,** 89–96.

Krinke G, Schaumburg HH, Spencer PS, Suter J, Thomann O, and Hess R (1980) Pyridoxine megavitaminosis produces degeneration of peripheral sensory neurons (sensory neuropathy) in the dog. *Neurotoxicology* **2,** 13–24.

Krishnamurthy K, Surolia N, and Adiga PR (1984) Mechanism of foetal wastage following immunoneutralization of riboflavin carrier protein in the pregnant rat: disturbances in flavin coenzyme levels. *FEBS Letters* **178,** 87–91.

Kumar MV, Sunvold GD, and Scarpace PJ (1999) Dietary vitamin A supplementation in rats: suppression of leptin and induction of UCP1 mRNA. *Journal of Lipid Research* **40,** 824–9.

Kumar R (2000) Tumor-induced osteomalacia and the regulation of phosphate homeostasis. *Bone* **27,** 333–8.

Kurlandsky SB, Duell EA, Kang S, Voorhees JJ, and Fisher GJ (1996) Auto-regulation of retinoic acid biosynthesis through regulation of retinol esterification in human keratinocytes. *Journal of Biological Chemistry* **271,** 15346–52.

Lakaye B, Makarchikov AF, Fernandes Antunes A, Zorzi W, Coumans B, De Pauw E, Wins P, Grisar T, and Bettendorff L (2002) Molecular characterization of a specific thiamine triphosphatase widely expressed in mammalian tissues. *Journal of Biological Chemistry* **277,** 13771–7.

Lakshmanan MR, Chansang H, and Olson JA (1972) Purification and properties of carotene 15,15′-dioxygenase of rabbit intestine. *Journal of Lipid Research* **13,** 477–82.

Lakshmi AV and Bamji MS (1974) Tissue pyridoxal phosphate concentration and pyridoxaminephosphate oxidase activity in riboflavin deficiency in rats and man. *British Journal of Nutrition* **32,** 249–55.

Lal KJ, Dakshinamurti K, and Thliveris J (1996) The effect of vitamin B_6 on the systolic blood pressure of rats in various animal models of hypertension. *Journal of Hypertension* **14**, 355–63.

Lancaster CR (2002) Succinate:quinone oxidoreductases: an overview. *Biochimica et Biophysica Acta* **1553**, 1–6.

Langlais PJ (1995) Pathogenesis of diencephalic lesions in an experimental model of Wernicke's encephalopathy. *Metabolic Brain Diseases* **10**, 31–44.

Langman LJ and Cole DE (1999) Homocysteine: cholesterol of the 90s? *Clinica Chimica Acta* **286**, 63–80.

Langsjoen PH and Langsjoen AM (1999) Overview of the use of CoQ10 in cardiovascular disease. *Biofactors* **9**, 273–84.

Lawson DE, Paul AA, Black AE, Cole TJ, Mandal AR, and Davie M (1979) Relative contributions of diet and sunlight to vitamin D state in the elderly. *British Medical Journal* **2**, 303–5.

Lee HC (1996) Modulator and messenger functions of cyclic ADP-ribose in calcium signaling. *Recent Progress in Hormone Research* **51**, 355–88; discussion 389.

Lee HC (1999) A unified mechanism of enzymatic synthesis of two calcium messengers: cyclic ADP-ribose and NAADP. *Biological Chemistry* **380**, 785–93.

Lee HC (2000) NAADP: an emerging calcium signaling molecule. *Journal of Membrance Biology* **173**, 1–8.

Lee HC (2001) Physiological functions of cyclic ADP-ribose and NAADP as calcium messengers. *Annual Reviews of Pharmacology and Toxicology* **41**, 317–45.

Lee SS and McCormick DB (1985) Thyroid hormone regulation of flavocoenzyme biosynthesis. *Archives of Biochemistry and Biophysics* **237**, 197–201.

Leklem JE (1990) Vitamin B-6: a status report. *Journal of Nutrition* **120 Suppl 11**, 1503–7.

Lemire JM, Adams JS, Sakai R, and Jordan SC (1984) 1 alpha,25-Dihydroxyvitamin D_3 suppresses proliferation and immunoglobulin production by normal human peripheral blood mononuclear cells. *Journal of Clinical Investigation* **74**, 657–61.

Lemoyne M, Van Gossum A, Kurian R, and Jeejeebhoy KN (1988) Plasma vitamin E and selenium and breath pentane in home parenteral nutrition patients. *American Journal of Clinical Nutrition* **48**, 1310–15.

Leo MA and Lieber CS (1985) New pathway for retinol metabolism in liver microsomes. *Journal of Biological Chemistry* **260**, 5228–31.

Leo MA, Lasker JM, Raucy JL, Kim CI, Black M, and Lieber CS (1989) Metabolism of retinol and retinoic acid by human liver cytochrome P450IIC8. *Archives of Biochemistry and Biophysics* **269**, 305–12.

Leong DK and Butterworth RF (1996) Neuronal cell death in Wernicke's encephalopathy: pathophysiologic mechanisms and implications for PET imaging. *Metabolic Brain Diseases* **11**, 71–9.

Lerner V, Miodownik C, Kaptsan A, Cohen H, Matar M, Loewenthal U, and Kotler M (2001) Vitamin B(6) in the treatment of tardive dyskinesia: a double-blind, placebo-controlled, crossover study. *American Journal of Psychiatry* **158**, 1511–14.

Levine M, Dhariwal KR, Welch RW, Wang Y, and Park JB (1995) Determination of optimal vitamin C requirements in humans. *American Journal of Clinical Nutrition* **62**, 1347S–56S.

Levine M, Conry-Cantilena C, Wang Y, Welch RW, Washko PW, Dhariwal KR, Park JB, Lazarev A, Graumlich JF, King J, and Cantilena LR (1996) Vitamin C pharmacokinetics in healthy volunteers: evidence for a recommended dietary allowance. *Proceedings of the National Academy of Sciences of the USA* **93**, 3704–9.

Levine M, Rumsey SC, Daruwala R, Park JB, and Wang Y (1999) Criteria and recommendations for vitamin C intake. *JAMA* **281**, 1415–23.

Li E and Norris AW (1996) Structure/function of cytoplasmic vitamin A-binding proteins. *Annual Reviews of Nutrition* **16**, 205–34.

Liang WJ, Johnson D, and Jarvis SM (2001) Vitamin C transport systems of mammalian cells. *Molecular Membrane Biology* **18**, 87–95.

Lima L, Obregon F, Cubillos S, Fazzino F, and Jaimes I (2001) Taurine as a micronutrient in development and regeneration of the central nervous system. *Nutritional Neuroscience* **4**, 439–43.

Lin FJ, Song W, Meyer-Bernstein E, Naidoo N, and Sehgal A (2001) Photic signaling by cryptochrome in the Drosophila circadian system. *Molecular and Cell Biology* **21**, 7287–94.

Lippman SM and Lotan R (2000) Advances in the development of retinoids as chemopreventive agents. *Journal of Nutrition* **130**, 479S–82S.

Liska DJ and Suttie JW (1988) Location of gamma-carboxyglutamyl residues in partially carboxylated prothrombin preparations. *Biochemistry* **27**, 8636–41.

Liu YY, Shigematsu Y, Bykov I, Nakai A, Kikawa Y, Fukui T, and Sudo M (1994) Abnormal fatty acid composition of lymphocytes of biotin-deficient rats. *Journal of Nutritional Science and Vitaminology* (Tokyo) **40**, 283–8.

Lombardo YB, Serdikoff C, Thamotharan M, Paul HS, and Adibi SA (1999) Inverse alterations of BCKA dehydrogenase activity in cardiac and skeletal muscles of diabetic rats. *American Journal of Physiology* **277**, E685–92.

Lopez-Lluch G, Blazquez MV, Perez-Vicente R, Macho A, Buron MI, Alcain FJ, Munoz E, and Navas P (2001) Cellular redox state and activating protein-1 are involved in ascorbate effect on calcitriol-induced differentiation. *Protoplasma* **217**, 129–36.

Louet JF, Le May C, Pegorier JP, Decaux JF, and Girard J (2001) Regulation of liver carnitine palmitoyltransferase I gene expression by hormones and fatty acids. *Biochemical Society Transactions* **29**, 310–16.

Ludwig ML and Matthews RG (1997) Structure-based perspectives on B_{12}-dependent enzymes. *Annual Reviews of Biochemistry* **66**, 269–313.

Luo G, Ducy P, McKee MD, Pinero GJ, Loyer E, Behringer RR, and Karsenty G (1997) Spontaneous calcification of arteries and cartilage in mice lacking matrix GLA protein. *Nature* **386**, 78–81.

Lykidis A and Jackowski S (2001) Regulation of mammalian cell membrane biosynthesis. *Progress in Nucleic Acid Research and Molecular Biology* **65**, 361–93.

Ma JJ and Truswell AS (1995) Wernicke-Korsakoff syndrome in Sydney hospitals: before and after thiamine enrichment of flour. *Medical Journal of Australia* **163**, 531–4.

Mackenzie RE and Baugh CM (1980) Tetrahydropterolypolyglutamate derivatives as substrates of two multifunctional proteins with folate-dependent enzyme activities. *Biochimica et Biophysica Acta* **611**, 187–95.

Maden M (2000) The role of retinoic acid in embryonic and post-embryonic development. *Proceedings of the Nutrition Society* **59,** 65–73.

Maeda T, Taguchi H, Minami H, Sato K, Shiga T, Kosaka H, and Yoshikawa K (2000) Vitamin B_6 phototoxicity induced by UVA radiation. *Archives of Dermatology Research* **292,** 562–7.

Maeda Y, Kawata S, Inui Y, Fukuda K, Igura T, and Matsuzawa Y (1996) Biotin deficiency decreases ornithine transcarbamylase activity and mRNA in rat liver. *Journal of Nutrition* **126,** 61–6.

Magboul BI and Bender DA (1983) The effects of a dietary excess of leucine on the synthesis of nicotinamide nucleotides in the rat. *British Journal of Nutrition* **49,** 321–9.

Maguire JJ, Wilson DS, and Packer L (1989) Mitochondrial electron transport-linked tocopheroxyl radical reduction. *Journal of Biological Chemistry* **264,** 21462–5.

Mahmoodian F and Peterkofsky B (1999) Vitamin C deficiency in guinea pigs differentially affects the expression of type IV collagen, laminin, and elastin in blood vessels. *Journal of Nutrition* **129,** 83–91.

Malfait P, Moren A, Dillon JC, Brodel A, Begkoyian G, Etchegorry MG, Malenga G, and Hakewill P (1993) An outbreak of pellagra related to changes in dietary niacin among Mozambican refugees in Malawi. *International Journal of Epidemiology* **22,** 504–11.

Malo C and Wilson JX (2000) Glucose modulates vitamin C transport in adult human small intestinal brush border membrane vesicles. *Journal of Nutrition* **130,** 63–9.

Mangelsdorf D and Evans R (1995) The RXR heterodimers and orphan receptors. *Cell* **83,** 841–50.

Manolagas SC, Provvedini DM, and Tsoukas CD (1985) Interactions of 1,25-dihydroxyvitamin D_3 and the immune system. *Molecular and Cellular Endocrinology* **43,** 113–22.

Manson JA and Carpenter KJ (1978a) The effect of a high level of dietary leucine on the niacin status of chicks and rats. *Journal of Nutrition* **108,** 1883–8.

Manson JA and Carpenter KJ (1978b) The effect of a high level of dietary leucine on the niacin status of dogs. *Journal of Nutrition* **108,** 1889–98.

Manwaring JD and Csallany AS (1988) Malondialdehyde-containing proteins and their relationship to vitamin E. *Lipids* **23,** 651–5.

Mardones P, Strobel P, Miranda S, Leighton F, Quinones V, Amigo L, Rozowski J, Krieger M, and Rigotti A (2002) Alpha-tocopherol metabolism is abnormal in scavenger receptor class B type I (SR-BI)-deficient mice. *Journal of Nutrition* **132,** 443–9.

Mark M, Ghyselinck NB, Wendling O, Dupe V, Mascrez B, Kastner P, and Chambon P (1999) A genetic dissection of the retinoid signalling pathway in the mouse. *Proceedings of the Nutrition Society* **58,** 609–13.

Marquet A, Bui BT, and Florentin D (2001) Biosynthesis of biotin and lipoic acid. *Vitamins and Hormones* **61,** 51–101.

Mason JB, Gibson N, and Kodicek E (1973) The chemical nature of the bound nicotinic acid of wheat bran: studies of nicotinic acid-containing macromolecules. *British Journal of Nutrition* **30,** 297–311.

Masse PG, van den Berg H, Duguay C, Beaulieu G, and Simard JM (1996) Early effect of a low dose (30 micrograms) ethinyl estradiol-containing Triphasil on vitamin B_6 status. A follow-up study on six menstrual cycles. *International Journal of Vitamin and Nutrition Research* **66,** 46–54.

Matthews RG and Daubner SC (1982) Modulation of methylenetetrahydrofolate reductase activity by *S*-adenosylmethionine and by dihydrofolate and its polyglutamate analogues. *Advances in Enzyme Regulation* **20,** 123–31.

Mawer EB, Taylor CM, Backhouse J, Lumb GA, and Stanbury SW (1973) Failure of formation of 1,25-dihydroxycholecalciferol in chronic renal insufficiency. *Lancet* **1,** 626–8.

May JM (1999) Is ascorbic acid an antioxidant for the plasma membrane? *FASEB Journal* **13,** 995–1006.

Mayson JS, Schumaker O, and Nakamura RM (1973) False-negative tests for urine glucose. *Lancet* **1,** 780–1.

McBee JK, Kuksa V, Alvarez R, de Lera AR, Prezhdo O, Haeseleer F, Sokal I, and Palczewski K (2000) Isomerization of all-*trans*-retinol to *cis*-retinols in bovine retinal pigment epithelial cells: dependence on the specificity of retinoid-binding proteins. *Biochemistry* **39,** 11370–80.

McCandless DW, Hanson C, Speeg KV Jr, and Schenker S (1970) Cardiac metabolism in thiamin deficiency in rats. *Journal of Nutrition* **100,** 991–1002.

McCormick DB (1989) Two interconnected B vitamins: riboflavin and pyridoxine. *Physiological Reviews* **69,** 1170–98.

McCreanor GM and Bender DA (1986) The metabolism of high intakes of tryptophan, nicotinamide and nicotinic acid in the rat. *British Journal of Nutrition* **56,** 577–86.

McChrisley B, Thye FW, McNair HM, and Driskell JA (1988) Plasma B6 vitamer and 4-pyridoxic acid concentrations of men fed controlled diets. *Journal of Chromatography* **428,** 35–42.

McCullough FS, Northrop-Clewes CA, and Thurnham DI (1999) The effect of vitamin A on epithelial integrity. *Proceedings of the Nutrition Society* **58,** 289–93.

McCully KS (1971) Homocysteine metabolism in scurvy, growth and arteriosclerosis. *Nature* **231,** 391–2.

McEntee WJ (1997) Wernicke's encephalopathy: an excitotoxicity hypothesis. *Metabolic Brain Diseases* **12,** 183–92.

McGrath JJ and Soares KV (2000) Cholinergic medication for neuroleptic-induced tardive dyskinesia. *Cochrane Database Systematic Reviews,* CD000207.

McGregor DO, Dellow WJ, Lever M, George PM, Robson RA, and Chambers ST (2001) Dimethylglycine accumulates in uremia and predicts elevated plasma homocysteine concentrations. *Kidney International* **59,** 2267–72.

McLachlan RS and Brown WF (1995) Pyridoxine dependent epilepsy with iatrogenic sensory neuronopathy. *Canadian Journal of Neurological Science* **22,** 50–1.

McMahon RJ (2002) Biotin in metabolism and molecular biology. *Annual Reviews of Nutrition* **22,** 221–39.

McNulty H, McKinley MC, Wilson B, McPartlin J, Strain JJ, Weir DG, and Scott JM (2002) Impaired functioning of thermolabile methylenetetrahydrofolate reductase is dependent on riboflavin status: implications for riboflavin requirements. *American Journal of Clinical Nutrition* **76,** 436–41.

Medical Research Council: Vitamin C Sub-committee of the Accessory Food Factor Committee (1948) Vitamin C requirements of human adults. *Lancet* **1,** 835–58.

Meeks RG, Zaharevitz D, and Chen RF (1981) Membrane effects of retinoids: possible correlation with toxicity. *Archives of Biochemistry and Biophysics* **207,** 141–7.

Meganathan R (2001) Biosynthesis of menaquinone (vitamin K_2) and ubiquinone (coenzyme Q): a perspective on enzymatic mechanisms. *Vitamins and Hormones* **61,** 173–218.

Mehta K and Cheema S (1999) Retinoid-mediated signaling pathways in CD_{38} antigen expression in myeloid leukemia cells. *Leukemia and Lymphoma* **32,** 441–9.

Meister A (1990) On the transamination of enzymes. *Annals of the New York Academy of Sciences* **585,** 13–31.

Meleady R and Graham I (1999) Plasma homocysteine as a cardiovascular risk factor: causal, consequential, or of no consequence? *Nutrition Reviews* **57,** 299–305.

Menniti FS, Knoth J, and Diliberto EJ Jr (1986) Role of ascorbic acid in dopamine beta-hydroxylation. The endogenous enzyme cofactor and putative electron donor for cofactor regeneration. *Journal of Biological Chemistry* **261,** 16901–8.

Meydani M (2000) Vitamin E and prevention of heart disease in high-risk patients. *Nutrition Reviews* **58,** 278–81.

Militante JD, Lombardini JB, and Schaffer SW (2000) The role of taurine in the pathogenesis of the cardiomyopathy of insulin-dependent diabetes mellitus. *Cardiovascular Research* **46,** 393–402.

Miller DA and DeLuca HF (1986) Biosynthesis of retinoyl-beta-glucuronide, a biologically active metabolite of all-*trans*-retinoic acid. *Archives of Biochemistry and Biophysics* **244,** 179–86.

Miller LT and Linkswiler H (1967) Effect of protein intake on the development of abnormal protein metabolism by men during vitamin B_6 depletion. *Journal of Nutrition* **93,** 53–9.

Miller LT, Leklem JE, and Shultz TD (1985) The effect of dietary protein on the metabolism of vitamin B_6 in humans. *Journal of Nutrition* **115,** 1663–72.

Miller RK, Hendrickx AG, Mills JL, Hummler H, and Wiegand UW (1998) Periconceptional vitamin A use: how much is teratogenic? *Reproductive Toxicology* **12,** 75–88.

Min H, Shane B, and Stokstad EL (1988) Identification of 10-formyltetrahydrofolate dehydrogenase-hydrolase as a major folate binding protein in liver cytosol. *Biochimica et Biophysica Acta* **967,** 348–53.

Miyamoto T, Kakizawa T, and Hashizume K (1999) Inhibition of nuclear receptor signalling by poly(ADP-ribose) polymerase. *Molecular and Cell Biology* **19,** 2644–9.

Miyoshi K, Egi Y, Shioda T, and Kawasaki T (1990) Evidence for in vivo synthesis of thiamin triphosphate by cytosolic adenylate kinase in chicken skeletal muscle. *Journal of Biochemistry* (Tokyo) **108,** 267–70.

Mizushima Y, Harauchi T, Yoshizaki T, and Makino S (1984) A rat mutant unable to synthesize vitamin C. *Experientia* **40,** 359–61.

Mizutani A, Nakagawa N, Hitomi K, and Tsukagoshi N (1997) Ascorbate-dependent expression of ubiquitin genes in guinea pigs. *International Journal of Biochemistry and Cell Biology* **29,** 575–82.

Mock DM (1991) Skin manifestations of biotin deficiency. *Seminars in Dermatology* **10,** 296–302.

Mock DM (1999) Biotin status: which are valid indicators and how do we know? *Journal of Nutrition* **129,** 498S–503S.

Mock DM and Mock NI (2002) Lymphocyte propionyl-CoA carboxylase is an early and sensitive indicator of biotin deficiency in rats, but urinary excretion of 3-hydroxypropionic acid is not. *Journal of Nutrition* **132,** 1945–50.

Mock DM, Baswell DL, Baker H, Holman RT, and Sweetman L (1985) Biotin deficiency complicating parenteral alimentation: diagnosis, metabolic repercussions, and treatment. *Journal of Pediatrics* **106,** 762–9.

Mock DM, Johnson SB, and Holman RT (1988a) Effects of biotin deficiency on serum fatty acid composition: evidence for abnormalities in humans. *Journal of Nutrition* **118,** 342–8.

Mock DM, Mock NI, Johnson SB, and Holman RT (1988b) Effects of biotin deficiency on plasma and tissue fatty acid composition: evidence for abnormalities in rats. *Pediatric Research* **24,** 396–403.

Mock DM, Quirk JG, and Mock NI (2002) Marginal biotin deficiency during normal pregnancy. *American Journal of Clinical Nutrition* **75,** 295–9.

Mock NI, Malik MI, Stumbo PJ, Bishop WP, and Mock DM (1997) Increased urinary excretion of 3-hydroxyisovaleric acid and decreased urinary excretion of biotin are sensitive early indicators of decreased biotin status in experimental biotin deficiency. *American Journal of Clinical Nutrition* **65,** 951–8.

Moestrup SK and Verroust PJ (2001) Megalin- and cubilin-mediated endocytosis of protein-bound vitamins, lipids, and hormones in polarized epithelia. *Annual Reviews of Nutrition* **21,** 407–28.

Momoi T, Hanaoka K, and Momoi M (1990) Spatial and temporal expression of cellular retinoic acid binding protein (CRABP) along the anteroposterior axis in the central nervous system of mouse embryos. *Biochemical and Biophysical Research Communications* **169,** 991–6.

Mori T, Itoh S, Ohgiya S, Ishizaki K, and Kamataki T (1997) Regulation of CYP1A and CYP3A mRNAs by ascorbic acid in guinea pigs. *Archives of Biochemistry and Biophysics* **348,** 268–77.

Moriguchi S and Muraga M (2000) Vitamin E and immunity. *Vitamins and Hormones* **59,** 305–36.

Morris DP, Stevens RD, Wright DJ, and Stafford DW (1995) Processive post-translational modification. Vitamin K-dependent carboxylation of a peptide substrate. *Journal of Biological Chemistry* **270,** 30491–8.

Morrissey PA and Sheehy PJ (1999) Optimal nutrition: vitamin E. *Proceedings of the Nutrition Society* **58,** 459–68.

Mortensen PB, Kolvraa S, and Christensen E (1980) Inhibition of the glycine cleavage system: hyperglycinemia and hyperglycinuria caused by valproic acid. *Epilepsia* **21,** 563–9.

Moss J, Balducci E, Cavanaugh E, Kim HJ, Konczalik P, Lesma EA, Okazaki IJ, Park M, Shoemaker M, Stevens LA, and Zolkiewska A (1999) Characterization of NAD:arginine ADP-ribosyltransferases. *Molecular and Cellular Biochemistry* **193,** 109–13.

Moss J, Zolkiewska A, and Okazaki I (1997) ADP-ribosylarginine hydrolases and ADP-ribosyltransferases. Partners in ADP-ribosylation cycles. *Advances in Experimental Medicine and Biology* **419,** 25–33.

Mowat C and McColl KE (2001) Alterations in intragastric nitrite and vitamin C levels during acid inhibitory therapy. *Best Practice Research in Clinical Gastroenterology* **15**, 523–37.

Moyers S and Bailey LB (2001) Fetal malformations and folate metabolism: review of recent evidence. *Nutrition Reviews* **59**, 215–24.

Mpofu C, Alani SM, Whitehouse C, Fowler B, and Wraith JE (1991) No sensory neuropathy during pyridoxine treatment in homocystinuria. *Archives of Disease in Childhood* **66**, 1081–2.

MRC Vitamin Study Research Group (1991) Prevention of neural tube defects: results of the Medical Research Council Vitamin Study. *Lancet* **338**, 131–7.

Mudd SH (1971) Pyridoxine-responsive genetic disease. *Federation Proceedings* **30**, 970–6.

Muller DP (1986) Vitamin E – its role in neurological function. *Postgraduate Medical Journal* **62**, 107–12.

Muller DP, Lloyd JK, and Wolff OH (1983) Vitamin E and neurological function. *Lancet* **1**, 225–8.

Mutucumarana VP, Stafford DW, Stanley TB, Jin DY, Solera J, Brenner B, Azerad R, and Wu SM (2000) Expression and characterization of the naturally occurring mutation L394R in human gamma-glutamyl carboxylase. *Journal of Biological Chemistry* **275**, 32572–7.

Myhre AM, Takahashi N, Blomhoff R, Breitman TR, and Norum KR (1996) Retinoylation of proteins in rat liver, kidney, and lung in vivo. *Journal of Lipid Research* **37**, 1971–7.

Nagao A, During A, Hoshino C, Terao J, and Olson JA (1996) Stoichiometric conversion of all *trans*-beta-carotene to retinal by pig intestinal extract. *Archives of Biochemistry and Biophysics* **328**, 57–63.

Nagata K and Okada M (1985) Characterization of mitochondrial aspartate aminotransferase from the liver of pyridoxine-deficient rats. *Journal of Biochemistry* (Tokyo) **97**, 501–7.

Nakagawa I, Takahashi T, Suzuki T, and Masana Y (1969) Effect in man of the addition of tryptophan or niacin to the diet on the excretion of their metabolites. *Journal of Nutrition* **99**, 325–30.

Nakahiro M, Fujita N, Fukuchi I, Saito K, Nishimura T, and Yoshida H (1985) Pantoyl-gamma-aminobutyric acid facilitates cholinergic function in the central nervous system. *Journal of Pharmacology and Experimental Therapeutics* **232**, 501–6.

Napoli J and Race K (1987) The biosynthesis of retinoic acid from retinol by rat tissues. *Archives of Biochemistry and Biophysics* **255**, 95–101.

Napoli JL (1996) Retinoic acid biosynthesis and metabolism. *FASEB Journal* **10**, 993–1001.

Napoli JL (2001) 17Beta-hydroxysteroid dehydrogenase type 9 and other short-chain dehydrogenases/reductases that catalyze retinoid, 17beta- and 3alpha-hydroxysteroid metabolism. *Molecular and Cellular Endocrinology* **171**, 103–9.

Napoli JL, Boerman MH, Chai X, Zhai Y, and Fiorella PD (1995) Enzymes and binding proteins affecting retinoic acid concentrations. *Journal of Steroid Biochemistry and Molecular Biology* **53**, 497–502.

Narisawa S, Wennberg C, and Millan JL (2001) Abnormal vitamin B_6 metabolism in alkaline phosphatase knock-out mice causes multiple abnormalities, but not the impaired bone mineralization. *Journal of Pathology* **193**, 125–33.

Narvaez CJ, Zinser G, and Welsh J (2001) Functions of 1alpha,25-dihydroxyvitamin D(3) in mammary gland: from normal development to breast cancer. *Steroids* **66,** 301–8.

Natadisastra G, Wittpenn JR, West KP Jr, Muhilal, and Sommer A (1987) Impression cytology for detection of vitamin A deficiency. *Archives of Ophthalmology* **105,** 1224–8.

National Institutes of Health (2000) Osteoporosis prevention, diagnosis, and therapy. *NIH Consensus Statement* **17,** 1–45.

National Research Council (1989) *Recommended Dietary Allowances,* 10th Edition. Washington, DC: National Academy Press.

Nau H (2001) Teratogenicity of isotretinoin revisited: species variation and the role of all-*trans*-retinoic acid. *Journal of the American Academy of Dermatology* **45,** S183–7.

Nelsestuen GL, Shah AM, and Harvey SB (2000) Vitamin K-dependent proteins. *Vitamins and Hormones* **58,** 355–89.

Nemere I and Farach-Carson MC (1998) Membrane receptors for steroid hormones: a case for specific cell surface binding sites for vitamin D metabolites and estrogens. *Biochemical and Biophysical Research Communications* **248,** 443–9.

Ness AR, Powles JW, and Khaw KT (1996) Vitamin C and cardiovascular disease: a systematic review. *Journal of Cardiovascular Risk* **3,** 513–21.

Neufeld EJ, Fleming JC, Tartaglini E, and Steinkamp MP (2001) Thiamine-responsive megaloblastic anemia syndrome: a disorder of high-affinity thiamine transport. *Blood Cells Molecules and Diseases* **27,** 135–8.

Nghiem HO, Bettendorff L, and Changeux JP (2000) Specific phosphorylation of Torpedo 43K rapsyn by endogenous kinase(s) with thiamine triphosphate as the phosphate donor. *FASEB Journal* **14,** 543–54.

Nishino K, Itokawa Y, Nishino N, Piros K, and Cooper JR (1983) Enzyme system involved in the synthesis of thiamin triphosphate. I. Purification and characterization of protein-bound thiamin diphosphate: ATP phosphoryltransferase. *Journal of Biological Chemistry* **258,** 11871–8.

Nishino T and Okamoto K (2000) The role of the [2Fe-2S] cluster centers in xanthine oxidoreductase. *Journal of Inorganic Biochemistry* **82,** 43–9.

Nittynen L, Nurminen ML, Korpela R, and Vapaatalo H (1999) Role of arginine, taurine and homocysteine in cardiovascular diseases. *Annals of Medicine* **31,** 318–26.

Nixon PF, Kaczmarek MJ, Tate J, Kerr RA, and Price J (1984) An erythrocyte transketolase isoenzyme pattern associated with the Wernicke-Korsakoff syndrome. *European Journal of Clinical Investigation* **14,** 278–81.

Nokubo M, Ohta M, Kitani K, and Nagy I (1989) Identification of protein-bound riboflavin in rat hepatocyte plasma membrane as a source of autofluorescence. *Biochimica et Biophysica Acta* **981,** 303–8.

Norman AW, Bishop JE, Collins ED, Seo EG, Satchell DP, Dormanen MC, Zanello SB, Farach-Carson MC, Bouillon R, and Okamura WH (1996) Differing shapes of 1alpha, 25-dihydroxyvitamin D$_3$ function as ligands for the D-binding protein, nuclear receptor and membrane receptor: a status report. *Journal of Steroid Biochemistry and Molecular Biology* **56,** 13–22.

Norman AW, Henry HL, Bishop JE, Song XD, Bula C, and Okamura WH (2001a) Different shapes of the steroid hormone 1alpha,25(OH)(2)-vitamin D(3) act as agonists for two different receptors in the vitamin D endocrine system to mediate genomic and rapid responses. *Steroids* **66,** 147–58.

Norman AW, Ishizuka S, and Okamura WH (2001b) Ligands for the vitamin D endocrine system: different shapes function as agonists and antagonists for genomic and rapid response receptors or as a ligand for the plasma vitamin D binding protein. *Journal of Steroid Biochemistry and Molecular Biology* **76**, 49–59.

North American Menopause Society (2001) The role of calcium in peri- and post-menopausal women: consensus opinion of The North American Menopause Society. *Menopause* **8**, 84–95.

Norum K, Blomhoff R, Green M, Green J, Wathne K-O, Gjoen T, Botilsrud M, and Berg T (1986) Metabolism of retinol and other retinoids. *Biochemical Society Transactions* **14**, 923–5.

Novotny JA, Dueker SR, Zech LA, and Clifford AJ (1995) Compartmental analysis of the dynamics of beta-carotene metabolism in an adult volunteer. *Journal of Lipid Research* **36**, 1825–38.

Noy N (2000) Retinoid-binding proteins: mediators of retinoid action. *Biochemical Journal* **348**(Pt 3), 481–95.

Obayashi M, Sato Y, Harris RA, and Shimomura Y (2001) Regulation of the activity of branched-chain 2-oxo acid dehydrogenase (BCODH) complex by binding BCODH kinase. *FEBS Letters* **491**, 50–4.

O'Connell MJ, Chua R, Hoyos B, Buck J, Chen Y, Derguini F, and Hammerling U (1996) Retro-retinoids in regulated cell growth and death. *Journal of Experimental Medicine* **184**, 549–55.

Oduho GW, Han Y, and Baker DH (1994) Iron deficiency reduces the efficacy of tryptophan as a niacin precursor. *Journal of Nutrition* **124**, 444–50.

Oka T (2001) Modulation of gene expression by vitamin B_6. *Nutrition Research Reviews* **14**, 257–65.

Oka T, Komori N, Kuwahata M, Suzuki I, Okada M, and Natori Y (1994) Effect of vitamin B_6 deficiency on the expression of glycogen phosphorylase mRNA in rat liver and skeletal muscle. *Experientia* **50**, 127–9.

Okamoto H (1999a) The CD38-cyclic ADP-ribose signaling system in insulin secretion. *Molecular and Cellular Biochemistry* **193**, 115–18.

Okamoto H (1999b) Cyclic ADP-ribose-mediated insulin secretion and Reg, regenerating gene. *Journal of Molecular Medicine* **77**, 74–8.

Olson J (1986) Metabolism of vitamin A. *Biochemical Society Transactions* **14**, 928–30.

Olson J (1987a) Recommended dietary intakes (RDI) of vitamin A in humans. *American Journal of Clinical Nutrition* **45**, 704–16.

Olson JA (1987b) Recommended dietary intakes (RDI) of vitamin K in humans. *American Journal of Clinical Nutrition* **45**, 687–92.

Olson JA and Hodges RE (1987) Recommended dietary intakes (RDI) of vitamin C in humans. *American Journal of Clinical Nutrition* **45**, 693–703.

Olson RE (1984) The function and metabolism of vitamin K. *Annual Reviews of Nutrition* **4**, 281–337.

Olson RE, Chao J, Graham D, Bates MW, and Lewis JH (2002) Total body phylloquinone and its turnover in human subjects at two levels of vitamin K intake. *British Journal of Nutrition* **87**, 543–53.

Omdahl JL and DeLuca HF (1971) Strontium induced rickets: metabolic basis. *Science* **174**, 949–51.

Omdahl JL, Bobrovnikova EA, Choe S, Dwivedi PP, and May BK (2001) Overview of regulatory cytochrome P450 enzymes of the vitamin D pathway. *Steroids* **66**, 381–9.

Omenn GS, Goodman G, Thornquist M, Barnhart S, Balmes J, Cherniack M, Cullen M, Glass A, Keogh J, Liu D, Meyskens F Jr, Perloff M, Valanis B, and Williams J Jr (1996a) Chemoprevention of lung cancer: the beta-Carotene and Retinol Efficacy Trial (CARET) in high-risk smokers and asbestos-exposed workers. *IARC Scientific Publication* **136**, 67–85.

Omenn GS, Goodman GE, Thornquist MD, Balmes J, Cullen MR, Glass A, Keogh JP, Meyskens FL, Valanis B, Williams JH, Barnhart S, and Hammar S (1996b) Effects of a combination of beta carotene and vitamin A on lung cancer and cardiovascular disease. *New England Journal of Medicine* **334**, 1150–5.

Onorato JM, Jenkins AJ, Thorpe SR, and Baynes JW (2000) Pyridoxamine, an inhibitor of advanced glycation reactions, also inhibits advanced lipoxidation reactions. Mechanism of action of pyridoxamine. *Journal of Biological Chemistry* **275**, 21177–84.

Overvad K, Diamant B, Holm L, Holmer G, Mortensen SA, and Stender S (1999) Coenzyme Q10 in health and disease. *European Journal of Nutrition* **53**, 764–70.

Packer L, Weber SU, and Rimbach G (2001) Molecular aspects of alpha-tocotrienol antioxidant action and cell signalling. *Journal of Nutrition* **131**, 369S–73S.

Page MG, Ankoma-Sey V, Coulson WF, and Bender DA (1989) Brain glutamate and gamma-aminobutyrate (GABA) metabolism in thiamin-deficient rats. *British Journal of Nutrition* **62**, 245–53.

Palczewski K and Saari JC (1997) Activation and inactivation steps in the visual transduction pathway. *Curent Opinions in Neurobiology* **7**, 500–4.

Palm D, Klein HW, Schinzel R, Buehner M, and Helmreich EJ (1990) The role of pyridoxal 5'-phosphate in glycogen phosphorylase catalysis. *Biochemistry* **29**, 1099–1107.

Paquin J, Baugh CM, and MacKenzie RE (1985) Channeling between the active sites of formiminotransferase-cyclodeaminase. Binding and kinetic studies. *Journal of Biological Chemistry* **260**, 14925–31.

Park JB and Levine M (2000) Intracellular accumulation of ascorbic acid is inhibited by flavonoids via blocking of dehydroascorbic acid and ascorbic acid uptakes in HL-60, U937 and Jurkat cells. *Journal of Nutrition* **130**, 1297–1302.

Parker RA, Pearce BC, Clark RW, Gordon DA, and Wright JJ (1993) Tocotrienols regulate cholesterol production in mammalian cells by post-transcriptional suppression of 3-hydroxy-3-methylglutaryl-coenzyme A reductase. *Journal of Biological Chemistry* **268**, 11230–8.

Parker RS (1989) Carotenoids in human blood and tissues. *Journal of Nutrition* **119**, 101–4.

Parker RS (1996) Absorption, metabolism, and transport of carotenoids. *FASEB Journal* **10**, 542–51.

Parker RS, Swanson JE, You CS, Edwards AJ, and Huang T (1999) Bioavailability of carotenoids in human subjects. *Proceedings of the Nutrition Society* **58**, 155–62.

Parsons WB (1961a) Studies on the use of nicotinic acid in hypercholesterolemia: changes in hepatic function, carbohydrate tolerance and uric acid metabolism. *Archives of Internal Medicine* **107**, 653–67.

Parsons WB (1961b) Treatment of hypercholesterolemia by nicotinic acid: progress report with review of studies regarding mode of action. *Archives of Internal Medicine* **107,** 639–52.

Parvin SG and Sivakumar B (2000) Nutritional status affects intestinal carotene cleavage activity and carotene conversion to vitamin A in rats. *Journal of Nutrition* **130,** 573–7.

Patel S, Churchill GC, and Galione A (2001) Coordination of Ca^{2+} signalling by NAADP. *Trends in Biochemical Science* **26,** 482–9.

Patrini C, Reggiani C, Laforenza U, and Rindi G (1988) Blood-brain transport of thiamine monophosphate in the rat: a kinetic study in vivo. *Journal of Neurochemistry* **50,** 90–3.

Paukert JL, Straus LD, and Rabinowitz JC (1976) Formyl-methyl-methylenetetrahydrofolate synthetase-(combined). An ovine protein with multiple catalytic activities. *Journal of Biological Chemistry* **251,** 5104–11.

Pauling L (1970) Evolution and the need for ascorbic acid. *Proceedings of the National Academy of Sciences of the USA* **67,** 1643–8.

Pawlik F, Bischoff A, and Bitsch I (1977) Peripheral nerve changes in thiamine deficiency and starvation. An electron microscopic study. *Acta Neuropathologica* (Berlin) **39,** 211–18.

Pekovich SR, Martin PR, and Singleton CK (1996) Thiamine pyrophosphate-requiring enzymes are altered during pyrithiamine-induced thiamine deficiency in cultured human lymphoblasts. *Journal of Nutrition* **126,** 1791–8.

Pekovich SR, Martin PR and Singleton CK (1998) Thiamine deficiency decreases steady-state transketolase and pyruvate dehydrogenase but not alpha-ketoglutarate dehydrogenase mRNA levels in three human cell types. *Journal of Nutrition* **128,** 683–7.

Pelliniemi TT and Beck WS (1980) Biochemical mechanisms in the Killmann experiment: critique of the deoxyuridine suppression test. *Journal of Clinical Investigation* **65,** 449–60.

Peracchi M, Bamonti Catena F, Pomati M, De Franceschi M, and Scalabrino G (2001) Human cobalamin deficiency: alterations in serum tumour necrosis factor-alpha and epidermal growth factor. *European Journal of Haematology* **67,** 123–7.

Perry TL, Yong VW, Kish SJ, Ito M, Foulks JG, Godolphin WJ, and Sweeney VP (1985) Neurochemical abnormalities in brains of renal failure patients treated by repeated hemodialysis. *Journal of Neurochemistry* **45,** 1043–8.

Peters R (1963) *Biochemical Lesions and Lethal Synthesis.* Oxford: Pergamon Press.

Petersen OH and Cancela JM (1999) New Ca^{2+}-releasing messengers: are they important in the nervous system? *Trends in Neuroscience* **22,** 488–95.

Peterson P, Nilsson S, Ostberg l, Rask L, and Vahlquist A (1974) Aspects of the metabolism of retinol-binding protein and retinol. *Vitamins and Hormones* **32,** 181–214.

Peto R, Doll R, Buckley J, and Sporn M (1981) Can dietary β-carotene materially reduce human cancer risk? *Nature* **290,** 201–8.

Phelps DL (1987) Current perspectives on vitamin E in infant nutrition. *American Journal of Clinical Nutrition* **46,** 187–91.

Phillips WE, Mills JH, Charbonneau SM, Tryphonas L, Hatina GV, Zawidzka Z, Bryce FR, and Munro IC (1978) Subacute toxicity of pyridoxine hydrochloride in the beagle dog. *Toxicology and Applied Pharmacology* **44,** 323–33.

Pieper AA, Verma A, Zhang J, and Snyder SH (1999) Poly(ADP-ribose) polymerase, nitric oxide and cell death. *Trends in Pharmacological Science* **20,** 171–81.

Pietrzik K, Hesse CH, Zur Wiesch ES, and Hotzel D (1975) Urinary excretion of pantothenic acid as a measurement of nutritional requirements. *International Journal of Vitamin and Nutrition Research* **45**, 153–62.

Pinto J, Huang YP, and Rivlin RS (1981) Inhibition of riboflavin metabolism in rat tissues by chlorpromazine, imipramine, and amitriptyline. *Journal of Clinical Investigation* **67**, 1500–6.

Pirie A, Werb Z, and Burleigh M (1975) Collagenase and other proteinases in the cornea of the retinol deficient rat. *British Journal of Nutrition* **34**, 297–309.

Plaitakis A, Nicklas WJ, Van Woert MH, Hwang EC, and Berl S (1981) Uptake and metabolism of serotonin and amino acids in thiamine deficiency. *Advances in Experimental Medicine and Biology* **133**, 391–416.

Poortmans JR and Francaux M (2000) Adverse effects of creatine supplementation: fact or fiction? *Sports Medicine* **30**, 155–70.

Poston JM (1984) The relative carbon flux through the alpha- and the beta-keto pathways of leucine metabolism. *Journal of Biological Chemistry* **259**, 2059–61.

Powers HJ (1995) Riboflavin-iron interactions with particular emphasis on the gastrointestinal tract. *Proceedings of the Nutrition Society* **54**, 509–17.

Powers HJ, Weaver LT, Austin S, Wright AJ, and Fairweather-Tait SJ (1991) Riboflavin deficiency in the rat: effects on iron utilization and loss. *British Journal of Nutrition* **65**, 487–96.

Prasad PD and Ganapathy V (2000) Structure and function of mammalian sodium-dependent multivitamin transporter. *Current Opinion in Clinical Nutrition and Metabolic Care* **3**, 263–6.

Prentice AM and Bates CJ (1981a) A biochemical evaluation of the erythrocyte glutathione reductase (EC 1.6.4.2) test for riboflavin status. 1. Rate and specificity of response in acute deficiency. *British Journal of Nutrition* **45**, 37–52.

Prentice AM and Bates CJ (1981b) A biochemical evaluation of the erythrocyte glutathione reductase (EC 1.6.4.2) test for riboflavin status. 2. Dose-response relationships in chronic marginal deficiency. *British Journal of Nutrition* **45**, 53–65.

Prigge ST, Mains RE, Eipper BA, and Amzel LM (2000) New insights into copper monooxygenases and peptide amidation: structure, mechanism and function. *Cellular and Molecular Life Sciences* **57**, 1236–59.

Proud VK, Rizzo WB, Patterson JW, Heard GS, and Wolf B (1990) Fatty acid alterations and carboxylase deficiencies in the skin of biotin-deficient rats. *American Journal of Clinical Nutrition* **51**, 853–8.

Pruthi S, Allison TG, and Hensrud DD (2001) Vitamin E supplementation in the prevention of coronary heart disease. *Mayo Clinic Proceedings* **76**, 1131–6.

Quandt L and Huth W (1984) Modulation of rat-liver mitochondrial acetyl-CoA acetyltransferase activity by a reversible chemical modification with coenzyme A. *Biochimica et Biophysica Acta* **784**, 168–76.

Quandt L and Huth W (1985) On the mechanism of the chemical modification of the mitochondrial acetyl-CoA acetyltransferase by coenzyme A. *Biochimica et Biophysica Acta* **829**, 103–8.

Rahman MM, Wahed MA, Fuchs GJ, Baqui AH, and Alvarez JO (2002) Synergistic effect of zinc and vitamin A on the biochemical indexes of vitamin A nutrition in children. *American Journal of Clinical Nutrition* **75**, 92–8.

Rajagopalan KV (1997) Biosynthesis and processing of the molybdenum cofactors. *Biochemical Society Transactions* **25,** 757–61.

Rajagopalan KV and Johnson JL (1992) The pterin molybdenum cofactors. *Journal of Biological Chemistry* **267,** 10199–202.

Ramaswamy K (1999) Intestinal absorption of water-soluble vitamins focus on "molecular mechanism of the intestinal biotin transport process." *American Journal of Physiology* **277,** C603–4.

Raskind JY and El-Chaar GM (2000) The role of carnitine supplementation during valproic acid therapy. *Annals of Pharmacotherapy* **34,** 630–8.

Raux E, Schubert HL, and Warren MJ (2000) Biosynthesis of cobalamin (vitamin B_{12}): a bacterial conundrum. *Cellular and Molecular Life Sciences* **57,** 1880–93.

Reddi A, DeAngelis B, Frank O, Lasker N, and Baker H (1988) Biotin supplementation improves glucose and insulin tolerances in genetically diabetic KK mice. *Life Sciences* **42,** 1323–30.

Reddy GS and Tseng KY (1989) Calcitroic acid, end product of renal metabolism of 1,25-dihydroxyvitamin D_3 through C-24 oxidation pathway. *Biochemistry* **28,** 1763–9.

Redmond TM, Gentleman S, Duncan T, Yu S, Wiggert B, Gantt E, and Cunningham FX Jr (2001) Identification, expression, and substrate specificity of a mammalian beta-carotene 15,15′-dioxygenase. *Journal of Biological Chemistry* **276,** 6560–5.

Refsum H, Ueland PM, Nygard O, and Vollset SE (1998) Homocysteine and cardiovascular disease. *Annual Reviews of Medicine* **49,** 31–62.

Reggiani C, Patrini C, and Rindi G (1984) Nervous tissue thiamine metabolism in vivo. I. Transport of thiamine and thiamine monophosphate from plasma to different brain regions of the rat. *Brain Research* **293,** 319–27.

Reiss J (2000) Genetics of molybdenum cofactor deficiency. *Human Genetics* **106,** 157–63.

Reynolds EH (1967) Effects of folic acid on the mental state and fit-frequency of drug-treated epileptic patients. *Lancet* **1,** 1086–8.

Ribaya-Mercado JD (2002) Influence of dietary fat on beta-carotene absorption and bioconversion into vitamin A. *Nutrition Reviews* **60,** 104–10.

Rindi G and Laforenza U (2000) Thiamine intestinal transport and related issues: recent aspects. *Proceedings of the Society for Experimental Biology and Medicine* **224,** 246–55.

Rindi G, Comincioli V, Reggiani C, and Patrini C (1984) Nervous tissue thiamine metabolism in vivo. II. Thiamine and its phosphoesters dynamics in different brain regions and sciatic nerve of the rat. *Brain Research* **293,** 329–42.

Ritchie HE, Webster WS, Eckhoff C, and Oakes DJ (1998) Model predicting the teratogenic potential of retinyl palmitate, using a combined in vivo/in vitro approach. *Teratology* **58,** 113–23.

Rivers JM (1987) Safety of high-level vitamin C ingestion. *Annals of the New York Academy of Sciences* **498,** 445–54.

Rivlin RS and Langdon RG (1966) Regulation of hepatic FAD levels by thyroid hormone. *Advances in Enzyme Regulation* **4,** 45–58.

Robinson BH, MacKay N, Chun K, and Ling M (1996) Disorders of pyruvate carboxylase and the pyruvate dehydrogenase complex. *Journal of Inherited Metabolic Disorders* **19,** 452–62.

Robishaw JD, Berkich D, and Neely JR (1982) Rate-limiting step and control of coenzyme A synthesis in cardiac muscle. *Journal of Biological Chemistry* **257**, 10967–72.

Rock CO, Calder RB, Karim MA, and Jackowski S (2000) Pantothenate kinase regulation of the intracellular concentration of coenzyme A. *Journal of Biological Chemistry* **275**, 1377–83.

Rodriguez-Martin JL, Qizilbash N, and Lopez-Arrieta JM (2001) Thiamine for Alzheimer's disease. *Cochrane Database Systematic Reviews*, CD001498.

Rodriguez-Melendez R, Cano S, Mendez ST, and Velazquez A (2001) Biotin regulates the genetic expression of holocarboxylase synthetase and mitochondrial carboxylases in rats. *Journal of Nutrition* **131**, 1909–13.

Roessner CA, Santander PJ, and Scott AI (2001) Multiple biosynthetic pathways for vitamin B_{12}: variations on a central theme. *Vitamins and Hormones* **61**, 267–97.

Rohde CM, Manatt M, Clagett-Dame M, and DeLuca HF (1999) Vitamin A antagonizes the action of vitamin D in rats. *Journal of Nutrition* **129**, 2246–50.

Rose DP (1966a) Excretion of xanthurenic acid in the urine of women taking progesterone-oestrogen preparations. *Nature* **210**, 196–7.

Rose DP (1966b) The influence of oestrogens on tryptophan metabolism in man. *Clinical Science* **31**, 265–72.

Rose DP, Leklem JE, Brown RR, and Linkswiler HM (1975) Effect of oral contraceptives and vitamin B_6 deficiency on carbohydrate metabolism. *American Journal of Clinical Nutrition* **28**, 872–8.

Rowe A and Brickell P (1993) Current status review: the nuclear retinoid receptors. *International Journal of Experimental Pathology* **74**, 117–26.

Rowe DW and Kream BE (1982) Regulation of collagen synthesis in fetal rat calvaria by 1,25-dihydroxyvitamin D_3. *Journal of Biological Chemistry* **257**, 8009–15.

Rowling MJ, McMullen MH, and Schalinske KL (2002) Vitamin A and its derivatives induce hepatic glycine *N*-methyltransferase and hypomethylation of DNA in rats. *Journal of Nutrition* **132**, 365–9.

Saareks V, Mucha I, Sievi E, and Riutta A (1999) Nicotinic acid and pyridoxine modulate arachidonic acid metabolism in vitro and ex vivo in man. *Pharmacology and Toxicology* **84**, 274–80.

Said HM (1999) Cellular uptake of biotin: mechanisms and regulation. *Journal of Nutrition* **129**, 490S–3S.

Said HM, Ortiz A, McCloud E, Dyer D, Moyer MP, and Rubin S (1998) Biotin uptake by human colonic epithelial NCM460 cells: a carrier-mediated process shared with pantothenic acid. *American Journal of Physiology* **275**, C1365–71.

Said HM, Ortiz A, Moyer MP, and Yanagawa N (2000) Riboflavin uptake by human-derived colonic epithelial NCM460 cells. *American Journal of Physiology Cell Physiology* **278**, C270–6.

Sakmar TP (1998) Rhodopsin: a prototypical G protein-coupled receptor. *Progress in Nucleic Acid Research and Molecular Biology* **59**, 1–34.

Saliba KJ, Horner HA, and Kirk K (1998) Transport and metabolism of the essential vitamin pantothenic acid in human erythrocytes infected with the malaria parasite *Plasmodium falciparum*. *Journal of Biological Chemistry* **273**, 10190–5.

Salter M, Bender DA, and Pogson CI (1985) Leucine and tryptophan metabolism in rats. *Biochemical Journal* **225,** 277–81.

Salter M, Knowles RG,and Pogson CI (1986) Quantification of the importance of individual steps in the control of aromatic amino acid metabolism. *Biochemical Journal* **234,** 635–47.

Salter M and Pogson CI (1985) The role of tryptophan 2,3-dioxygenase in the hormonal control of tryptophan metabolism in isolated rat liver cells. Effects of glucocorticoids and experimental diabetes. *Biochemical Journal* **229,** 499–504.

Sancar A (2000) Cryptochrome: the second photoactive pigment in the eye and its role in circadian photoreception. *Annual Reviews of Biochemistry* **69,** 31–67.

Sano T, Vajda S, and Cantor CR (1998) Genetic engineering of streptavidin, a versatile affinity tag. *Journal of Chromatography B Biomedical Science Applications* **715,** 85–91.

Saransaari P and Oja SS (2000) Taurine and neural cell damage. *Amino Acids* **19,** 509–26.

Sato A, Nishioka M, Awata S, Nakayama K, Okada M, Horiuchi S, Okabe N, Sassa T, Oka T, and Natori Y (1996) Vitamin B_6 deficiency accelerates metabolic turnover of cystathionase in rat liver. *Archives of Biochemistry and Biophysics* **330,** 409–13.

Sato M, Shirota M, and Nagao T (1995) Pantothenic acid decreases valproic acid-induced neural tube defects in mice (I). *Teratology* **52,** 143–8.

Sato P and Udenfriend S (1978) Studies on ascorbic acid related to the genetic basis of scurvy. *Vitamins and Hormones* **36,** 33–52.

Sauberlich H, Dowdy RP, and Skala JH (1974a) *Laboratory Tests for the Assessment of Nutritional Status.* Cleveland: CRC Press.

Sauberlich H, Hodges R, Wallace D, Kolder H, Canham J, Hood J, Raica N, and Lowry L (1974b) Vitamin A metabolism and requirements in the human studied with the use of labelled retinol. *Vitamins and Hormones* **32,** 251–75.

Sauberlich HE (1975) Human requirements and needs. Vitamin C status: methods and findings. *Annals of the New York Academy of Sciences* **258,** 438–50.

Sauberlich HE, Canham JE, Baker EM, Raica N Jr, and Herman YF (1972) Biochemical assessment of the nutritional status of vitamin B 6 in the human. *American Journal of Clinical Nutrition* **25,** 629–42.

Savage DG and Lindenbaum J (1995) Neurological complications of acquired cobalamin deficiency: clinical aspects. *Baillieres Clinics in Haematology* **8,** 657–78.

Saxena SP, Israels ED, and Israels LG (2001) Novel vitamin K-dependent pathways regulating cell survival. *Apoptosis* **6,** 57–68.

Scalabrino G, Tredici G, Buccellato FR, and Manfridi A (2000) Further evidence for the involvement of epidermal growth factor in the signaling pathway of vitamin B_{12} (cobalamin) in the rat central nervous system. *Journal of Neuropathology and Experimental Neurology* **59,** 808–14.

Schaeppi U and Krinke G (1982) Pyridoxine neuropathy: correlation of functional tests and neuropathology in beagle dogs treated with large doses of vitamin B_6. *Agents and Actions* **12,** 575–82.

Schatz DA and Bingley PJ (2001) Update on major trials for the prevention of type 1 diabetes mellitus: the American Diabetes Prevention Trial (DPT-1) and the European Nicotinamide Diabetes Intervention Trial (ENDIT). *Journal of Pediatrics Endocrinology and Metabolism* **14**(Suppl 1), 619–22.

Schaumburg H, Kaplan J, Windebank A, Vick N, Rasmus S, Pleasure D, and Brown MJ (1983) Sensory neuropathy from pyridoxine abuse. A new megavitamin syndrome. *New England Journal of Medicine* **309,** 445–8.

Schenk G, Duggleby RG, and Nixon PF (1998) Properties and functions of the thiamin diphosphate dependent enzyme transketolase. *International Journal of Biochemistry and Cell Biology* **30,** 1297–1318.

Schneider G and Lindqvist Y (2001) Structural enzymology of biotin biosynthesis. *FEBS Letters* **495,** 7–11.

Schoental R (1983) Mycotoxins and nutritional deficeincies. *Nutrition and Health* **2,** 147–52.

Schorah CJ, Habibzadeh N, Hancock M, and King RF (1986) Changes in plasma and buffy layer vitamin C concentrations following major surgery: what do they reflect? *Annals of Clinical Biochemistry* **23**(Pt 5), 566–70.

Schwerdt G and Huth W (1993) Turnover and transformation of mitochondrial acetyl-CoA acetyltransferase into CoA-modified forms. *Biochemical Journal* **292,** 915–19.

Scientific Committee for Food (1993) *Nutrient and Energy Intakes for the European Community.* Luxemburg: Commission of the European Communities.

Scientific Committee on Food (2000) *Opinion of the Scientific Committee on Food on the Tolerable Upper Intake Level of Vitamin B_6.* Brussels: European Commission Health and Consumer Protection Directorate-General.

Scott JM (1999) Folate and vitamin B_{12}. *Proceedings of the Nutrition Society* **58,** 441–8.

Sedrani SH (1988) Correlation between concentrations of humoral antibodies and vitamin D nutritional status: a survey study. *European Journal of Nutrition* **42,** 243–8.

Seetharam B (1999) Receptor-mediated endocytosis of cobalamin (vitamin B_{12}). *Annual Reviews of Nutrition* **19,** 173–95.

Seetharam B and Li N (2000) Transcobalamin II and its cell surface receptor. *Vitamins and Hormones* **59,** 337–66.

Segal I, Ou Tim L, Demetriou A, Paterson A, Hale M, and Lerios M (1986) Rectal manifestations of pellagra. *International Journal of Colorectal Diseases* **1,** 238–43.

Seitz G, Gebhardt S, Beck JF, Bohm W, Lode HN, Niethammer D, and Bruchelt G (1998) Ascorbic acid stimulates DOPA synthesis and tyrosine hydroxylase gene expression in the human neuroblastoma cell line SK-N-SH. *Neuroscience Letters* **244,** 33–6.

Selhub J, Jacques PF, Wilson PW, Rush D, and Rosenberg IH (1993) Vitamin status and intake as primary determinants of homocysteinemia in an elderly population. *JAMA* **270,** 2693–8.

Selhub J, Jacques PF, Bostom AG, Wilson PW, and Rosenberg IH (2000) Relationship between plasma homocysteine and vitamin status in the Framingham study population. Impact of folic acid fortification. *Public Health Review* **28,** 117–45.

Semba RD (1999) Vitamin A as "anti-infective" therapy, 1920–1940. *Journal of Nutrition* **129,** 783–91.

Semba RD and Tang AM (1999) Micronutrients and the pathogenesis of human immunodeficiency virus infection. *British Journal of Nutrition* **81,** 181–9.

Shankar AH (2000) Nutritional modulation of malaria morbidity and mortality. *Journal of Infectious Diseases* **182**(Suppl 1), S37–53.

Shaw NJ and Pal BR (2002) Vitamin D deficiency in UK Asian families: activating a new concern. *Archives of Disease in Childhood* **86,** 147–9.

She QB, Hayakawa T, and Tsuge H (1995) Alteration in the phosphatidylcholine biosynthesis of rat liver microsomes caused by vitamin B$_6$ deficiency. *Bioscience Biotechnology and Biochemistry* **59,** 163–7.

Shearer MJ, Bach A, and Kohlmeier M (1996) Chemistry, nutritional sources, tissue distribution and metabolism of vitamin K with special reference to bone health. *Journal of Nutrition* **126,** 1181S–6S.

Shin DJ and McGrane MM (1997) Vitamin A regulates genes involved in hepatic gluconeogenesis in mice: phosphoenolpyruvate carboxykinase, fructose-1,6-bisphosphatase and 6-phosphofructo-2-kinase/fructose-2,6-bisphosphatase. *Journal of Nutrition* **127,** 1274–8.

Shiratori T (1974) Uptake, storage and excretion of chylomicra-bound ^{3}H-alpha-tocopherol by the skin of the rat. *Life Sciences* **14,** 929–35.

Shoulson I (1998) DATATOP: a decade of neuroprotective inquiry. Parkinson Study Group. Deprenyl and tocopherol antioxidative therapy of parkinsonism. *Annals of Neurology* **44,** S160–6.

Shrimpton D (1997) *Vitamins and Minerals: A Scientific Evaluation of the Range of Safe Intakes*. Thames Ditton Survey. Brussels: European Federation of Health Product Manufacturers Associations.

Sidell N, Sawatsri S, Connor MJ, Barua AB, Olson JA, and Wada RK (2000) Pharmacokinetics of chronically administered all-*trans*-retinoyl-beta-glucuronide in mice. *Biochimica et Biophysica Acta* **1502,** 264–72.

Simbulan-Rosenthal CM, Rosenthal DS, Hilz H, Hickey R, Malkas L, Applegren N, Wu Y, Bers G, and Smulson ME (1996) The expression of poly(ADP-ribose) polymerase during differentiation-linked DNA replication reveals that it is a component of the multiprotein DNA replication complex. *Biochemistry* **35,** 11622–33.

Sirotnak FM and Tolner B (1999) Carrier-mediated membrane transport of folates in mammalian cells. *Annual Reviews of Nutrition* **19,** 91–122.

Smirnoff N (2000) Ascorbate biosynthesis and function in photoprotection. *Philosophical Transactions of the Royal Society of London B Biological Sciences* **355,** 1455–64.

Smith J, McDaniel E, Fan F, and Halsted J (1973) Zinc, a trace element essential in vitamin A metabolism. *Science* **181,** 945–55.

Snell EE (1990) Vitamin B$_6$ and decarboxylation of histidine. *Annals of the New York Academy of Sciences* **585,** 1–12.

Snell K, Baumann U, Byrne PC, Chave KJ, Renwick SB, Sanders PG, and Whitehouse SK (2000) The genetic organization and protein crystallographic structure of human serine hydroxymethyltransferase. *Advances in Enzyme Regulation* **40,** 353–403.

Sokoll LJ and Sadowski JA (1996) Comparison of biochemical indexes for assessing vitamin K nutritional status in a healthy adult population. *American Journal of Clinical Nutrition* **63,** 566–73.

Solomon LR and Cohen K (1989) Erythrocyte O$_2$ transport and metabolism and effects of vitamin B$_6$ therapy in type II diabetes mellitus. *Diabetes* **38,** 881–6.

Solomons N and Russell R (1980) The interaction of vitamin A and zinc: implications for human nutrition. *American Journal of Clinical Nutrition* **33,** 2031–40.

Sonneveld E, van den Brink CE, van der Leede BM, Schulkes RK, Petkovich M, van der Burg B, and van der Saag PT (1998) Human retinoic acid (RA) 4-hydroxylase (CYP26) is highly specific for all-*trans*-RA and can be induced through RA receptors in human breast and colon carcinoma cells. *Cell Growth and Differentiation* **9**, 629–37.

Soprano DR and Soprano KJ (1995) Retinoids as teratogens. *Annual Reviews of Nutrition* **15**, 111–32.

Soprano DR, Smith JF, and Goodman DS (1982) Effect of retinol status on retinol-binding protein biosynthesis rate and translatable messenger RNA level in rat liver. *Journal of Biological Chemistry* **257**, 7693–7.

Speck WT, Chen CC, and Rosenkranz HS (1975) In vitro studies of effects of light and riboflavin on DNA and HeLa cells. *Pediatric Research* **9**, 150–3.

Stadtman TC (1996) Selenocysteine. *Annual Reviews of Biochemistry* **65**, 83–100.

Stahl W and Sies H (1996) Lycopene: a biologically important carotenoid for humans? *Archives of Biochemistry and Biophysics* **336**, 1–9.

Stahl W, von Laar J, Martin HD, Emmerich T, and Sies H (2000) Stimulation of gap junctional communication: comparison of acyclo-retinoic acid and lycopene. *Archives of Biochemistry and Biophysics* **373**, 271–4.

Stamler JS and Slivka A (1996) Biological chemistry of thiols in the vasculature and in vascular-related disease. *Nutrition Reviews* **54**, 1–30.

Stanley JS, Griffin JB, and Zempleni J (2001) Biotinylation of histones in human cells. Effects of cell proliferation. *European Journal of Biochemistry* **268**, 5424–9.

Stanulovic M and Chaykin S (1971a) Aldehyde oxidase: catalysis of the oxidation of N^1-methylnicotinamide and pyridoxal. *Archives of Biochemistry and Biophysics* **145**, 27–34.

Stanulovic M and Chaykin S (1971b) Metabolic origins of the pyridones of N^1-methylnicotinamide in man and rat. *Archives of Biochemistry and Biophysics* **145**, 35–42.

St-Arnaud R (1999) Targeted inactivation of vitamin D hydroxylases in mice. *Bone* **25**, 127–9.

Stayton PS, Nelson KE, McDevitt TC, Bulmus V, Shimoboji T, Ding Z, and Hoffman AS (1999) Smart and biofunctional streptavidin. *Biomolecular Engineering* **16**, 93–9.

Steinberg SE, Campbell CL, and Hillman RS (1979) Kinetics of the normal folate enterohepatic cycle. *Journal of Clinical Investigation* **64**, 83–8.

Steiner M (1999) Vitamin E, a modifier of platelet function: rationale and use in cardiovascular and cerebrovascular disease. *Nutrition Reviews* **57**, 306–9.

Stenflo J, Ferlund P, Egan W, and Roepstorff P (1974) Vitamin K dependent modifications of glutamic acid residues in prothrombin. *Proceedings of the National Academy of Sciences of the USA* **71**, 2730–33.

Stephens NG, Parsons A, Schofield PM, Kelly F, Cheeseman K, and Mitchinson MJ (1996) Randomised controlled trial of vitamin E in patients with coronary disease: Cambridge Heart Antioxidant Study (CHAOS). *Lancet* **347**, 781–6.

Stites TE, Mitchell AE, and Rucker RB (2000a) Physiological importance of quinoenzymes and the O-quinone family of cofactors. *Journal of Nutrition* **130**, 719–27.

Stites TE, Sih TR, and Rucker RB (2000b) Synthesis of [(14)C]pyrroloquinoline quinone (PQQ) in *E. coli* using genes for PQQ synthesis from *K. pneumoniae*. *Biochimica et Biophysica Acta* **1524**, 247–52.

Stocker A and Azzi A (2000) Tocopherol-binding proteins: their function and physiological significance. *Antioxidants and Redox Signaling* **2,** 397–404.

Stryer L (1986) Cyclic GMP cascade of vision. *Annual Reviews of Neuroscience* **9,** 87–119.

Sugden MC, Bulmer K, Augustine D, and Holness MJ (2001a) Selective modification of pyruvate dehydrogenase kinase isoform expression in rat pancreatic islets elicited by starvation and activation of peroxisome proliferator-activated receptor-alpha: implications for glucose-stimulated insulin secretion. *Diabetes* **50,** 2729–36.

Sugden MC, Bulmer K, and Holness MJ (2001b) Fuel-sensing mechanisms integrating lipid and carbohydrate utilization. *Biochemical Society Transactions* **29,** 272–8.

Sugiura I, Furie B, Walsh CT, and Furie BC (1997) Propeptide and glutamate-containing substrates bound to the vitamin K-dependent carboxylase convert its vitamin K epoxidase function from an inactive to an active state. *Proceedings of the National Academy of Sciences of the USA* **94,** 9069–74.

Suh JR, Herbig AK, and Stover PJ (2001) New perspectives on folate catabolism. *Annual Reviews of Nutrition* **21,** 255–82.

Sun AY and Chen YM (1998) Oxidative stress and neurodegenerative disorders. *Journal of Biomedical Sciences* **5,** 401–14.

Sundaram KS, Fan JH, Engelke JA, Foley AL, Suttie JW, and Lev M (1996) Vitamin K status influences brain sulfatide metabolism in young mice and rats. *Journal of Nutrition* **126,** 2746–51.

Suter PM and Vetter W (2000) Diuretics and vitamin B_1: are diuretics a risk factor for thiamin malnutrition? *Nutrition Reviews* **58,** 319–23.

Sutor AH, von Kries R, Cornelissen EA, McNinch AW, and Andrew M (1999) Vitamin K deficiency bleeding (VKDB) in infancy. ISTH Pediatric/Perinatal Subcommittee. International Society on Thrombosis and Haemostasis. *Thrombosis and Haemostasis* **81,** 456–61.

Suttie JW (1995) The importance of menaquinones in human nutrition. *Annual Reviews of Nutrition* **15,** 399–417.

Suttie JW, Mummah-Schendel LL, Shah DV, Lyle BJ, and Greger JL (1988) Vitamin K deficiency from dietary vitamin K restriction in humans. *American Journal of Clinical Nutrition* **47,** 475–80.

Tabet N, Birks J, and Grimley Evans J (2000) Vitamin E for Alzheimer's disease. *Cochrane Database Systematic Reviews,* CD002854.

Tahiliani AG and Beinlich CJ (1991) Pantothenic acid in health and disease. *Vitamins and Hormones* **46,** 165–228.

Takahashi K, Kukimoto I, Tokita K, Inageda K, Inoue S, Kontani K, Hoshino S, Nishina H, Kanaho Y, and Katada T (1995) Accumulation of cyclic ADP-ribose measured by a specific radioimmunoassay in differentiated human leukemic HL-60 cells with all-*trans*-retinoic acid. *FEBS Letters* **371,** 204–8.

Takahashi N, De Luca LM, and Breitman TR (1997) Decreased retinoylation in NIH 3T3 cells transformed with activated Ha-ras. *Biochemical and Biophysical Research Communications* **239,** 80–4.

Tannenbaum SR and Wishnok JS (1987) Inhibition of nitrosamine formation by ascorbic acid. *Annals of the New York Academy of Sciences* **498,** 354–63.

Tanumihardjo S, Barua A, and Olson J (1987) Use of 3,4-didehydroretinol to assess vitamin A status in rats. *International Journal of Vitamin and Nutrition Research* **57,** 127–32.

Tanumihardjo SA (2002) Factors influencing the conversion of carotenoids to retinol: bioavailability to bioconversion to bioefficacy. *International Journal of Vitamin and Nutrition Research* **72,** 40–5.

Tappel AL and Dillard CJ (1981) In vivo lipid peroxidation: measurement via exhaled pentane and protection by vitamin E. *Federation Proceedings* **40,** 174–8.

Taylor EW (1995) Selenium and cellular immunity. Evidence that selenoproteins may be encoded in the +1 reading frame overlapping the human CD4, CD8, and HLA-DR genes. *Biological Trace Element Research* **49,** 85–95.

Teicher VB, Kucharski N, Martin HD, van der Saag P, Sies H, and Stahl W (1999) Biological activities of apo-canthaxanthinoic acids related to gap junctional communication. *Archives of Biochemistry and Biophysics* **365,** 150–5.

Thaller C and Eichele G (1987) Identification and spatial distribution of retinoids in the developing chick limb bud. *Nature* **327,** 625–8.

Thaller C and Eichele G (1990) Isolation of 3,4-didehydroretinoic acid, a novel morphogenetic signal in the chick wing bud. *Nature* **345,** 815–19.

Theriault A, Chao JT, Wang Q, Gapor A, and Adeli K (1999) Tocotrienol: a review of its therapeutic potential. *Clinical Biochemistry* **32,** 309–19.

Thijssen HH and Drittij-Reijnders MJ (1996) Vitamin K status in human tissues: tissue-specific accumulation of phylloquinone and menaquinone-4;. *British Journal of Nutrition* **75,** 121–7.

Thijssen HH, Drittij-Reijnders MJ, and Fischer MA (1996) Phylloquinone and menaquinone-4 distribution in rats: synthesis rather than uptake determines menaquinone-4 organ concentrations. *Journal of Nutrition* **126,** 537–43.

Thomas SR and Stocker R (2000) Molecular action of vitamin E in lipoprotein oxidation: implications for atherosclerosis. *Free Radicals in Biology and Medicine* **28,** 1795–1805.

Thomas SR, Neuzil J, Mohr D, and Stocker R (1995) Coantioxidants make alpha-tocopherol an efficient antioxidant for low-density lipoprotein. *American Journal of Clinical Nutrition* **62,** 1357S–64S.

Thomas SR, Witting PK, and Stocker R (1999) A role for reduced coenzyme Q in atherosclerosis? *Biofactors* **9,** 207–24.

Thompson JN and Scott ML (1970) Impaired lipid and vitamin E absorption related to atrophy of the pancreas in selenium-deficient chicks. *Journal of Nutrition* **100,** 797–809.

Thony B, Auerbach G, and Blau N (2000) Tetrahydrobiopterin biosynthesis, regeneration and functions. *Biochemical Journal* **347** (Pt 1), 1–16.

Thurnham D and Northrop-Clewes C (1999) Optimum nutrition: vitamin A and the carotenoids. *Proceedings of the Nutrition Society* **58,** 449–57.

Thurston JH and Hauhart RE (1992) Amelioration of adverse effects of valproic acid on ketogenesis and liver coenzyme A metabolism by cotreatment with pantothenate and carnitine in developing mice: possible clinical significance. *Pediatric Research* **31,** 419–23.

Tibaduiza EC, Fleet JC, Russell RM, and Krinsky NI (2002) Excentric cleavage products of beta-carotene inhibit estrogen receptor positive and negative breast tumor cell growth in vitro and inhibit activator protein-1-mediated transcriptional activation. *Journal of Nutrition* **132**, 1368–75.

Traber MG (2001) Does vitamin E decrease heart attack risk? Summary and implications with respect to dietary recommendations. *Journal of Nutrition* **131**, 395S–7S.

Traber MG and Arai H (1999) Molecular mechanisms of vitamin E transport. *Annual Reviews of Nutrition* **19**, 343–55.

Trakatellis A, Dimitriadou A, and Trakatelli M (1997) Pyridoxine deficiency: new approaches in immunosuppression and chemotherapy. *Postgraduate Medical Journal* **73**, 617–22.

Trang HM, Cole DE, Rubin LA, Pierratos A, Siu S, and Vieth R (1998) Evidence that vitamin D_3 increases serum 25-hydroxyvitamin D more efficiently than does vitamin D_2. *American Journal of Clinical Nutrition* **68**, 854–8.

Tsai MY, Yang F, Bignell M, Aras O, and Hanson NQ (1999) Relation between plasma homocysteine concentration, the 844ins68 variant of the cystathionine beta-synthase gene, and pyridoxal-5′-phosphate concentration. *Molecular Genetics and Metabolism* **67**, 352–6.

Tsaioun KI (1999) Vitamin K-dependent proteins in the developing and aging nervous system. *Nutrition Reviews* **57**, 231–40.

Tschanz CL and Noy N (1997) Binding of retinol in both retinoid-binding sites of inter-photoreceptor retinoid-binding protein (IRBP) is stabilized mainly by hydrophobic interactions. *Journal of Biological Chemistry* **272**, 30201–7.

Tsuchiya H and Bates CJ (1997) Vitamin C and copper interactions in guinea-pigs and a study of collagen cross-links. *British Journal of Nutrition* **77**, 315–25.

Tsuge H, Hotta N, and Hayakawa T (2000) Effects of vitamin B-6 on (n-3) polyunsaturated fatty acid metabolism. *Journal of Nutrition* **130**, 333S–4S.

Tully DB, Allgood VE, and Cidlowski JA (1994) Modulation of steroid receptor-mediated gene expression by vitamin B_6. *FASEB Journal* **8**, 343–9.

Tweto J and Larrabee AR (1972) The effect of fasting on synthesis and 4′-phosphopantetheine exchange in rat liver fatty acid synthetase. *Journal of Biological Chemistry* **247**, 4900–4.

Ubbink JB (1997) The role of vitamins in the pathogenesis and treatment of hyperhomocyst(e)inaemia. *Journal of Inherited Metabolic Disorders* **20**, 316–25.

Ubbink JB, Vermaak WJ, van der Merwe A, Becker PJ, Delport R, and Potgieter HC (1994) Vitamin requirements for the treatment of hyperhomocysteinemia in humans. *Journal of Nutrition* **124**, 1927–33.

Underwood B (1990) Methods for assessment of vitamin A status. *Journal of Nutrition* **56**, 1459–63.

Upston JM, Terentis AC, and Stocker R (1999) Tocopherol-mediated peroxidation of lipoproteins: implications for vitamin E as a potential antiatherogenic supplement. *FASEB Journal* **13**, 977–94.

Valk EE and Hornstra G (2000) Relationship between vitamin E requirement and polyunsaturated fatty acid intake in man: a review. *International Journal of Vitamin and Nutrition Research* **70**, 31–42.

van den Berg H (1993) Flair concerted action no 10 status papers: vitamin B$_{12}$. *International Journal of Vitamin and Nutrition Research* **63,** 282–9.

van den Veyver IB (2002) Genetic effects of methylation diets. *Annual Reviews of Nutrition* **22,** 255–82.

Van Gossum A, Shariff R, Lemoyne M, Kurian R, and Jeejeebhoy K (1988) Increased lipid peroxidation after lipid infusion as measured by breath pentane output. *American Journal of Clinical Nutrition* **48,** 1394–99.

van Leeuwen JP, van den Bemd GJ, van Driel M, Buurman CJ, and Pols HA (2001) 24,25-Dihydroxyvitamin D(3) and bone metabolism. *Steroids* **66,** 375–80.

van Vliet T, van Vlissingen MF, van Schaik F, and van den Berg H (1996) beta-Carotene absorption and cleavage in rats is affected by the vitamin A concentration of the diet. *Journal of Nutrition* **126,** 499–508.

Vas A, Gachalyi B, and Kaldor A (1990) Pantothenic acid, acute ethanol consumption and sulphadimidine acetylation. *Inetrnational Journal of Clinical Pharmacology Therapeutics and Toxicology* **28,** 111–14.

Vaxman F, Olender S, Lambert A, Nisand G, Aprahamian M, Bruch JF, Didier E, Volkmar P, and Grenier JF (1995) Effect of pantothenic acid and ascorbic acid supplementation on human skin wound healing process. A double-blind, prospective and randomized trial. *European Surgery Research* **27,** 158–66.

Vaxman F, Olender S, Lambert A, Nisand G, and Grenier JF (1996) Can the wound healing process be improved by vitamin supplementation? Experimental study on humans. *European Surgery Research* **28,** 306–14.

Vaz FM and Wanders RJ (2002) Carnitine biosynthesis in mammals. *Biochemical Journal* **361,** 417–29.

Veitch K, Draye JP, Van Hoof F, and Sherratt HS (1988) Effects of riboflavin deficiency and clofibrate treatment on the five acyl-CoA dehydrogenases in rat liver mitochondria. *Biochemical Journal* **254,** 477–81.

Velazquez A, Zamudio S, Baez A, Murguia-Corral R, Rangel-Peniche B, and Carrasco A (1990) Indicators of biotin status: a study of patients on prolonged total parenteral nutrition. *European Journal of Nutrition* **44,** 11–16.

Vermeer C, Jie KS, and Knapen MH (1995) Role of vitamin K in bone metabolism. *Annual Reviews of Nutrition* **15,** 1–22.

Vermeer C, Gijsbers BL, Craciun AM, Groenen-van Dooren MM, and Knapen MH (1996) Effects of vitamin K on bone mass and bone metabolism. *Journal of Nutrition* **126,** 1187S–91S.

Vidal J, Fernandez-Balsells M, Sesmilo G, Aguilera E, Casamitjana R, Gomis R, and Conget I (2000) Effects of nicotinamide and intravenous insulin therapy in newly diagnosed type 1 diabetes. *Diabetes Care* **23,** 360–4.

Vieth R (1999) Vitamin D supplementation, 25-hydroxyvitamin D concentrations, and safety. *American Journal of Clinical Nutrition* **69,** 842–56.

Villalba JM and Navas P (2000) Plasma membrane redox system in the control of stress-induced apoptosis. *Antioxidants and Redox Signaling* **2,** 213–30.

Villarroya F, Giralt M, and Iglesias R (1999) Retinoids and adipose tissues: metabolism, cell differentiation and gene expression. *International Journal of Obesity and Related Metabolic Disorders* **23,** 1–6.

Vogel S, Mendelsohn CL, Mertz JR, Piantedosi R, Waldburger C, Gottesman ME, and Blaner WS (2001) Characterization of a new member of the fatty acid-binding protein family that binds all-*trans*-retinol. *Journal of Biological Chemistry* **276,** 1353–60.

Volpe JJ and Vagelos PR (1973) Fatty acid synthetase of mammalian brain, liver and adipose tissue. Regulation by prosthetic group turnover. *Biochimica et Biophysica Acta* **326,** 293–304.

Voziyan PA, Metz TO, Baynes JW, and Hudson BG (2002) A post-Amadori inhibitor pyridoxamine also inhibits chemical modification of proteins by scavenging carbonyl intermediates of carbohydrate and lipid degradation. *Journal of Biological Chemistry* **277,** 3397–403.

Wald G (1968) The molecular basis of visual excitation. *Nature* **219,** 800–7.

Wang J, Chai X, Eriksson U, and Napoli JL (1999) Activity of human 11-*cis*-retinol dehydrogenase (Rdh5) with steroids and retinoids and expression of its mRNA in extraocular human tissue. *Biochemical Journal* **338,** 23–7.

Wang JJ, Martin PR, and Singleton CK (1997) A transketolase assembly defect in a Wernicke-Korsakoff syndrome patient. *Alcohol Clinical and Experimental Research* **21,** 576–80.

Wang X and Quinn PJ (1999) Vitamin E and its function in membranes. *Progress in Lipid Research* **38,** 309–36.

Wang X and Quinn PJ (2000) The location and function of vitamin E in membranes (review). *Molecular Membrane Biology* **17,** 143–56.

Wang X-D and Russell R (1999) Procarcinogenic and anticarcinogenic effects of β-carotene. *Nutrition Reviews* **57,** 263–72.

Wang XD, Tang GW, Fox JG, Krinsky NI, and Russell RM (1991) Enzymatic conversion of beta-carotene into beta-apo-carotenals and retinoids by human, monkey, ferret, and rat tissues. *Archives of Biochemistry and Biophysics* **285,** 8–16.

Wang XD, Krinsky NI, Tang GW, and Russell RM (1992) Retinoic acid can be produced from excentric cleavage of beta-carotene in human intestinal mucosa. *Archives of Biochemistry and Biophysics* **293,** 298–304.

Wei S, Lai K, Patel S, Piantedosi R, Shen H, Colantuoni V, Kraemer FB, and Blaner WS (1997) Retinyl ester hydrolysis and retinol efflux from BFC-1beta adipocytes. *Journal of Biological Chemistry* **272,** 14159–65.

Weimann BI and Hermann D (1999) Studies on wound healing: effects of calcium D-pantothenate on the migration, proliferation and protein synthesis of human dermal fibroblasts in culture. *International Journal of Vitamin and Nutrition Research* **69,** 113–19.

Weir DG and Scott JM (1995) The biochemical basis of the neuropathy in cobalamin deficiency. *Baillieres Clinics in Haematology* **8,** 479–97.

Welch GN and Loscalzo J (1998) Homocysteine and atherothrombosis. *New England Journal of Medicine* **338,** 1042–50.

Welch RW, Wang Y, Crossman A Jr, Park JB, Kirk KL, and Levine M (1995) Accumulation of vitamin C (ascorbate) and its oxidized metabolite dehydroascorbic acid occurs by separate mechanisms. *Journal of Biological Chemistry* **270,** 12584–92.

Wellik DM and DeLuca HF (1995) Retinol in addition to retinoic acid is required for successful gestation in vitamin A-deficient rats. *Biology and Reproduction* **53,** 1392–7.

Wellik DM, Norback DH, and DeLuca HF (1997) Retinol is specifically required during midgestation for neonatal survival. *American Journal of Physiology* **272**, E25–9.

Wendling O, Chambon P, and Mark M (1999) Retinoid X receptors are essential for early mouse development and placentogenesis. *Proceedings of the National Academy of Sciences of the USA* **96**, 547–51.

Werner ER, Werner-Felmayer G, and Wachter H (1993) Tetrahydrobiopterin and cytokines. *Proceedings of the Society for Experimental Biology and Medicine* **203**, 1–12.

Werner ER, Werner-Felmayer G, and Mayer B (1998) Tetrahydrobiopterin, cytokines, and nitric oxide synthesis. *Proceedings of the Society for Experimental Biology and Medicine* **219**, 171–82.

West CE and Castenmiller JJ (1998) Quantification of the "SLAMENGHI" factors for carotenoid bioavailability and bioconversion. *International Journal of Vitamin and Nutrition Research* **68**, 371–7.

White HB 3rd (1996) Sudden death of chicken embryos with hereditary riboflavin deficiency. *Journal of Nutrition* **126**, 1303S–7S.

White HB 3rd, Orth WH 3rd, Schreiber RW Jr, and Whitehead CC (1992) Availability of avidin-bound biotin to the chicken embryo. *Archives of Biochemistry and Biophysics* **298**, 80–3.

Whitehead CC, Bannister DW, Evans AJ, Siller WG, and Wight PA (1976) Biotin deficiency and fatty liver and kidney syndrome in chicks given purified diets containing different fat and protein levels. *British Journal of Nutrition* **35**, 115–25.

WHO (1996) *Trace Elements in Human Nutrition and Health.* Geneva: World Health Organization.

Wickramasinghe SN (1995) Morphology, biology and biochemistry of cobalamin- and folate-deficient bone marrow cells. *Baillieres Clinics in Haematology* **8**, 441–59.

Wickramasinghe SN (1999) The wide spectrum and unresolved issues of megaloblastic anemia. *Seminars in Hematology* **36**, 3–18.

Wiegand UW, Hartmann S, and Hummler H (1998) Safety of vitamin A: recent results. *International Journal of Vitamin and Nutrition Research* **68**, 411–16.

Wikvall K (2001) Cytochrome P450 enzymes in the bioactivation of vitamin D to its hormonal form (review). *International Journal of Molecular Medicine* **7**, 201–9.

Wilgus H and Roskoski R Jr (1988) Inactivation of tyrosine hydroxylase activity by ascorbate in vitro and in rat PC12 cells. *Journal of Neurochemistry* **51**, 1232–9.

Williams EA, Powers HJ, and Rumsey RD (1995) Morphological changes in the rat small intestine in response to riboflavin depletion. *British Journal of Nutrition* **73**, 141–6.

Williams EA, Rumsey RD, and Powers HJ (1996) Cytokinetic and structural responses of the rat small intestine to riboflavin depletion. *British Journal of Nutrition* **75**, 315–24.

Wilson HL and Galione A (1998) Differential regulation of nicotinic acid-adenine dinucleotide phosphate and cADP-ribose production by cAMP and cGMP. *Biochemical Journal* **331**, 837–43.

Winbauer AN, Pingree SS, and Nuttall KL (1999) Evaluating serum alpha-tocopherol (vitamin E) in terms of a lipid ratio. *Annals of Clinical and Laboratory Science* **29**, 185–91.

Wittpenn JR, Tseng SC, and Sommer A (1986) Detection of early xerophthalmia by impression cytology. *Archives of Ophthalmology* **104**, 237–9.

Wittwer CT, Beck S, Peterson M, Davidson R, Wilson DE, and Hansen RG (1990) Mild pantothenate deficiency in rats elevates serum triglyceride and free fatty acid levels. *Journal of Nutrition* **120,** 719–25.

Wittwer CT, Burkhard D, Ririe K, Rasmussen R, Brown J, Wyse BW, and Hansen RG (1983) Purification and properties of a pantetheine-hydrolyzing enzyme from pig kidney. *Journal of Biological Chemistry* **258,** 9733–8.

Wolf B and Feldman GL (1982) The biotin-dependent carboxylase deficiencies. *American Journal of Human Genetics* **34,** 699–716.

Wolf B and Heard GS (1991) Biotinidase deficiency. *Advances in Pediatrics* **38,** 1–21.

Wolf G (2001) Retinoic acid homeostasis: retinoic acid regulates liver retinol esterification as well as its own catabolic oxidation in liver. *Nutrition Reviews* **59,** 391–4.

Wolosker H, Sheth KN, Takahashi M, Mothet JP, Brady RO Jr, Ferris CD, and Snyder SH (1999) Purification of serine racemase: biosynthesis of the neuromodulator D-serine. *Proceedings of the National Academy of Sciences of the USA* **96,** 721–5.

Wood RJ and Fleet JC (1998) The genetics of osteoporosis: vitamin D receptor polymorphisms. *Annual Reviews of Nutrition* **18,** 233–58.

Wright AJ, Finglas PM, and Southon S (1998) Erythrocyte folate analysis: a cause for concern? *Clinical Chemistry* **44,** 1886–91.

Wu JM, DiPietrantonio AM, and Hsieh TC (2001) Mechanism of fenretinide (4-HPR)-induced cell death. *Apoptosis* **6,** 377–88.

Wyatt KM, Dimmock PW, Jones PW, and Shaughn O'Brien PM (1999) Efficacy of vitamin B-6 in the treatment of premenstrual syndrome: systematic review. *British Medical Journal* **318,** 1375–81.

Wyss A, Wirtz GM, Woggon WD, Brugger R, Wyss M, Friedlein A, Riss G, Bachmann H, and Hunziker W (2001) Expression pattern and localization of beta,beta-carotene 15,15′-dioxygenase in different tissues. *Biochemical Journal* **354,** 521–9.

Yamada K, Yamada Y, Fukuda M, and Yamada S (1999) Bioavailability of dried asakusanori (*Porphyra tenera*) as a source of cobalamin (vitamin B_{12}). *International Journal of Vitamin and Nutrition Research* **69,** 412–18.

Yamada K, Chen Z, Rozen R, and Matthews RG (2001a) Effects of common polymorphisms on the properties of recombinant human methylenetetrahydrofolate reductase. *Proceedings of the National Academy of Sciences of the USA* **98,** 14853–8.

Yamada S, Yamamoto K, Masuno H, and Choi M (2001b) Three-dimensional structure-function relationship of vitamin D and vitamin D receptor model. *Steroids* **66,** 177–87.

Yamada Y, Merrill AH Jr, and McCormick DB (1990) Probable reaction mechanisms of flavokinase and FAD synthetase from rat liver. *Archives of Biochemistry and Biophysics* **278,** 125–30.

Yeum KJ and Russell RM (2002) Carotenoid bioavailability and bioconversion. *Annual Reviews of Nutrition* **22,** 483–504.

Yeum KJ, dos Anjos Ferreira AL, Smith D, Krinsky NI, and Russell RM (2000) The effect of alpha-tocopherol on the oxidative cleavage of beta-carotene. *Free Radicals in Biology and Medicine* **29,** 105–14.

Ylikomi T, Laaksi I, Lou YR, Martikainen P, Miettinen S, Pennanen P, Purmonen S, Syvala H, Vienonen A, and Tuohimaa P (2002) Antiproliferative action of vitamin D. *Vitamins and Hormones* **64,** 357–406.

Young AJ and Lowe GM (2001) Antioxidant and prooxidant properties of carotenoids. *Archives of Biochemistry and Biophysics* **385**, 20–7.

Youssef JA, Song WO, and Badr MZ (1997) Mitochondrial, but not peroxisomal, beta-oxidation of fatty acids is conserved in coenzyme A-deficient rat liver. *Molecular and Cellular Biochemistry* **175**, 37–42.

Yu W, Simmons-Menchaca M, Gapor A, Sanders BG and Kline K (1999) Induction of apoptosis in human breast cancer cells by tocopherols and tocotrienols. *Nutrition and Cancer* **33**, 26–32.

Yubisui T, Matsuki T, Tanishima K, Takeshita M, and Yoneyama Y (1977) NADPH-flavin reductase in human erythrocytes and the reduction of methemoglobin through flavin by the enzyme. *Biochemical and Biophysical Research Communications* **76**, 174–82.

Yun M, Park CG, Kim JY, Rock CO, Jackowski S, and Park HW (2000) Structural basis for the feedback regulation of *Escherichia coli* pantothenate kinase by coenzyme A. *Journal of Biological Chemistry* **275**, 28093–9.

Zannoni VG, Flynn EJ, and Lynch M (1972) Ascorbic acid and drug metabolism. *Biochemical Pharmacology* **21**, 1377–92.

Zeisel SH (2000) Choline: an essential nutrient for humans. *Nutrition* **16**, 669–71.

Zempleni J and Mock DM (1999a) Advanced analysis of biotin metabolites in body fluids allows a more accurate measurement of biotin bioavailability and metabolism in humans. *Journal of Nutrition* **129**, 494S–7S.

Zempleni J and Mock DM (1999b) Bioavailability of biotin given orally to humans in pharmacologic doses. *American Journal of Clinical Nutrition* **69**, 504–8.

Zempleni J and Mock DM (2000a) Marginal biotin deficiency is teratogenic. *Proceedings of the Society for Experimental Biology and Medicine* **223**, 14–21.

Zempleni J and Mock DM (2000b) Utilization of biotin in proliferating human lymphocytes. *Journal of Nutrition* **130**, 335S–7S.

Zempleni J and Mock D (2001) Biotin homeostasis during the cell cycle. *Nutrition Research Reviews* **14**, 45–63.

Zempleni J, Galloway JR, and McCormick DB (1996) Pharmacokinetics of orally and intravenously administered riboflavin in healthy humans. *American Journal of Clinical Nutrition* **63**, 54–66.

Zempleni J, Steven Stanley J, and Mock DM (2001) Proliferation of peripheral blood mononuclear cells causes increased expression of the sodium-dependent multivitamin transporter gene and increased uptake of pantothenic acid. *Journal of Nutritional Biochemistry* **12**, 465–73.

Zhang DQ and McMahon DG (2000) Direct gating by retinoic acid of retinal electrical synapses. *Proceedings of the National Academy of Sciences of the USA* **97**, 14754–9.

Zhang F, Thottananiyil M, Martin DL, and Chen CH (1999) Conformational alteration in serum albumin as a carrier for pyridoxal phosphate: a distinction from pyridoxal phosphate-dependent glutamate decarboxylase. *Archives of Biochemistry and Biophysics* **364**, 195–202.

Zhang H, Osada K, Sone H, and Furukawa Y (1997) Biotin administration improves the impaired glucose tolerance of streptozotocin-induced diabetic Wistar rats. *Journal of Nutritional Science and Vitaminology* (Tokyo) **43**, 271–80.

Zhou B, Westaway SK, Levinson B, Johnson MA, Gitschier J, and Hayflick SJ (2001) A novel pantothenate kinase gene (PANK2) is defective in Hallervorden-Spatz syndrome. *Nature Genetics* **28**, 345–9.

Zhyvoloup A, Nemazanyy I, Babich A, Panasyuk G, Pobigailo N, Vudmaska M, Naidenov V, Kukharenko O, Palchevskii S, Savinska L, Ovcharenko G, Verdier F, Valovka T, Fenton T, Rebholz H, Wang ML, Shepherd P, Matsuka G, Filonenko V, and Gout IT (2002) Molecular cloning of CoA synthase. The missing link in CoA biosynthesis. *Journal of Biological Chemistry* **277**, 22107–10.

Zolfaghari R and Ross AC (2000) Lecithin:retinol acyltransferase from mouse and rat liver. cDNA cloning and liver-specific regulation by dietary vitamin A and retinoic acid. *Journal of Lipid Research* **41**, 2024–34.

Zubaran C, Fernandes JG, and Rodnight R (1997) Wernicke-Korsakoff syndrome. *Postgraduate Medical Journal* **73**, 27–31.

Index